T0201606

The Anthropocene as a Geological Time Unit

The Anthropocene, a term launched into public debate by Nobel Prize winner Paul Crutzen, has been used informally to describe the time period during which human actions have had a drastic effect on the Earth and its ecosystems. This book, written by the high-profile international team analysing the Anthropocene's potential addition to the Geological Time Scale, presents evidence for defining the Anthropocene as a geological epoch. The evidence ranges from chemical signals arising from pollution to landscape changes associated with urbanisation and biological changes associated with species invasion and extinctions. Global environmental change is placed within the context of planetary processes and deep geological time, allowing the reader to appreciate the scale of human-driven change and compare the global transition taking place today with major transitions in Earth history. This is an authoritative review of the Anthropocene for graduate students and academic researchers across scientific, social science and humanities disciplines.

The editors and contributing authors, most being part of the Anthropocene Working Group (AWG), have been involved in the geological analysis of the Anthropocene from the beginning, playing a central role in characterising and defining it as a unit of geological time and contributing to its wider multidisciplinary study.

Jan Zalasiewicz is Professor of Palaeobiology at the University of Leicester and Chair of the Anthropocene Working Group. His research interests include mudrock processes; early Paleozoic and Quaternary stratigraphy and sedimentology; and stratigraphic analysis, notably the study of the Anthropocene concept.

Colin N. Waters is Honorary Professor in the School of Geography, Geology and the Environment at the University of Leicester and Secretary of the Anthropocene Working Group with a central role in coordinating activities of the Working Group members. He recently retired as Principal Mapping Geologist at the British Geological Survey, where he specialised in geological mapping of the United Kingdom and parts of the Sahara Desert, as well as stratigraphic analysis, principally of the Carboniferous and Anthropocene.

Mark Williams is Professor of Paleobiology at the University of Leicester. He is interested in the evolution of the biosphere over geological timescales, with an emphasis on understanding the rate and degree of current biological change. He was the first secretary of the Anthropocene Working Group from 2009 to 2011.

Colin P. Summerhayes is an Emeritus Associate of the Scott Polar Research Institute at the University of Cambridge. He is a marine geologist and oceanographer with expertise in the role of climate in forming marine sediments and in interpretation of the history of climate from sedimentary records. He is a former manager of the Stratigraphy Branch of the Exploration Division of BP Research and former director of the Institute of Oceanographic Sciences Deacon Laboratory, Wormley, United Kingdom.

The Anthropocene as a Geological Time Unit

A Guide to the Scientific Evidence and Current Debate

Edited by

JAN ZALASIEWICZ
University of Leicester

COLIN N. WATERS
University of Leicester

MARK WILLIAMS
University of Leicester

COLIN P. SUMMERHAYES
Scott Polar Research Institute

CAMBRIDGE
UNIVERSITY PRESS

University Printing House, Cambridge CB2 8BS, United Kingdom

One Liberty Plaza, 20th Floor, New York, NY 10006, USA

477 Williamstown Road, Port Melbourne, VIC 3207, Australia

314–321, 3rd Floor, Plot 3, Splendor Forum, Jasola District Centre, New Delhi – 110025, India

79 Anson Road, #06–04/06, Singapore 079906

Cambridge University Press is part of the University of Cambridge.

It furthers the University's mission by disseminating knowledge in the pursuit of education, learning, and research at the highest international levels of excellence.

www.cambridge.org
Information on this title: www.cambridge.org/9781108475235
DOI: 10.1017/9781108621359

First published 2019

Printed in the United Kingdom by TJ International Ltd. Padstow Cornwall

A catalogue record for this publication is available from the British Library.

Library of Congress Cataloging-in-Publication Data
Names: Zalasiewicz, J. A., editor.
Title: The anthropocene as a geological time unit : a guide to the scientific evidence and current debate / edited by Jan Zalasiewicz (University of Leicester) [and three others].
Description: Cambridge : Cambridge University Press, 2019. | Includes bibliographical references and index.
Identifiers: LCCN 2018020098 | ISBN 9781108475235 (hardback)
Subjects: LCSH: Human ecology. | Nature–Effect of human beings on. | Geology, Stratigraphic.
Classification: LCC GF75 .A644 2018 | DDC 304.2–dc23
LC record available at https://lccn.loc.gov/2018020098

ISBN 978-1-108-47523-5 Hardback

Paul Crutzen first developed the term Anthropocene during work with many of his colleagues on the International Geosphere-Biosphere Programme, when the term Holocene seemed inappropriate to describe the scale and rate of recent change to the physics, chemistry and biology of the Earth System. Since then, he has seen the term grow and develop hugely, within a wide range of studies. This book summarises research on the geological context of the Anthropocene as geology, work that might see the Anthropocene become part of the Geological Time Scale. We dedicate this book to Paul, who is glad to see these studies proceeding, on a topic that affects us all.

CONTENTS

CONTRIBUTORS

David Aldridge
University of Cambridge

Zhisheng An
Institute of Earth Environment,
Chinese Academy of Sciences

Anthony Barnosky
Stanford University

Valentin Bault
University of Lyon

Sharon A. Billings
University of Kansas

Alejandro Cearreta
University of the Basque Country UPV/EHU

Erich Draganits
University of Vienna

Matt Edgeworth
University of Leicester

Ian J. Fairchild
Birmingham University

Agnieszka Gałuszka
Jan Kochanowski University

Phil Gibbard
University of Cambridge

Jacques Grinevald
Graduate Institute of International and
Development Studies

Peter Haff
Duke University

Irka Hajdas
ETH Zurich

Trevor Hancock
University of Victoria

Robert M. Hazen
Carnegie Institution for Science

Martin J. Head
Brock University

Juliana Assunção Ivar do Sul
Leibniz Institute for Baltic Sea Research, Warnemünde

Catherine Jeandel
LEGOS University of Toulouse, CNES, CNRS, IRD

Reinhold Leinfelder
Free University of Berlin

John McNeill
Georgetown University

Eric Odada
University of Nairobi

Naomi Oreskes
Harvard University

Simon Price
University of Cambridge

Daniel deB. Richter
Duke University

Neil Rose
University College London

Andy Smith
University of Derby

Will Steffen
Australian National University

Colin P. Summerhayes
University of Cambridge

James Syvitski
University of Colorado

Davor Vidas
Fridtjof Nansen Institute

Michael Wagreich
University of Vienna

Colin N. Waters
University of Leicester

Ian Wilkinson
University of Leicester

Mark Williams
University of Leicester

Scott Wing
Smithsonian Institution

Jan Zalasiewicz
University of Leicester

FIGURE CREDITS

The following institutions, publishers and authors are gratefully acknowledged for their kind permission to use figures based on illustrations in journals, books and other publications for which they hold copyright. We have cited the original sources in our figure captions. We have made every effort to obtain permissions to make use of copyrighted materials and apologise for any errors or omissions. The publishers welcome errors and omissions being brought to their attention.

Institutions and publishers	Figure number(s)
American Association of the Advancement of Science	
Science	5.2.2, 5.2.3, 5.8.4, 6.1.8, 6.2.2, 6.3.3, 7.6.2
American Association of Petroleum Geologists	
North American Commission on Stratigraphic Nomenclature	2.7.5
American Chemical Society	
Environmental Science and Technology	5.1.2
American Geophysical Union Publications	
Earth's Future	4.2.1
Water Resources Research	2.8.2
Andy Lee Robinson, Haveland Robinson Associates	6.2.5
Cambridge University Press	
Annals of Glaciology	6.2.4
The Surface Waters Acidification Programme	7.8.4
Carbon Dioxide Information Analysis Center	7.4.2
Centre for the Study of Provincial interdependence in Classical Antiquity	2.5.3
Copernicus Publications (EGU)	
Geoscience Model Development Discussions	5.5.2, 5.5.3
Elsevier	
A Geological Time Scale 2012	1.3.6, 1.3.7
Anthropocene	4.3.1, 4.3.2, 7.7.1
Earth and Planetary Science Letters	7.8.2c
Earth-Science Reviews	2.4.3, 2.8.4, 5.4.2, 5.8.7, 7.8.1, 7.8.5, 7.8.6, 7.8.7

(*cont.*)

Institutions and publishers	Figure number(s)
Encyclopedia of the Anthropocene	2.2.1
Geochimica et Cosmochimica Acta	5.3.2, 5.5.1
Global and Planetary Change	6.3.2
Journal of Environmental Radioactivity	5.8.5
Journal of Marine Systems	2.6.2
Marine Micropalaeontology	5.4.4
Nuclear Instruments and Methods in Physics Research	5.8.6b
European Geophysical Union	
Climate of the Past	6.1.7
Geological Society of America	
GSA Today	1.3.2, 6.1.2
Geological Society of America Bulletin	2.7.1
Geological Society of America, Special Paper	1.3.5
Geological Society of London	
Geological Society, London, Special Publications	1.3.3, 2.1.1, 2.1.2, 2.5.4, 3.1.2, 4.2.2, 5.2.1, 5.8.6c
Quarterly Journal of Engineering Geology and Hydrogeology	2.5.6b
ICOLD Committee on Reservoir Sedimentation	
ICOLD Bulletin preprint 147	2.8.1
International Commission on Stratigraphy	
The Geological Time Scale	1.3.1
International Union of Geological Sciences	
Episodes	1.3.4, 1.3.5
Inter-Research	
Marine Ecology Progress Series	5.8.6a
Keck Geology Consortium	2.8.3
MIT Press	
Energy Transitions: History, Requirements, Prospects	7.4.1
NASA Earth Observatory	5.4.3
National Academies Press, Washington, D.C.	
National Academy of Sciences	6.2.3
Proceedings of the National Academy of Sciences	7.8.2b
National Museums of Kenya, Nairobi	7.2.1

Nature	
Nature	5.3.1, 6.1.3, 6.1.5, 6.1.6, 6.2.1
Nature Climate Change	6.1.10, 7.6.3
Nature Communications	1.3.9, 6.1.1
Scientific Reports	2.3.4a, 2.4.2
National Oceanic and Atmospheric Administration (NOAA)	
Global analysis: Annual 2016	7.6.1
Office of Ocean Exploration and Research	2.8.5
The Royal Society	
Philosophical Transactions of the Royal Society	5.4.1, 7.8.3
Sage	
Bulletin of the Atomic Scientists	5.8.1, 5.8.2
The Holocene	2.3.4b
Soil Science Society of America	
Soil Science Society of America Journal	2.7.4
Springer	
Developments in Paleoenvironmental Research	2.3.2
Journal of Paleolimnology	2.3.4b, c
Space Science Reviews	6.1.9
Sustainability Science	6.3.1
United Nations	
United Nations Scientific Committee on the Effects of Atomic Radiation	5.8.3
University of Chicago Press	
Current Anthropology	1.1.1
United States Forest Service	2.7.2
Westermann Gruppe	
Diercke Three Universal Atlas Niederlande	2.5.7
Wiley	
Geophysical Research Letters (AGU)	2.4.1, 7.8.2a
Global Biogeochemical Cycles (AGU)	2.4.1
Journal of Quaternary Science	1.3.8
Authors	
Bault, V.	3.3.2
Fairchild, I. J.	5.1.1, 5.5.4
Gałuszka, A.	5.6.1, 5.6.2, 5.6.3, 5.6.4, 5.6.5, 5.7.1, 5.7.2, 5.7.3, 5.7.4, 5.7.5

(*cont.*)

Institutions and publishers	Figure number(s)
Hou, X.	3.1.1
Leinfelder, R.	3.4.1, 3.4.2, 3.4.3
Lovejoy, S.	6.1.4
Migaszewski, Z. M.	2.3.5, 2.3.6, 2.3.7
Rose, N.	2.3.1, 2.3.3, 2.3.4a
Smith, A.	2.2.2
Steffen, W.	7.5.1, 7.5.2, 7.5.3
Waters, C. N.	2.2.3, 2.5.1, 2.5.2, 2.5.5, 2.5.6a, 2.6.1, 2.7.3
Wilkinson, I. P.	3.2.1
Williams, M.	3.3.3
Ysebaert, T.	3.3.1

We thank Lisa Barber for kindly redrawing figures 1.3.4, 1.3.5, 1.3.9 and 2.8.2.

1 History and Development of the Anthropocene as a Stratigraphic Concept

CONTENTS

We introduce here the concept of the Anthropocene as a potential geological unit of time, while noting that antecedents of this concept were sporadically present in previous literature prior to its effective inception, by Paul Crutzen, in 2000 CE. We describe how the Anthropocene compares with examples from the Geological Time Scale throughout Earth history, and demonstrate the extent to which the term has practical utility in the field of geology, in the field of natural science generally, and to the wider academic community. In this book we describe the geological Anthropocene, while this definition does not exclude other, different, interpretations of the Anthropocene that have appeared in recent years amongst other scholarly communities, particularly in the humanities. We explain here how this book will help to inform the process of producing a formal proposal for the Anthropocene as a geological time unit. Examples from the beginning of the Cenozoic Era, the Cambrian, Silurian and Quaternary periods, and the Eocene and Holocene epochs are used to demonstrate how chronostratigraphic boundaries are defined and what lessons from these can be applied to defining the Anthropocene.

1.1 A General Introduction to the Anthropocene

Jan Zalasiewicz, Colin N. Waters, Mark Williams, Colin P. Summerhayes, Martin J. Head and Reinhold Leinfelder

The Anthropocene, launched as a concept by Paul Crutzen in 2000 (Crutzen & Stoermer 2000; Crutzen 2002), has in less than two decades grown astonishingly in its range and reach amongst different academic communities. Fundamentally, it was coined to crystallise the growing realisation that human activities – or, more often, the unintended consequences of human activities – had fundamentally changed the Earth System. Hence, the patterns of behaviour of the oceans, atmosphere, land (i.e., the geosphere's terrestrial surface), cryosphere, biosphere and climate are no longer those that over 11 millennia characterised the great bulk of the epoch that we still formally live in, the Holocene. The accent on planetary processes reflected the character of the scientific community that Paul Crutzen was working in, that of the Earth System science (ESS) community, concerned most acutely with contemporary global change.

Nevertheless, the Anthropocene was explicitly described as a geological time interval, as an epoch in direct comparison to – and different from – the Holocene because of the inferred geological significance of the altered Earth System processes. The implicit hypothesis was that the Holocene had terminated, perhaps about when the Industrial Revolution started. This improvised proposal chimed with the conclusions on the nature, scale and speed of global change being reached by the ESS community, and the term soon began to be widely used in publications, matter-of-factly, as if it were already part of accepted geological time terminology. It was not formal, though, having gone through none of the extensive formal analysis, debate, agreement (via an established pattern of voting amongst appropriate stratigraphic bodies) and ratification that formal geological time terms require (and which are described fully in Section 7.8.1).

A few years after Crutzen's intervention, increasing use of the term began to be noticed by the geological community, and a preliminary analysis by a national body, the Stratigraphy Commission of the Geological Society of London, suggested that the term had merit and should be studied further with respect to any potential formalisation. This conclusion was in sharp contrast to the general response by the geological community to sporadic earlier suggestions of a 'human era', which had indeed been made since the late 18th century (Stoppani 1873; Buffon 2018). These suggestions had always been generally rejected, on the basis that the great forces of nature that drove Earth's geology were considered to operate on a vaster and longer-term scale than any kind of human impact, which by comparison was widely considered 'too puny'. The realisation, even amongst geologists, that humans could indeed significantly affect not only the Earth System parameters but, as a consequence of this, also the course of Earth's geological evolution, led to an invitation from the Subcommission of Quaternary Stratigraphy (SQS) of the International Commission on Stratigraphy (ICS) to set up a formal Anthropocene Working Group (AWG); to examine the case for formalisation; and ultimately to make recommendations to the SQS, ICS and the latter's parent body, the International Union of Geological Sciences (IUGS).

This book is the outcome of the work of the Anthropocene Working Group since 2009 in developing and testing the general case for the Anthropocene as a formal geological time unit. This work was a necessary prelude to preparing any specific formalisation proposal to the SQS, ICS and IUGS (a task that is underway). It summarises the evidence gathered in the intervening time, both by AWG members and others, for what we may here call the 'geological Anthropocene' or perhaps 'stratigraphic Anthropocene'. This distinguishes it from other interpretations of the Anthropocene that have emerged in these last few years as a range of communities, including those within the social sciences, humanities and arts, have explored this term and concept through the prisms of their own disciplines.

Thus, in our discussions of the Anthropocene to follow, there are a few things to bear in mind. Firstly, its interpretation here is non-exclusive – it does not in any way restrict (or seek to restrict) the potential use of the word in other meanings, by other communities, as has indeed been the case in the last decade (e.g., Edgeworth et al. 2015; Ruddiman et al. 2015a). Many words have more than one meaning – the word 'mantle', for instance, can be applied to part of the Earth beneath the crust, to an item of clothing, to a type of tissue on a mollusc or to part of an old-fashioned gas lamp. Sometimes the meaning of the word is clear from the context, and sometimes an appropriate qualifier needs to be used to ensure precision of communication; we suggest that such care in communication now needs to apply to the term 'Anthropocene' too.

We recognise that accepting the various material signals of the geological Anthropocene as a valid scientific outcome of stratigraphic analysis may lead, as a corollary, to analysis of the societal, cultural and political causes and consequences of the existence of a geological Anthropocene. Such a broader level of analysis is potentially of considerable importance and would involve extensive cooperation of the sciences, the humanities, the arts and society. However, it goes beyond the mandate of the Anthropocene Working Group and the scope of this book. One might use a medical metaphor, in that the characterisation and definition of a geological Anthropocene may be said to be diagnosing the condition of a planet through a particular set of symptoms, against the background of a very long family history. Such analysis of the geological Anthropocene does not, though, investigate the causes of the condition too deeply, nor does it offer any treatment plan or much in the way of a prognosis.

In a geological context, the Anthropocene is here considered as a unit of Earth history and, more than this, as a potentially *formal* unit that might become part of the ICS-produced International Chronostratigraphic Chart (which informs the Geological Time Scale). It would thus comprise a potential Anthropocene Epoch and, as its essential material counterpart and alter ego, simultaneously an Anthropocene Series, which is a unit of strata that can be dug into, sampled and – in a few cases, despite its geological youth – hit with a hammer. The value of such a designation is to make the most effective comparison between present processes and those of the deep geological past: to, as far as possible, compare like with like in making such comparisons. As the history of the Earth prior to human documentation can *only* be inferred from the rock record, this focus on material, stratal evidence is critical to comparing the modern and ancient histories of this planet and therefore to gauging the relative scale and rate of human-driven perturbation. The geological Anthropocene, therefore, has to be considered within the established rules and guidelines that apply to all other units of the Geological Time Scale. For instance, it is important that, as far as possible, its beginning (and its base, when applied to strata) is synchronous around the world (see Section 7.8).

The geological Anthropocene is not a diachronous unit of human cultural history like the Iron Age and Palaeolithic, which unfolded in mosaic fashion across the planet, or like the Renaissance (though other social science interpretations of the Anthropocene may approximate to such units). More generally, descriptions of it as a 'human epoch' are in some respects misleading. The Anthropocene is here considered as an epoch of Earth time, just like all Earth's previous epochs. It so happens that its distinctive characteristics have up until now been driven largely by a variety of human actions. But if these characteristics (such as sharply increased atmospheric carbon dioxide levels, global carbon isotope and nitrogen isotope anomalies, a biosphere modified by species extinctions and invasions, and so on – Figure 1.1.1) were driven by any other means – such as by a meteorite impact, volcanic eruptions or the actions of another species – then they would have exactly the same importance geologically.

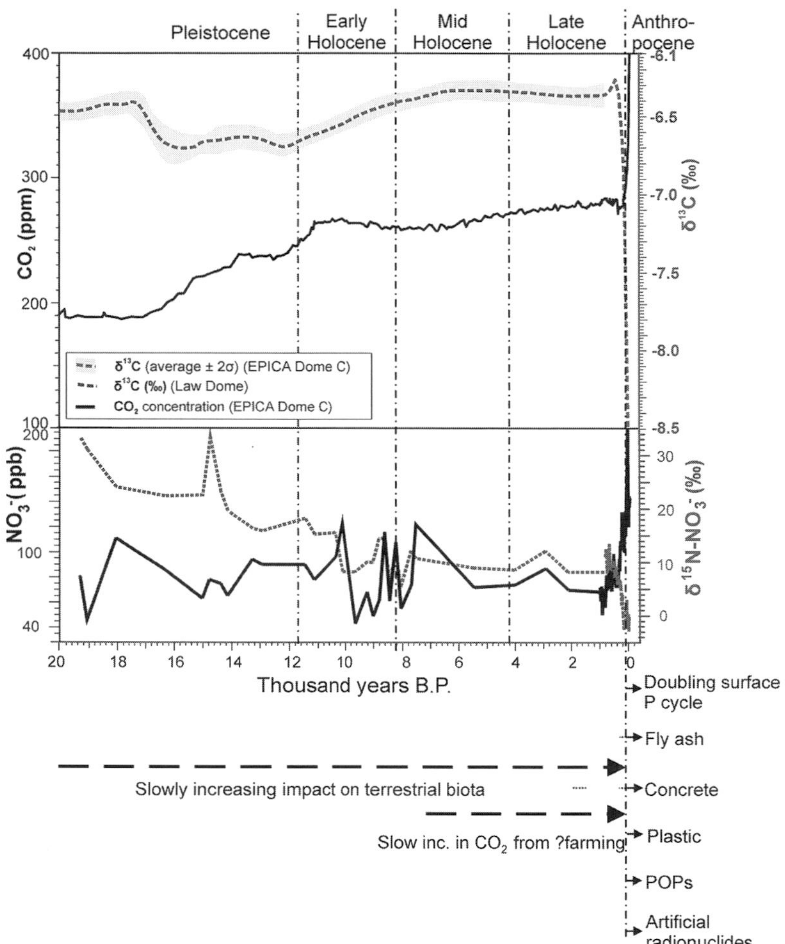

Figure 1.1.1 Trends in key stratigraphic indicators from the Late Pleistocene to the present time. Note the largely gradual change (at this scale) across the Pleistocene-Holocene boundary, the general stability through the Holocene, and the marked inflections and incoming of novel indicators that clearly demarcate a changed trajectory from the mid-20th century, identified as the Anthropocene. From Zalasiewicz et al. (2018). (POPs = persistent organic pollutants.)

Therefore, setting out these preliminary constraints of what we consider the stratigraphic Anthropocene to be and also *not* to be (constraints that are placed upon all of the units of the Geological Time Scale) helps explain the particular content and emphases that we place in this book. The Anthropocene represents a remarkable episode in the history of the Earth, a narrative that is unfinished but that has emphatically begun, and one that is of no little consequence for present and future communities. Examining it in classical geological terms will, we hope, be useful to geologists and non-geologists alike.

1.2 History of the Anthropocene Concept

Jacques Grinevald, John McNeill, Naomi Oreskes, Will Steffen, Colin P. Summerhayes and Jan Zalasiewicz

Is the modern scientific concept of the Anthropocene an old idea, dating back a century or more yet retaining its meaning and perspective? Or is it a new, paradigm-shifting conceptual novelty? This question is rendered more complicated by the diversity of the perspectives from which the Anthropocene and related ideas have been addressed, their varied interpretations and the problems inherent in making historical retrospectives (e.g., Uhrqvist & Linnér 2015).

The notion that collective human action (or 'mankind', in older parlance) is a geomorphological and geological agent altering the Earth is certainly not new in Western thought (Glacken 1956), with ideas developed by such thinkers as René Descartes and Francis Bacon around the domination or transformation of nature by humankind. But the extent to which this notion has been embedded within a context of geological and biospheric processes and deep-time Earth history – and, more specifically, in the stratigraphic nomenclature for classifying Earth history – has varied, as has scientific appreciation of our home planet as a specific and remarkable element within the solar system. The history, and indeed prehistory, of the Anthropocene concept and related ideas is still an emerging and debated topic, but it has received attention after Crutzen's (2002) early suggestions of historical antecedents in both concise (e.g., Steffen et al. 2011) and more comprehensive (Grinevald 2007; Davis 2011) accounts.

An in-depth study has yet to be written. The history of science and the development of knowledge are connected in intricate and reciprocal ways, so the appearance of a conceptual novelty and new scientific terminology is often bedevilled by misunderstanding. The new 'big idea' of the Anthropocene, as first coined by Paul Crutzen and Eugene Stoermer (2000) in the context of the IGBP (International Geosphere-Biosphere Programme) and by Crutzen (2002) and then considered by Zalasiewicz et al. (2008) in the geological context of stratigraphy, is no exception (Hamilton & Grinevald 2015).

The ancients sometimes pondered how humans relate to their world, as in Lucretius' suggestion of an Earth made weary through the weight of a growing human population. But perhaps the first significant reference in the Western world is within an influential work in which the Earth's history was, for the first time, systematically chronologically described on the basis of empirical geological evidence. This is Buffon's *Les Époques de la Nature*, published in 1778 (Roger 1962; Buffon 2018; see also Heringman 2015). In this pioneering book, the seven 'epochs' represent distinct phases in Earth history, ranging from its initial cooling to the formation of the oceans and the lowering of sea level, the weathering of primordial rocks and the deposition of sedimentary strata, and the origin and progression of successive, different forms of life. The 'seventh and last epoch – When the power of Man assisted the operation of nature' is described as one in which humans not only are present but, as 'civilised humans' (placed by Buffon in overt opposition to 'savages'), are modifying key Earth processes such as regional temperature and precipitation by altering vegetation patterns and burning coal. In attempting to describe how key planetary mechanisms (crust formation, sea level, volcanism and so on) might be interlinked and how they can evolve through time, Buffon was a pioneer of Earth history, and the late (in Buffon's chronology) addition of human participation in Earth history is placed within this same intellectual framework. Buffon, like James Hutton, Joseph Black, Adam Smith and James Watt, was a natural philosopher of pre-industrial Europe, a man of the 'Age of Enlightenment' and one of many thinkers considering the place of humans in Earth history (see Rudwick 2005, 2008).

The idea of 'man' as a geographical and geological agent arose in a succession of geological and related naturalist publications in the mid- to late 19th century. The Welsh geologist and theologian Thomas Jenkyn (1854a, b; mentioned by Lewis & Maslin 2015) also wrote of a 'human epoch' that he referred to as an 'Anthropozoic' that would leave a future fossil record. The term Anthropozoic was also used by Haughton (1865) and the Italian abbot and geologist Stoppani (1873; quoted by the US ambassador in Italy; Marsh 1874) and was rediscovered by William Clark in the 1980s (Clark 1986, quoted by Crutzen). Stoppani observed that humans, since the rise of Christianity, were changing not only the present but also the future of the Earth. The roles of humans and environmental change in the geology of the recent past were later to be conflated with the classification of geologically recent strata as the Holocene (a term proposed to replace Lyell's 'Recent' by Paul Gervais in the 1860s and adopted after the Third International Geological Congress of 1885), in which the geologically defining forces were seen to be marked by post-Pleistocene glacial warming and sea-level rise, but in which it was recognised that locally abundant human activities and traces formed part of the characterisation.

The entire Quaternary Period (Gibbard & Head 2009), broadly representing the Ice Ages (see Section 1.3.1.5), was recognised as the time when the human genus diversified (albeit mostly remaining ecologically and geologically insignificant) and was termed the Anthropogene (sometimes transcribed as Anthropocene) by some early- to mid-20th-century Soviet geologists and geochemists. While the Anthropogene was essentially a synonym for the Quaternary (Gerasimov 1979), Piruzyan et al. (1980; quoted in Grinevald 2007) noted the following:

> The notion that mankind was becoming a power of geological scale was, by the beginning of the 20th century, clearly expressed by A. P. Pavlov in Moscow and, independently, by C. Schuchert in New Haven. They interpreted in a new way long-known facts on the changes in the environment caused by human activities, coming to the conclusion that their manifestations characterised the beginning of a new geological era. Ideas on the new geological era – 'Psychozoic' according to Schuchert, 'anthropozoic' according to Pavlov – were developed in detail by V. I. Vernadsky.

A focus on the changes that humans specifically were making had been first documented by George Perkins Marsh in his classic book *Man and Nature* (1864), which was retitled as *The Earth as Modified by Human Action* in the second edition of 1874. Marsh's study was couched in environmental or geographical rather than geological (or stratigraphic) terms, reflecting his posthumous status as 'North America's first conservationist' or 'Prophet of Conservation' (Lowenthal 2000). But his themes and influence were overtly restated and examined in later meetings and publications (Thomas 1956; Nir 1983; Orio & Botkin 1986; Turner et al. 1990; Naredo & Gutiérrez 2005). A classically geological analysis by Sherlock (1922) systematically documented the lithostratigraphic dimension driven by mining, building and related activities, assembling statistics on different types of mineral production and rock and earth movement and considering not only the effects in sedimentological and geomorphological terms but also geochemical effects, not least following Arrhenius (see the next paragraph) in linking coal burning to previsaged climate warming (see also Shaler 1905).

While Marsh and others, including Thomas Jefferson, had realised that human changes to Earth's plant cover led to changes in the temperature of the air, John Tyndall had demonstrated in the 1860s that the minor gases of the air, like water vapour, carbon dioxide, methane and ozone, had the power to absorb and re-emit long-wave radiation, meaning that fluctuations in their abundance could change the climate (Tyndall 1868). Arrhenius had calculated 30 years later that doubling the amount of CO_2 in the air would warm the planet by about 6°C (Arrhenius 1896). By 1908 he had modified that figure to 4°C and noted that the burning of coal by industry would emit enough CO_2 to measurably warm the atmosphere (Arrhenius 1908). He thought that would be no bad thing – humans would benefit from living in a warmer, more equable climate, and rising warmth and CO_2 would stimulate plant growth, providing more food for a larger population and even preventing the occurrence of another glacial period. This kind of human impact on the planet was well beyond that envisaged by the likes of Marsh or Sherlock. But it was not until the mid-20th century that scientists were able to build on Arrhenius's findings and become fully aware of the growing human impact of changing atmospheric chemistry, not least because the technology to provide us with the full spectrum of CO_2 in the atmosphere was not available until the mid-1950s (Plass 1961). For more on CO_2 and climate, see Section 6.1.

More or less simultaneously, influential conceptual developments under the same terms of 'biosphere' and 'nöosphere' were made by two French Catholic visionary thinkers: Pierre Teilhard de Chardin, then professor of geology, and Édouard Le Roy, a mathematician turned philosopher and Bergson's successor at the Collège de France. Another significant contributor was the remarkable Russian geoscientist, Vladimir I. Vernadsky, a hugely influential member of the Saint Petersburg Academy of Sciences, who was then staying in Paris. The nöosphere (or anthroposphere, including the technosphere) denoted accelerating human transformation of 'the face of the Earth' (a term derived from the massive and widely read early-20th-century geological synthesis of Eduard Suess). These various ideas of Teilhard, Le Roy and Vernadsky generated a range of interpretations (and confusions) in subsequent years, mainly after the Second World War (WWII) and the birth of the Nuclear Age. Teilhard disagreed with Vernadsky's meaning of the 'biosphere', which both took from Suess. Teilhard's evolutionary view of life and man on Earth was ignorant of Vernadsky's biogeochemical perspective, and he probably never read *La Biosphère*, the 1929 French translation of Vernadsky (1926, in Russian) – at least, he never quoted it in his writings. In general, Vernadsky's biogeochemical teachings and his own ambitious concept of the Earth's biosphere in the cosmos were commonly ignored (Vernadsky 1998).

The term 'nöosphere' was adopted by Vernadsky only after Le Roy's books of 1927 and 1928 (Vernadsky 1945, 1997). It was originally seen as a direct offshoot of the biosphere, a term and notion briefly coined by Suess in his 1875 book *Die Entstehung der Alpen* (The Origin of the Alps) and restated in 1909 in the final chapter, 'Das Leben' (Life), of his great work *Das Antlitz der Erde* (The Face of the Earth). The term 'biosphere' was adopted by Teilhard and Le Roy, with a restricted biological meaning, and developed in a global biogeochemical perspective by Vernadsky (1926; see 1998) to represent not just the sum total of living matter (or biota, according to Teilhard) on the Earth's rocky crust, but an evolving complex system representing the dynamic interaction and co-evolution of life, crustal mineral matter, ocean, atmosphere and energy (mainly from the Sun). It was this geobiological system that Vernadsky viewed as being changed and perturbed by growing human activities, particularly technical and scientific development (Vernadsky 1924, 1945, 1997).

Vernadsky's ideas foreshadowed many of those developed by James Lovelock and Lynn Margulis (1974) in the 'Gaia hypothesis', specifically that life acts together as a system to modify and regulate surface conditions on Earth. Lovelock, like most Western scientists, only became aware of Vernadsky after he had developed his own ideas (Grinevald 1987, 1988). As in the case of Plass (1961) and the measurement of the spectrum of CO_2 in Earth's atmosphere in the 1950s, Lovelock's Gaia concept also depended on the development of a new technology, in his case for the measurement of gases in the atmospheres of other planets, in the search for signs of life. The atmosphere of a planet with life would contain a cocktail of gases out of equilibrium with one another, much like Earth's, while a planet without life would contain an atmosphere dominated by gases like CO_2, as on Mars and Venus (Lenton 2016). In due course, Lynn Margulis was instrumental in the United States for the publication in New York of a first 'complete annotated edition' of Vernadsky's *The Biosphere* (Vernadsky 1998), significantly cited by Crutzen and Stoermer (2000) and Crutzen (2002).

Over the 20th century, the epic scale of Earth history (e.g., Hazen et al. 2008; Lenton & Watson 2011; Zalasiewicz & Williams 2012) was becoming progressively clearer – not just its multi-billion[1]-year duration, as resolved by radiometric dating, which allowed the time necessary for the evolution of many successive life forms by Darwinian evolution, but also the profound nature of geological change. The plate tectonics revolution (Oreskes 1999; Oreskes 2003) showed that even ocean basins and mountain ranges were ephemeral features on a planetary timescale, while detailed geological studies showed that rare, extraordinary volcanic outbursts (far greater than anything in recorded human history) and meteorite impacts could fundamentally perturb the Earth System and lead to mass extinctions. Geologists also came to understand that the evidence of the last few million years, of the Ice Ages, revealed that present-day temperate landscapes were formerly buried under kilometre-thick sheets of ice, while global sea-level changes reached amplitudes of ~130 m, roughly twice the amount of sea-level rise that would happen if all of the Earth's present ice were melted (see Chapter 6 for a fuller discussion).

Small wonder that, until recently, the great majority of geologists thought human impact on the geology of the planet (if they thought of it at all) to be trivial and fleeting by comparison with these more obvious large-scale geological events. Collations of the physical impact on the Earth's geology (in terms of such things as volumes of raw material excavated) by such as Sherlock (1922) were impressive, but the resulting constructions were generally regarded as temporary, easily erodible structures that (once humans were no longer present) would simply be recycled back into the Earth by processes of erosion and sedimentation. There was also a tendency to regard geology as ending as human history began and giving way to disciplines such as anthropology, archaeology and written history (cf. Finney 2014).

One might take the opinions of the influential North American geologist Edward Wilber Berry (1925) on the Psychozoic as typical of widely held opinion in the international geological community through much of the 20th century. While admitting the 'magnitude and multifarious effects of human activity', he said that these were 'scarcely of geological magnitude' and that the Psychozoic was 'not only a false assumption, but altogether wrong in principle, and is really nurtured as a surviving or atavistic idea from the holocentric philosophy of the Middle Ages'.

Widespread acceptance that humans could profoundly alter the course of the Earth's geological evolution – and that geology (particularly stratigraphy) as a discipline reached into the present – emerged only slowly and fitfully, in the post-WWII years. Significant change in opinion was associated with such developments as the emergence of Earth System science, closely associated with the development of atmospheric science and the rise of biogeochemistry, and the ambitious International Geosphere-Biosphere Programme (IGBP) in the later part of the 20th century (see the 'Reflections on Earth System Science' by IGBP's leaders published in *Global Change*, Rosswall et al. 2015). These had built on earlier developments in the post-WWII years. Fairfield Osborn's book *Our Plundered Planet* wrote of 'man as now becoming for the first time a large-scale geological force' (Osborn 1948, p. 29) and included a chapter on this theme, with explicit reference to Vernadsky's work. The role of the early debate on the first Meadows report to the Club of Rome, *The Limits to Growth* (see Georgescu-Roegen 1975), was significant here, too, as illustrated by the emergence of Georgescu-Roegen's bioeconomic paradigm, in which he suggested that natural resources are irreversibly degraded once they are exploited in economic activity, and in which he developed concepts of ecological economics and industrial ecology.

These developments led to a growing appreciation of human impact (e.g., Turner et al. 1990), not so much upon the physical structures of the planet but rather on its chemical and biological fabric, with such phenomena as climate change and biodiversity loss

[1] Billion is used throughout this book as a thousand million.

coming to the fore. As a further factor, both the United States and the USSR started paying much more attention to the 'environment' as a theatre of warfare and pouring large amounts of funding into atmospheric and oceanic sciences. Given that such processes could be geologically long-lived (as regards climate change) or even permanent (as regards species extinctions), realisation grew of the scale and potentially lasting nature of human-driven perturbations.

Suggestions of 'geological' terms to describe this global change reappeared. Andrew Revkin published the term 'Anthrocene' in a 1992 book on global warming (Revkin 1992). The biologist Michael Samways (1999) coined the term 'Homogenocene' to encompass the increasing global homogenisation of animal and plant communities through widespread species invasions. The oceanographer Daniel Pauly (2010) came up with the term 'Myxocene' to describe his projection of future oceans dominated by jellyfish and microbial slime.

However, it was the term Anthropocene that began to take hold, initially within the Earth System science community. In February 2000, the term was offered on the spur of the moment by Paul Crutzen, the Nobel Prize–winning atmospheric chemist, at a meeting of the IGBP Scientific Committee in Cuernavaca, Mexico. Becoming progressively impatient at discussion of global change in the Holocene, he broke into the discussion, saying that we were no longer in the Holocene but in (and here he improvised) . . . the Anthropocene. Part of the rest of the meeting was taken in discussion of this idea; afterwards, Crutzen researched the term, found that it had been used for some years informally by a lake ecologist, Eugene Stoermer, and invited him to join him in publishing the term (though the two men never met). It was published in 2000, in the *IGBP Newsletter*; the article was invited and edited by IGBP executive director and newsletter editor Will Steffen, who had been present at the Mexico meeting. Two years later, Crutzen published a brief, vivid one-page article on the term in *Nature* in 2002, which gave the term wide visibility. He suggested that the Anthropocene began with the Industrial Revolution.

Continued research within the IGBP community led to the recognition that the time since ~1950 CE has without doubt seen the most rapid transformation of the human relationship with the natural world in the history of humankind (Steffen et al. 2004). At a 2005 Dahlem Conference on the history of the human-environment relationship, in which Crutzen participated, the sharp upward inflection of many trends of global significance in the mid-20th century was recognised as the 'Great Acceleration' (Hibbard et al. 2006). That term was first used in a journal article in 2007 (Steffen et al. 2007), in which it was regarded as a 'second stage' of the Anthropocene, following the Industrial Revolution.

The term Anthropocene began to be widely used and further analysed, particularly within the IGBP-based community (e.g., Steffen et al. 2004). In publications, the term began to be used as if it were a formal part of the Geological Time Scale, without inverted commas or other such qualifications – but it was not formal, and to this time it remains informal.

In response to the growing visibility and use of the term, the Stratigraphy Commission of the Geological Society of London considered the Anthropocene as a potential addition to the Geological Time Scale. Although it is a national body, not an international one, and has no power of formalisation, it published a discussion paper (Zalasiewicz et al. 2008) signed by a majority of commission members (21 out of 22) suggesting that there was geological evidence to support the term and that it should be examined further with respect to potential formalisation.

There followed an invitation from the Subcommission on Quaternary Stratigraphy, a component body of the International Commission on Stratigraphy (the body responsible for maintaining the Geological Time Scale, more technically known as the International Chronostratigraphic Chart), to set up an Anthropocene Working Group (AWG) to examine the case for formalisation. The AWG has been working since 2009 and has published two volumes of evidence (Williams et al. 2011; Waters et al. 2014), together with a number of individual papers on particular aspects (e.g., Edgeworth et al. 2015; Waters et al. 2015, 2016, 2018), as well as responses to emerging critiques of the Anthropocene from both the stratigraphic (Zalasiewicz et al. 2017d and references therein) and other communities (e.g., Zalasiewicz et al. 2018). This book represents a summary of these and related studies on the Anthropocene.

The AWG process was (and remains) in many ways novel as regards the assessment and determination of stratigraphic units – particularly in view of its inverted sequence of evidence and deductions (Barnosky 2014). Instead of stratigraphic names (such as the Cambrian, Cretaceous and so on) emerging from prolonged study of ancient strata, the Anthropocene Working Group was considering a concept that had emerged from another (albeit related) field of science and then determining whether it could work in both geohistorical terms (for example, as an Anthropocene Epoch) and stratal terms (to enable a time-based material unit of strata – an Anthropocene Series – to be recognised and correlated across the Earth) (see Section 1.3 for explanation of this distinction). The group also had to consider human phenomena and timescales as well as non-human, geological ones – and hence needed to include representatives of archaeology, ecology, oceanography, history, law and so on. There was also the matter of the very short timescale as compared with the million-year-scale units normally considered by stratigraphers (although the establishment of the Holocene had already provided an epoch-scale unit measured in centuries and millennia rather than in millions of years).

The stratigraphic examination of the Anthropocene has taken place in tandem with its exploration as a key concept by a wide variety of other disciplines, many from outside the Earth sciences and including the social sciences, humanities and arts (e.g., Hansen 2013; Chakrabarty 2014; Davies & Turpin 2015; Latour 2015; Angus 2016; Bonneuil & Fressoz 2016; Davis 2016; McNeill & Engelke 2016; Clark & Yusoff 2017; Hamilton 2017; see also McNeill 2001). The Anthropocene has been seen both as providing some measure of, and deep-time context to, human 'environmental' change to the planet and as integrating the effects of a wide variety of environmental change that are commonly considered more or less separately (such as climate change, biodiversity loss, ocean acidification). That integration is made via extension of the use of the 'multi-proxy' approach typical of modern stratigraphic studies, and it may be related to such compilations of global environmental change as in the 'indicator graphs' of Steffen et al. (2007, 2015) and the planetary boundaries concept (Rockström et al. 2009; Steffen et al. 2016).

Following several years' work, the AWG provided its initial findings and recommendations to the 2016 International Geological Congress held at Cape Town (Zalasiewicz et al. 2017d). It found, overall, that the Anthropocene possesses geological reality consistent with a potential formal time unit and that a proposal towards formalisation should be made, at the hierarchical level of epoch/series with a boundary to be defined by a GSSP (Global Boundary Stratotype Section and Point) at some level at or around the mid-20th century (Wolfe et al. 2013; Steffen et al. 2015; Zalasiewicz et al. 2015b; Waters et al. 2016). 'Bomb test' radionuclides were suggested as the primary marker.

Support for formalisation has not been unanimous within the stratigraphic community, and detailed and searching questions have been asked as to whether it is appropriate to consider a unit so geologically brief and with so many novel features as a part of the Geological Time Scale (e.g., Finney 2014; Gibbard & Walker 2014; Smil 2015; Finney & Edwards 2016; for responses see Zalasiewicz et al. 2017d). And there have been suggestions that the Anthropocene should not be defined in geological terms but should become a term of the social sciences – or be suppressed because it is inappropriate to other disciplines (Ellis et al. 2017;

Bauer & Ellis 2018; for responses see Zalasiewicz et al. 2017b; Zalasiewicz et al. 2018).

As these debates proceed, the current focus of the AWG is on identifying potential GSSP candidate sites within suitable kinds of sedimentary archive (such as annually laminated lake, marine or polar ice deposits; see Section 7.8 herein and Waters et al. 2018) and also on exploring the utility of a potential formal Anthropocene unit both to the Earth sciences and to other fields of study, all in preparation for a formal proposal to the ICS. Meanwhile, the use of the Anthropocene continues to expand into areas where Earth history once did not venture. Its future status may in general be regarded as secure as regards concept but uncertain in formal terms.

1.3 Stratigraphy and the Geological Time Scale

Jan Zalasiewicz, Colin P. Summerhayes, Martin J. Head, Scott Wing, Phil Gibbard and Colin N. Waters

Earth history spans in excess of 4,500 million years and so is of the order of a million times longer than recorded human history. Geologists cope in practical terms with this enormous time span by resolving the main episodes of Earth history and representing these as named units of the Geological Time Scale[2] (Figure 1.3.1).

Essentially all of Earth history is gleaned from biological, chemical or physical evidence preserved in rocks, particularly within strata (because strata, being laid down successively one on top of another, can preserve a detailed record of successive events through time – and from this is derived the discipline of *stratigraphy*: the inference of geological history from the rock record).

The primacy of this strata-based evidence has led to there being two parallel means of classifying Earth history. There is a *geochronological* classification, simply of time intervals within which certain events and processes took place (for example, one might speak of the Quaternary Period in which we live, comprising the last 2.6 million years, approximately since major glaciations began occurring in both the Northern and Southern hemispheres). Together with this, there is a parallel time-based *chronostratigraphic* classification of the material record (i.e., of strata) that preserves the evidence of that history (thus, the Quaternary System is made up of all the strata laid down during the Quaternary Period).

The units are exactly parallel in scope, and if the definition of any of a pair of parallel units is changed, they are changed in lockstep with the other. Thus, when the Pleistocene was recently redefined to begin at 2.6 million years ago, to formally replace an older definition of 1.8 million years ago (Gibbard et al. 2010; and see Section 1.3.1.5), this change simultaneously affected both the Pleistocene Epoch and the Pleistocene Series. If the Anthropocene is to be defined as a formal geological time unit, as an Anthropocene Epoch, say, then that in current practice must have a material counterpart in the form of an Anthropocene Series.

The Anthropocene might be set at another rank (see discussion in Section 7.7). The Geological Time Scale is hierarchical, and smaller-scale units are grouped together into large ones (Figure 1.3.1). Thus, the largest geochronological units are eons (with eonothems as their material chronostratigraphic counterpart). We currently live in the Phanerozoic Eon, so far ~541 million years in duration, the beginning of which is tied to the emergence and diversification of metazoan organisms (see discussion in Section 1.3.1.1). Eons are divided into eras (and eonothems into erathems). We currently live in the Cenozoic Era, which began when a large meteorite

[2] Technically, this is the *International Chronostratigraphic Chart* of the International Commission on Stratigraphy, but in this book we use the more widely understood general term Geological Time Scale which it informs.

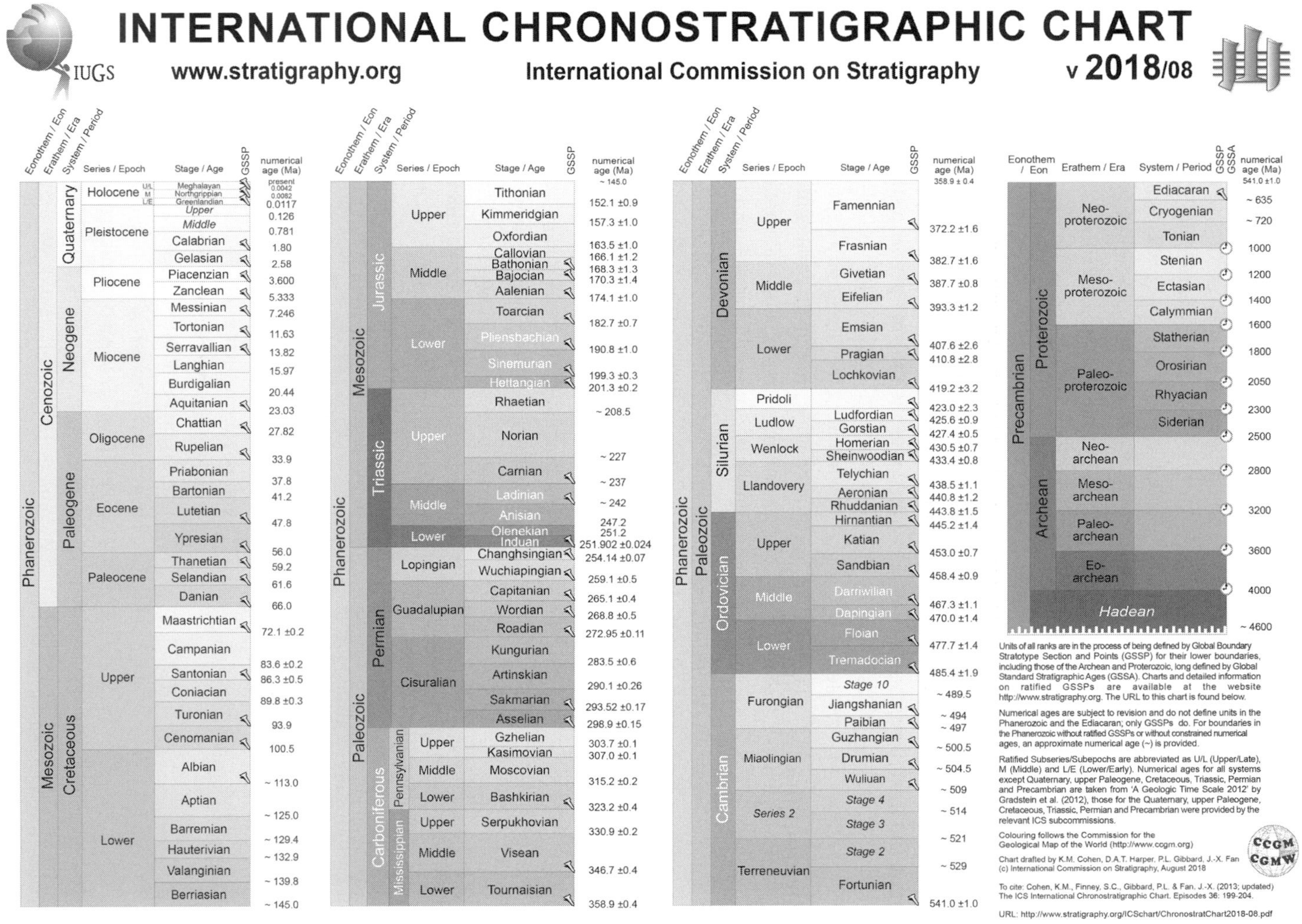

Figure 1.3.1 The Geological Time Scale of the International Commission on Stratigraphy (http://stratigraphy.org/index.php/ics-chart-timescale). Reproduced by permission © ICS International Commission on Stratigraphy 2018. (A black-and-white version of this figure appears in some formats. For the colour version, please refer to the plate section.)

strike ended (or gave the coup de grâce to, following other environmental perturbations) the Mesozoic world of non-avian dinosaurs on land and ammonites and belemnites in the seas, ~66 million years ago. Within this, we live in the Quaternary Period (often on landscapes underlain by deposits of the Quaternary System). Within the Quaternary Period, the last of many warm intervals that alternated with successive cold glacial phases is separated off (from the Pleistocene Epoch that makes up about 99.5% of the Quaternary) as the Holocene Epoch, within which we still live. The Holocene Epoch, though brief at ~11,700 years duration (see Section 1.3.1.6), is justifiable and generally unquestioned by geologists, because its deposits (of the Holocene Series) have largely formed our soils, river floodplains, deltas and coastal plains – and hence a good deal of our most fertile and productive terrains, while its deposits may be distinguished from those of previous interglacials by its rich archaeological record.

The characterisation and definition of these and other units of the Geological Time Scale are carried out not directly but as the end results of detailed study and classification of the strata by a range of stratigraphic means. Thus, the strata may be divided on the basis of their physical characters into *lithostratigraphic* units (bodies of rock or unconsolidated materials characterised by their lithologies and stratigraphic context). A lithostratigraphic unit may be very nearly of the same age, i.e., almost *synchronous*, throughout its extent: one based on a volcanic ash layer, for instance. Or it may be of substantially different ages in different places, i.e., *diachronous* – such as a fossilised beach deposit, progressively deposited across different parts of a landscape as sea level slowly rose or fell.

Strata (particularly those of the Phanerozoic Eon) may be divided up on the basis of the fossils they contain in the discipline of *biostratigraphy*. Biostratigraphic units (biozones) help establish the relative age of strata and facilitate *correlation* (i.e., they demonstrate age equivalence) between stratal successions in different places. They are thus *proxies* for time, often very good ones; but they are never perfect in this respect, because any species cannot appear (or disappear) everywhere simultaneously around the world and hence cannot define a single global time plane. Nevertheless, fossils are very often an excellent *guide* to a time boundary. Chapter 3 addresses the extent to which such biostratigraphic signals provide a useful means of correlating strata within the Anthropocene.

There is a variety of other means of classifying strata. One is through *chemostratigraphy*, exploiting different chemical patterns within strata. Particularly effective chemostratigraphic patterns are provided by ratios of stable isotopes of certain elements such as carbon and oxygen, as these may reflect global environmental changes and so can provide useful means of correlation of strata. This topic forms the basis for Chapter 5 in the context of the Anthropocene. Other correlatable patterns are provided by magnetostratigraphy, exploiting magnetic patterns preserved within rocks (see Section 2.6), notably patterns of reversals of the Earth's magnetic field. Yet others are based on changes in global sea level (sequence stratigraphy; see Section 6.3) or upon abrupt regional or global *events*, most notoriously the dusting of the Earth's surface with iridium-rich particles following the end-Mesozoic meteorite impact. Correlation is also helped by numerical calibration of such stratigraphic patterns, by means of radiometric dating or by the analysis of astronomically forced (Milankovitch) patterns in strata.

Several of these stratigraphic methods establish good to excellent correlation between stratal successions in different areas so that a detailed history of the world can be built up, with events taking place in different parts of the world being placed in their correct time order relative to each other. None of these methods provides *perfect* worldwide time planes (magnetic reversals come close to providing perfect time planes, though there have been none in late Quaternary time; see Section 2.6). So to provide a stable and reliable geological time framework, geologists use stable

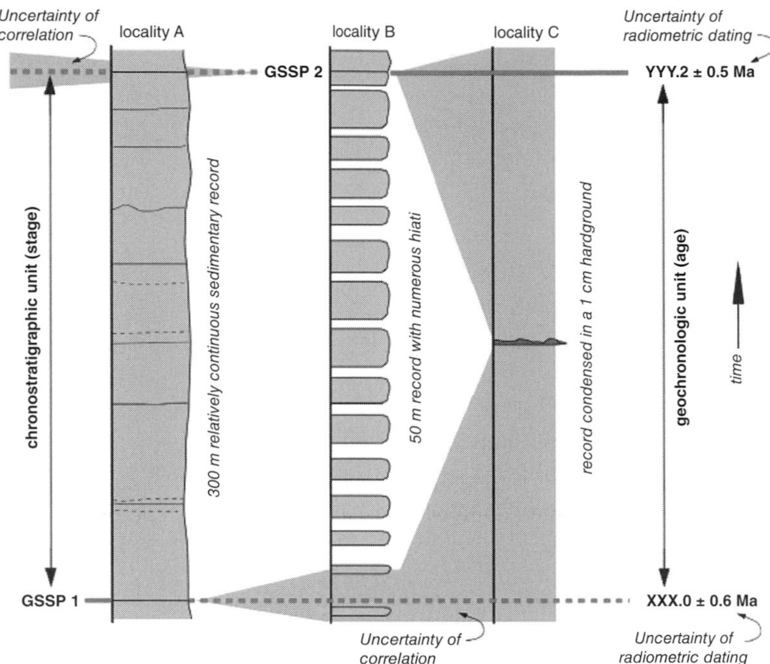

Figure 1.3.2 Diagram to illustrate GSSPs, correlation and uncertainties in correlation. Reproduced from figure 1 of Zalasiewicz et al. (2013). ©2013 GSA.

reference points within time. These are of two kinds: Global Boundary Stratotype Sections and Points (GSSPs), more commonly known as 'golden spikes'; and Global Standard Stratigraphic Ages (GSSAs).

Global Boundary Stratotype Section and Point (GSSP): To establish a GSSP, a single level is selected within a stratal succession (Figure 1.3.2), often close to where a common and distinctive fossil first appears or where there is a marked chemical change. This level is taken to have been deposited at the instant when the time interval began (Remane et al. 1996). Then geologists try to trace this level within strata all around the world, by any means possible. Importantly, the exact level chosen remains the reference point, even if the key fossil is later found to have appeared lower down in strata (i.e., earlier) at the same location (which has indeed sometimes happened; see Section 1.3.1.1). The ability to trace (i.e., to correlate) this level across the world varies greatly depending on how well the evidence is preserved in any particular case (Figure 1.3.2). For instance, in deep-ocean floor strata, the classic end-Mesozoic boundary, with its iridium-rich layer, can probably be traced to within a few millennia (in relative terms) or less. Individual tephra (volcanic ash) layers can provide similar stratigraphic resolution, as can Heinrich layers in the late Quaternary deposits of the North Atlantic, which represent debris layers dropped by sporadic iceberg 'armadas' (see Section 6.2.2). However, in strata that may represent, for example, desert dunes of the same general age, where the iridium dust would have been blown away, the degree of uncertainty in locating the boundary may be several million years.

Global Standard Stratigraphic Age (GSSA): For the older stratigraphic record, where fossils are scarce rendering unambiguous correlation by means of relative dating between stratal successions more difficult, boundaries are mostly defined in terms of numerical ages (GSSAs). For instance, the boundary between the Archean and Proterozoic eons is defined at 2.5 billion years ago exactly. To locate this boundary as precisely as possible, numerical means of dating such as radiometric methods are needed (though once the boundary is located in any particular succession by this means, then it may also be traced elsewhere around the world by any type of relative dating – by chemostratigraphy, for instance). Until recently, the beginning of the Holocene was in practice taken as a GSSA (see Section 1.3.1.6), and this has also been suggested as a potential means to establish an Anthropocene beginning (Zalasiewicz et al. 2015b; see Section 7.8 herein).

In defining the chronostratigraphic units of the Geological Time Scale, there are some further features of importance that will be of significance to consideration of the Anthropocene. Firstly, the determination of such geological units hinges much more on *effect* than on *cause*, not least because of the importance of strata, which are the physical archives of elapsed Earth processes, in their definition. One might illustrate this with the case for the Cretaceous-Palaeogene boundary (discussed more fully in Section 1.3.1.3), where the defining iridium layer, shocked quartz and mass extinction function as effective boundary markers regardless of whether they were the result of asteroid impact or extraordinary volcanic eruption, as has been debated (Alvarez et al. 1984; Officer et al. 1987). Thus, debates about the driving forces of the Anthropocene and the role of different modes of human social, technological and political behaviour (e.g., Chakrabarty 2014; Angus 2016; Hamilton 2017) are scientific questions of deep importance, just as are studies into the dynamics and wider effects of bolide impacts and volcanic eruptions. Yet it is the inherent pattern of strata and how well their particular characters can be recognised and correlated between different geographical places that act as the primary empirical basis for the Anthropocene as a geological unit. This is, of course, a basis that can then also help to inform scientific inquiry into the causes, processes and dynamics of the Anthropocene.

Then there is the significance of the particular name of the Anthropocene in this context. It is named after the ancient Greek term for human (*anthropos*) and *cene*, from *kainos*, the ancient Greek for 'new' or 'recent' time. However, as with all geological time terms (see Section 1.3.1), it has *no* particular significance or symbolic character – except that it is the name that has in practical terms clearly won out as regards global scientific recognition amongst the various other terms suggested for the phenomenon of a planet's geology deeply impacted by humans – the Anthrocene (Revkin 1992), the Homogenocene (Samways 1999), the Myxocene (Pauly 2010), the Plasticene, the Pyrocene, the Plantationocene, the Capitalocene (see Haraway 2015) and others, the names either reflecting a chosen part of the set of diagnostic characters or providing a suggested explanation for the causes of the epoch's existence. The Anthropocene is a name, a practical label, just like that of other geological time units considered below, such as Silurian, Triassic and Quaternary. The Silurian was originally named after the Silures, an ancient Welsh tribe; the Triassic was named because the strata where it was first described (but by no means everywhere) are made of three main rock types; and the Quaternary is a holdover from the times of Primary, Secondary, Tertiary and Quaternary geological time units (the Primary and Secondary have long been in disuse, while the Tertiary is no longer a formal unit). Within the contexts of epochs of the Cenozoic, as denoting 'human new', it might be said to strike a note little different from earlier Cenozoic epochs: thus, there is the 'old new' (Palaeocene), an 'early new' (Eocene), a 'little new' (Oligocene), a 'weak new' (Miocene), a 'more new' (Pliocene), a 'still more new' (Pleistocene) and a 'fully new' (Holocene).

As a geological unit, therefore, attempts to 'design' a name that might better symbolise its essence (e.g., the Chthulucene of Haraway 2015) would have little significance – even if such a name could be devised and agreed upon. There is considerable congruence between the meaning of the Anthropocene as originally devised and used in the Earth System science community and the Anthropocene as considered geologically, as a chronostratigraphic unit (Steffen et al. 2016; Zalasiewicz et al. 2017a). This, together with the way that the name has become quickly established in the literature, suggests that the term Anthropocene should be retained with this meaning – with appropriate qualifications as needed when it is necessary to distinguish it from other meanings and interpretations of the word.

Another feature of the chronostratigraphic units concerns the definition of their beginnings (as formal time units of geochronology) and bases (as the parallel formal time-rock units of chronostratigraphy; see Section 7.8). Once it is considered that there *is* a need to establish a beginning/base of a unit (because there are two distinct units of time and of strata that need to be separated), then the boundary between these two units is established pragmatically, for maximum ease of recognition worldwide – that is, to allow the best correlation between the strata and the events and processes that they represent in different regions. An additional corollary is that as this is a *time* boundary (whether 'abstractly' or in rock), then this boundary must be established so that – as far as is reasonably possible – it can be placed synchronously around the world.

This need not be the case for other kinds of boundaries in geology. A boundary between rock units (of what is known in geology as lithostratigraphy) follows changes in rock properties and can commonly be markedly time transgressive – that is, it can be of different ages in different places. Even the boundaries between palaeontological zones (of biostratigraphy) – although commonly used as guides to the relative age of the enclosing strata – are in reality generally time transgressive to some degree, reflecting the time it took for assemblages of animals and plants to migrate from one part of the world to another (see Section 3.3). But the aim is for chronostratigraphic boundaries to be synchronous, to provide clear separation between what used to be known as the 'holy trinity' of rocks, fossils and time. The Anthropocene, if it is to be considered as a geological time unit, must follow the same pattern.

Selection of such an Anthropocene boundary would thus firstly seek maximum time-correlation potential, with less emphasis placed upon factors interpreted to have most geohistorical significance. At a few established geological time boundaries in the ancient record, the two (correlative potential and geohistorical significance) coincide – as arguably with the Cretaceous-Palaeogene boundary (see Section 1.3.1.3). *Much* more often there are extended boundary intervals reflecting an array of complex changes in time and space, as one Earth state – and hence one pattern of strata and fossils – gives way to another, and decisions need to be made as to which event, in such a prolonged interval, provides the best time marker (see, e.g., Zalasiewicz & Williams 2014; Williams et al. 2014; discussed in more detail in Sections 1.3.1.1 and 1.3.1.2).

Within such a context, an effective Anthropocene boundary does not need to be based, say, on the earliest significant traces of human activity (for example, the wave of large mammal extinctions beginning in the Late Pleistocene) or even those that may be regarded as of most transformative significance (some 10,000 years ago, for instance, as agriculture started). Instead – and especially as the geological Anthropocene is in essence Earth centred (and strata based) rather than human centred – it should provide the clearest, most recognisable, most nearly synchronous geological division. The boundary, indeed, need not be based on a human-made signal. Had there been, say, a globally recognisable volcanic ash layer from some particularly violent single eruption somewhere within the boundary interval (if the 1815 Tambora event had been even larger, for instance; cf. Zalasiewicz et al. 2008), then that might have served admirably as a candidate boundary. Similarly, it is more important that the boundary allows the best tracing of a single time plane around the world than that it exactly coincides with the timing of greatest global change, and there are a number of boundaries of the

Geological Time Scale where the two (time plane and time of greatest change) are significantly offset (e.g., Zalasiewicz & Williams 2014). In the case of the Anthropocene, there is in fact reasonably close congruence between the boundary considered most optimal in this volume (see Figure 1.1.1) and the change in trajectory of major parts of the Earth System (perturbations to the carbon and nitrogen cycles, for instance).

Examples from the ancient record described below present a selection of chronostratigraphic boundaries, from ancient to geologically recent. They demonstrate the kind of evidence that has been used to divide the geological column into sensible and pragmatically recognisable time units, the kind of decisions and compromises that needed to be taken, and the creativity deployed to provide a clear and unambiguous framework for navigation within a complex and variable succession of both strata and planetary history. Establishing a proper definition for the Anthropocene will need similar decisions and compromises and comparable creativity.

1.3.1 Defining Units of the Geological Time Scale: Some Examples

1.3.1.1 Beginning of the Phanerozoic Eon (and Base of the Phanerozoic Eonothem)

This is arguably the most important geological boundary on Earth. It reflects the puzzlement of, amongst others, Charles Darwin when he contemplated the difference between the very old rocks of the Earth that we now call Precambrian (now an informal term), which seemed unfossiliferous, with the younger strata above that teemed with the fossils of arthropods, molluscs, worms and many other multicellular organisms.

This geologically rapid appearance and radiation of essentially all of the many animal groups is still something of an enigma, but its course is now better understood (Erwin et al. 2011) – and we also know that the Precambrian rocks do in fact include fossils, representing various forms of microbe or microbial colony and, in the later Precambrian, multicellular organisms too. Nevertheless, the evolution of animals represents a state shift on Earth at a more fundamental level than just providing a range of easily visible fossils to date and correlate strata with (important though that is). Via their complex trophic networks, they fundamentally changed the cycles of carbon, phosphorus and nitrogen; by burrowing through the seafloor, they disrupted the microbial mats that had held sway for over three billion years; and by filter feeding, they cleaned the ocean waters of fine particulate organic matter, allowing their easier oxygenation (Butterfield 2011). Hence, in terms of both Earth System function and stratal distinctiveness – producing bioturbated, macrofossil-bearing rocks – this is fully consistent with an eon-scale difference.

As larger-scale boundaries also define smaller-scale ones, the beginning of the Phanerozoic Eon also aligns with that of the Paleozoic Era, the Cambrian Period, the Terreneuvian Epoch and the Fortunian Age (Figure 1.3.1). The problem, though, is to pick a precise boundary.

Examined more closely, the 'Cambrian explosion' of animal groups appears as a complex, stepped event (Erwin et al. 2011; Hou et al. 2017). Some 580 million years ago, enigmatic, extinct multicellular organisms called the Ediacaran biota appeared (Figure 1.3.3). These represent a novel set of metazoan morphologies that appear to have gone extinct after some 40 million years at the end of the Ediacaran Period. Some 550 million years ago, muscular wormlike organisms appeared and began leaving burrows in sediment layers. At some 549 million years ago, the earliest biomineralised (i.e., shelly) fossils are found. At ~541 million years ago, a distinctive type of burrow that has been given the name *Treptichnus pedum* appeared. Small shelly fossils, representing the skeletons of many metazoans, became widespread some 526 million years ago. About 521 million years ago, the trademark fossil of the Cambrian, the trilobite, appeared (Hou et al. 2017). There is other evidence in strata, such as organic-walled microfossils that show changes across this interval,

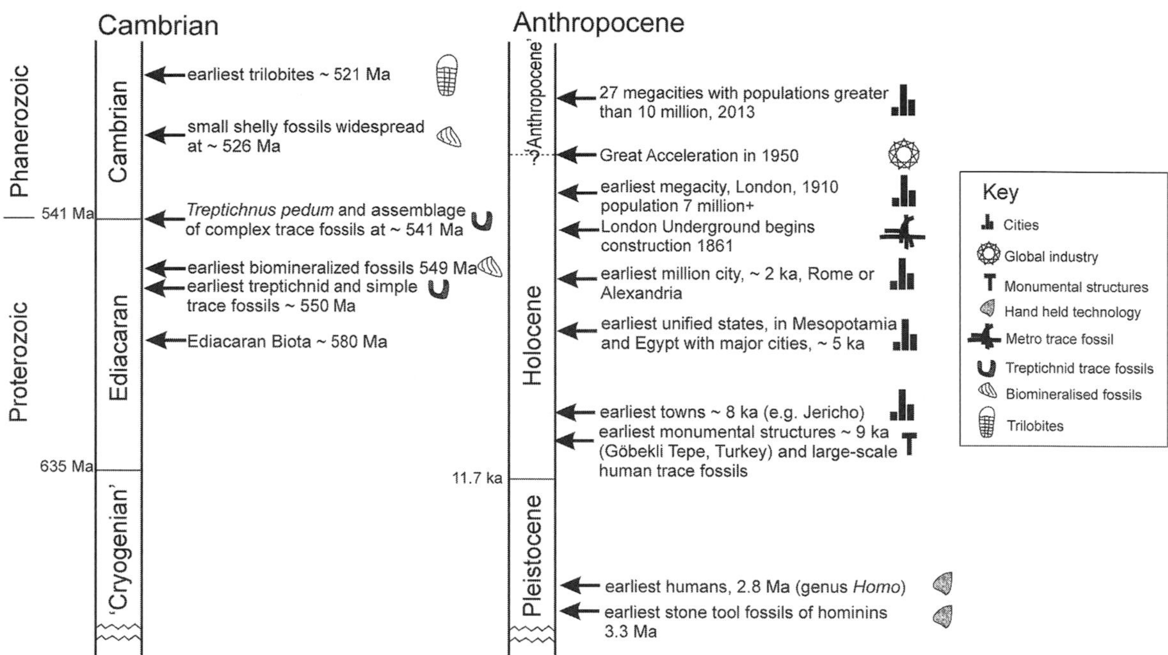

Figure 1.3.3 Cambrian boundary-related events compared with events related to a putative Anthropocene boundary; from Williams et al. (2014, figure 2) with amendments.

and there are global changes in ratios of the 'light' ^{12}C and 'heavy' ^{13}C isotopes of carbon, too. Which of these significant events serves best as a boundary?

The various possibilities were extensively debated by the working group tasked with this question. The main criterion was not which event 'started' this global transformation or which was deemed to be most important. Rather, it was which of them, pragmatically, provided the best time marker to enable correlation within rock strata around the world. Traditionally, the appearance of trilobites marked the boundary, but it turned out that these fossils appeared as representatives of two separate families on two separate continents (with uncertain relations between them and no common ancestor yet found), and so suggested definitions later focused more on the small shelly fossils at older levels. In 1992, the boundary was formally decided and ratified at a yet older level, with the lowest occurrence of the distinctive *Treptichnus pedum* burrows (Figure 1.3.3) in a stratal section at Fortune Head, in Newfoundland, coinciding with the GSSP (Landing 1994).

Since 1992, the boundary has not proved ideal. Firstly, *T. pedum* was later found *lower* than the designated GSSP level at Fortune Head, by up to 4 m. This does not move the boundary level (that is fixed) – but it does make it more difficult to use. More problematic, just below the (revised) lowest appearance of *T. pedum*, there was found to be a geological fault – a tectonic dislocation of strata – that made it impossible to work out precisely how much further down *T. pedum* may actually range at the GSSP section. The problems with the boundary deepened when subsequent work showed not only that *T. pedum* was confined to a rather limited range of shallow marine environments (it is not found in deep water or in terrestrial strata) but that it seemed to spread rather slowly, over several million years, across the world, within the environments that it did inhabit. There were taxonomic problems too (which, in general, are all too familiar to palaeontologists) in clearly separating *T. pedum* from related species of fossilised burrow.

A call has come (Babcock et al. 2014) to re-examine the whole question of the beginning of the Cambrian (and hence of the Paleozoic and the Phanerozoic), and it was suggested that all options should be open, from keeping the current boundary, with all its problems, to considering using the 'traditional' boundary of the appearance of trilobites (some 20 million years later than the first *T. pedum*) to using a prominent global change in carbon isotope ratios now often used as a de facto Cambrian base within strata (but which cannot be recognised at Fortune Head, because the strata there have been too strongly heated during metamorphism).

This tale provides an example of the complexities involved in defining a boundary, even a very major one. While formal time boundaries are *meant* to be permanent, to provide stability in communication between geologists, in practice they may (and often do) evolve to fit new data and new interpretations of Earth history. The rules simply stipulate that a boundary, once ratified, cannot be changed for a minimum of ten years.

Whatever its problems, the succession of events involved in the transition from the Precambrian (technically, from the Proterozoic Eon) to the Phanerozoic Eon has been used as an analogue of the succession of events that has been described in terms of a change from a 'Holocene' to an 'Anthropocene' world (Williams et al. 2014). The latter transition is compressed to centuries and decades rather than millions of years, but there are some similarities in the difficult decisions to be made regarding the choice of a single time or event that may be selected as a boundary within a complex transition and some parallels between the emergence of bioturbation (burrowing) by animals as an important process on Earth (in this case to a maximum of a few metres depth) and the development by humans of widespread 'anthroturbation' via mines, tunnels and boreholes (now commonly to kilometres depth; see Zalasiewicz et al. 2014b, Waters et al. 2018 in press and Chapter 4 herein).

1.3.1.2 Beginning of the Silurian Period (and Base of the Silurian System)

By comparison with the currently embattled Proterozoic–Phanerozoic boundary, the Ordovician-Silurian boundary (Figure 1.3.1), at an estimated 443.8 million years ago in the early Paleozoic (Melchin et al. 2012), is, for now at least, settled, effective in practice, widely accepted and uncontroversial.

The boundary exists largely because of the fundamental difference between Ordovician and Silurian fossil faunas. Within the Ordovician, for example, there lived such organisms as distinctive trinucleid trilobites, with no eyes but a remarkable pitted fringe around the head, and also a number of pelagic, free-swimming trilobites, which in the plankton were joined by a variety of multiple-branched graptolites (extinct animal colonies widely used to date and correlate the rocks). In the Silurian, by contrast, there were no more trinucleid or pelagic trilobites, and the graptolites were dominated by single-branched forms.

This major reorganisation of faunas and ecosystems came about because of a major biological crisis (one of the 'Big Five' mass extinctions in Earth history) that in turn coincided with the growth and subsequent collapse of a short-lived but intense glacial phase (Hammarlund et al. 2012). As ice rapidly grew (on what is now South America and northwest Africa, then conjoined and lying over the South Pole), sea level dropped precipitously, exposing much of the continental shelves, sweeping sediment into deep water and driving the first phase of the mass extinction. Less than a million years later, there came rapid deglaciation: water flooded back into the seas, which became deeper and extensively anoxic at the seafloor, causing the second phase of the mass extinction event. Global changes in carbon isotope ratios accompanied these perturbations of the Earth's biology. Following this, the surviving species evolved and radiated to recover overall diversity over the next few million years, but into patterns and taxa different from the Ordovician ones.

As with the Proterozoic–Phanerozoic boundary interval, there is a number of possible candidates where a precise boundary level might be defined. These include either of the biological extinction events, stratigraphic signals reflecting the major eustatic sea-level fall associated with the glacial acme of the Hirnantian Stage 445 million years ago (such as the sweeping of sediment from shallow into deep water), stratigraphic signals associated with the subsequent sea-level rise (such as the change from carbon-poor to carbon-rich sediment as seafloors became anoxic) or associated signals (such as the carbon isotope changes). In the end, it was decided to select an event that post-dated all of these: the appearance of a distinctive species of fossil graptolite, *Akidograptus ascensus* (in practice a couple of related graptolite species that appear quasi-simultaneously), in an early part of the biological recovery event, within an anoxic deepwater succession of rocks at Dob's Linn, in southern Scotland (Melchin et al. 2012 and references therein).

This was chosen pragmatically as a marker level because of the wide distribution of the boundary-determining species and the inference that they spread around the world sufficiently quickly to provide the nearest approximation to a traceable time plane. Their appearance is globally a trivial event compared with the major events that drove the Ordovician-Silurian transition, and their distribution is not completely worldwide (they are rare or absent in shallow-water strata and absent from terrestrial strata; to try to find the Ordovician-Silurian boundary within such deposits, other means of correlation must be used, and the placing of the boundary in these circumstances can be very approximate, with wide error bars).

The Ordovician-Silurian boundary provides a useful comparison for the stratigraphic changes associated with the Anthropocene (Zalasiewicz & Williams 2014). It represented a rapid change from icehouse to greenhouse conditions that involved substantial glacio-eustatic sea-level rise and a mass extinction of overwhelmingly shallow-marine biota in association with extensive anoxic conditions. But ultimately the Ordovician-Silurian boundary is marked by a minor biological event that, though it both postdates and is of far smaller scale than the main palaeoenvironmental changes, is nevertheless regarded as a useful time marker (Zalasiewicz & Williams 2014). As for the Ordovician-Silurian boundary, any putative Holocene-Anthropocene boundary could be defined using a plethora of criteria; similarly, the relevant signals in many cases are not concurrent and are evolving at different rates. Biological extinctions, sea-level changes and oceanic anoxia are in early stages of development in what may become extensive features of the near-future Earth. The concentration of carbon dioxide has increased in the atmosphere and become enriched in the light isotope of carbon as a result of fossil-fuel combustion since the Industrial Revolution. Novel materials such as plastic and novel chemicals such as artificial radionuclides have appeared and circulated the planet remarkably quickly. Ultimately, a signature will need to be chosen that provides the nearest approximation to a globally traceable time plane that represents the array of changes observed across this boundary (see discussion in Section 7.8).

1.3.1.3 The Mesozoic-Cenozoic Era Boundary

There is a great difference between the animals of the Mesozoic world – with non-avian dinosaurs on land and ammonites and belemnites abundant in the seas – and those of the succeeding 'modern' world of the Cenozoic, where these were no longer present but were replaced by mammals and a proliferation of bivalves and gastropods. This was one of the first great changes in the Earth System to be noticed by the early geologists, and the 'death of the dinosaurs' (and of many other organisms, as this event was further anatomised) became, for over a century, one of the great geological mysteries. Its analysis, and the hypotheses put forward to explain it, had considerable implications for how this mass extinction event might be used to precisely define the boundary between these two eras and hence also of the time units lower in the hierarchy (Cretaceous, Paleogene and so on; Figure 1.3.1).

Important and relevant questions included how rapid the extinction event was and what was the kill mechanism. Determining whether the disappearance of

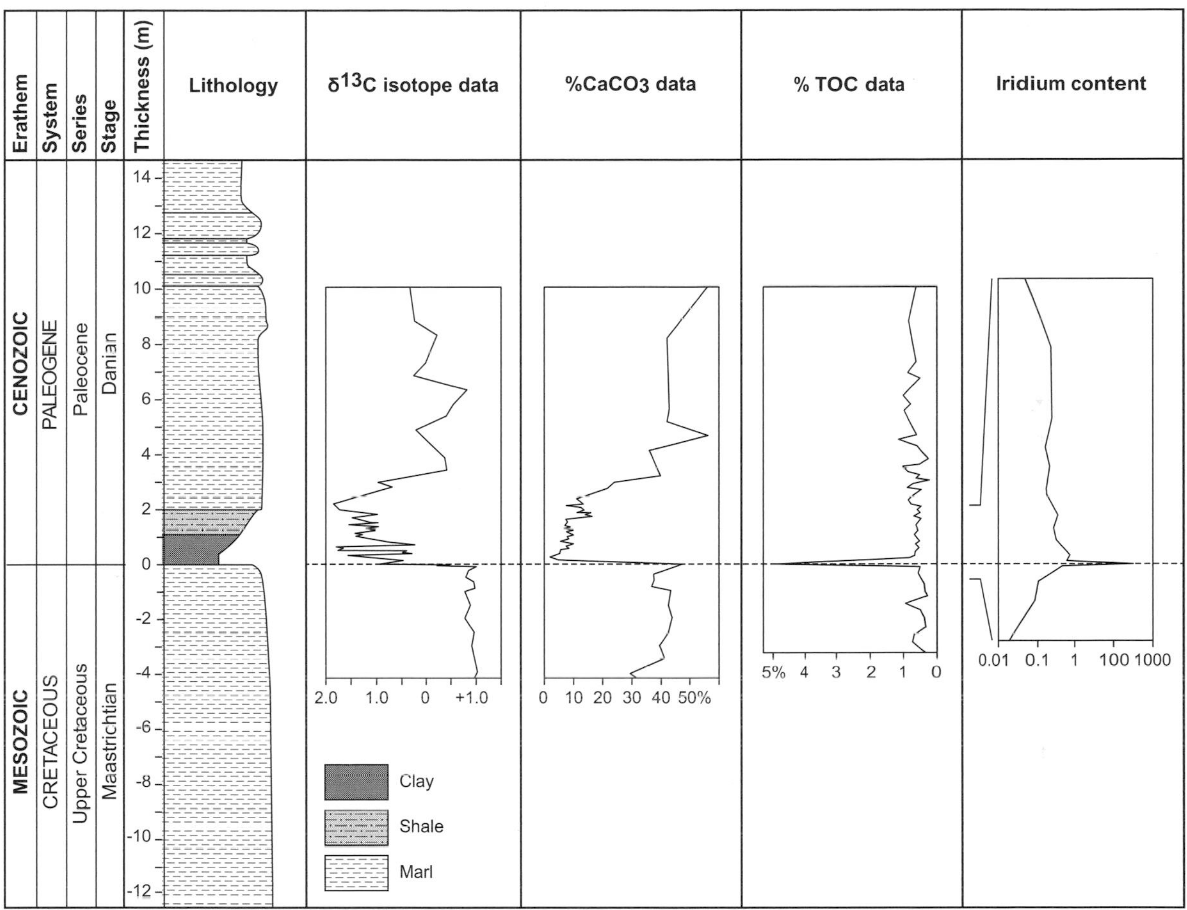

Figure 1.3.4 Stratigraphy of the El Kef GSSP section for the Cretaceous–Paleogene boundary; from figure 2 of Molina et al. (2006).

many species is effectively a single, globally synchronous event or a protracted and geographically varied decline over very many thousands of years is not a trivial task. It needs the painstaking collection and analysis of many fossil assemblages, through a number of stratal successions that represent diverse palaeoenvironments in different parts of the Earth. In any one of these individual rock successions, the geologist is plagued by what has become known as the Signor-Lipps effect (named after the two palaeontologists who formulated it), which in essence says that, at any one place, the geologist will not find all of the species then extant, partly because of the hit-and-miss of collecting and partly because of the patchy distribution of any species around the Earth, even within its favoured environment. This is all the more true for fossils that are rare anyway – such as dinosaurs.

The detailed pattern of species extinctions and appearances can thus be frustratingly difficult to pin down precisely, and for a long while it was unclear whether the species-extinction pattern here did represent a sudden crash or a slow decline. This made the question of causal mechanism harder to assess, with many ideas (climate change, vegetation change, volcanism and so on) being suggested yet remaining unconstrained in the absence of further evidence.

The discovery of an iridium anomaly at exactly the extinction level (Figure 1.3.4) both focused further

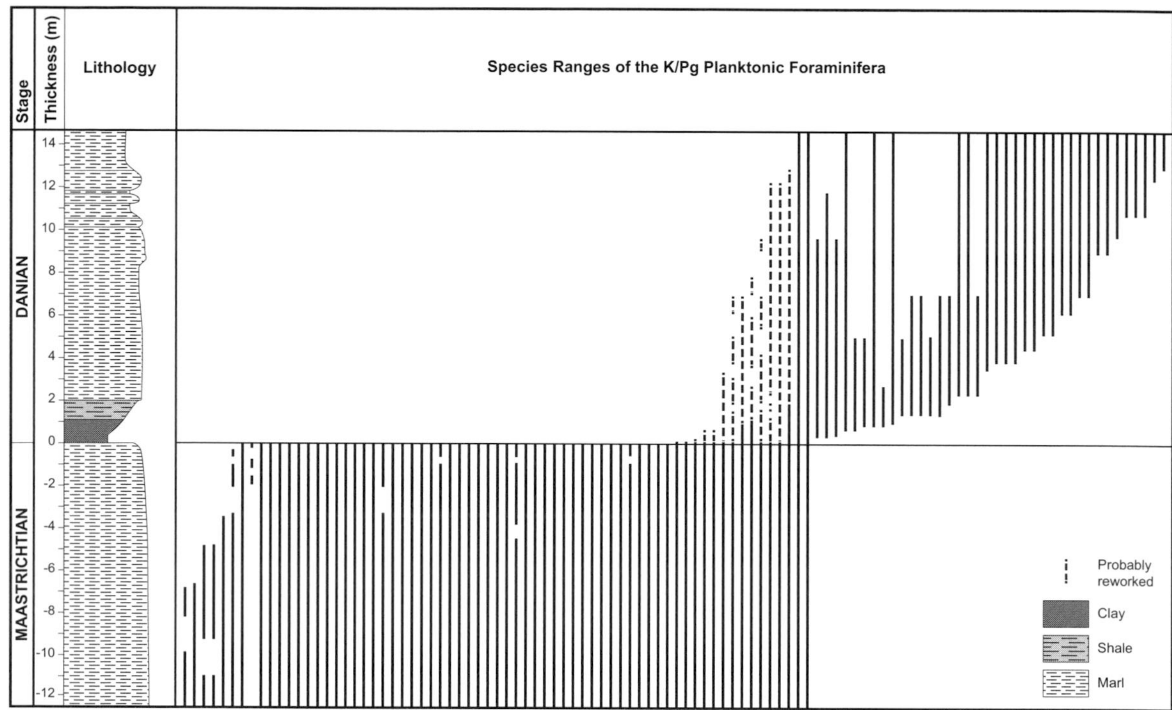

Figure 1.3.5 Planktonic foraminifera in the El Kef section. After figure 5 of Molina et al. (2006) and Arenillas et al. (2002).

palaeontological study and, for the first time, provided concrete, testable evidence of a kill mechanism – a major meteorite impact (Schulte et al. 2010). More detailed palaeontological investigations worldwide around this level, particularly of marine microfossils (Figure 1.3.5), indicated abrupt extinction, followed by a low-diversity 'survival' association. The iridium anomaly, too, was found to be commonly associated with a physical layer: a dark 'boundary clay' with a basal millimetric 'rusty layer' showing the maximum iridium enrichments, both interpreted as far-flung impact debris. This unit thickened towards a site in Mexico, where a 200 km diameter impact crater was discovered.

An impact-driven mass extinction has not been universally accepted, particularly by researchers who noted the coincidence with another event of potential global environmental impact: the outpouring of enormous amounts of basaltic lavas with attendant toxic gases on what is now the Indian subcontinent. Furthermore, the 'expanded' stratal sections close to the impact site (which, counter-intuitively, are harder to interpret than more distal sites) appear to show that the impact may have been separated from the mass extinction event by as much as 300,000 years (Keller et al. 2004; though see Schulte et al. 2010).

Nevertheless, many researchers now regard the impact as coincident with, and the primary cause of, the mass extinction event (or at least as a powerful and geologically instantaneous coup de grâce for a global ecosystem perhaps weakened by climate change and enhanced volcanic activity). This provides a rationale for using the impact event and its debris layer as a means to define the Mesozoic-Cenozoic boundary. The working group involved (see Molina et al. 2006) indeed almost unanimously chose this as the boundary level. The question then was at which precise level to place the boundary and where geographically to site it. Some votes were cast for levels such as the iridium maximum and the lowest occurrence of a microfossil species, but most (11 out of 19) opted for the base of the rusty layer at the bottom of the boundary clay. A number of geographically separated stratal sections exposing this same interval worldwide were considered as GSSP candidates. Their relative merits – such as perceived completeness of preservation, fossil content above and below the impact layer and accessibility – were assessed and voted on. The candidate obtaining most votes (26 out of 35 votes, with 4 nil responses) was a section at El Kef, in Tunisia. The choice was supported by the voting membership of the International Commission on Stratigraphy (winning 80% of votes). It has not been seriously questioned since, though there have been concerns about the degradation through weathering of the rocks at the site and its current accessibility.

This level, of course, marked the first incoming of meteoritic debris onto the seafloor at the spot that was to become El Kef, this debris having travelled from the Yucatan Peninsula, Mexico (then some 7,000 km away). This is probably the best-known geological time boundary in the world. Yet it is unique in that it is represented by a major event marker that closely approximates to being both global in scale and instantaneous in development; hence, the choice of boundary level is relatively straightforward. Most other major geological time boundaries (including those examples discussed in this chapter) are based on transitions that are considerably more protracted and variable in time and space, with potential boundary candidates needing to be assessed in detail in order to select the optimum level.

The debates concerning the definition of the Mesozoic-Cenozoic era boundary have particular relevance to the current Anthropocene debate. Firstly, the voting preferences of the working group indicate that such decisions are taken by supermajority, as required under ICS statutes; rarely can unanimous consensus be achieved amongst stratigraphers with diverse lines of research. Secondly, the meteorite impact that marked the end of the Cretaceous, though potentially vastly more destructive, might be considered analogous to the use of nuclear weapons in the mid-20th century, raising the question as to whether the first appearance of a fallout signature in a GSSP in a stratigraphic section or the timing of the actual first nuclear detonation is the most suitable signature (Zalasiewicz et al. 2015b).

1.3.1.4 The Paleocene-Eocene Epoch Boundary

Even in the early days of organised geology, the strata of the Cenozoic Era (Figure 1.3.1) were seen to include fossils that represented successively more modern animals and plants, and this interval was subdivided into a succession of epochs on this basis. The earliest of these epochs was originally the Eocene (the 'new dawn' following the time of dinosaurs), though in the late 19th century an even earlier unit, the Paleocene, was separated from this based on plant fossils from western Europe (Schimper 1874). The distinctiveness of the Paleocene was long disputed (Vandenberghe et al. 2012), but studies over many decades showed that this epoch is indeed clearly separable from the Eocene and furthermore by events that might be said to foreshadow, in some ways, the development of the Anthropocene.

The 'classical' Paleocene-Eocene boundary was recognised on the basis of fossils in a marine sedimentary succession near Ypres, in Belgium. Correlation of this level to other parts of the world proved problematic, though, with miscorrelations, particularly between land and sea, of up to 1.5 million years (Aubry et al. 2007). Study of biostratigraphy and isotopic chemistry in deep-marine borehole cores revealed evidence of a profound global perturbation at a level nearly a million years older than the classical one. Sedimentary and palaeontological signs of this perturbation could be not only recognised in deep-ocean cores but also correlated to environments ranging from the nearshore marine to river floodplains in continental interiors. These events form the basis of a revised boundary.

The chief event now used to recognise the beginning of the Eocene is a geologically sudden, global increase in the proportion of the light isotope of carbon (^{12}C). This shift, seen in strata that were deposited on both land and sea, was caused by a massive injection of isotopically light carbon into the atmosphere/ocean system from sources that are still disputed (McInerney & Wing 2011; Reynolds et al. 2017). The carbon release also provoked global warming (the Paleocene-Eocene Thermal Maximum, or PETM) and ocean acidification (Koch et al. 1992; Zachos et al. 2003, 2005; Penman et al. 2014; see also Section 6.1 herein). The effects of the carbon release on Earth's carbon cycle and climate lasted for nearly 200,000 years (Röhl et al. 2007; Murphy et al. 2010). The biotic effects of the PETM varied by ecosystem, with about 50% species extinction in some deep-sea groups (Thomas 2003), rapid evolution and poleward range expansion in surface-ocean plankton (Gibbs et al. 2006), major intercontinental migration and dwarfism amongst mammals (Gingerich 2006) and a combination of extinction and migration amongst plants (Wing & Currano 2013).

This perturbation provided several possible signals that were considered as primary markers, including some of the palaeontological events (including the extinction, appearance or sudden spread of particular microfossil species) and preserved changes to the Earth's magnetic field, although this last was quite unrelated to the perturbation. In the end, the GSSP level chosen and subsequently agreed to and ratified was the beginning of the marked change in carbon isotope values, at a well-exposed stratal section at Dababiya, in Egypt (Figure 1.3.6 herein; Vandenberghe et al. 2012).

It was something of a revolutionary decision, being the first Phanerozoic GSSP to be based on chemical signals in the strata, rather than on fossils. Nevertheless, it has subsequently found wide favour, not least because this particular signal is preserved in and provides a correlatory link between terrestrial and marine strata. This GSSP is considered one of the most successful of geochronological boundaries because its correlatability around the globe and across most environments has permitted studies at high temporal resolution (Miller & Wright 2017). This precedent has significant implications for the potential use of chemostratigraphic signatures to define the base of the Anthropocene (see Chapters 5 and 7).

The precise nature and timing of the rise in atmospheric/oceanic carbon has been disputed, with estimates ranging from essentially instantaneous (Wright & Schaller 2013) to as much as 20,000 years (Cui et al. 2011), though the best recent estimates are in the range of 3,000–5,000 years (Bowen et al. 2015; Zeebe et al. 2014). Part of the uncertainty in the duration of the onset arises because in many deep-marine sequences the acidification associated with the PETM destroyed calcareous microfossil evidence. Recent work in continental sections suggests there may have been additional carbon isotope excursions prior to the main one that is used to recognise the base of the Eocene (Bowen et al. 2015), demonstrating that even the best correlation tools are complex when investigated at fine scale.

1.3.1.5 The Neogene-Quaternary Period Boundary

The recognition by Leonardo da Vinci, ~1500 CE, that hard rocks were overlain by loose 'earth' has been taken as the beginnings of stratigraphy (Vai 2007), and over the next few centuries, that 'earth' formed the topmost part of later classifications, notably the 'fourth order' (*Quarto ordine*) of Giovanni Arduino (1760), later converted into 'Quaternary' (*Quaternaire*) by others, including Desnoyers (1829).

The realisation, in the mid- to late 19th century, that much of this 'earth' had been laid down or influenced by ice in a glacial climate strongly influenced subsequent attempts to define the term (see Pillans & Naish 2004). Closer analysis included, for instance, the recognition of cold-climate fossils (such as the mollusc *Arctica islandica*) within strata in the currently warm Mediterranean region. This kind of evidence was crucial in the decision in 1948 to place the Pliocene–Pleistocene boundary at the first indication of climate deterioration in the

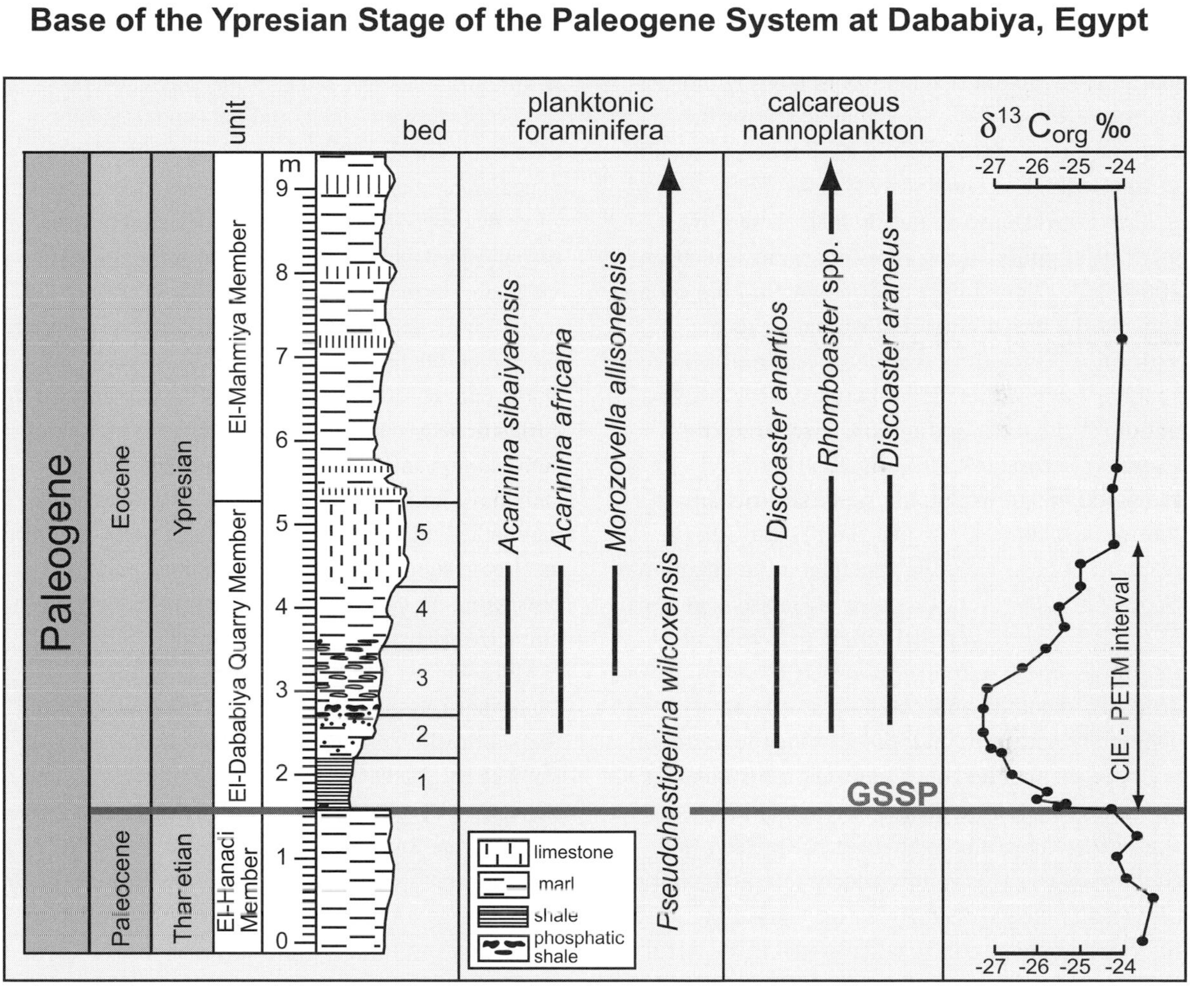

Base of the Ypresian Stage of the Paleogene System at Dababiya, Egypt

Figure 1.3.6 The Paleocene-Eocene GSSP at Dababiya, Egypt. (from figure 28.5 in Vandenberghe et al. 2012, based on Aubry et al. 2007). ©2012, with permission from Felix Gradstein

Italian succession (where recent tectonics had thrust magnificent, virtually continuous marine successions up on to dry land), at a level that was later determined to be ~1.8 million years old, and this level was formalised four decades later (Aguirre & Pasini 1985).

Subsequent research showed that this level was neither the beginning of the Cenozoic Ice Age (which had begun more than 30 million years earlier, when Antarctica became widely glaciated, at a level subsequently chosen to coincide with the beginning of the Oligocene Epoch) nor that of the intensification of Northern Hemisphere glaciation that fully established the bipolar glaciation that persists until today, which took place at about 2.7 million years ago. A date centred on ~2.6 million years ago was seen as a much more 'natural' and practically recognisable beginning to the Pleistocene (and hence Quaternary) by many workers than the ~1.8 Ma level, and it became a frequently, though not universally, used de facto boundary (see Gibbard et al. 2010).

Hence, there was tension between groups that used the unofficial 'natural' boundary and those (particularly amongst scientists working with deep-marine records) who were happy to see the 1.8 Ma boundary retained, citing the importance of stability of the Geological Time Scale.

A crisis was precipitated in 2004, when two influential publications appeared (Gradstein et al. 2004, 2005) that simply omitted the Quaternary and showed the Neogene Period extending to the present and including the Pleistocene and Holocene epochs. The reasons given included the 'archaic' nature of the term Quaternary, given that the Primary, Secondary and (arguably) Tertiary were no longer formal stratigraphic units. Intense debate followed, including not just the ICS but other relevant bodies, including the International Union of Quaternary Research (the latter a hierarchical equal to the IUGS and overwhelmingly in favour of retaining the Quaternary). A variety of solutions was proposed, debated and voted on. Out of this ultimately emerged, in 2009, a ratified lowering of the base of the Pleistocene and the reinstatement of the Quaternary, both sharing the same GSSP with an age of 2.6 Ma (Gibbard & Head 2010; Head & Gibbard 2015). This is now widely accepted – but not universally so, particularly within certain factions of the Neogene community (e.g., Hilgen et al. 2012).

What, then, is the Quaternary GSSP? Ratified in 2009 as the base of the Gelasian Stage, which had formerly belonged to the of the Pliocene Series, it is on a steep hillside at Monte San Nicola in Sicily that exposes rhythmically bedded marine strata, in which the rhythms represent astronomically driven ('Milankovitch') alternations in lithology (Pillans & Gibbard 2012, figure 30.3). The GSSP level is at the termination of Marine Isotope Stage 103[3] (Head & Gibbard 2015; Figure 1.3.7 herein), which can be correlated into the deep-ocean succession penetrated by the Ocean Drilling Program and its successors. In itself this isotope event is unremarkable, though it is within an interval of intensifying glacial events (Lisiecki & Raymo 2005), and palaeontologically, it does not coincide with either the extinction or the appearance of any of the marine microfossils used in these deposits. However, it coincides closely with a major palaeomagnetic reversal, from the Gauss magnetic chron of normal Earth polarity (Figure 1.3.7) to the Matuyama magnetic chron of reversed polarity (i.e., when the Earth's magnetic field 'flipped', the North Pole becoming the South). This can be used to correlate between any deposits (volcanic lavas and tuffs, some mudrocks) that, at the time of their formation, can preserve a signal of the Earth's magnetic field, with the reversal itself being regarded as effectively geologically synchronous and possessing the capacity to be preserved in both marine and terrestrial strata.

More widely, the basal boundary of the Quaternary System approximately coincides with the southward spread of ice-rafted debris onto the floor of the North Atlantic, with aridification and the spread of savannah in East Africa (and the slow rise of *Homo* spp., though for most of that time as a succession of rare, ecologically insignificant primate species) and with the beginning of substantial loess deposition in China (Pillans & Naish 2004; Gibbard et al. 2005; Figure 1.3.7 herein). It also coincides with a major shift in the position of the North Atlantic Current, with implications of a major climatic reorganisation of the Northern Hemisphere almost precisely at the beginning of the Quaternary (Hennissen et al. 2014, 2015). Hence, it is a generally effective level for a boundary that is substantial, albeit often transitional over hundreds of thousands of years, as regards global change. The best means of pinpointing the boundary is via the associated magnetic reversal and cyclostratigraphy; where these are not recorded (for instance, in some shallow marine or terrestrial sandy strata), boundary recognition may be associated with uncertainties.

[3] Even numbers are major glacial episodes, and the odd numbers in between are interglacials.

Base of the Gelasian Stage of the Pleistocene Series of the Quaternary System at Monte San Nicola, Italy

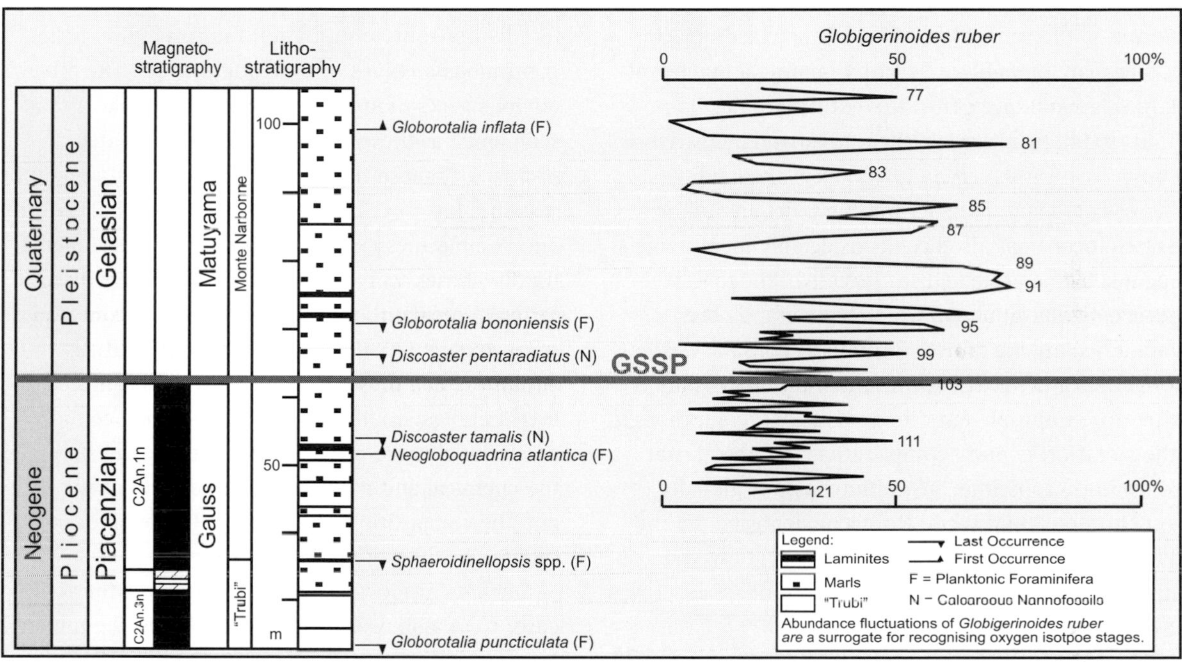

Figure 1.3.7 The base Quaternary GSSP at Monte San Nicola (from Pillans & Gibbard 2012). The succession of glacial and interglacial phases over the past five million years (Lisiecki & Raymo 2005) showing the Quaternary boundary at 2.6 Ma within a ~0.5 Myr transition from less to more intense climate oscillations. ©2012, with permission from Felix Gradstein.

The Neogene-Quaternary boundary may be analogous to the proposed Holocene–Anthropocene boundary in that no prominent palaeontological extinction or appearance event is recognisable. Criteria proposed to mark the base of the Quaternary, such as Milankovitch cyclicity and geomagnetic reversals, are not available to help define the base of the Anthropocene, but again this emphasises the need to identify the most suitable signature from a wide array of potential markers.

1.3.1.6 The Pleistocene-Holocene Epoch Boundary

Most of the geological time boundaries associated with the Quaternary are linked to climate change. The beginning of the Quaternary was recently repositioned to better reflect the intensification of Northern Hemisphere glaciation (see Section 1.3.1.5), while its division into Pleistocene and Holocene epochs is based on the warming (and associated major ~120 m sea-level rise) from the last of many glacial phases into the current interglacial phase, less than 12,000 years ago.

There are a number of reasons to separate off this current interglacial as an epoch in its own right, even though it is more than three orders of magnitude briefer than the average epoch (and more than two orders of magnitude briefer than the next shortest epoch, the Pliocene, which is ~2.7 million years in duration; Figure 1.3.1). Holocene deposits make up much of our landscape: soils, river floodplains and coastal plains, deltas and so forth; and in contrast with the preceding glacial-phase deposits, they are usually relatively easily distinguishable. The Holocene is the epoch in which humans made the transition from low numbers of hunter-gatherers, with relatively little wider impact (other than, probably, efficiently hunting a number of large mammal species towards extinction, e.g., Barnosky et al. 2014), to a farming, then urban and industrialised species present in very large numbers and having increasingly larger impacts on the wider environment. The Holocene has also been considerably more stable as regards temperature and sea level than the time of the preceding glacial phase, which has certainly been a factor helping the growth of human civilisation.

Defining the Holocene and tracing its deposits is therefore unproblematic in general. In detail, though, the situation is more complicated. The postglacial warming was neither instantaneous nor globally synchronous, and while Holocene deposits and the Pleistocene-Holocene transition interval are widely preserved, finding an appropriate boundary level was not straightforward.

Until 2008, the Pleistocene-Holocene boundary was in practice identified numerically, essentially as a GSSA, at 10,000 radiocarbon years before present (present being then defined as 1950 AD). However, this was known to be a rather poor approximation for the start of the current interglacial. In the Northern Hemisphere, the picture is of abrupt warming from the height of the last glacial, warmth persisting for some two millennia before abrupt cooling that brought back glacial conditions for a millennium in an interval called the Younger Dryas (after *Dryas octopetala*, an arctic and high-alpine flower that spread widely across European lowlands at this time), and then abrupt warming into the current interglacial (see Section 6.1.1). In the Northern Hemisphere, the obvious boundary to take for the Holocene was the abrupt transition at the end of the Younger Dryas, which took place about 11,700 years ago. But where was the best place to define this level as a GSSP, and how precisely could one recognise this level elsewhere?

It is normal to site GSSP boundaries in marine strata, but in the case of a boundary as recent as the Holocene, where one might seek very fine time resolution, most marine strata are too incomplete and too disrupted by bioturbation (burrowing) to form continuous archives of time and process. The other obvious types of strata considered were lake sediments, as these tend to be less bioturbated, particularly those that are varved, showing annual or seasonal layering; but even here problems of dating and completeness were encountered. The sediments finally chosen were deeply buried ice layers on central Greenland, representing virtually continuous snow accumulation, sampled by the NGRIP borehole (Monnin et al. 2001). Although the ice layers are essentially unfossiliferous, they contain a detailed archive of environmental change through the chemical and physical characteristics of the ice and the composition of air trapped in bubbles in the ice.

The succession chosen, represented in the NGRIP core, clearly shows via oxygen isotopes the regional temperature trend (Figure 1.3.8a herein; Walker et al. 2009). Other climate proxies include dust content, ice-layer thickness (the thinner the ice layers and the more dust, the more arid the climate over Greenland) and 'excess deuterium' levels (inferred as indicating distance from source of moisture).

The sharp temperature step at gross level resolves, in detail, as a change in most parameters over several decades from a glacial pattern (cold, dusty, arid) into a warm interglacial pattern (warm, moist); the sharpest change is seen in the excess deuterium (Figure 1.3.8b), interpreted as representing a reorganisation of North Atlantic ocean/atmospheric circulation in only a few years (Steffensen et al. 2008). The beginning of this was the level chosen, lying about midway in the local overall multi-decadal warming/moistening trend. Its position within the detailed oxygen isotope and dust record suggests it lies at the beginning of a small cooling

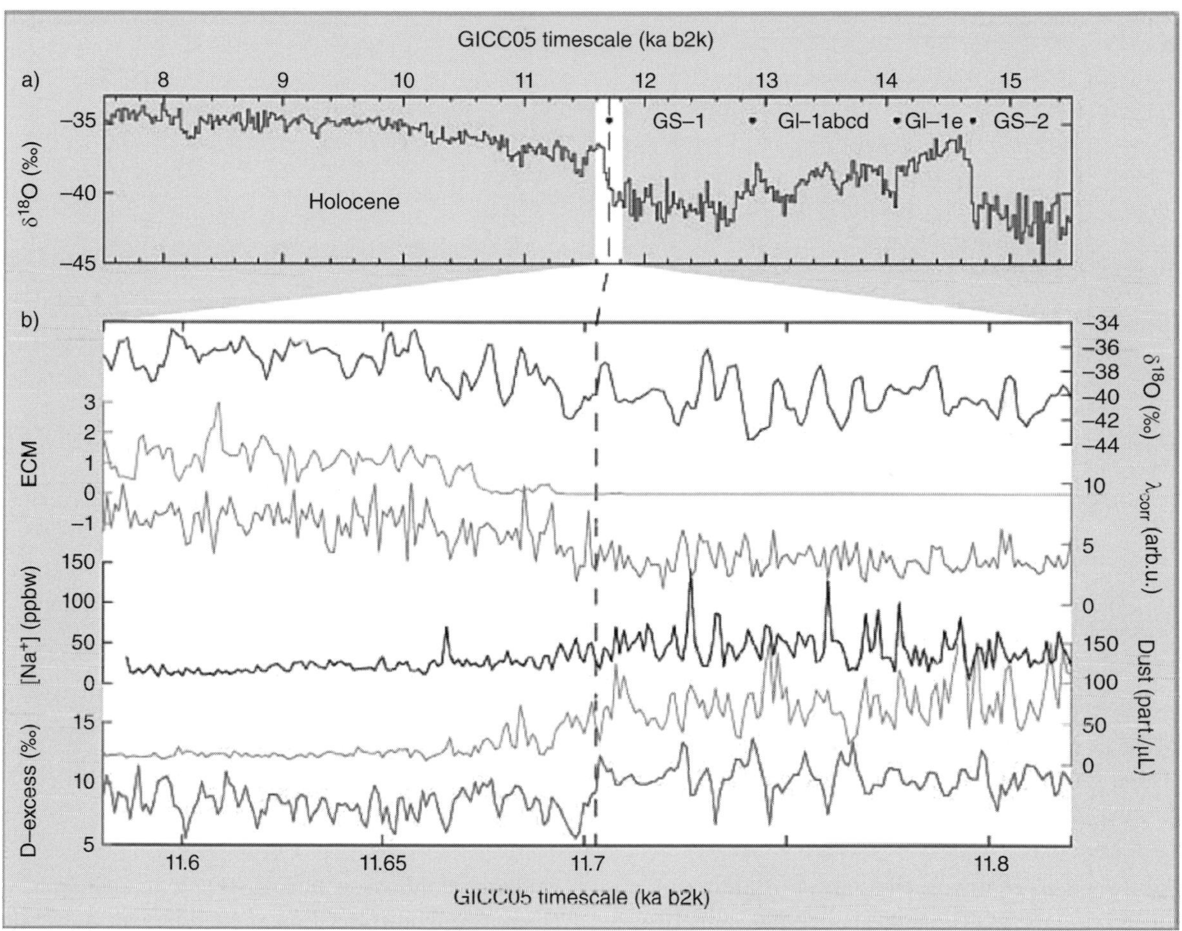

Figure 1.3.8 The Late Pleistocene-Holocene transition of the NGRIP ice core, Greenland, and the location of the GSSP. From Walker et al. (2009). Used with permission from Wiley.

oscillation lasting a few years, but this is one of several such oscillations within the marked overall warming trend (Figure 1.3.9).

How long ago did this happen? At such depths, ice layers are so compressed that annual layers generally cannot be clearly distinguished, but the estimate obtained was 11,700 years b2k (i.e., before 2000 CE) ± 99 years at 2 sigma (Walker et al. 2009), meaning that there is a 95% probability that the GSSP level occurs within the interval 11,601–11,799 years b2k.

How well can this GSSP be correlated? This might be illustrated by five 'auxiliary stratotypes' that were established together with the NGRIP GSSP (Walker et al. 2009). These, in five very closely studied sections (four lacustrine and one marine) from around the world, were correlated by a variety of means, including radiocarbon dating (though this is hampered around the Pleistocene-Holocene boundary by an approximately half-millennium 'plateau' in which it is difficult to tell dates apart, caused by fluctuations in radiocarbon production).

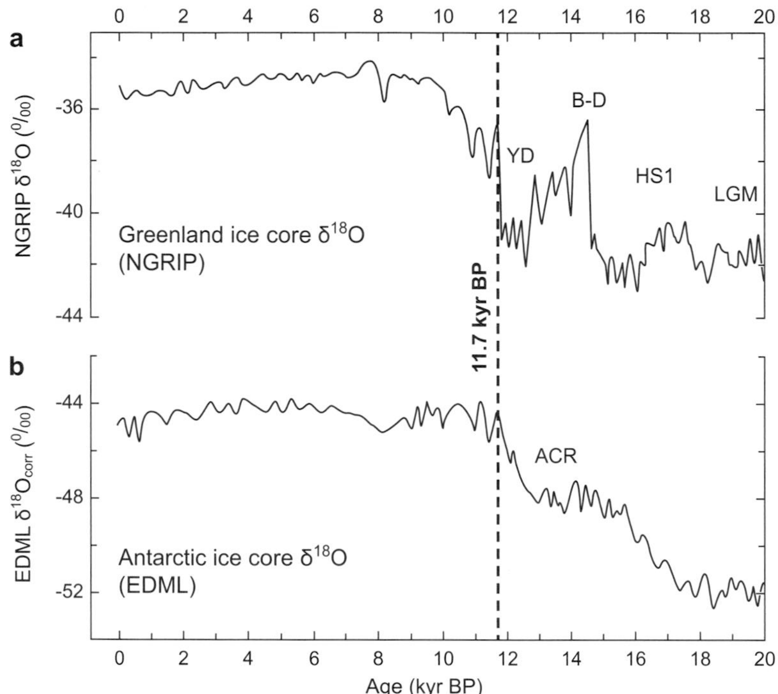

Figure 1.3.9 Contrasting climate histories in the Northern and Southern hemispheres. Modified and simplified after Felis et al. (2012). Abbreviations: Antarctic Cold Reversal (ACR); Bølling-Allerød (B-A); Heinrich Stadial 1 (HS1); Last Glacial Maximum (LGM); Younger Dryas (YD).

Figure 1.3.10 compares the published dating of these sections, which suggests optimal numerical correlation to the GGSP is to within a couple of centuries' error. With other, less well-studied sections, it will commonly exceed that.

In particular, it is difficult to translate precisely the GSSP into the Southern Hemisphere, because that had a different climate history to the Northern Hemisphere. Because of 'seesaw' redistribution of heat between the Northern and Southern Hemispheres, caused by changes in ocean circulation, the Younger Dryas interval and its clear termination are generally not directly reflected in the south. Rather, there was a more diffuse Antarctic Cold Reversal as an interruption to the warming, which began before the Younger Dryas interval and also had largely terminated before the abrupt end–Younger Dryas warming, so the south was in interglacial warmth while the north was still cold. This is an additional complication in correlation and serves to demonstrate the difficulty of establishing fixed time boundaries in successions of deposits or rocks that accumulated under rapidly changing environmental conditions. For instance, moraines show that glacial advances took place in the northern Andes, up into equatorial latitudes, during the Antarctic Cold Reversal rather than during the Younger Dryas interval as earlier thought (Jomelli et al. 2014).

In summary, therefore, the Holocene is a generally effectively defined epoch, in globally distributed, well-preserved and geologically very recent strata, with a thoughtfully and optimally chosen GSSP level. Nevertheless, correlation to this boundary commonly has an uncertainty of the order of a couple of centuries even in very well-studied, information-rich sections, illustrating the difficulty of locating

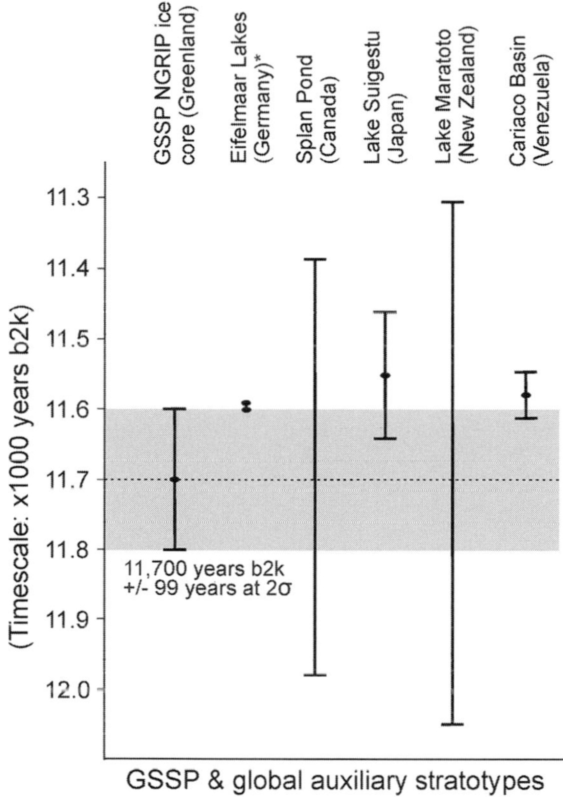

Figure 1.3.10 Comparison of republished numerical dates of the Holocene GSSP and auxiliary stratotypes. Data from Walker et al. (2009). * = dates given without error bar.

geological time boundaries *precisely* in sedimentary successions. The processes followed, involving initially recognising a specific age for the start of the epoch, followed by adoption of glacial core as the GSSP and a range of marine and lake successions as auxiliary stratotypes, can be a guide to the process by which the base of the Anthropocene may be determined (see Section 7.8).

1.3.1.7 Lessons

The lessons to be learned from this summary of some of the key boundaries defined in the Geological Time Scale are that for all the immense amount of work that has been achieved over the last two centuries, time boundaries remain difficult to define. In essence, all boundaries are a compromise, since the Earth System is so complex, and few if any of the many variables respond at the same time instant and at the same rate throughout the world. The identification of a boundary is a matter of consensus and ultimately convenience, since the boundaries defined in the Geological Time Scale have to be practical and as far as possible reflect real events in Earth history. The critical point is that no boundary is perfect, but if it is defined as carefully and as precisely as possible, based on the evidence available at the time, it forms an essential tool for understanding and communicating about the evolution of our changing planet.

1.4 The Utility of Formalisation of the Anthropocene for Science

Davor Vidas, Jan Zalasiewicz, Will Steffen, Trevor Hancock, Anthony Barnosky, Colin P. Summerhayes and Colin N. Waters

Since its proposal in natural science (Crutzen & Stoermer 2000; Crutzen 2002), the Anthropocene concept has become increasingly used also in the social sciences and humanities (e.g., Chakrabarty 2009; Vidas 2010; Latour 2015) to designate the time when humans began to decisively influence the state, dynamics and future of the Earth System. Earth's geological trajectory has already been profoundly and demonstrably modified. Yet there remains no formal acknowledgement, through appropriate scientific analysis, that we now live in a new and distinct geological time interval, the Anthropocene.

The mandate of the Anthropocene Working Group (AWG) consists of two tasks. Its first task is comparable to the process regarding any other proposed geological time unit: to identify and assess *geological evidence* on whether the Anthropocene is scientifically *justified* in stratigraphy, in having a sufficiently large, clear, distinctive and persistent 'geological signal' already preserved in strata. The second task of the AWG is specific to this proposed geological time unit: to explain the *usefulness* of formalisation of the Anthropocene for both geological and wider scientific communities, which in this case include those beyond the physical sciences.

Ongoing work by the AWG relates to both tasks. There is, however, a difference in the scope that can be expected in each. Evidence presented in a formal proposal by the AWG regarding the first task will have to be *conclusive* – and the study of the AWG in that respect must necessarily be comprehensive. For the second task – the study of scientific usefulness – the arguments by the AWG will have to be *convincing*, even if remaining introductory only. This area of study necessarily exceeds the confines of the AWG and even of geology, offering potential for broader, transdisciplinary impacts extending to other natural sciences as well as to social sciences and humanities. This, in turn, leads to a major engagement in explaining the transformative dimensions of the Anthropocene, once formalised, for science as a whole.

To take into account broader scientific-community interest in the formalisation of a geological time interval is a special, although not entirely unique, feature of the Anthropocene and the AWG mandate (compare, for example, the 2004–2009 debate on the status and the base of the Quaternary; Head & Gibbard 2015). What is, however, unique here is the diversity and the extent to which a formal Anthropocene may inform different branches of science and their scientific communities.

The purpose of this section is to briefly present the main threads in this 'scientific utility' debate pertaining to formalisation of the Anthropocene. A key consideration here is to distinguish the scientific utility for geology (and Earth sciences more generally) from the potential scientific utility to other branches of natural science – as well as extending beyond these into, e.g., the life sciences or the social sciences and humanities (which we will illustrate below with reference to international law theory and public health science). The organisation into three main sections – on geology, on natural sciences and beyond natural sciences, respectively – follows this trichotomy of utility.

1.4.1 Utility for Geology

It has been questioned whether the formalisation of the Anthropocene would bring any utility for geology, and it has even been suggested that there would be no utility whatsoever (cf. Autin & Holbrook 2012; Gibbard & Walker 2014; Klein 2015; Finney & Edwards 2016; see response by Zalasiewicz et al. 2017d). However, that blanket statement ignores several relevant observations.

On a general level, names are given in science to label distinct phenomena that are clearly separable from other phenomena, in order to enable and facilitate scientific discussion. Therefore, the initial question is whether the Anthropocene has demonstrable reality as a stratigraphic unit (both as a geochronological unit of Earth history and as a material chronostratigraphic unit of strata). As discussed throughout this book (see also Waters et al. 2016), the Anthropocene is characterised by an array of widespread signatures that are lithostratigraphic, biostratigraphic and chemostratigraphic in nature and that may be traced across most of the Earth's surface. Some of these signatures have counterparts in older stratigraphic units (e.g., carbon isotope anomalies); others are novel, with no such counterparts (e.g., artificial radionuclides, plastics, and industrially-produced fly ash and glass microbead particles within sediments).

These signatures reflect a demonstrably distinct phase in Earth history, which departed from the overall Earth System stability of the Holocene at around the time of the Industrial Revolution and intensified markedly during the 'Great Acceleration' of the mid-20th century, which seems to be the optimum chronostratigraphic boundary level (see Sections 7.5 and 7.8), with marked perturbation of the carbon, nitrogen, phosphorus and other cycles and increasingly marked effects on Earth's biota (Zalasiewicz et al. 2017d, Figure 1.1.1). While humans had a succession of impacts on Earth's environments and ecology during the Holocene (and indeed from Late Pleistocene times; Section 7.2), which have left a rich archaeological record, these have largely been local to regional in nature and are also highly diachronous in time from one region to another. A slight rise in atmospheric CO_2 levels began about 7,000 years ago, a global change that was perhaps (though this is controversial) due to early agriculture (Ruddiman 2003, 2013; cf. Elsig et al. 2009), which may have forestalled the return to a glacial phase (Ganopolski et al. 2016; see also Clark et al. 2016) and so prolonged Holocene interglacial conditions (for further discussion, see Section 6.1).

While the changes we here associate with the Anthropocene are indeed geologically brief so far, their consequences have led to a clear shift of Earth history into a new trajectory, with effects that will variously persist from centuries to millennia to millions of years (and, in many cases, will likely intensify in the short to medium term). Some of the changes are irreversible, even if humanity and its environmental-forcing effects were somehow to disappear tomorrow. The nature of change in such a case determines the duration of the effects:

The physical effects ('artificial ground'/'urban strata'; see Section 2.5) will either be eroded away or be preserved as a stratigraphic event layer, depending on the geomorphological and tectonic situation (with much eroded material likely to include physical, chemical or biological traces of an urban provenance).

The climatic/oceanic acidification effects of carbon release to date (see Section 5.2) will be largely dissipated in ~50,000 years (Clark et al. 2016; Ganopolski et al. 2016), though with only modest further additions it will persist for ~100,000 years (Ganopolski et al. 2016), based on modelling studies and on ancient analogues. Hence, the beginning of a long-lived climate-event layer of distinct stratigraphic character is being produced, with some of the same signals, such as changes in carbon isotope ratios, as for the Paleocene-Eocene Thermal Maximum (see Section 1.3.1.4), the beginning of which defines the start of the Eocene Epoch.

The biological effects of extinctions, invasions and species redistributions (already considerable and in some respects without close analogues in Earth history; see Chapter 3 and Williams et al. 2015b, 2016) will be permanent, as future evolution on any part of the Earth will take place from the surviving/transplanted biological communities. Already, this has given rise to a recognisable Anthropocene biota and, in the far future, this will appear in the rock record as a geologically sharp and substantial palaeontological break between distinct pre-Anthropocene and Anthropocene strata.

The Anthropocene, hence, can already be reasonably said to be of long-term significance to the geological record and is therefore clearly geologically 'real' both as a unit of time and process and as a distinctive stratal unit across a large range of environments (see Section 7.8).

The process of formalisation of the Anthropocene includes the need to precisely fix the boundary level and thereby stabilise the meaning of this term for geology. If the mid-20th-century date is adopted, then the Anthropocene would not impact drastically upon the Holocene as a stratigraphic unit (other than terminating it at a level ~70 years ago) nor interfere with the tripartite subdivision of the Holocene (Walker et al. 2012), recently ratified. The degree of disruption to established stratigraphic nomenclature would arguably be considerably less than some recent changes to the Geological Time Scale, such as the lowering of the Pliocene-Pleistocene (and hence Neogene-Quaternary) boundary (Gibbard et al. 2010) or the wholesale restructuring of the Ordovician series/epochs and stages/ages (Webby 1998).

Is there any utility in mapping distinct Anthropocene deposits? The bulk composition of artificial deposits has changed markedly over the past 70 years. Such deposits are commonly rich in concrete (Waters & Zalasiewicz 2018) and plastic (Zalasiewicz et al. 2016a). Not only do these materials act as a means of dating the deposits, but also they impart engineering properties that are distinct from those typical of Holocene anthropogenic deposits. So on the local scale this may become an important factor when developing the urban cityscape, in which foundation design will require knowledge of the age of the underlying anthropogenic strata. A deposit rich in concrete-masonry debris will behave markedly differently to one where natural stone-building debris or bricks are abundant. Furthermore, the types of contaminants will vary markedly between Anthropocene and pre-Anthropocene deposits (see Sections 5.6 and 5.7). Polycyclic aromatic hydrocarbons (PAHs) sourced from a 19th-century coal-tar works will have different organic contaminants to those from a modern petrochemical works. Heavy metals from industrial activity during the Industrial Revolution will be markedly different from rare-earth elements widely used in the manufacture of electronic equipment. There is clear advantage in knowing which pollution species one may encounter as one excavates deeper through anthropogenic strata. These geochemical signals extend into what may be termed 'natural deposits' but which are clearly strongly influenced by human interaction.

On the global scale, the utility in producing a world map of Anthropocene deposits would need to be considered, not least because of the complexity of the processes involved: the Roman deposits of the English city of Leicester, for instance, are still an active deposit, as they get reworked time and time again by current redevelopment of the urban landscape (Edgeworth 2014). So old cities will be both temporarily and spatially highly complex (perhaps akin to strongly bioturbated deposits, but on a larger scale), while new parts of cities (e.g., much of modern Shanghai) will be essentially wholly Anthropocene, whereas the ancient centre of Shanghai will show a complex interworking of Holocene and Anthropocene anthropogenic deposits (Zalasiewicz et al. 2014c).

Furthermore, the process of investigating the definition of the Anthropocene, given the vast array of available signals and potential sites and the presence of highly resolvable successions, may help geologists better navigate the process of defining GSSPs in deep-time successions.

1.4.2 Scientific Utility beyond Geology

Here, the utility may be regarded within the context of the sciences connected with the study of recent and ongoing global change, most notably Earth System science (ESS), where the Anthropocene as a term originated (Crutzen & Stoermer 2000; Crutzen 2002).

For the ESS community, the significance of the Anthropocene is clear, as the term encapsulates many of the phenomena with which this community is concerned (climatic, oceanic, biotic change, etc.), and it provides an integrating concept in Earth System study. The paradigmatic-shift effect of the Anthropocene can explain the rapid spread of this term through the ESS community (Steffen et al. 2016). The chronostratigraphic aspects here are of lesser significance, as most of the data used in these studies come from direct observations of one kind or another, rather than from analysis of strata, though geological analysis of the Anthropocene has highlighted data that may be useful for Earth System science studies, such as the use of fly ash within recent deposits as a proxy for airborne particulates (Zalasiewicz et al. 2017a). It seems clear that use of the term will continue in this community, but formalisation may nevertheless bring benefits, such as stabilising the term with a meaning that is consistent with the way that it is understood in the ESS.

Importantly, ESS is concerned not only with contemporary changes in Earth System structure and functioning but also with past changes. Understanding the long-term dynamics of the Earth System is crucial to understanding the changes driven by human activities in the Anthropocene. Thus, ESS has relied strongly on the Geological Time Scale and its interpretations by the geological community to infer changes that have occurred in the past (Steffen et al. 2016). This synergistic relationship, crucial to ESS, would be continued into the future with the formalisation of the Anthropocene by the stratigraphic community.

The archaeological community, like the geological community, deals with material stratigraphies. In this community, though, the geological Anthropocene as a stratigraphic concept in some ways conflicts with preferred archaeological usage of the term, because of the inherent need for a chronostratigraphic unit to have a globally synchronous base/beginning. Seemingly similar archaeological time terms, such as Neolithic, Bronze Age and so on, and the newer concept of the 'archaeosphere' (Edgeworth 2014; Edgeworth et al. 2015) are all inherently diachronously bounded, especially when considered interregionally, and hence are more akin to the lithostratigraphic and biostratigraphic units of geology. Calendar time serves as a framework in archaeology to establish levels of synchroneity and diachroneity – something that was long impossible for geological successions and that remains difficult for these, even following the advent of different means of radiometric dating.

Nevertheless, some Anthropocene deposits – particularly artificial ground – may be analysed and dated highly successfully by what are essentially archaeological techniques, such as applying typological studies to artefacts/technofossils, and indeed the potential for this kind of study could encourage multidisciplinary studies between archaeological, geological, urban-study and related communities (e.g., de Beer et al. 2012). A formal and stabilised Anthropocene boundary within this context should help analyse and understand the profound transformation of urban material culture that took place globally with the advent of the Great Acceleration in the mid-20th century.

1.4.3 Scientific Utility beyond Natural Sciences

The Anthropocene hypothesis has already passed beyond the boundaries of natural science, emerging as a new way of understanding the human role in environmental transformation and the implications of our actions for the world we live in. The relevance of the Anthropocene to many users helps explain why it has become such a popular term, already widely used to embrace a variety of meanings, from extreme human-driven modification of the planet to loss of biodiversity, modification of the landscape, pollution and climate change. Acceptance of this wider use of the Anthropocene concept has a further implication: it can no longer be expected that our global environmental background will remain stable, as was the case for much of the Holocene. The resultant impacts of this realisation on law, insurance, urban resilience, ability to ensure adequate food and water supplies and so on therefore need to be addressed.

Raising awareness of these issues is societally important – but this is quite separate from consideration of the stratigraphically based criteria for formalisation of the term. Confusing the possible wider use of a formalised Anthropocene with the evidence used to support formalisation would risk politicising the analysis, as noted by Finney and Edwards (2016).

Similarly, the potential societal relevance of a formal Anthropocene must be distinguished from its potential use for science.

Formalising the Anthropocene would impact on a wide range of communities in the life sciences, social sciences, humanities and arts. Amongst the many societal consequences (see, e.g., Dalby 2009; Tickell 2011) are its potential implications for interstate relations as regulated by international law (Vidas et al. 2015a), while its significance for public health science has also been recognised (Whitmee et al. 2015; Hancock et al. 2015). These two examples will be used here to illustrate several key distinctions that must be made in considering whether and how formalisation in geology, a (potential) outcome of scientific analysis by the relevant stratigraphic bodies, may be reflected in the sphere of social and life sciences.

1.4.3.1 The Example of International Law Scholarship

The first conceptualisation of how the Anthropocene may be linked to international law, specifically to the International Law of the Sea, took place shortly after the AWG was established in 2009 (Vidas 2010; Vidas 2011). Subsequently, the concept of the Anthropocene became more broadly discussed in international law, initially at international academic conferences from the early 2010s, in academic debates such as within the International Law Association (ILA Committee Interim Report 2016) and, increasingly over the course of the past few years, in scholarly writings on international law. Over this time, a trend can be observed from the initial focus on implications for the law of the sea, as well as on questions of international environmental law (Robinson 2012; Scott 2013; Kim & Bosselmann 2013; Ebbesson 2014; Kotzé 2014), to a much broader inquiry, involving the exploration of the potential relevance of the Anthropocene for international law more generally (Vidas 2015; Vidas et al. 2015b; Biber 2016; Hey 2016; Torres Camprubi 2016; Vinuales 2016).

Here, the Anthropocene poses some deep-lying conceptual questions. In today's international law, a fundamental notion is stability, which operates at two levels. One level concerns the conscious objective of working towards legally guaranteed stability in international relations, which themselves are vulnerable to frequent political change. The other level of stability is implied: it is based on human experience of the generally stable environmental (including geographical) conditions of the Late Holocene. Changes in that underlying element of stability, into the conditions of the Anthropocene, will bring about a fundamental shift of the context in which international law operates. This is a shift in which the challenges are increasingly recognised as the consequences of natural, not only political, change.

Throughout recent human history, an underlying stable condition of the Earth System has been taken as a given. This is the premise upon which our legal and political structures have been created over the past several centuries. In the relationship between international law and observed geographical features of the Earth – and indeed as regards the overall geological dimension of our planet – there has been an implicit assumption that current conditions form an objective and unchanging reality that has surrounded us since time immemorial. The definition of current international law may, in many respects, be said to be that of a system of rules resting on foundations that evolved under the circumstances of the Late Holocene, which are assumed to be everlasting. International law takes the observed conditions of the Holocene for granted, and on that premise a huge edifice of international law has been constructed over the past several centuries.

However, it is now becoming widely recognised that these underlying conditions are changing. For instance, the onset of a significant change in the ratio between sea, ice and land is already inbuilt due to ocean-atmosphere interplay and the delayed thermal response time of the oceans (DeConto & Pollard 2016). The removal of that underlying element of stability – and that is what the transition from the Holocene to the Anthropocene represents – contains the potential for an unprecedented new type of tension in the relations between states. This can spill over to, and aggravate, existing tensions between the territorial integrity of states and territorial claims – tensions which are already difficult to resolve because of the immense geopolitical differences between different states, on the one hand, and the sovereign equality of states as the founding postulate of international law, on the other. With the progressive onset of changing conditions in the Earth System and the possible formalisation of a new geological epoch as scientific response to this change, international law is set to become a subject of particular scrutiny (Vidas 2011, 2015).

The Anthropocene contains the potential for profound implications as regards international law in two main ways. The first is a shorter-term perspective: the formalisation of the Anthropocene as a new geological time unit in the history of the Earth, ratified through due scientific process in stratigraphy, may bring increased focus on the implications of such formalisation within the academic international law community. The second is directly related to the political consequences of the changing conditions in the Anthropocene, as these changes become ever more evident and seriously impinge upon daily life. Here the perspective is a longer-term one, although not restricted to some far theoretical future. Some of the changing conditions on the horizon, such as sea-level rise, may already become serious over this current century. The potential for interplay between those two types of implication is where a timely formalisation of the Anthropocene could play a crucial role for international law scholarship and its development of theory to meet the emerging challenges.

The first distinction to be made here concerns the difference between the reality of geological and Earth System change ascribed to the Anthropocene, on the one hand, and the formalisation of the Anthropocene in stratigraphy, on the other. While formalisation in itself will not alter any of the underlying geological realities, it contains the potential for shifting the focus in international law scholarship towards these issues, thereby contributing to the timely development of expertise for the elaboration of appropriate legal mechanisms and rules. Hence, as Anthropocene-related changes intensify and cause larger societal and political issues, the proposals for such mechanisms could already be in place, instead of needing to be improvised belatedly, once conflicting interests have already emerged and have become acute.

Such a perspective has not been present until very recently. For instance, it was absent during the negotiations of the United Nations Convention on the Law of the Sea (1973–1982), which codified the existing architecture of the law of the sea. This architecture of law was based on an assumption of the general stability of the coastal baselines, and it was upon these baselines that limits of all other maritime zones are now determined. As these coastal baselines are now set to change profoundly, acknowledgement of the need for progressive development of international law to take this profound change into account becomes of key importance, in order to facilitate the avoidance of future conflicts – or at least to contribute to diminishing the risk of such conflicts. The formalisation of the Anthropocene, based on objective geological evidence, could here play an important role in giving focus to the international law scholarship that will be required to facilitate the legal developments.

A relevant ongoing study is the work of the ILA International Committee on International Law and Sea Level Rise, which was established by the Executive Council of the ILA in November 2012, has since adopted its interim report (ILA Committee Interim Report 2016; Freestone et al. 2017) and presented its final report at the 78th ILA Conference in Sydney, in August 2018. The mandate of this international committee of legal scholars is ‘to study the possible impacts of sea-level rise and the implications under international law of the partial and complete inundation of state territory, or depopulation thereof, in particular of small island and low-lying states’ and ‘to develop proposals for the progressive development of international law in relation to the possible loss of all or of parts of state territory and maritime zones due to sea-level rise, including the impacts on statehood, nationality, and human rights’. The wider context for the proposal for this committee in 2012 was provided by scientific findings regarding the profound changes that have been taking place in the Earth System, especially since the second half of the 20th century. This included various lines of scientific evidence showing that the Earth may already be undergoing a shift from the conditions of the current officially accepted geological time interval, the Holocene, to a new planetary state (ILA Committee Interim Report 2016).

The formal stratigraphic analysis leading to potential formalisation of the Anthropocene may also have direct, more imminent effects in other spheres of international law, such as regarding treaty-interpretation theory and the application of the rules of the law of international treaties. The cornerstone of the law of treaties is contained in a general rule of law, codified in the Vienna Convention on the Law of Treaties (VCLT 1969), according to which every treaty in force is binding upon the parties to it and must be performed by them in good faith – the basic rule known as *pacta sunt servanda* (Article 26, VCLT). A fundamental change of circumstances, however, which has occurred with regard to those existing at the time of the conclusion of a treaty and which was not foreseen by the parties, may in some situations be invoked as grounds for terminating or withdrawing from the treaty, as well as grounds for suspending the operation of the treaty. For such termination or withdrawal from a treaty to happen legally, though, this fundamental change of circumstances must relate to those circumstances that constituted an essential basis of the consent of the parties to be bound by the treaty, and the effect of the change must be such to radically transform the extent of obligations still to be performed under the treaty (Article 62, VCLT). This rule, which is known as *clausula rebus sic stantibus*, could – being exposed to the progressively changing conditions of the Anthropocene – lead to increasing exculpation of the parties for unilaterally suspending the operation of international treaties on the grounds of unforeseeable changes.

This type of argument and such exculpation could, however, be difficult to invoke with the Anthropocene being a formally ratified geological time interval, since treaty parties could not argue that the change, being a manifestation of a formally ratified geological time interval – itself an epochal decision presumed to become a part of common public knowledge – was unforeseeable. Should the Anthropocene, in contrast, be seen as a cultural narrative, informal metaphor or the like, it would remain wide open to different interpretations, including those that are treaty related. In that context, the formalisation of the Anthropocene in geology could be seen in the light of providing legal certainty under international law, which is the ultimate goal of a legal order; and while at present this aspect may belong to a scholarly debate, its normative effects and links with the rules for treaty interpretation may become tested over time.

1.4.3.2 The Example of Public Health Science

The field of public health has always been concerned with the relationship between the environments where people live and their health. In the 19th century and well into the 20th century the focus was on issues such as clean water, sanitation, housing quality, air pollution and, more recently, persistent organic pollutants and urban design. But in the late 20th century, growing concern with the global environment led to apprehension about the health implications of this, perhaps best crystallised by McMichael (1993).

In 2013 Richard Horton, the editor of *The Lancet*, proposed the concept of 'planetary health' and tied the idea to the Anthropocene and to planetary boundaries, noting that 'the way we organise society's actions in the face of threats is more important than the threats themselves' (Horton 2013). This led to the creation of a Commission on Planetary Health, which defined planetary health in its final report, 'Safeguarding Human Health in the Anthropocene Epoch' (Whitmee et al. 2015), as follows:

> 'The achievement of the highest attainable standard of health, wellbeing, and equity worldwide through judicious attention to the human systems – political,

economic, and social – that shape the future of humanity and the Earth's natural systems that define the safe environmental limits within which humanity can flourish. Put simply, planetary health is the health of human civilisation and the state of the natural systems on which it depends.'

The *Lancet* report noted the evidence indicating fundamental and ongoing change to the Earth System, including large perturbations of the carbon, phosphorus and nitrogen cycles and changes to land use, erosion, climate and biosphere. It took this as evidence that humanity had become 'a primary determinant of Earth's biophysical conditions, giving rise to a new term for the present geological epoch, the Anthropocene'. As in discussions on the significance of the Anthropocene for international law, a key message of the *Lancet* report was that this proposed new epoch has brought about conditions generally characterised by unpredictability and uncertainty, for which systems of governance and the organisation of human knowledge with respect to human health are currently inadequate.

The link between health and the Anthropocene was made at about the same time in a discussion document and background paper prepared for the Canadian Public Health Association (Hancock et al. 2015) and in a number of other publications since then (Butler 2016; Hancock 2016, 2017; Hancock et al. 2016). Landrigan et al. (2017) noted the Anthropocene as context in their study of global pollution-related mortality, a phenomenon where a number of the pollutants involved may be monitored and assessed through stratigraphic proxy indicators (e.g., black carbon, heavy metals, persistent organic pollutants) as well as through direct environmental measurement.

Clearly, public health scientists and professionals see the utility of the Anthropocene and are beginning to use it in their work to safeguard and improve the health of the population. In this respect, too, the best solution is to plan for prevention and to encourage the kinds of development that will prevent health problems from arising (Summerhayes 2010).

1.4.4 The Utility of the Formalisation of the Anthropocene for Science: Key Distinctions

There is a key distinction to be made between a 'broader societal relevance' of the formalisation of the Anthropocene and its 'scientific usefulness' in the sphere of social sciences. With this distinction absent or not fully appreciated, the Anthropocene concept has sometimes been criticised as a political agenda or ideology under the guise of proposed geological epoch (see, e.g., Baskin 2015). Thus, the phrase 'the scientific and societal utility' of formalising the Anthropocene refers in fact to two profoundly different matters: One is the potential usefulness for science, involving or facilitating a paradigm shift (and this is the matter to which the mandate of the AWG study is limited). The second is a broader societal relevance due to enhanced awareness raising (and therefore stretching into the sphere of political perception of the Anthropocene), and this is a fundamentally different consideration. Why and how could the formal Anthropocene in geology be useful for science (including social science)? – that is the question of *utility for science*, which the AWG is addressing and aims at providing some clarification towards. What is the point of the formalisation exercise for the society at large? – that is the question of *societal relevance*, which is beyond the scope of, and independent of, the AWG mandate.

This distinction is perhaps not always easily appreciated from the perspective of a bona fide broader interested public and indeed often becomes blurred in some criticisms targeting the Anthropocene concept. The distinction can also be illustrated by the example of international law described above. A formalised Anthropocene, thus, can be of utility to the scholarly legal discipline within a broader social science spectrum, while a generalised Anthropocene concept (formal or informal) can be of political relevance in matters such as interstate relations.

It can also be illustrated in the case of public health, as briefly described above, as being both scientifically useful and societally relevant. For public health science, the paradigm shift resulting from the formalisation of the Anthropocene is to see global ecological change as a fundamental and vitally important determinant of health. But since public health is also political, in that it seeks to influence public policy and the market in favour of health (Rudolf Virchow famously stated in 1848: 'Medicine is a social science, and politics but medicine writ large'), the Anthropocene as a concept is societally relevant, pointing to the need to create social awareness and seek a policy response.

Political implications are sometimes alluded to with respect to a formal Anthropocene. However, it is important to be aware that *any* decision in the formalisation process – be it positive or negative to formalising the Anthropocene as a geological time interval – will have certain political implications. Decision either way, be it 'Holocene preserving' or 'Anthropocene introducing', can be expected to have political resonance. An explicit decision denying formalisation of the Anthropocene and resulting in the formal continuation of the Holocene would be as much a politically relevant statement as would be the inclusion of the Anthropocene as a new time interval in the Geological Time Scale.

The final consideration here relates to the responsibility of stratigraphers in specific and scientists in general, when faced with geologically relevant evidence of change, to record that change and, if appropriate, to formalise it. Geologists, thus, would be in error if they saw a scientifically demonstrable, significant and substantial change and did not give it commensurate recognition.

Thus, it is important to appreciate that there is potential utility to other scientific disciplines of the outcome of the formalisation process by the relevant stratigraphic bodies. It is the stratigraphic consideration and its outcome for the geological sciences that is the primary one for the Anthropocene concept: this would provide 'official' confirmation of a new geological time interval in the (ongoing) history of the Earth. In formal stratigraphy, there is a tightly regulated and rigorous process of formalisation applied in accordance with stratigraphic rules, representing due scientific process and procedurally involving a hierarchy of competent stratigraphic bodies legitimating the outcome and ultimately leading to ratification in the case of a positive decision. The outcome of this process will necessarily result in a spillover to other scientific disciplines, some of which may appear as distant as international law theory and public health. From this perspective, however, the features of a well-regulated, rigorous process of stratigraphic formalisation are invaluable and profoundly different from a situation in which the Anthropocene remains part of an informal scientific vocabulary or cultural narrative.

2 Stratigraphic Signatures of the Anthropocene

CONTENTS

We explore here the broad stratigraphic expression of the Anthropocene. Novel 'minerals' and novel human-made 'rocks', novel particulates, human reshaping of the landscape resulting in accumulations of artificial deposits, local magnetostratigraphic signals and modifications of soil development, have caused a broad change in patterns and compositions of sediments deposited across land surfaces, rivers, lakes, deltas and seafloors. We describe how, while some 5,300 natural minerals have evolved with time, they have recently been augmented by more than 180,000 new human-made mineral-like compounds. These novel 'minerals' form many new human-made rocks (e.g., concrete, brick, ceramics and asphalt), which in turn contribute to the development of diverse and widespread bodies of artificially modified ground. Novel particulate materials are widely distributed by the atmosphere, including glass microspheres, fly-ash, black carbon and organic carbon associated with fossil fuel and biomass burning. Mineral magnetic records in sediments can discern such particulate pollution as well as changes caused by deforestation and soil erosion. While deforestation for early agriculture increased the supply of sediments to the ocean, the subsequent construction of vast numbers of dams then severely reduced that supply.

2.1 Rock Components – Synthetic Mineral-Like Compounds

Robert M. Hazen and Jan Zalasiewicz

Each of Earth's more than 5,300 recognised mineral species is defined as a naturally occurring solid substance with a unique combination of crystal structure and chemical composition. Many mineral species are known to form exclusively via abiological processes (e.g., the majority of rock-forming igneous and metamorphic minerals). However, most species – perhaps two-thirds of the total number – are indirectly mediated by biological processes, most notably as a consequence of oxygenic photosynthesis (Hazen et al. 2008). In addition, dozens of minerals are the direct consequence of biology, either through biomineralisation (e.g., the calcite of mollusc shells and hydroxylapatite of teeth and bones) or through the decay of biomass (e.g., dozens of organic mineral species that occur in guano, coal, or decaying plant matter).

By contrast, human-made crystalline compounds of fixed chemical compositions are not generally supposed to be considered minerals, according to the rules of the International Mineralogical Association (Nickel 1995a, b; Nickel & Grice 1998). However, these taxonomic formalisms are complicated by the existence of dozens of human-mediated minerals that were approved prior to current rules – for example, minerals discovered coating bronze or tin artefacts, lining mine tunnel walls, or clogging geothermal piping systems (Hazen et al. 2017). Other similar compounds discovered in more recent years have not been approved as mineral species and have been referred to as 'synthetic mineral-like compounds' (Hazen et al. 2017).

In spite of these formalities regarding the official definition of a 'mineral', there is no question that the human contribution to the synthesis and distribution of mineral-like crystalline compounds is substantial. Here we consider human-influenced synthetic crystalline compounds within the context of the stratigraphy of terrestrial mineralogy.

2.1.1 Deep-Time Context

Hazen et al. (2008) outlined the growth in mineral species during Earth's 4.5-billion-year evolution in terms of mineral eras and stages (Figure 2.1.1). A first era began with clouds of interstellar mineral dust, the aftermath of supernova explosions, comprising about a dozen 'ur-minerals' – diamond, graphite and a few carbides, nitrides, oxides and silicates. With formation of the solar nebula, the heat of the early Sun melted and recondensed the circling dust into primitive chondritic meteorites, where ~50 additional minerals formed, such as the first pyroxenes and feldspars. Larger meteorites and planetesimals grew through collisions and interacted with water from comets: new minerals such as calcite, gypsum, chlorite and the first clays appeared and can be seen in carbonaceous chondrites that fall to Earth, taking

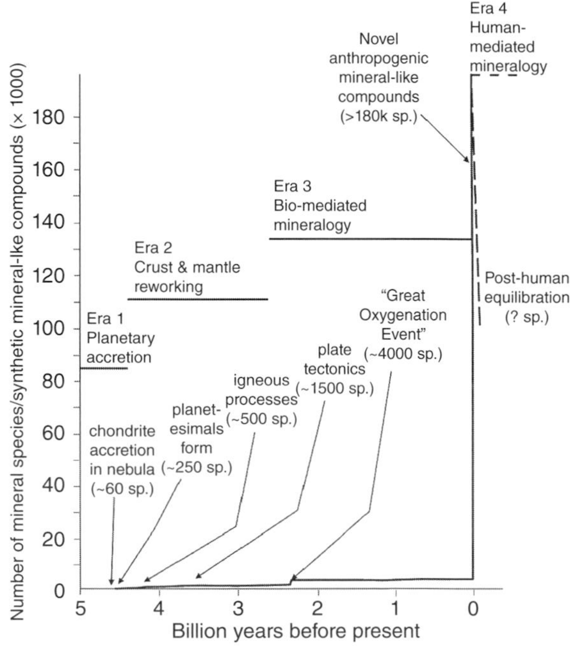

Figure 2.1.1 Earth's mineral evolution (after Hazen et al. 2008) modified to show the human addition (from Zalasiewicz et al. 2014a), the scale of which is here shown to be much greater than originally figured (see Section 2.1.2 and Hazen et al. 2017).

the total number of minerals to ~150. As larger planetesimals grew, they began to melt and differentiate into metal-rich cores and silicate shells, where further minerals such as quartz and potassium feldspar formed. By then, the ~250 minerals that can be found generally in meteorites were present.

As Earth formed, the second mineral era began. Classical igneous processes operating early on took the number of minerals to ~500 (roughly the same as has been estimated for Mars, whereas the anhydrous Moon has an estimated 350 mineral species; Hazen & Ferry 2010). Crust-related processes, such as the formation of granite pegmatites enriched in incompatible elements, saw further mineral diversity. Plate tectonics, with associated volcanism, deep metamorphism and ore-forming fluids, facilitated yet more mineral diversification. On such a dynamic (though still biologically dead) Earth, some 1,500 minerals may have been present. The origin of life on Earth, probably sometime before 3.5 billion years ago, though a momentous step, initially had little impact on mineralogy.

Then, about 2.4 billion years ago, the rise of oxygenic photosynthesis began to drive a new burst of mineral diversification and the third mineral era began. The 'Great Oxygenation Event' restricted the distribution of some mineral types, such as detrital pyrite or uraninite in terrestrial surface deposits (Rasmussen & Buick 1999). But in creating oxides and hydroxides of primary minerals, it more than doubled the number of minerals on Earth to greater than 4,000 (though most of these species are vanishingly rare). Further stages of this era modified the mineral inventory of Earth, but with little further increase in new minerals. However, old minerals took a multitude of new morphologies as biomineralised structures, independently acquired across several different animal groups from the Late Proterozoic to Early Phanerozoic (Porter 2011), including structures such as the siliceous skeletons of sponges, the carbonate skeletons of bivalve shells and corals, and ultimately the apatite bones of humans. Their first appearance in rock strata marks a fundamental change in Earth's biosphere from one dominated by microbes to one enriched by complex metazoan life.

The next major, ongoing development in Earth's mineralogy is driven by humans. The current, rapid phase of global mineralogical innovation is, indeed, a distinctive and robust feature of the Anthropocene (Zalasiewicz et al. 2014a; Hazen et al. 2017; Heaney 2017).

2.1.2 Novel Human-Made 'Minerals'

Hazen et al. (2017) identified two main kinds of anthropogenic mineral formation. Firstly, there are 'human-mediated minerals' – those that have arisen as inadvertent by-products of human activities, often associated with mining – for instance, by the weathering of mineral slags, crystallisation from mine drainage systems, or precipitation on tunnel walls, as well as corrosion products around archaeological artefacts. Some 206 of these human-mediated phases (~6% of the total IMA-approved mineral total of ~5,300) were recognised and named prior to the tighter IMA guidelines that excluded human-influenced crystalline inorganic compounds as minerals, but these have been retained nevertheless, as the guidelines are not retrospective.

In addition, Hazen et al. (2017) tabulated a number of representative synthetic mineral-like compounds that are directly manufactured for a wide variety of purposes. Most significant from a volumetric perspective are mass-produced ubiquitous building materials such as Portland cement, which is the basis of the widespread artificial 'rock' known as concrete, and clay-fired products such as porcelain and bricks (see Section 2.2). Less voluminous but equally widespread are innumerable technological crystals, including those used in semiconductor devices, magnets, phosphors and other electronic applications.

Also amongst the human additions to terrestrial mineralogy are pure or alloyed metals, rare on a prehuman Earth, where gold and (less commonly) copper and iron were found to naturally occur in amounts that could be exploited. Commencing early in the Holocene (see Section 7.3), humans have isolated metals by smelting from their compounds, beginning with lead, silver and tin (most copper and iron, too, had to be extracted from compound ores). In a burst of innovation from the late 18th to mid-20th century, most metals were isolated, including some never known to have existed previously in native form, such as magnesium, calcium, sodium, vanadium, and molybdenum and some that only occur rarely and in minuscule amounts, such as aluminium, titanium and zinc (Zalasiewicz et al. 2014a, figure 2). Novel metal alloys have also been produced (e.g., bronze, brass, pewter), as well as steel (iron-carbon alloys, often with chromium, molybdenum and other metals). Annual world production of aluminium now exceeds 35 million tons[1], and historical world-production estimates (USGS 2010) indicate that since the mid-20th century about half a billion tons of aluminium have been produced globally, sufficient to place the entire land area of the USA under a layer of standard kitchen aluminium foil (Figure 2.1.2). Annual world production of iron now exceeds a billion tons, and the total produced to date is probably of the order of 15 billion tons. Such amounts represent significant change to the Earth's surface mineral assemblage.

A wide range of other synthetic minerals, including novel forms of garnet for use in lasers (including yttrium-aluminium garnets, or 'YAG crystals') and as artificial gems; other synthetic crystalline materials used in lasers (including potassium titanyl phosphate and lithium niobate); boron nitride (Borazon), an industrial abrasive that rivals diamond in hardness; boron carbide, used in tank armour and bulletproof vests; tungsten carbide, used as the balls in ballpoint pens; various artificial zeolites; graphene; and many others. Hazen et al. (2017) note that the Inorganic Crystal Structure Database (http://icsd.fiz-karlsruhe.de) now lists more than 180,000 different types of 'synthetic mineral-like compounds'.

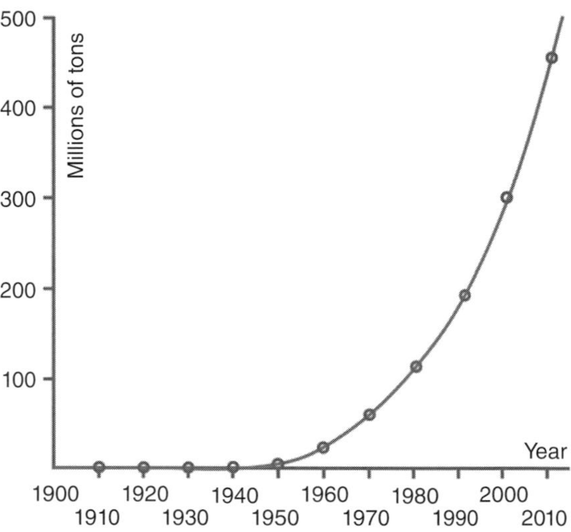

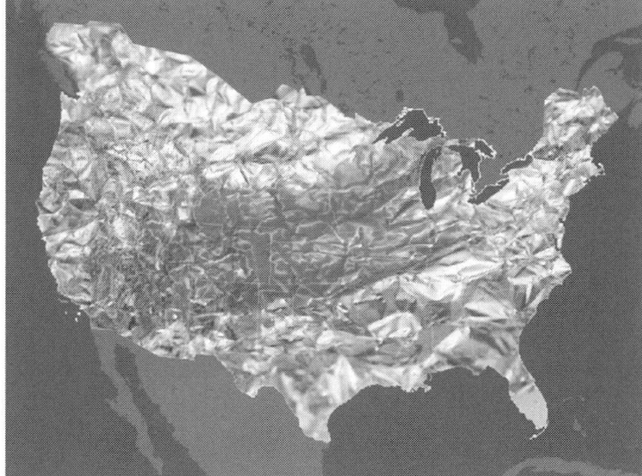

Figure 2.1.2 Global cumulative aluminium production expressed as a graph, and the same amount as a cover of standard kitchen aluminium foil over North America (image by Yesenia Thibault-Picazo). From Zalasiewicz et al. (2014a).

2.1.3 Mineral Textures and Habits

When considering unique aspects of the stratigraphic record of our time, one should consider novel textures and habits, in addition to the distinctive identities of synthetic minerals. Comparison may be made with

[1] Tons, as used throughout this book, refers to metric tons (or tonnes), equivalent to 1,000 kg.

the origin, increasing complexity and sheer range of biomineralised structures in Phanerozoic organisms. Biomineralisation may be regarded as a major mineral-forming process in its own right, given the complex architectures (down to the molecular scale) of shells and bones, not least where organisms have combined two or more minerals in one structure, as where bivalves and bryozoans may each combine aragonite and calcite in 'bimineralic' skeletons. New human-made minerals commonly also take on new and distinctive forms, or mineral habits, in their incorporation in artefacts or 'technofossils' (see Section 4.2). Anthropogenic mineral fabrication and combination is reaching near-biological states of sophistication, for example in the extraordinary, rapidly evolving miniaturisation of computer chips and central processing units. This patterning of crystalline productions is a significant elaboration of mineral habit in Earth history, the latest culmination in a trend where primordial magmatic textures began to be supplemented with metamorphic and hydrothermal textures as plate tectonics began, and later by biogenic textures, notably as a metazoan skeleton-building biota radiated following the so-called Cambrian explosion of life with hard parts. The new mineral habits created in human laboratories and factories represent an elaboration of mineral patterning as distinctive as and much more rapidly developed than that associated with the emergence of metazoan life.

2.1.4 Significance to Stratigraphic Definition and Wider Aspects

This abundant and varied synthetic production of mineral-like compounds indicates that a significant new epoch in Earth's mineral evolution has begun. Considered simply in terms of the range of new minerals *sensu lato*, it is the most striking new development since – and greatly exceeds the diversification rate and scale of – the Great Oxygenation Event of the Precambrian (Hazen et al. 2017). It represents, in effect, an explosive radiation of mineral-like species on Earth, some of which may never have been produced elsewhere in the universe (Hazen et al. 2017; Heaney 2017). Human action is hence accentuating the chemical disequilibrium on Earth – a disequilibrium that became a signature of this planet as soon as life evolved on it.

This change to Earth's near-surface mineral composition is likely to form a global-scale, robust and permanent signal within Earth's future stratigraphic record for at least two reasons. A major contributor, as outlined above, arises from the simple fact of 'mineral' diversification, with literally hundreds of thousands of new crystalline compounds, some of them of worldwide distribution in a distinctive stratigraphic horizon. In addition, as Hazen et al. (2017) emphasise, humans are also responsible for the widespread redistribution of naturally-occurring minerals, in applications from the large-scale use of anthropogenic 'xenoliths' employed as building and ornamental stone to gemstone and precious metals in jewellery to the decidedly unnatural juxtaposition of varied mineral species in countless amateur and museum collections around the world.

This new phase in Earth's mineral evolution may be short lived as an active phenomenon, either piecemeal as man-made 'minerals' become extinct (i.e., cease to form) when they come to be obsolete for humans or ultimately as humans themselves become extinct. However, many of the minerals already formed likely will be preserved for billions of years in the rock record, while others may alter into as-yet-unseen novel forms during diagenesis and metamorphism (Hazen et al. 2017).

Nevertheless, the new anthropogenic 'minerals' already provide a high-resolution stratigraphy (e.g., as noted for plastics above) that can help characterise Anthropocene strata today, augmented by the yet higher-resolution stratigraphy provided by the technofossils that the new minerals are shaped into (Section 4.2). Overall, these new human-made mineral-like species add a distinctive mineralogical character to the Anthropocene strata of the present and also of the geological future.

2.2 Anthropogenic Rock Types

Colin N. Waters and Andy Smith

Natural rocks comprise minerals and when present with associated rocks can be combined to form lithostratigraphic units. The same relationships are present within anthropogenic rocks, which comprise minerals that may be artificial (see Section 2.1) or natural, and these rocks in turn can be aggregated into mappable geological units known as artificial ground ('made ground' of earlier terminology; see Section 2.5), which can include both anthropogenic and natural rocks, with a consistent relationship of being emplaced by human action. As a recent deposit, artificial ground typically comprises unconsolidated aggregates, as the deposits have not undergone the deep burial, diagenesis and recrystallisation that convert natural sediments to sedimentary rocks. But in some circumstances human activity carries out the necessary imposition of heat and/or pressure to convert materials to anthropogenic rocks. The commonest novel rock types, which are described in this section, include concrete, brick, ceramics and asphalt.

2.2.1 Concrete

Concrete is the modern building material of choice (see Section 2.5.2), making it the most abundant anthropogenic sedimentary rock on the planet (Waters & Zalasiewicz 2018). The production process requires the transport of component cement and aggregates on scales far greater than natural geological processes, with production requiring the quarrying of vast amounts of limestone, clay, shale, sand and gravel, as well as crushed rock aggregates. It is estimated that in 2015 some 4.1 billion tons of Portland cement were produced (USGS 2016), some two orders of magnitude greater than the amount produced at the end of WWII (Figure 2.2.1). From

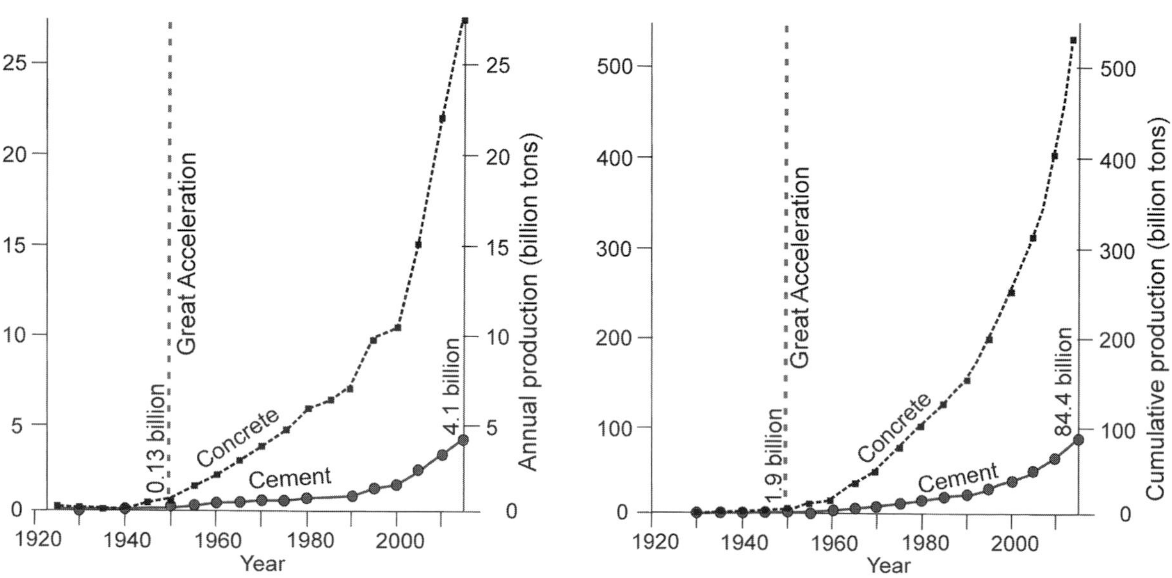

Figure 2.2.1 Global production of cement over the last century, derived from United States Geological Survey global cement-production statistics (from figure 6 of Waters & Zalasiewicz 2018). Approximate global production of concrete is derived by assuming most cement goes into concrete and that ~15% of average concrete mass is cement. Both annual and cumulative production amounts are shown (left and right, respectively). Note that the mid-20th-century 'Great Acceleration' is clearly seen. ©2018, with permission from Elsevier.

cement production statistics it is possible to estimate annual concrete production in 2015 at 27.3 billion tons (Waters & Zalasiewicz 2018; Figure 2.2.1 herein). It is estimated that annually 1 billion tons of construction and demolition waste, a significant component of which is concrete, finds its way into refuse sites or is recycled into road bases (Mehta & Monteiro 2006), into artificial ground, or back into concrete as recycled aggregate (Figure 2.2.2). Furthermore, the annual production of ~4 billion tons of cement has a significant influence on the carbon cycle (see Section 5.2), accounting for about 7% of all human-generated CO_2 emissions (Mehta & Monteiro 2006), through the burning of fuel needed to heat the kilns and the release of CO_2 from the baked limestone.

Concrete as a building material was first widely used over two millennia ago by the Romans. Roman concrete used volcanic sand, reacting with burnt lime and water to produce an exceptionally durable cement (Jackson et al. 2014), although the vast majority of Roman concretes were less robust, as they used a simple air-lime binder that gains strength slowly by combining with atmospheric carbon dioxide.

Figure 2.2.2 Example of a processed and size-graded recycled aggregate containing a blend of natural aggregates: quartzite gravel, sandstone, granite, basalt and limestone, along with recycled aggregate composed of construction and demolition waste, including crushed concrete, crushed clay brick and crushed roof tile.

Following the fall of the Roman Empire, the method of concrete production was lost, being reinvented in 1824 with the patenting of Portland cement. This cement is produced from a mixture of finely ground limestone and clay, heated to produce a clinker that is ground to a powder that hardens rapidly in the presence of water (Waters & Zalasiewicz 2018). Roman concrete has less silica and more alumina than Portland cement, the latter also containing subordinate additional components such as iron oxide (Fe_2O_3), hydrated aluminium ($Al(OH)_3$) and gypsum ($CaSO_4{\cdot}2H_2O$); modern cements contain higher concentrations of alumina and sulphate than traditional Portland cement (Scrivener & Kirkpatrick 2008).

Portland cement was used in many Victorian engineering projects in the United Kingdom, as the binder for both concrete and mortar, but it has become ubiquitous globally as the main reactive component in concrete production. The invention of reinforced concrete with iron/steel rebars (reinforcing bars) in the mid-19th century facilitated the construction of the first high-rise buildings at the beginning of the 20th century, which subsequently have become a typical feature of the modern cityscape (see Section 2.5.2.1). Ready-mix concrete, first developed in 1913, permits concrete to be transported as a liquid slurry, facilitating its much wider use in major construction schemes, including dams and roads. The invention in the 1930s of lightweight cinder/breeze blocks (also known as dense and lightweight aggregate concrete-masonry units), which incorporate blast-furnace slag and power-station fly-ash wastes, a distinctive novel particulate material (see Section 2.3.1), has radically changed the construction of load-bearing walls in most modern houses (Waters & Zalasiewicz 2018). Modern concretes also commonly contain up to 2% fibres (e.g., steel, glass, polypropylene, cellulose and, formerly, asbestos) and silica microspheres (silica 'fume'), calcium chloride and polycarboxylate ethers in order to modify specific properties of the concrete (Scrivener & Kirkpatrick 2008), producing a distinct Anthropocene mineralogical and geochemical fingerprint.

Most modern Portland cements comprise the powdered form of the mineral-like phases alite, aluminate, belite and ferrite (Mehta & Monteiro 2006; Hazen et al. 2017). Following hydration, the cement is modified to produce secondary minerals, such as ettringite, hillebrandite, jennite, portlandite and Al-tobermorite, all of which are relatively rare in nature but are now common as components of the hydrated cement-paste part of concrete (Waters & Zalasiewicz 2018). Concrete is prone both to physical weathering through frost action, abrasion, erosion, cavitation and marked temperature changes and to chemical weathering, especially in the presence of sulphate- or chloride-rich groundwaters and within marine conditions (Mehta & Monteiro 2006) or through corrosion of steel reinforcements (Figure 2.2.3a). Hence, it is likely that concrete will be robust over millennial scales (dependent on environmental conditions) and effectively be preserved as a 'man-made conglomerate'. Over millions of years and through normal geological processes, the cement paste and/or the aggregate will recrystallise to common natural minerals and disaggregate, losing much mineralogical evidence of human origin, though commonly retaining aspects of human-made texture and form.

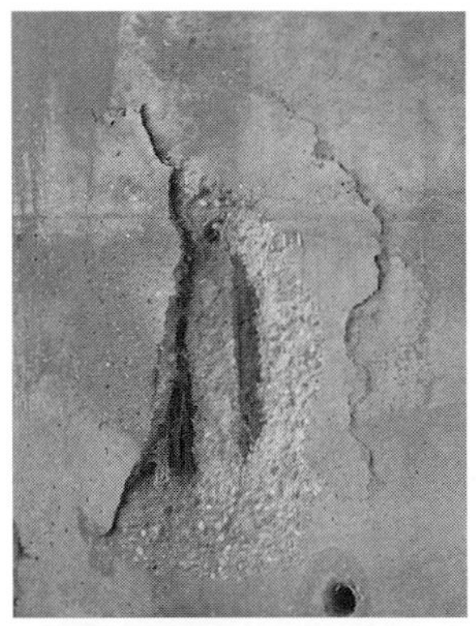

Figure 2.2.3 Examples of novel materials. (a) Flaking of modern concrete due to oxidation and expansion of steel rebars (reinforced bars); (b) bitumen used as a binder for seawall defences on Guernsey starting to show gravitational flow.

2.2.2 Brick

The early construction of buildings commonly used unfired, sun-dried mud bricks, commonly referred to as adobe. Their first usage is recorded from some 7500 BCE at Tell Aswad in Syria (Stordeur & Khawam 2007) and at Jericho in the Palestinian territories (Wright 1985), as well as within the Indus Valley civilisation by 7000 BCE (Khan & Lemmen 2014). Such bricks have low durability, being bound by straw, and deteriorate quickly through rainfall. Adobe bricks are still used today for constructing temporary or cheap structures. Heating of the mud to produce baked or fired bricks significantly increases the preservation potential. This technology was first employed within Indus Valley civilisation cities from about 2800–1500 BCE, including for building foundations, linings for sewerage systems and flood prevention systems, all prone to water damage of mud bricks, as well as city walls that must be resilient (Khan & Lemmen 2014), but mud bricks still continued to dominate much of the city construction. This civilisation standardised brick ratios to 4:2:1 (length:width:height). The Romans also produced standardised fired brick, held together by mortar, which as a common building material expanded across much of the empire. Although hand-moulded fired-brick technology was widely available over succeeding centuries, it was during the Industrial Revolution that fired-brick manufacturing expanded greatly in production, partly through improved mechanical manufacturing processes commencing in the 1850s but also to meet the need for a cheap building material that could be produced rapidly during phases of urban expansion and new infrastructure construction (see Section 2.5.2). During the urban expansions of the late 20th century and a growing reliance on concrete, fired brick was mainly limited in use to small and medium-sized buildings and is now used less in areas prone to earthquakes, where reinforced structural concrete is preferred.

Fired bricks are composed of shale, siltstone, mudstone, or clay, with some additional components that are used to help drying (sand or fired ceramic 'grog') and prevent cracking or shrinkage during firing, to act as a flux to enable melting of silicate components, and to provide colour. While raw-material composition varies widely, the generic composition of a brick-making raw material is 50%–60% silica (from sand, silt and clay minerals), 20%–30% aluminium oxide (from the clay minerals – typically combinations of illite, kaolinite, chlorite and smectite), 2%–5% CaO (which may occur as a calcareous component of the mud), 5%–6% iron oxide, and less than 1% MgO (Punmia et al. 2004), with average compositions for United Kingdom bricks provided by Dunham (1992). The brick clay is artificially metamorphosed at high temperatures (typically 900°C–1150°C) to drive out moisture and other volatiles and produce a harder 'ceramic-bonded' material through high-temperature mineral transformations, but the resultant mineralogy will depend not only upon the composition of the source brick clay but also on the length and temperature of firing and the atmosphere in which that firing occurs (Dunham 1992). New minerals formed during brick making include mullite (from the breakdown of clay minerals); anhydrite, diopside, wollastonite and melilite (from the Ca-rich component); and cristobalite (a high-temperature silica polymorph) (Dunham 1992). These are all naturally occurring minerals, although their association within a natural rock, as a mineral assemblage, would be rare. The composition of bricks has not varied systematically over the last 200 years, as many of the raw-material sources have remained the same. However, with the advent of the mass transportation of construction materials during the latter half of the 20th century, brick types that were once restricted to a defined geographical or regionalised market have been distributed farther afield, both nationally and in many cases internationally, resulting in a much greater heterogeneity in brick compositions at any one location. The morphology and particularly the dimensions of bricks vary with time, but over recent decades these dimensions have tended to be become standardised geographically, with different countries having their own preferences.

Bricks are variably porous and may be susceptible to physical degradation in the presence of water and changes in temperature (Ford et al. 2014). As brick clays are heated rapidly and for short durations, as opposed to the slow thermal processes attained naturally, bricks may not contain geologically stable mineral assemblages (Dunham 1992). Durable over millennia, bricks are unlikely to be preserved in their current state over millions of years but are likely to leave fossil imprints.

2.2.3 Ceramics

Traditionally, ceramics include clay minerals (mainly illite and kaolinite), which are fired to produce earthenware pottery, stoneware, porcelain, or china. Ceramics are a common component of archaeological sites, with the evolution of pottery styles and compositions of particular importance in providing the dating of successions. Perhaps the earliest ceramic is a baked-clay figurine (Venus of Dolní Věstonice) from 27,000 to 31,000 years ago found in the Czech Republic (Vandiver et al. 1989). The earliest functional pottery vessels are from ~20,000 years ago in Jiangxi Province, China, but represent a low-volume, high-prestige material at that time (Cohen 2013). During the Holocene, pottery became the principal means of storage, transport and consumption of food and drink, and it is commonly an abundant component of archaeological deposits (see Section 2.5.2.3). Despite some of that functionality being overtaken by plastics, ceramics remain abundant and will be a common contributor to Anthropocene technofossils, where the morphology of designs, rather than compositional variations, will be diagnostic.

The types of ceramics will depend on the composition of the source clays, the additives and the firing temperature; ceramics are typically fired at higher temperatures than used for bricks. The principal constituents of porcelain, including 'china', fired at 1,200°C–1,400°C, are mullite and cristobalite (Hazen et al. 2017), the original clay source typically being kaolinite rich. Stoneware is fired at about 1,200°C, producing a vitrified, non-porous texture. Earthenware is fired at a lower temperature of 1,000°C–1,150°C and typically comprises quartz and feldspar (Hazen et al. 2017), producing a porous medium, which commonly is glazed during a second firing. Traditional clay ceramics tend to be hard and corrosion resistant but brittle, and potsherds comprising broken pottery fragments are the norm in many archaeological excavations. Those ceramics fired at higher temperatures may be considered more resilient to weathering.

More recent 'technical or advanced' ceramic materials, developed during the Anthropocene, are not sourced from clay but are based on aluminium oxide and zirconium oxide. Other modern advanced ceramic materials include abrasion-resistant silicon carbide and tungsten carbide, boron carbide used in armour, and molybdenum disilicide used in heating elements. Other diverse types have found applications in the aerospace and automotive industries, power generation, medicine and biomedical implants, electronics and semiconductors. These new ceramics may provide a near-permanent marker for the Anthropocene, particularly abrasives, which are especially hard, durable and resistant to chemical weathering.

2.2.4 Asphalt/Bitumen

Asphalt/bitumen is a black highly viscous liquid to semisolid substance that can occur naturally, forming pitch lakes, as found in Trinidad, or forming seeps in bodies of water such as the Dead Sea. It is also a product of coal/petroleum/oil refining and has been used by many ancient civilisations as a waterproofing or adhesive agent. Historically, bitumen was extracted directly from bituminous sands, e.g., the Athabasca oil sands of Canada. From the early 20th century through to the 1970s, distillation of coal to produce town gas produced a similar 'coal tar', but most gas is now refined from crude oil. In Europe during the 1830s, there was a sudden expansion in the use of asphalt for pavement construction. Modern roads commonly have a surface of asphalt ('tar macadam'/ blacktop), in which the asphalt is used to bind a hard aggregate, with crushed rock or gravel typically making up 95% of the content, to produce 'asphalt concrete' (see Section 2.5.2.4). Annual production of bitumen is estimated at 102 million tons, with ~85% used to produce road asphalt (Razali et al. 2016), but asphalt is also used as a waterproof roofing felt or as a sealant. Asphalt, in the form of road metal, is unlikely to become a long-lasting distinctive rock type, as the bitumen component will be prone to mobilisation (Figure 2.2.3b) or combustion over short timescales.

2.2.5 Anthropogenic Rock Types as Marker for the Anthropocene

The scale of production (some 90% of a cumulative production of about 500 billion tons; Waters et al. 2016) and mineralogical and geochemical distinctiveness of concrete produced since the mid-20th century make concrete a prominent marker for the Anthropocene within terrestrial and, increasingly, subterranean environments. While ready-mix concrete is typically used within a defined delivery area, normally based on transportation time to site, its components, the cement binder and the aggregate, often come from different places. Pre-1950s, both cement and aggregates would typically be sourced locally to the concrete batching plant. However, more recently, cement is traded as a commodity and therefore over the last 25 years has been shipped globally, whereas the aggregates are often still sourced much closer to the concrete batching plant. Modern concrete therefore has two different geographical-origin indicators, based on the chemical and mineralogical markers of its constituent parts.

Concrete types used in Victorian and earlier Roman engineering are both compositionally quite distinct from modern concretes and occur in structures that are recognisably pre-Anthropocene in origin (see Section 2.5). The need for rapid construction of defensive structures during WWII and subsequently for rebuilding Europe saw a rapid expansion in the production and use of concrete, coinciding with the 'Great Acceleration'. However, there is likely to be a decadal-scale lag between the concrete production figures (see Figure 2.2.1) and concrete's appearance within sedimentary successions either as a waste material or through erosion of infrastructure (Waters & Zalasiewicz 2018). Other novel rock types, such as ceramics and bricks, may be more recognisable as Anthropocene markers through their morphology as distinct technofossils (Section 4.2).

The long-term burial histories of natural stone in a range of burial environments for geological durations of time are now well constrained, but those of materials such as bricks, concrete, glass and ceramic represent novel planetary experiments. They will alter to various degrees as temperatures and pressures rise and as groundwater chemistry changes (Ford et al. 2014). Contemporary degradation is already locally resulting in new geological materials, such as modern speleothems associated with the breakdown of concrete or of limekiln waste (e.g., Field et al. 2017), though many constituents of brick and concrete, such as clay minerals, sand and gravel, have considerable geological durability.

2.3 Novel Materials as Particulates

Neil Rose and Agnieszka Gałuszka

In addition to the wide range of novel anthropogenic minerals and rock types discussed earlier in this chapter, some microscopic particulates can be identified in part by a morphology and/or composition not found in nature. Such particulates can be dispersed widely to form a potentially suitable marker for the Anthropocene. The two considered here are fly ash, sourced from the combustion of coal and oil fossil fuels, and glass microspheres, principally sourced from road markings and signs.

2.3.1 Fly Ash

When fossil fuels such as coal and fuel oil are burned at high temperatures, they produce flue gases including carbon, nitrogen and sulphur oxides. The former contributes to global warming, while the latter two have been linked with environmental concerns such as nitrate leaching and surface-water acidification (Monteith et al. 2014). Because these fuels contain some material that is non-combustible, and because power plants are never 100% efficient, the burning of these fuels also generates particulate matter, known as fly ash (Rose 2001).

There are two main components to fly ash. The part of the fuel that cannot be burned (for example, mineral inclusions within coal) melts in the power-plant furnace and then recoalesces as it cools into spherical particles known as inorganic ash spheres (IASs) (Rose 1996). These have a chemical composition dependent on that of the original mineral. The second component is left over when the coal and oil are not fully burned. Because fuel oil is sprayed into the furnace, and because coal is finely pulverised before it is burned, the resulting particulates are usually less than 100 μm in diameter. They are mainly composed of elemental carbon and are often porous due to volatile gases being emitted during the combustion process (Figure 2.3.1). They are known as spheroidal carbonaceous particles (SCPs). Because SCPs have no natural sources and are relatively easy to extract and identify from environmental archives, they are more widely used than IASs in environmental studies. SCPs therefore provide an unambiguous indicator of one of the key drivers of anthropogenically induced change – high-temperature fossil-fuel combustion.

The burning of fossil fuels such as petroleum and natural gas produces neither SCPs nor IASs. The black material resulting from this combustion (for example, from vehicle exhaust) is composed of amorphous agglomerations of submicron-sized particles and is often known as soot. In fact, both soot and SCPs form part of a 'black carbon continuum', which stretches from large particles of charred biomass produced at low combustion temperatures up to these small, high-temperature condensates (Rose & Ruppel 2015). SCPs lie near the higher-temperature end of this continuum (Figure 2.3.2). Black carbon is described separately as a potential marker for the Anthropocene in Section 2.4.

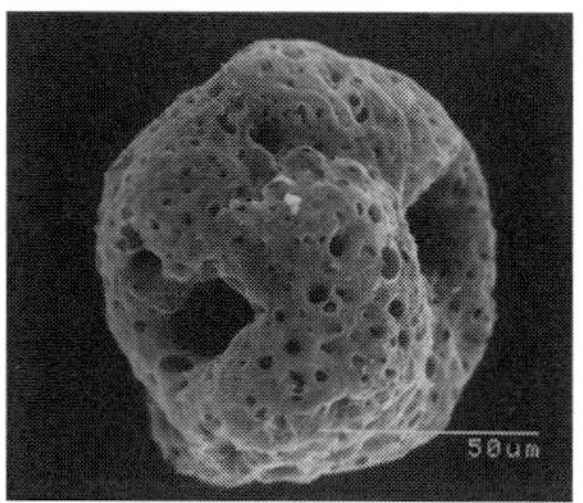

Figure 2.3.1 Scanning electron micrograph of a spheroidal carbonaceous particle (SCP).

Fly-ash particle size is very important, as it determines both capture efficiency and the distance a particle can potentially travel once it is emitted to the atmosphere. Most modern industrial combustion plants are fitted with particle-arrestor technology such as electrostatic precipitators, which are >99% efficient (Hart & Lawn 1977), and so most fly-ash particles (although preferentially the larger ones) are captured before they can be released to the atmosphere with the flue gases. Despite this efficiency, the scale of fossil-fuel combustion is such that estimates for fly-ash-particle emission from a single 2,000 MW coal-fired power plant can reach 25 trillion SCPs and over 450 trillion IASs each day of full operation (Rose in press). The emission height for an industrial chimney is typically around 200 m, and while this does not reduce the number of particles entering the environment, it results in their considerable geographical distribution. Atmospheric transport and deposition of fly ash are dependent on the meteorological conditions that the diffusing chimney plume encounters, and particles can be efficiently removed by rain and snow. Despite this,

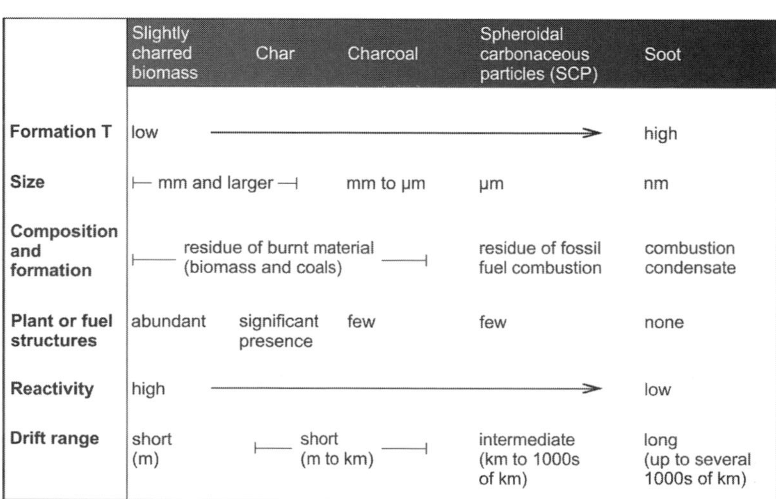

	Slightly charred biomass	Char	Charcoal	Spheroidal carbonaceous particles (SCP)	Soot
Formation T	low	→	→	→	high
Size	⊢ mm and larger ⊣		mm to μm	μm	nm
Composition and formation	⊢ residue of burnt material (biomass and coals) ⊣			residue of fossil fuel combustion	combustion condensate
Plant or fuel structures	abundant	significant presence	few	few	none
Reactivity	high	→	→	→	low
Drift range	short (m)	⊢ short (m to km) ⊣		intermediate (km to 1000s of km)	long (up to several 1000s of km)

Figure 2.3.2 The position of SCPs in the black-carbon continuum. (from Rose & Ruppel 2015)

fly-ash particles are known to be transported over very long distances and have been identified in polar regions (Rose et al. 2012) and other remote locations (Rose 2015).

2.3.1.1 Fly Ash in Natural Archives

Fly-ash-particle records were first widely used in the 1980s as supporting evidence in the debate on the causes of surface-water acidification (Battarbee 1990), when lake sediment records of SCPs were compared with historical pH reconstructed from changing diatom assemblages. Diatoms are single-celled algae with siliceous cell walls, which generally preserve well in sediments. As diatom species have well-defined preferences for the chemistry of the water in which they live (pH; nutrients), changes to species assemblages can be used to infer changing conditions. Due to their provenance, SCPs are emitted from the same sources as the gases (SO_2; NO_x), which are converted to the acidifying ions SO_4^{2-} and NO_3^-. The coincidence of the timing of pH change as recorded by diatoms and the increasing sediment concentrations of SCPs and other indicators of industrial emissions, such as trace metals, therefore enabled a convincing case to be made for atmospheric deposition ('acid rain') being the source of lake acidification (Battarbee et al. 1996).

As there are an estimated 117 million lakes greater than 2,000 m^2 on Earth (Verpoorter et al. 2014), present on every continent, lake sediments provide an invaluable natural archive of environmental change, including the storage of a record of atmospheric contamination deposited onto their surfaces and their surrounding catchments. As SCPs are only produced from industrial combustion of coal and oil fuels, their earliest record in lake sediments is from the mid-19th century (Rose 2001). Their first observable presence in lake sediments varies considerably from region to region and depends on a number of parameters, including the timing and scale of regional industrial development, as well as more site-specific factors such as sediment accumulation rate (Rose & Appleby 2005). In the mid-20th century, following WWII, there was a major increase in the demand for electricity, and so the scale of coal combustion increased dramatically around the world. Fuel oil was also first used to generate electricity at around this time. This resulted in a rapid increase in SCP accumulation in lake sediments globally at a time coincident with the 'Great Acceleration' (see Section 7.5).

A synthesis of lake-sediment records shows that this rapid increase in SCP accumulation appears to occur in all areas of the world where such records exist (Rose 2015). In regions with long histories of industrialisation (e.g., Europe, North America), this increase is observable as a marked upturn in SCP accumulation rate. In areas where industrial development has been more recent, it is observable as the start of the record as SCP sediment concentrations move from below to above the sedimentary analytical limit of detection for the first time. Combining all these data, normalised to the peak SCP accumulation rate to compensate for the scale of regional industrialisation, provides a global picture of SCP accumulation (Figure 2.3.3). While the overall trend is biased towards Europe due to data availability, the mid-20th-century increase is apparent for all regions (Rose 2015). There is also a clear decline in SCP accumulation rate from the 1970s to the 1990s and onwards due to the introduction of particle-arrestor technology, changes in fuel use for electricity production (i.e., to natural gas), and for some regions, declines in heavy industry. As with the start of the SCP record, the timings of these changes are also regional, and only the mid-20th-century increase appears to be truly global.

While the vast majority of data for fly ash in natural archives are for lake sediments, it has also been reported in peat sequences (Yang et al. 2001, Kuoppamaa et al. 2009, Swindles et al. 2015) and ice cores (Hicks & Isaksson 2006), although data are scarce. Figure 2.3.4a, b shows two examples, both from the United Kingdom, where SCP analysis has been undertaken on both lake-sediment cores and peat sequences from the lake's catchment. In each case, the historical patterns are remarkably similar,

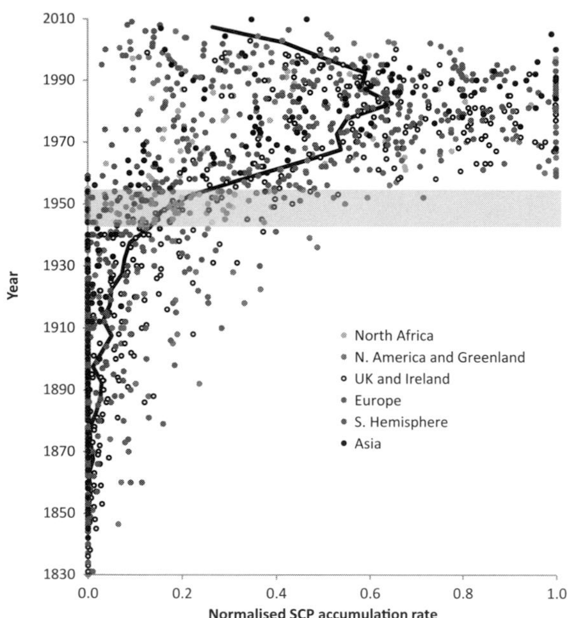

Figure 2.3.3 Global synthesis of SCP lake-sediment accumulation-rate data. Historical SCP data are normalised to peak accumulation rate (=1.0). Each region is coloured separately. Grey horizontal bar is 1950 ± 5 representing the Great Acceleration. Data from Rose (2015). (A black-and-white version of this figure appears in some formats. For the colour version, please refer to the plate section.)

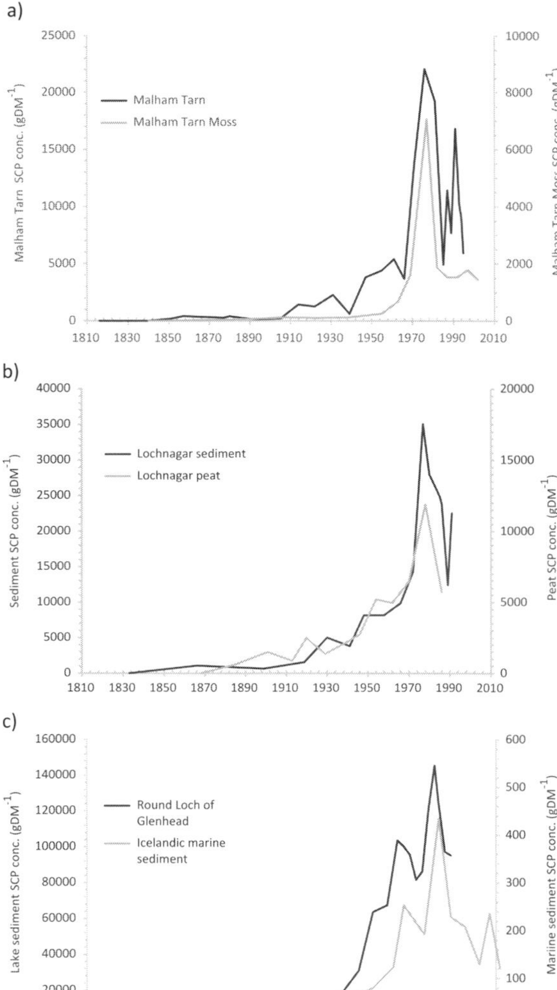

Figure 2.3.4 Comparison of SCP profiles from lake sediments and peat sequences from adjacent sites in (a) Malham Tarn, United Kingdom (data from Swindles et al. 2015; Rose unpublished); (b) Lochnagar (adapted from Yang et al. 2001, and data from Rose & Appleby 2005). (c) This compares the historical records of a lake-sediment core from the Round Loch of Glenhead in Galloway, Scotland (Rose & Appleby 2005), and a marine-sediment core from south of Iceland (unpublished data). In each case, the lake-sediment profile is shown in black.

not only for the timing of the mid-20th-century increase in SCP contamination, but also for the start of the record and concentration peaks. Given the consistency in the global record for the mid-20th-century SCP increase, perhaps this should not be a surprise, but from the point of view of using SCPs as a stratigraphic indicator, it is useful to know that records can be matched between different types of archive. Similarly, the comparison between historical trends of SCPs in marine and lake sediments shows good agreement. Figure 2.3.4c shows the records from a marine core taken from a location approximately 200 km off the south coast of Iceland and from a lake-sediment core from Galloway in southwest Scotland. While there is a considerable distance (c. 1,200 km) between the sites, the United Kingdom is likely to have been a historically important source of contamination to this area of the North Atlantic. This comparison provides further support for the agreement between different archives and indicates the presence of a very extensive potential natural archive within which to explore SCP accumulation as a stratigraphic marker. As regards other archives, while SCP data from ice cores are scarce, the record for other black-carbon components in ice is more extensive (e.g., Lavanchy et al. 1999, McConnell et al. 2007), suggesting considerable potential for SCP storage. No data exist for SCPs in corals, although it is known that other anthropogenic particles such as microplastics do become incorporated within them (Hall et al. 2015). Further exploration of these archives is warranted.

2.3.1.2 Fly Ash as a Marker for the Anthropocene

The historical record of fly ash in lake sediments, and in particular the rapid increase in SCP accumulation, appears to provide a consistent globally synchronous marker for the mid-20th century. Furthermore, while data are currently scarce, they would appear to be consistent between different natural archives, including lake and marine sediments and peat sequences. The scale and extent of fossil-fuel combustion from the mid-20th century onwards have made this a key driving force for environmental change (see Section 7.5), and it seems appropriate that stratigraphic markers associated with this force can be used to define the start of the Anthropocene.

It is not clear how well SCPs may be preserved over very long periods, as the earliest known particles were only produced in the mid-19th century. However, particles of a similar nature, which may have been produced by the low-temperature (300°C–800°C) expansion but not combustion of volatiles within organic-rich particulates, have been observed both in Bronze Age peats (Rose & Ruppel 2015) and possibly even at the Cretaceous-Paleogene boundary ~66 million years ago (Harvey et al. 2008) associated with a major meteorite impact (Section 1.3.1.3). The earliest 'true' SCPs show no signs of decomposition or deterioration, and if they are as robust as these 'precursors', then the record in natural archives could well be considered permanent. Of course, SCPs are only one indicator of industrial-scale fossil-fuel combustion, and a number of other indicators may also be used to support or verify the record. These include an unprecedented increase in carbon dioxide in Greenland and Antarctic ice cores (Barnola et al. 1995, Rubino et al. 2013; see Section 5.2 herein); increases in concentrations of trace metals (such as mercury, lead and zinc) in lake and marine sediments, peats and ice (e.g., Wolff & Suttie 1994; Shotyk et al. 1998; Yang et al. 2002; see Section 5.6 herein); increases in polycyclic aromatic hydrocarbons (Smith & Levy 1990, Muri et al. 2006; see Section 5.7 herein); changes in nitrogen deposition exemplified by $\delta^{15}N$ records in lake sediments (Holtgrieve et al. 2011; see Section 5.4 herein); and biological responses such as changes to diatom assemblages as a result of acid deposition (Battarbee 1990) and in response to the eutrophying effects of nitrogen deposition (Pla et al. 2009). Between them, these indicators may provide a wealth of stratigraphic markers for the start of a mid-20th-century Anthropocene Epoch. However, while promising, greater exploration is still required of marine sediments, ice cores and coral records to determine whether their records may also be used. This is especially the case in the Southern Hemisphere, where SCP records remain sparse.

2.3.2 Glass Microspheres

Commercially produced tiny spherical glass particles with a typical diameter range of 1–1,000 μm are known as glass microspheres. They include solid glass microspheres (glass microbeads) and hollow glass microspheres (microballoons or microbubbles). Industrial-scale production of the first solid glass microspheres started in 1914, while hollow glass microsphere production began in the 1950s. The number of applications of glass microspheres has grown tremendously since the mid-20th century (Gałuszka & Migaszewski 2017). They are most widely used as fillers in composite polymer materials, as additives in paints and coatings, and in pavement-marking materials (Amos & Yalcin 2015). The last mentioned application is of special interest because the glass microspheres are embedded in road safety markings only while the paint

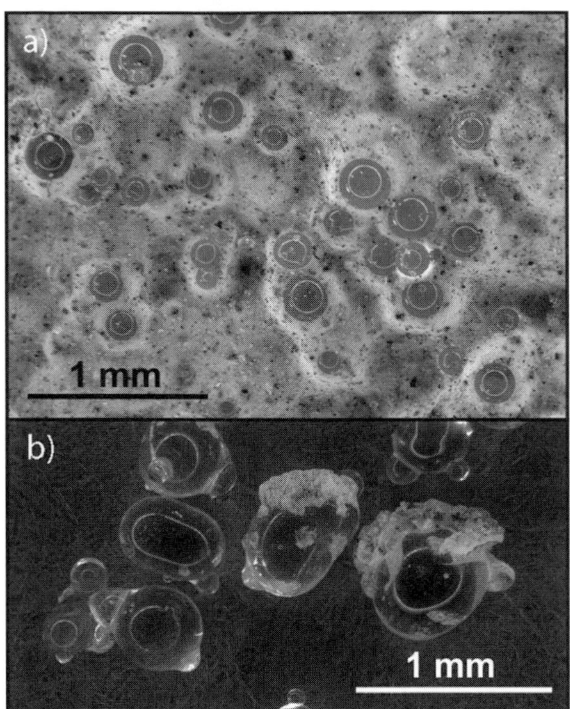

Figure 2.3.5 (a) Glass microspheres embedded in pavement-marking paint; (b) individual microspheres separated from river sediments. (microphotographs taken by Z. M. Migaszewski)

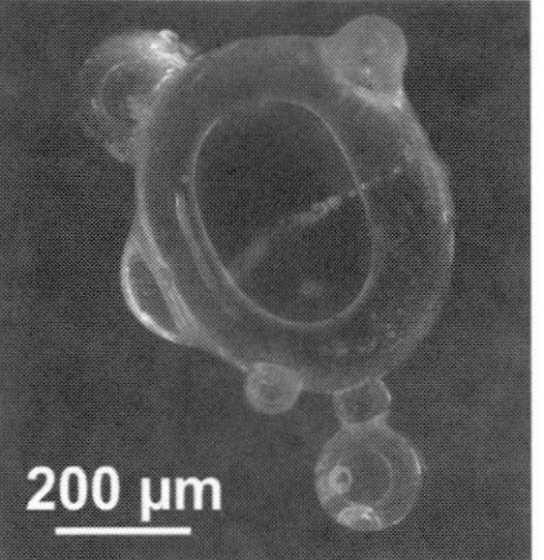

Figure 2.3.6 An aggregate of glass microspheres found in road dust. (microphotograph taken by Z. M. Migaszewski)

lasts (typically 6–24 months; Burghardt et al. 2016), after which they are transported by wind to highway surroundings and by storm-water drainage to river systems. Glass microspheres embedded in traffic paint are shown in Figure 2.3.5a, and individual glass microspheres separated from river sediments are shown in Figure 2.3.5b.

The glass microspheres of the highest quality are perfectly spherical. Depending on the conditions during production in a high-temperature fluidised bed, glass microspheres of different shapes can be formed, including splash-form shapes (such as teardrop, pear, peanut and dumbbell) and aggregates (Figure 2.3.6).

The chemical composition of glass microspheres depends on the type of glass used for their production. The most often used glass types are recycled soda-lime silicate and alumino-borosilicate glasses. Their major components are shown in Table 2.3.1 together with the physical properties of glass microspheres.

Apart from degradation of traffic paints, an additional source of glass microspheres is their unintentional release to the roadside environment during traffic-paint application, especially when the glass microspheres are sprayed onto wet paint. Road-marking paints with typical durabilities of 6–24 months are used for marking horizontal traffic signs, spaces in parking lots, loading zones, airport runways and so on. The presence of glass microspheres in road-dust samples was first documented by Zannoni et al. (2016), whereas the occurrence of glass microspheres in river sediments receiving storm-water runoff was first reported by Gałuszka and Migaszewski (2017).

The similarity of glass microspheres to natural microtektites was noted by Marini (2003), who studied the physical properties and chemical composition of solid glass microspheres and found that the best method of distinguishing glass microspheres from microtektites was by chemical analysis for trace elements such as Pb, As and Sb. Quartz grains can also be mistaken for glass microspheres. While there is no conclusive method for discriminating between glass microspheres and microimpactites, because their physical features and often chemical composition are very similar (Marini 2003), Gałuszka and Migaszewski (2017) show that examination of samples under a polarising microscope can discriminate between glass microspheres (isotropic)

Table 2.3.1: **Selected properties of glass microspheres**

Parameter	Values/Features	Reference
Particle size	12–300 µm (hollow glass microspheres) 1–1,000 µm	Wood 2008 Marini 2003
Mohs hardness	6.0–6.5	Wypych 2016
Chemical composition	Soda-lime silicate glass: SiO_2: 72%–73%; Na_2O: 13.3%–14.3%; CaO: 7.2%–9.2%; MgO: 3.5%–4.0%; Al_2O_3: 0.8%–2.0%; K_2O: 0.2%–0.6%; Fe_2O_3: 0.08%–0.2% Alumino-borosilicate glass: SiO_2: 52.5%; CaO: 22.5%; B_2O_3: 8.6%; MgO: 1.2%; Na_2O: 0.3%; K_2O: 0.2%; Fe_2O_3: 0.2%	Wypych 2016
pH of water suspension	7–10	Wypych 2016
Density	Solid glass microspheres: 2.23–2.54 g/cm^3 Hollow glass microspheres: 0.12–1.1 g/cm^3	Wypych 2016
Floaters on water	89%–99.9%	Budov 1994
Thermal conductivity	0.029–0.115 W/m·K	Budov 1994

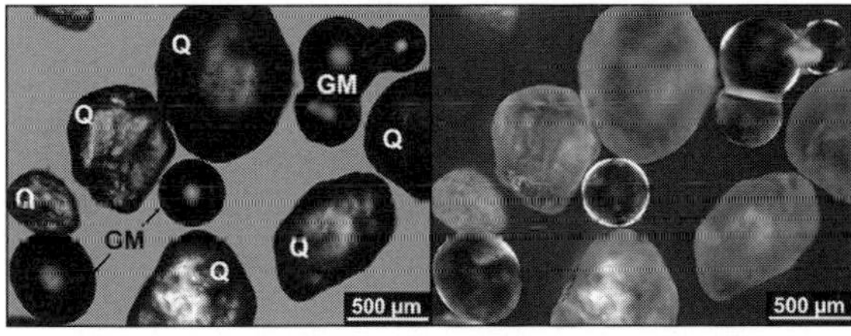

Figure 2.3.7 Optical-microscope images of isotropic glass microspheres (GM) and anisotropic quartz grains (Q): transmitted light, crossed nicols (left picture); reflected light, crossed nicols. (right picture) (microphotographs taken by Z. M. Migaszewski)

and quartz grains (anisotropic) (Figure 2.3.7). Another feature of glass microspheres that can be used for identification is that many have bubble inclusions.

2.3.2.1 Glass Microspheres as a Signal of the Anthropocene

Glass microspheres were first used in road-marking paints in North America in the 1930s, but very soon their use spread globally. Currently, the amount of glass microspheres used for pavement marking in the USA every year is about 250,000 tons (Jahan et al. 2011). Widespread use of glass microspheres has caused them to be widely distributed. A study of glass microspheres in the city of Kielce, Poland, showed that the numbers of glass microspheres decreased away from local sources: e.g., river sediments close to storm-water discharge (mean content: 5,100 microspheres/kg); road dust (mean content: 1,500 microspheres/kg); river sediments several hundred metres from the nearest road with reflective traffic paints (mean content: 300 microspheres/kg). The durability and high abundance of glass microspheres in different terrestrial surface deposits and probably in marine sediments suggest their potential use as a stratigraphic indicator of the Anthropocene.

The characteristic features of glass microspheres encompass their physical and chemical resistance, low density and very high thermal stability (Table 2.3.1). These properties contribute to the very high persistence of glass microspheres in the environment. The ideal spheroidal shape of most particles prevents significant corrosion. It seems very likely that their preservation potential is at least comparable to that of other proposed stratigraphic indicators of the Anthropocene, including organic pollutants, plastics, organic fossils, elemental aluminium and long-lived radioisotopes.

2.4 Black Carbon and Primary Organic Carbon from Combustion

Colin N. Waters and An Zhisheng

Black carbon (BC), or elemental carbon, is char or soot that is co-emitted with primary organic carbon (POC) during incomplete combustion of fossil fuels or anthropogenic and natural biomass burning (Bond et al. 2007, Han et al. 2017). BC is the inert, refractory part of the emitted organic carbon, which absorbs heat in the atmosphere and so contributes to global warming. POC has no heat-absorbing characteristics and is labile, or likely to change. Soot forms as a condensate at relatively high temperatures of pyrolysis compared with char, which often retains its original structure (Han et al. 2017). Fossil-fuel combustion, especially from motor-vehicle emissions, typically produces higher soot/char ratios than biomass burning (Han et al. 2016). The soot fraction, comprising smaller particulates, can be transported in the atmosphere and hence can be dispersed widely, though BC and POC particulates are considered to have relatively short atmospheric lifetimes of days to weeks. Once accumulated in sediment or ice, BC is resistant to degradation and may exist as a stratigraphic signal for tens of millions of years (Han et al. 2016, 2017).

BC is commonly identified in soil and in lacustrine and marine sediments. Glacial-ice cores from remote high-latitude glaciers are particularly useful in recording atmospheric BC (Xu et al. 2009), as they preserve primary aerosol deposition. Lake and marine sediments are prone to containing reworked BC if they include river systems within their catchment. Maar lakes provide a useful exception; as they lack fluvial reworking, they can provide atmospheric signals closer to industrial and urban sources of BC (Han et al. 2016).

Analysis of fuel usage and combustion emission factors can be used to provide approximations of the historical atmospheric load of BC from fossil fuels (Novakov et al. 2003) or BC and POC sourced from combustion of fossil fuels and biofuels (Bond et al. 2007). BC increased towards the end of the 19th century, especially from initially ~1950 and subsequently from ~1970 as China and India developed, with a peak estimated at 6.7 million tons per year in ~1990 (Novakov et al. 2003; Figure 2.4.1 herein). Lower estimates provided by Bond et al. (2007) suggest BC and POC production broadly increased linearly to 2000, with a value of 4.4 million tons of BC and 8.7 million tons of POC (Figure 2.4.1). The changes in combustion of an additional source of BC and POC, biomass burning, cannot be readily estimated. Biofuel accounts for the greater part of POC emissions, and it did also for BC before 1880 CE, whereas coal was the

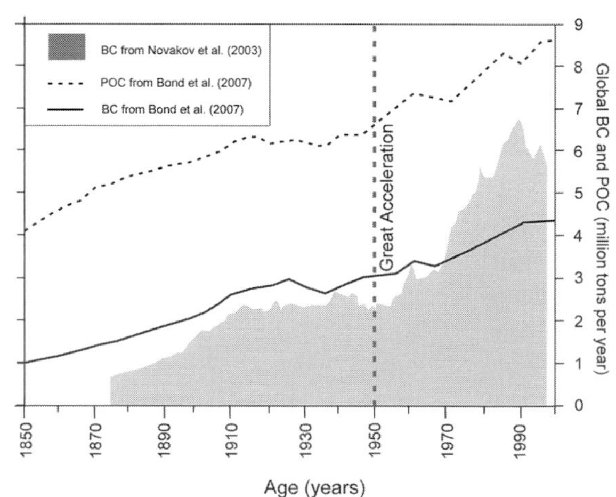

Figure 2.4.1 Global black carbon for available annual fossil-fuel-consumption data of 1875–1999. (redrawn from Novakov et al. 2003, figure 3) and black-carbon and primary-organic-carbon data of 1850–2000. (data from figure 6 of Bond et al. 2007) © by the American Geophysical Union & Wiley

main source from 1880 to 1975, after which diesel was the primary source (Bond et al. 2007).

Polycyclic aromatic hydrocarbons (PAHs) tend to be produced along with the BC during combustion, thus providing a potential dual signal for the Anthropocene (see Section 5.7.1). This is well expressed in Huguangyan Lake, eastern China, in which concentrations of BC and PAHs increased sharply during the late 1940s and early 1950s and again in the late 1970s, with a BC peak from 2004 to 2006 and subsequent rapid decrease not replicated in the PAH signal (Han et al. 2016; Figure 2.4.2 herein), consistent with BC production from China (Bond et al. 2007).

BC concentrations in a Greenland ice core measured from 1788 to 2002 CE vary in response to boreal forest fires and industrial activity (McConnell et al. 2007, McConnell & Edwards 2008). Prior to 1850, the main source of BC was from conifer fires, but from about 1850, industrial emissions resulted in a sevenfold increase in ice-core BC concentrations (Figure 2.4.3), with most change occurring in winter (McConnell et al. 2007). After reaching a peak in 1906 to 1910, an abrupt fall in BC concentrations occurring after about 1951 has persisted (McConnell et al. 2007, McConnell & Edwards 2008). This does not conform to the global production figures (Figure 2.4.1) or regional data from China (Figure 2.4.2), but it is consistent with predicted reductions in emissions from North America. That suggests that the BC signal is regional on a continental scale. In an ice core from near Mount Everest, in contrast, BC concentrations have increased approximately threefold from 1975 to 2000 relative to 1860 to 1975 (Kaspari et al. 2011), suggesting regional transport from South Asia and the Middle East.

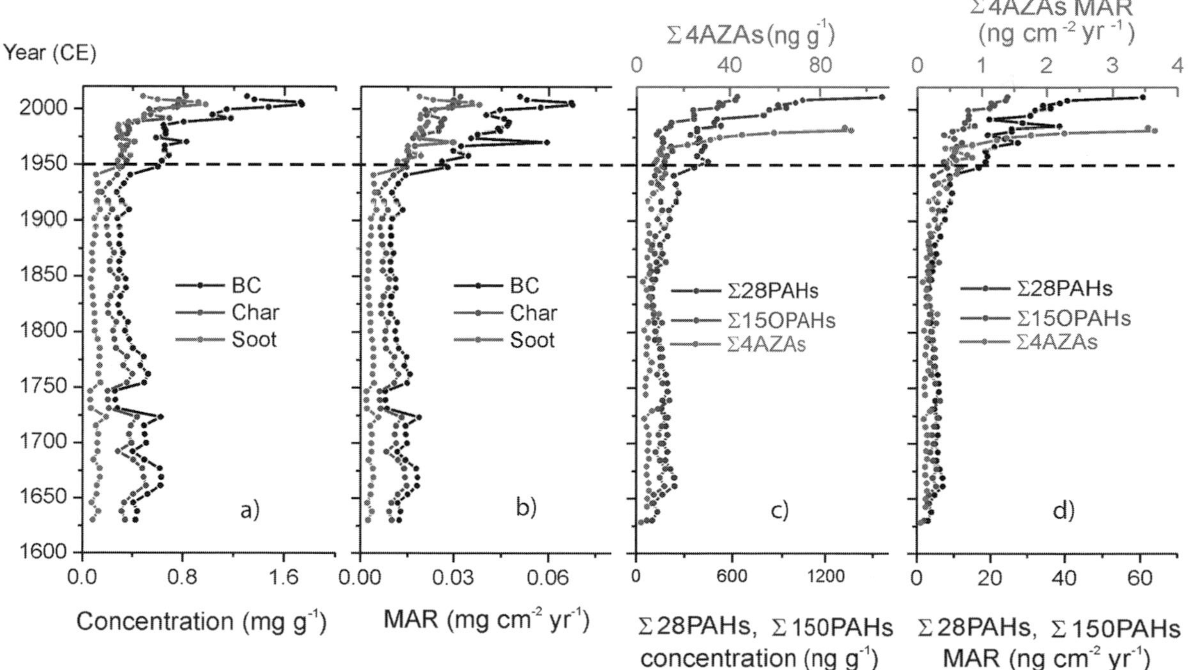

Figure 2.4.2 Historical variations of concentrations and mass-accumulation rates of black carbon (BC), char, soot, parent-PAHs, oxygenated PAHs (OPAHs) and azaarenes (AZAs) in the Huguangyan Maar Lake. (adapted from figure 2 of Han et al. 2016) (A black-and-white version of this figure appears in some formats. For the colour version, please refer to the plate section.)

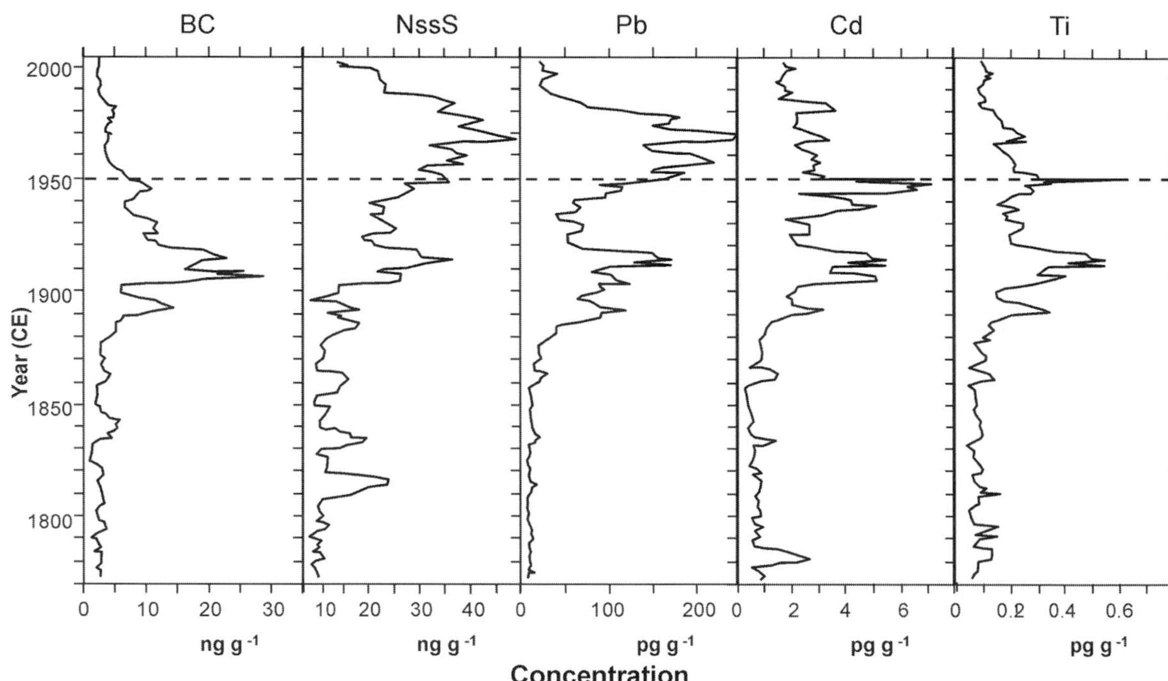

Figure 2.4.3 Concentrations (five-year running means) in Greenland ACT2 ice core from 1772 to 2003 CE of black carbon (BC), sulphur (NssS), lead (Pb), cadmium (Cd) and thallium (Tl). from McConnell and Edwards (2008), redrafted as figure 26 in Waters et al. (2018). ©2018, with permission from Elsevier

In summary, BC provides an inert signal found in lakes and ice cores, and probably in more extensive environments, that can be recognised on continental scales, but it does not show a single global pattern that can be used to consistently identify an Anthropocene boundary in the mid-20th century.

2.5 Artificial Ground, or Ground Modified by Humans

Colin N. Waters, Simon Price and Jan Zalasiewicz

Anthropogenic activities that deliberately and directly modify the landscape were referred to as producing 'novel sedimentary environments' by Zalasiewicz et al. (2011a). They are formed by the excavation, transport and deposition of natural geological materials and the accumulation of novel materials (see Section 2.2) related to urban development, mineral exploitation, waste dumping and land reclamation to produce artificial ground (Ford et al. 2014).

The deliberate, annual global flow of anthropogenic sediments through the mineral-extraction process alone is ~57,000 Mt (million tons), exceeding that of the transport of natural sediments to the world's oceans by almost a factor of two (Douglas & Lawson 2001). Anthropogenic sediments are moved across the surface of the planet via artificial processes that may be contrasted with those of natural sediment-transport systems. Natural transport processes are controlled by gravitational gradients in the case of rivers and glaciers, or through ocean or atmospheric circulation in marine and aeolian deposits. The artificial flux of anthropogenic materials follows no consistent natural pathway. However, urban areas tend to represent the main sinks for anthropogenic deposits, as a by-product of the development of the city landscape and infrastructure, while rural areas represent the main source for much of the material destined for urban and industrial conurbations (Ford et al. 2014). They supply the aggregates and the ingredients of cement and mudrock used to generate the concrete and bricks, the ores to produce the metals that form the superstructures, and the silica to create the glass or the building stones that clad city buildings. The cities are fuelled by resources such as coal, oil and gas and now, increasingly, are powered by renewable sources such as nuclear, hydro, wind and solar derived from rural areas.

Hooke et al. (2012) estimated that humans have modified >50% of the land surface. However, the distribution of this modification is uneven, with urban areas forming between 1% and 3% of the land surface (3.7 million km^2 according to Hooke et al. 2012) and much of the remainder associated with imposition of agriculture, pasture, or forestry on formerly pristine environments. This modification started in many parts of the world thousands of years ago (see Section 7.3), but through first the Industrial Revolution (Section 7.4) and subsequently the Great Acceleration (Section 7.5), both the rate and diversity of change have increased dramatically. The resulting anthropogenic landscape has been described through the various lenses of geology, archaeology, geomorphology and other disciplines, its characters reflecting functions such as habitation, burial and worship, disposal of domestic refuse, agriculture, pasture, forestry, transport, mineral extraction, industrial development, water supply, sewerage, flood defences, coastal reclamation and warfare. This chapter provides examples of the main types of contemporary and historical accumulations of anthropogenic sediments and of the common processes that shape them (Table 2.5.1).

2.5.1 Classification Schemes

Artificial ground is not in itself a marker for the Anthropocene, unless the Anthropocene is defined by the first physical traces of humans as

Table 2.5.1: **Selected types of anthropogenically modified ground**

Anthropogenic sediment or landform type	Process	Preservation potential/ stratigraphic significance
Habitation	Direct deposition and excavation	Moderate
Burial/worship sites	Direct deposition and excavation	High
Domestic refuse	Direct deposition and excavation	High
Transportation	Direct deposition and excavation	Moderate
Mineral extraction waste	Direct deposition and excavation	High
Industrial development	Direct and indirect deposition	Moderate
Water supplies and sewerage	Direct deposition and excavation	Moderate
Flood defence and coastal reclamation	Direct deposition	High
Warfare	Direct deposition and excavation	Low/moderate
Agriculture, pasture and forestry	Direct and indirect deposition and excavation	Low

geomorphological agents in the geological record, as discussed by Edgeworth et al. (2015). In such an interpretation, the Anthropocene would be not a unit of geological time (of chronostratigraphy) but one that reflects the classification of physical deposits (i.e., of lithostratigraphy). In this book the Anthropocene is considered as a chronostratigraphic unit, but anthropogenic deposits are important to its characterisation.

Anthropogenically modified deposits in an archaeological context have been referred to as the 'archaeosphere' (Capelotti 2010; Edgeworth 2014; Edgeworth et al. 2015). The markedly diachronous lower boundary of the archaeosphere, which marks the division between anthropogenically modified layers and underlying natural geological deposits, has been referred to as Boundary A (Edgeworth 2014; Edgeworth et al. 2015) or, in Japan, as the Jinji unconformity or discontinuity (Nirei et al. 2012). Assemblages of artefacts, their abundance and their first and last appearance within a sequence of archaeological deposits may be ordered and classified using stratification and seriation charts (Carver 1987). Complex erosional and non-depositional surfaces are of great value in the local description of individual excavations but are difficult to correlate regionally. However, artefact-laden layers (as opposed to surfaces) can be correlated across wide distances on the basis of the datable artefacts/novel materials they contain (see Section 4.2). The range of correlation is wider in younger deposits. In classical times, for example, the same types of pottery could be found in deposits at either end of the Athenian, Macedonian, or Roman empires but did not have worldwide distribution. Now, plastics can be found in strata on opposite sides of the Earth and even on the Moon, enabling correlation between deposits across much larger distances.

Many classification schemes for anthropogenic strata have been proposed. Time-based discrimination of materials and boundaries was proposed by Nirei et al. (2012), but for detailed site investigations, not for the widespread classification of artificial ground as undertaken by geologists (Ford et al. 2014). Chemekov (1983) recognised technogenic deposits as subaerial, subaqueous, or subterranean, with genesis, composition and morphology distinguishing different deposit types, including both entirely artificial deposits and natural deposits that have been anthropogenically modified. Peloggia et al. (2014) modified this scheme, recognising four main technogenic categories: aggraded ground, degraded ground, modified ground and mixed ground, each subdivided into specific genetic types. Howard (2014) proposed a twofold classification system of *anthrostratigraphic units* (ASU) and *technostratigraphic units* (TSU). ASUs are stratiform or irregular bodies of anthropogenic origin distinguished by lithological characteristics and/or bounding disconformities. TSUs are stratiform bodies of anthropogenic origin defined on the basis of human artefacts, with *technozones* defined by the commercial ranges of certain artefacts, comparable to the ranges shown for 'technofossils' as illustrated by Ford et al. (2014), Williams et al. (2016) and Zalasiewicz et al. (2016a).

Humanity's influence on geological processes was outlined by Sherlock (1922), with excavation regarded as a kind of denudation, complementing accumulation of anthropogenic debris as 'made ground'. From this concept, a stratigraphic classification scheme evolved for the geological mapping of artificially modified ground in the United Kingdom, largely based on morphogenetic attributes (McMillan & Powell 1999; Rosenbaum et al. 2003; Ford et al. 2010; Price et al. 2011). Five main classes are defined (worked, made, infilled, landscaped and disturbed ground) of artificially modified ground, each subdivisible into units reflecting different genetic types. This scheme emphasised the recognition of potential geohazards such as contamination, subsidence and instability, especially of industrial workings and deposits.

This was not a truly lithostratigraphic approach as used for natural strata, but an example of how the latter might be achieved was outlined by Ford et al.

(2014) for Swansea, Wales. There, formations were distinguished of 'pre-Industrial', Industrial Revolution and post-WWII deposits, each category being radically different lithologically (Ford et al. 2014, Terrington et al. 2018). Members there comprised deposits of broadly similar genetic origin within formations, while single, lithologically distinctive deposits were identified as lithostratigraphic beds. This approach showed how urban areas have grown laterally and vertically through time.

Artificial ground generally refers to landforms and sediments at or near the ground surface. But artificial ground is also created at deeper levels. Human disturbance of the subsurface may be considered as 'anthroturbation', comparable to bioturbation (Ford et al. 2014; Zalasiewicz et al. 2014b; Waters et al. 2018, in press). It might include such phenomena as boreholes, buried waste or resource repositories, underground mining, tunnels and even underground nuclear detonations.

2.5.2 Environments for Accumulation of Artificial Ground

Anthropogenic sediments, like natural sediments, show genetic relationships, with distinct bedforms and compositions that can be attributed to the environment of deposition. In this section, the characteristics of the fabric of deposits within each of the regimes identified within Table 2.5.1 are outlined, with the exception of agriculture, pasture and forestry, which are included in discussions about agricultural soils (Section 2.7.2).

2.5.2.1 Habitations

Landscaping and earthworks associated with building construction and demolition are included in the definition of artificially modified ground (see Section 2.5.1), whereas 'extant' buildings are generally excluded (Ford et al. 2014, Terrington et al. 2018). This is an arbitrary distinction, as buildings, their foundations and associated earthworks are in physical continuity and continually evolving. Buildings represent ephemeral sinks and sources of material in the rock cycle, with limited preservation potential on geological timescales (Ford et al. 2014), though on shorter timescales, the archaeological record includes many preserved buildings, not just the foundations.

Temporary settlements: The earliest habitations are associated with Palaeolithic cave deposits, in which food waste, stone tools, early artistic expression and bodies of the inhabitants are complexly interrelated (Figure 2.5.1a). As global population increased slowly, early human-constructed habitations associated with a hunter-gatherer way of life were mainly of wood, reeds, grass and clay, as in wattle-and-daub constructions. Such habitations, still common in many parts of the world (Figure 2.5.1b), typically have little surficial impact other than the presence of backfilled postholes. Organic materials are commonly oxidised in all but the most anaerobic of environments, such as peat mires.

Early cities: The shift towards construction of buildings in stone and mud bricks (see Section 2.2.2) reflects creation of more fixed settlements, which often grew into more substantial conurbations. Here, the amount of materials moved to form the foundations (i.e., earthworks) are typically much smaller than the built structures themselves. So it is the building materials that tend to be preserved in archaeological excavations. The earliest such buildings include Wadi Faynan 16, Jordan, with circular buildings used by early farmers between 11,600 and 10,200 years ago (Finlayson et al. 2011). The stratified remains at the Abu Hureyra mound in Syria (Moore et al. 2000) record 4,000 years of occupation from 11,000 years ago, spanning the transition from hunter-gathering to farming associated with domestication of plants and animals. The 600,000 m^3 settlement mound comprises the accumulated debris from the building, subsequent erosion and rebuilding of houses, with pottery appearing for the first time towards the end of the occupation of the site, 6–7 m up from the base of the initial habitation surface (Moore et al. 2000).

(a)

(b)

(c)

(d)

Figure 2.5.1 Habitation types: (a) recreation of cave habitation, Cango Caves, South Africa; (b) traditional straw hut still used by nomadic peoples in Mauritania; (c) Roman building from Pompeii; and (d) modern high-rise buildings from New York.

The earliest towns, such as Ur in Mesopotamia, show elaborate urban planning of non-linear alleys (Lay 1992), with the grid plan of wider perpendicular streets being adopted separately around 4,700 years ago at Harappa in the Indus Valley and Giza in Egypt and subsequently in Babylon and China from 3,700 to 3,500 years ago. The Romans too developed a grid plan, clearly seen in preserved remains such as Pompeii (Figure 2.5.1c), which they exported across large parts of the Mediterranean and western Europe. Reinvented in medieval Europe, this style of urban planning spread widely across the planet as colonisation proceeded and is common in modern cities, e.g., New York (Figure 2.5.1d). The habitation structures in ancient cities such as imperial Rome became complex, with subterranean hypocausts providing central heating. Settlements constructed during the Roman occupation of Britain saw an increase in the use of bricks and mortar, crushed rock, dimension stone and roofing slates (Ford et al. 2014), increasing their preservation potential. The Romans were also the first to use concrete and glass.

Modern developments: Urban areas, modified for residential and commercial purposes, were estimated by ~2007 CE to cover 3.7 ± 1.0 million km^2, about 2.8 ± 0.8% of the Earth's ice-free land surface, with an additional 4.2 million km^2 covered by rural housing (Hooke et al. 2012), which may amount to something like 17 trillion tons of buildings and foundation deposits (Zalasiewicz et al. 2016b). The rapid outward growth of urban areas is one of the characteristics of the 19th and 20th centuries, a consequence of expanding global populations and the trend towards urban living, with >50% of humanity now living in cities. Twenty-two metropolitan areas (megacities) now have populations of >10 million (https://en.wikipedia.org/wiki/Megacity), such as Shanghai (Zalasiewicz et al. 2014c). Cities are being continually rebuilt, whether for upgrade or to repair destruction from natural disasters or warfare. They are typically rebuilt upon the debris of older constructions, raising the land surface and so helping to preserve the remains of earlier urban strata as artificial deposits (e.g., Rivas et al. 2006). Rural areas have generally thinner, though still extensive, artificial deposits (Zalasiewicz et al. 2016b, table 1).

The building materials of choice since the mid-20th century are reinforced concrete, cinder/breeze-block (Waters & Zalasiewicz 2018; Section 2.2.1 herein), glass, and plastics (Zalasiewicz et al. 2016a; Section 4.3.1 herein). Such materials, now used worldwide, allow rapid construction, and the light but strong structures have permitted the trend to build skyward, as tower blocks, commonly reinforcing the grid plan of street networks (Figure 2.5.1d). Williams et al. (2017) describe a mid-20th-century transition from distinct regional building styles, developed over millennia, to marked convergence on an 'international style' – and lithological composition – of architecture. Such modern cities will leave a common global footprint, their distinct compositions and textures characterising post-WWII urban deposits.

Modern towns and cities have a complex geometry extending commonly downward to depths of ~20 m (Evans et al. 2009). This subsurface zone of human interaction includes subsurface containers and car parks; complex networks of public-utility pipelines/cables supplying electricity, gas, water, sewerage extraction and telecommunication; and transport facilities (Zalasiewicz et al. 2014b). As buildings get taller, the foundations of concrete and steel get deeper (Rivas et al. 2006); e.g., the Burj Khalifa in Dubai has foundations greater than 50 m deep.

Over geological timescales, the materials used in modern urban construction will, if buried long term, undergo diagenesis and, eventually, local metamorphism (see Section 2.2). Whether considerably altered or largely preserved, the resultant urban rock fabrics will, if buried in stratal successions, preserve distinctive structures, mineralogies and textures, and their nature and composition will provide a tool for distinguishing constructions from the Anthropocene from those that are older.

2.5.2.2 Burial and Ceremonial Sites

As nomadic hunter-gatherers gave way to sedentary cultivators, fixed sites for burials and religious worship were developed.

The organisation of skeletal and cremated human remains is a consistent feature of human development, the nature of burial customs varying in time and between cultures. Around settlements, human skeletal remains are often concentrated at specific locations, commonly with boundaries marked by earthworks, such as Neolithic long barrows. The burial ground may become an earthwork in itself, as ground is reworked over many generations (Edgeworth 2013). Burial patterns include rows of bodies aligned in grids, typical of most modern Christian and Muslim burials (Figure 2.5.2a), radial rows, satellite burials around a central mound or marker, or bodies stacked vertically in graves over time to save space (Edgeworth et al. 2015). Burials are common archaeological remains, but the shallow burial, typically <2 m depth, mean that both corpse and artefacts occupy oxic, unsaturated environments in which preservation potential, particularly of bone, is reduced (Burns et al. 2017).

Ossuaries, the stored remains of bones, are a common historical feature. In imperial Rome, inhumation in underground catacombs (Figure 2.5.2b) became fashionable from the second century CE, cremation being typical before then (Toynbee 1996). The 40 known catacombs, occupying 2.4 km^2, are up to 19 m below ground. Many of the contents are now lost, but these underground constructions in rock are likely to be preserved for considerable time.

(a)

(b)

(c)

(d)

Figure 2.5.2 Types of burial and ceremonial sites: (a) a modern Muslim burial site in United Arab Emirates with mosque in background; (b) Saint Paul's Roman catacombs from fourth century CE, Mdina-Rabat, Malta; (c) the Great Pyramid at Giza, Egypt; (d) the London Olympic Park from 2012.

The ultimate tomb constructions are the ancient Egyptian pyramids. The Great Pyramid of Khufu at Giza (Figure 2.5.2c), built between 2589 and 2566 BCE, with a height of 147 m (Herz-Fischler 2009), remained the tallest human-made free-standing structure until ~1300 CE. It comprises ~2.3 million blocks, averaging 2.27 tons in weight. The exotic limestone and granite blocks, sourced from up to 800 km away, are an early example of anthropogenic mass transfer of materials at a rate of ~0.26 million tons/yr. The construction, using natural stone arranged in a stable geometry with minimal internal voids, has excellent preservation potential, of at least many millennia. Contrast this with the Pyramid of the Sun at Moche (over 340 m long, 160 m wide and 40 m high) in northern coastal Peru (Shimada & Cavallaro 1985), constructed from 550 to 1532 CE of ~143 million adobe bricks. Despite the arid setting, this construction is affected by heavy rain damage and has little long-term preservation potential.

As global populations increase exponentially, the annual numbers of the deceased rise too; in 2011, the figure is estimated at ~55 million people (www.ecology.com/birth-death-rates/). Over recent decades the proportion of organised burials has steadily decreased and since 2015 has been outnumbered by the number of cremations (NFDA 2016). But this still results in huge numbers of interred bodies each year. In countries such as the United Kingdom there is pressure to reuse burial plots in urban cemeteries after 75 years, causing greater disturbance with reinterment of older remains at greater depth. The modern and widespread trend towards cremation leaves only charred fragments but a potential chemical signal from disperal of polychlorinated dibenzo-p-dioxins and dibenzofurans (PCDD/Fs) and mercury (Mari & Domingo 2010).

Related modern ceremonial sites include cathedrals and churches. Arguably, they also include developments such as sports facilities and music halls, as specific parts of urban conurbations. The scale of operations can be very large: in development for the London 2012 Olympic Park (Figure 2.5.2d), ~1.7 million m^3 of soil was washed to remove industrial contaminants (http://atkinsglobal.co.uk/en-GB/media-centre/features/enabling-olympic-park).

2.5.2.3 Domestic Refuse

The composition of domestic waste materials deposited as landfill, including historical middens, has changed greatly over time. The earliest middens relate to food foraging, with shell or kitchen middens found along coastlines, estuaries, lakes and rivers during the Late Mesolithic of Europe. The oldest such middens date from about 140,000 years ago, e.g., Blombos Cave in South Africa, but they can range up to recent times. From approximately 10,000 to 4,000 years ago, populations across much of the planet were still broadly hunter-gatherers but were reducing their range and selecting favourite sites to forage for shellfish, resulting in locally impressive shell middens. The deposits can include objects such as animal bones and pottery sherds associated with human occupation.

As humans developed the ability to preserve, store and transport foods, containers made initially from pottery became important debris. A spectacular example is the ~24,750,000 amphorae containers, used to transport olive oil from Spain, discarded from the early first century CE to post-217 CE to form the entire hill of Monte Testaccio in Rome (Figure 2.5.3). Some 50 m high, of 550,000 m^3 volume, and with a mass of 742,500 tons, the western accumulation of amphorae shows a clear stratigraphy (http://ceipac.ub.edu/MOSTRA/u_expo.htm). Pottery and glass have been the storage containers of choice for at least the last two millennia and are commonly found in anthropogenic deposits, the evolution of differing morphotypes being an important tool in dating such archaeological deposits. Since the mid-20th century, plastics have rapidly taken over many of the functions of these older materials and have increasingly become a significant component in waste materials (Zalasiewicz et al. 2016a).

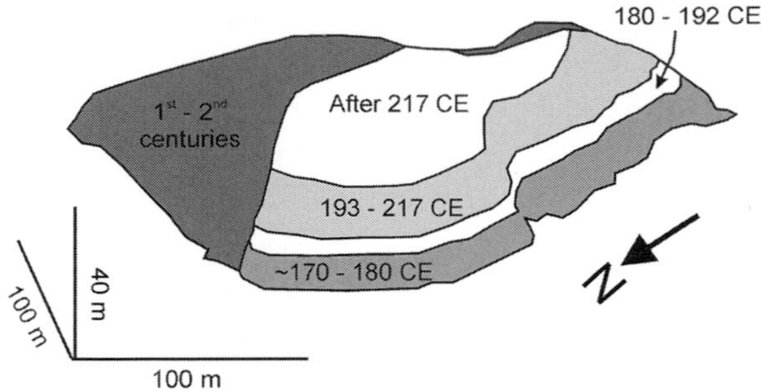

Figure 2.5.3 Perspective view of Monte Testaccio in Rome, formed entirely from amphorae deposited from the first to third century CE. Adapted from III. The Birth of Mount Testaccio http://ceipac.ub.edu/MOSTRA/u_expo.htm (Reproduced with permission of Centre for the Study of Provincial Interdependence in Classical Antiquity).

Landfill: As towns and cities grew, refuse began to be concentrated at sites outside them, to minimise proximity to pollution and vermin. Landfill materials vary geographically across the world as well as temporally, reflecting both local uses of materials and local legislation. Through much of their history, they have been uncontrolled tips of inert to biodegradable and potentially toxic wastes, while much of the organic waste was incinerated on site. However, as populations have soared over recent decades, the increasing risk of contamination of the air and water led to an emerging trend of sealing the landfill material, often in successive plastic-sheet-lined 'strata', to prevent seepage of leachate into groundwater. This slows decay, resulting for instance in the 'mummification' of food wastes (Rathje & Murphy 2001) and so increases long-term preservation potential, where landfill material is not eroded and reworked (Spencer & O'Shea 2014).

Landfills have changed composition, mirroring changes in urban living, hence producing variations that allow time correlation between sites. For example, in the United Kingdom prior to 1800, volumes of waste were small and dominated by ash, wood, bone and vegetable wastes, with little metal due to reuse and recycling (Waste Online 2004). By the early 20th century, most houses in the United Kingdom were fuelled by coal fires, much of the resultant ash going to landfill (seen as 'fines' in Figure 2.5.4). However, following the Clean Air Act in 1956, the use of domestic coal and hence the proportion of ash and cinder in landfill waste began to decline (Ford et al. 2014; Figure 2.5.4 herein). In the 1960s, with burgeoning plastics use and with increased consumerism and waste production, the volume and composition of wastes changed radically in the United Kingdom (Bridgewater 1986; Burnley 2007; Ford et al. 2014). Subsequent legislation and changes in recycling patterns resulted in greater segregation of waste and corresponding changes in waste volume and composition, a development throughout developed nations.

During the second half of the 20th century, as volumes of domestic refuse escalated, many urban areas developed large, long-lived landfill sites. The Fresh Kills Landfill, the repository for New York City refuse for over 50 years, opened in 1947 and now covers 890 ha and is up to 70 m high. The peak influx of garbage was 29,000 tons a day in 1986–1987 (http://nyc.gov/html/dcp/pdf/fkl/about_fkl.pdf; downloaded 22 February 2016); the total is ~150

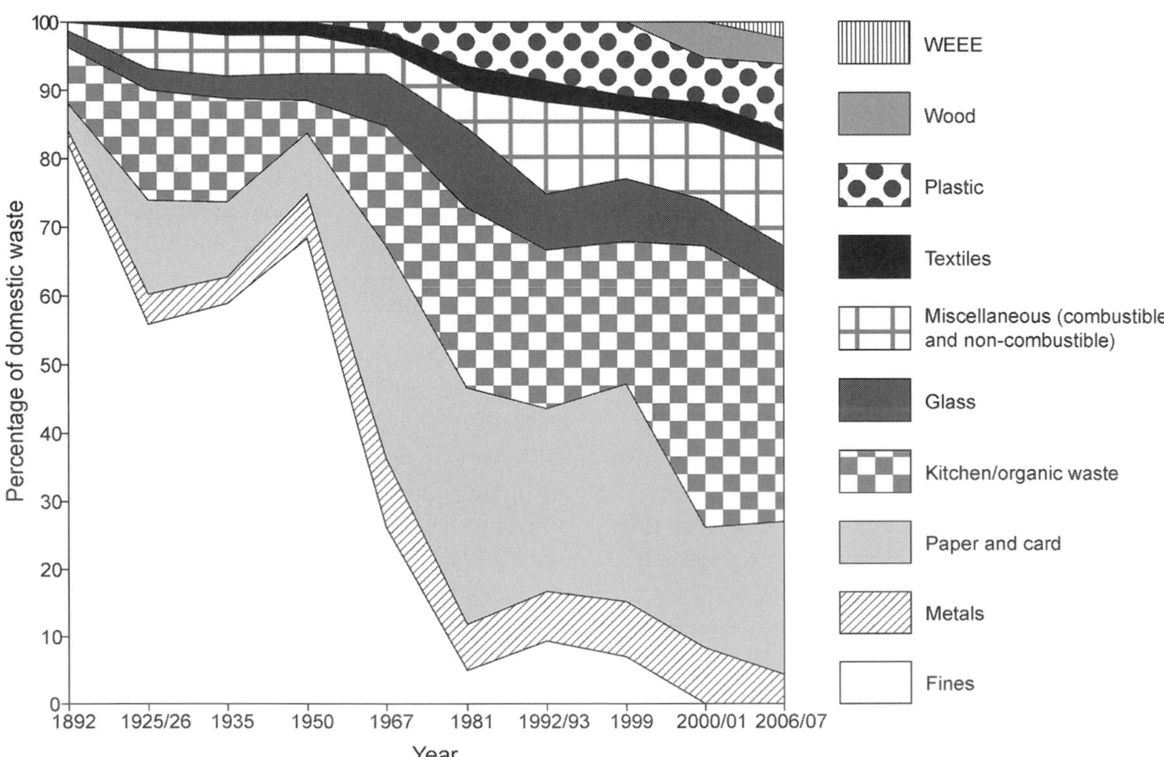

Figure 2.5.4 The changing composition of household municipal solid waste in Great Britain. WEEE-waste electrical and electronic equipment. Sourced from figure 11 of Ford et al. (2014).

million tons, including 1.2 million tons of debris from the terrorist attacks on the World Trade Center marking the final contribution.

Within landfill strata, the changing patterns of waste materials through time and the nature of the artefacts they are shaped into enable fine-scale 'technostratigraphic' age dating and therefore effective chronostratigraphic discrimination of 'Holocene' vs. 'Anthropocene' landfill material. However, the materials are not emplaced in a simple layer-by-layer process typical of natural sediments, forming a complex sedimentary architecture.

2.5.2.4 Transport Routes

Transport networks (outside urban areas) such as roads and railways are estimated to cover 0.5 million km^2 and 0.03 million km^2, respectively (Hooke et al. 2012), with an estimated 0.25 trillion tons of deposits for roads and 0.02 trillion tons of deposits for railways (Zalasiewicz et al. 2016b; excludes earthworks associated with cuttings and tunnels). For roads, railways, tunnels and most airports, the amount of materials moved to form the infrastructure (i.e., earthworks) is typically far greater than the built structures, such as the road surfaces and rail lines. Flat, flood-prone areas may require construction of concrete viaducts or embankments. The invention of dynamite by Alfred Nobel, patented in the USA in 1867, greatly helped expansion of major construction schemes, facilitating more rapid excavation of tunnels and cuttings. Further major advances in tunnel construction include the development of sprayed concrete lining (shotcrete/gunite), which became commonly used in the 1970s, and subsequently enhanced tunnel-boring machines (TBMs), which both excavate the tunnels and support the walls while precast concrete cylinder linings are inserted behind the TBM (Williams et al. 2019, in press). The effects of ship transport are discussed elsewhere (see Section 2.8.5).

Roads: Early roads developed to help travel by foot or packhorse; in South America this was the sole usage until Spanish colonisation. Transport of goods using ox and sleds began about 7,000 years ago, requiring smoother roadways (Lay 1992). The invention of the wheel at least 5,000 years ago (Lay 1992) was a technological revolution that helped transport of metal ores, stone for construction, food and widened trade networks (Hooke 2000). Early wheeled vehicles needed graded tracks with minor movement of earth or bedrock.

Construction using multilayered sand, gravel, crushed stone and setts fixed in mortar greatly improved road networks, and during the Roman Empire, some 300,000 km of such roads were

(a)

(b)

Figure 2.5.5 (a) Example of a stone-built Roman road, Pompeii; (b) section of a tarmac road exposed by a landslip, Mam Tor, England.

constructed in ~400 years (Forbes 1958; Figure 2.5.5a herein), moving some 5.35 billion tons of earth in the process (Hooke 2000).

There has been a major expanse of road networks since the 1940s coinciding with greatly increased use of motor vehicles. Modern roads commonly have a surface of asphalt (used since the 19th century; see Section 2.2.4), concrete, or road metal ('tar macadam'/blacktop) about 5 cm thick, resting upon a compacted base of sand, gravel, or crushed rock aggregate (Figure 2.5.5b). These construction materials vary greatly in thickness, but 1.5 m for a high-traffic road is not unusual. Mountain roads focus runoff and soil erosion and tend to initiate debris slides, whereas lowland roads can seriously impact natural flood patterns (Tarolli & Sofia 2016). Many roads display distinctive safety markings. Reflective paint, used to demarcate lanes and provide signals, contains hollow glass microspheres that may become dispersed within the local environment of the roads and may be a distinctive marker of the Anthropocene (Gałuszka & Migaszewski 2017; see Section 2.3.2 herein). Cat's eyes, fixed reflective studs also used to demarcate lanes, were first patented in 1934, but their use expanded widely after WWII in the United Kingdom, and variations on the invention are now used globally and provide a further distinctive characteristic of roads and their eroded debris constructed during the Anthropocene.

Canals: The Egyptians built a canal linking the Mediterranean and Red Seas, some 160 km long, which by 600 BCE was 60 m wide and 13 m deep (Forbes 1958). The Beijing-Hangzhou Grand Canal of China, the longest artificial waterway, is 1,776 km long, with the earliest construction commencing in 486 BCE. The Canal du Midi in southern France, completed in 1681, was the inspiration for the early development of canals in the United Kingdom, initiated with construction of the Bridgwater Canal (completed in 1776). The network evolved over half a century to facilitate transport of large volumes of high-bulk goods, such as raw materials, coal and finished products, during the early phase of the Industrial Revolution, but it was subsequently superseded by the railway network. Outside western Europe, where the Industrial Revolution arrived later, bulk transport by railways was established early, and there was no need to construct elaborate canal networks. However, canals are still being constructed where they can open up new and shorter transoceanic trade networks, e.g., the ~200 km Suez Canal opened in 1869 linking the Mediterranean and Red seas, and the 77 km long Panama Canal opened in 1914, providing a connection between the Atlantic and Pacific Oceans. The construction of canals has required large volumes of rock and deposits to be excavated to create cuttings, with the waste materials used to construct embankments in order to maintain even gradients. Their construction can significantly modify drainage on floodplains, particularly if associated with the canalisation of existing river channels (see Section 2.8.4). As an old transport technology, the nature of canals has not varied greatly during the transition into the Anthropocene.

Trains: The earliest rail transport comprised wagonways developed in Germany in the 16th century to haul ores from mines. Subsequently, a coincidence of inventions related to the Industrial Revolution (Watt's stationary steam engine in 1776, Trevithick's self-propelled steam engine in 1804) and increased demand for haulage of bulk materials led to rapid expansion of the rail network in Great Britain. The first public railway, Stephenson's Stockton and Darlington Railway in northeast England, opened in 1825. Innovations in the design of both locomotives and rails facilitated rapid expansion of networks from the late 1860s.

During the era of steam locomotives, railway lands (both depots/carriage works and track beds) included much timber, coal debris, oils and solvents and so were prone to spontaneous combustion (Waters et al. 1996). Modern rail tracks continue to be made of steel, now mainly resting on concrete sleepers rather than traditional wooden ones, resting in turn on a gravel base at least 0.3 m thick. With a significant decrease in the use of steam locomotives over recent decades, modern railway lines lack the deposits of ash and cinders and unburnt coal debris that formerly accumulated.

Railways need low gradients, necessitating the construction of cuttings, tunnels and embankments. The resultant earth movements are substantial: some 20 million tons of spoil was excavated and used for coastal reclamation in the Channel Tunnel project (Price et al. 2011). In cities, underground railways provide the most rapid access route from the suburbs. Rapid expansion of London during the Industrial Revolution led to the world's first underground railway system in 1863, seen by Williams et al. (2014) as key to the technological evolution of cities. By 2014, 148 cities had metro systems, with Shanghai the longest network of 548 km (Williams et al. 2019, in press). Railways in total now cover about 30,000 km^2 across the globe (Hooke et al. 2012) and associated deposits with a mass of ~20 billion tons (Zalasiewicz et al. 2016b).

2.5.2.5 Mineral-Extraction Waste

Mining and quarrying have affected about 0.4 million km^2 of the land surface (Hooke et al. 2012). There are estimates that some 50 billion tons (Jennings 2011), 57 billion tons (Douglas & Lawson 2001), or 316 billion tons (Cooper et al. 2018, in press) of industrial minerals, including sand, gravel, clay, metal ore, coal, etc., including spoil and waste, is quarried or mined annually. Humans presently mine coal at 8.5 billion tons per year (Gt/yr) (2010–2015 CE), iron ore at 3.2 Gt/yr (2015 CE), and aggregate at ~28 Gt/yr (2012 CE) (USGS online data), and together with other mining activities, data indicate that humans now move more sediment in this fashion than all natural processes combined (26 Gt/yr). Since 1945, mineral output and related waste production have risen to about 45 km^3 per year, a trend that continues (Ford et al. 2014). In Great Britain alone over 66.5 billion tons of sediment has been moved through mineral extraction and processing since the start of the Industrial Revolution (Price et al. 2011).

Mineral extraction generates large masses of artificial deposits and also produces, via what is in effect highly selective erosion of certain elements and compounds, a geochemical signature that can be dispersed widely as wind-distributed dust or via river water, groundwater and ocean currents (see Section 5.6). Surface extraction leads to increased erosion and consequently to increased sedimentation rates in neighbouring areas, especially of finer-grained deposits (Rivas et al. 2006). Together, mining activity and urban development increase denudation rates by typically two to four orders of magnitude over natural rates (Rivas et al. 2006, Tarolli & Sofia 2016). Associated subsidence of up to 50 mm/yr and lowering of the water table can accompany mineral extraction, particularly by underground mining (Tarolli & Sofia 2016). Landslides, debris flows and rockfalls are enhanced, too, in areas of extensive surface mining, within both the spoil debris and adjacent natural deposits (Tarolli & Sofia 2016).

Stone tools: Stone tools may have been used as early as 3.4 million years ago (e.g., McPherron et al. 2010), with more complex Palaeolithic stone-blade technologies from ~50,000 years ago (Bar-Yosef 2002; see Section 7.2 herein). Associated extensive landscape modification occurred from at least 50,000 years ago at the escarpment of Messak Settafet, Libya, where some 75 million stone-tool artefacts per km^2 were extracted by hominins from the silicified-sandstone bedrock and discarded, leaving a carpet of deposits and many shallow extraction pits (Foley & Lahr 2015). Shallow mining using rudimentary tools, such as antlers, has taken place from the Neolithic, as in shallow pits into chalk deposits to work flint at Grime's Grave, near Thetford, England, and elsewhere, worked from 3000 BCE. This is almost exclusively a pre-Anthropocene technology.

Metalliferous minerals: The earliest known mining activity, at least 42,000 years ago, is at Ngwenya Mine, Swaziland. The site shows ~30,000 years of continued workings for iron ore (Vogel 1970) to produce new tools and weapons. By about 400 CE, smelting of iron took place at the site, where there is now a modern opencast iron-ore mine, which opened in 1964 (http://whc.unesco.org/en/tentativelists/5421/).

This shows how mineral extraction has evolved from small-scale, shallow mining through to deep mining enhanced by mechanisation and now to the development of super opencasts able to quarry minerals by surface excavation from depths previously only possible from subsurface mines.

In both the Mediterranean region and China, copper began to be mined in the Late Mesolithic (>5,000 years ago), with the addition of tin to copper to make the harder alloy bronze beginning about two millennia later in both regions. The scale of metal-tool manufacture expanded significantly with the start of the Iron Age in Europe ~800 BCE (in China ~600 BCE, though iron was locally used there several centuries earlier). Iron occurs in much commoner metal ores, and the tools are tougher, so this development increased rates of mining (Hooke 2000). This technological development broadly coincides with the first widespread trace-metal pollution in the Northern Hemisphere at 3,100–2,800 years ago (see Section 7.3.5). Despite the importance of such mining to the development of civilisation, it was neither globally extensive nor volumetrically large when compared with modern extraction rates.

The Industrial Revolution saw great increase in metalliferous mining in Europe during the 19th century, due in part to increased demand and in part to improved mining technologies. By the late 19th century, mining in Europe declined relative to that in North America (ICMM 2012).

After WWII, the United States saw a similar dramatic decline, with mining shifting to developing countries following depletion of the easily extractable minerals, while bulk-transport ships made transfer of metals to the main markets more economic (ICMM 2012). Hence, the pattern associated with the Anthropocene is for smelter and refinery production, and the wastes associated with them, to still largely remain located in the developed countries (ICMM 2012). The scale of production is now, for example, 2,000 million tons of iron ore each year, a little less than 20 million tons of copper, 2,000 tons of gold and 200 tons of platinum-group metals (ICMM 2012).

Nickel, manganese, chromite and other alloys have been increasingly extracted for stainless- and specialist-steel production, while various rare-earth elements, though still volumetrically small, have become critical to high-tech industries.

Most deep mining takes place at depths of a few hundred metres, but currently the deepest gold mines in South Africa extend to >4 km (http://mining-technology.com/features/feature-top-ten-deepest-mines-world-south-africa/). Deep mining is only economic where the metals are of high value and occur in veins, particularly at depths below which an open pit is feasible. The greater cost of subsurface-mine extraction results in much less production of surface waste compared with strip mines, targeting higher-grade ores, and using waste rock to backfill old workings, both to stabilise the workings and to reduce haulage costs. The general trend is towards strip mining/surface quarrying, with a decline in ore grade (ICMM 2012), both driving an increased output of spoil. The Bingham Canyon mine, 0.97 km deep and 4 km wide, is one of the largest human-made excavations on the planet (Figure 2.5.6a).

Mine tailings (slag heaps) can be reused, like the tailings from Cornish tin mines proved to contain uraninite, which was mined from them during WWII to provide a British source of uranium for the Manhatten Project. These tailings tended to wash out to sea, forming locally tin- (cassiterite) rich coastal sediments.

Mining takes place not only on land but also from beaches – for example, for glass sand or titanium (in the heavy mineral rutile) in New Zealand and elsewhere. Tin is also mined offshore, for example, from coastal sediments off Malaysia, while diamonds are mined from beaches and drowned river valleys offshore along the coast of Namibia.

Coal: The start of the Industrial Revolution in the mid-18th-century United Kingdom saw the first large-scale working of coal to fuel steam engines (Section 7.4.2). In the United Kingdom, by the 1910s coal extraction peaked at about 300 million tons per annum (Price et al. 2011). This expansion in coal mining was

(a)

(b)

Figure 2.5.6 Types of mineral extraction. (a) Surface opencast (strip mine) at Bingham Canyon, USA; (b) pillar-and-stall working of coal. (from Waters et al. 2018, in press)

enabled by successive improvements in mining technology: steam engines were used to pump water from mine workings; improvements were made in ventilation; and the development of mining techniques, such as the longwall method, allowed mining operations to go to greater depths and expand into new areas (Waters et al. 1996, Zalasiewicz et al. 2014c).

Underground coal mining produces large surface spoil tips. Historically, uncontrolled deposition of these tips resulted in acid leachates in groundwaters and rivers, spontaneous combustion of carbonaceous material, and collapse of unstable steep slopes, as in the Aberfan (Wales) disaster of 1966. This resulted in UK legislation to control the disposal of colliery waste and led to reengineering of existing tips and their reclamation and 'greening' post-use.

Subsurface coal mining also leads to subsidence, the nature of which varies depending on working methods (illustrated by Waters et al. 1996). Coal was historically worked in small, shallow (up to 10 m) bell pits, which would collapse to form small circular depressions. The pillar-and-stall method of working, involving a grid of narrow roadways through a coal seam leaving pillars of coal to support the roof, allowed somewhat deeper extraction of coal (up to 30 m), with subsequent failure of pillars causing variable surface subsidence (Figure 2.5.6b). Subsequently, the longwall mining method became the norm for most 20th-century deep-mine coal extraction. It involves the total extraction of coal from a panel with the roof being temporarily supported by props, which are moved forward as the face is advanced. Wholesale subsidence occurs very rapidly after extraction of the coal and removal of the props, typically in days, and is complete after a few years. The amount of such regional surface subsidence can reach several metres (Bowell et al. 1999).

Changes in technology and in the economics of mineral extraction have led to a significant shift to surface mining since the mid-20th century. Global production of coal and lignite was 7,269 million tons in 2016, with China controlling 44.5% of production (IEA 2017). The combustion of coal widely disperses metals, including arsenic, cadmium, chromium and mercury, and fly ash as stratigraphic signatures (see Sections 2.3.1 and 5.6).

Aggregates and limestone: World cement and aggregate (mainly sand and gravel) production is currently associated with around 40 km^3 of limestone, shale, and aggregate excavated and moved each year (Ford et al. 2014). Output has increased greatly in tandem with the concrete production that provides the main Anthropocene signal of this activity (see Section 2.2.1). These bulk materials are mainly worked from quarries, in the case of aggregates either directly from sand and gravel deposits or as crushed bedrock, though also commonly through dredging of rivers and coastal zones. Such materials are largely inert and hence do not generate a significant pollution signal, other than the CO_2 created through concrete production, though quarries may produce fines (clay and silt) as a waste product, and offshore dredging destroys the local seabed ecology. Disused quarries and gravel pits may subsequently leave a secondary Anthropocene signature via their use for landfill.

Hydrocarbons: Most oil and gas extraction produces no significant waste deposit, although their extraction can result in ground-surface subsidence, and combustion produces greenhouse gases and fuel-ash particulates that form a widespread anthropogenic signature (see Sections 2.3 and 5.2). The main physical evidence of former hydrocarbon extraction will be the drilling infrastructure left in the subsurface, but offshore the deposition of oil-stained drill cuttings around oil and gas rigs usually takes the form of a halo up to 10 km wide in which levels of pollution decrease away from the rig site. Oil sands differ in being extracted in large strip-mine operations. The Athabasca oil sands of Canada comprise 14,000 km^2 of potentially extractable deposits; large-scale mining began in 1967, removing less than 75 m overburden and 40–60 m of oil sands. One operation (Syncrude Mine) has alone removed and processed 30 billion tons of sediment (Syvitski & Kettner 2011), with other nearby mines processing >1.5 billion tons/yr of tar sand (2012 CE). The sand worked via strip mines is rinsed with hot water to separate the oil, and sand and wastewater are then stored in tailings ponds, which are reforested when dry. Deeper deposits are worked by injecting hot water via wells and extracting the liquefied bitumen.

2.5.2.6 Industrial Development

Coordination of manufacturing within large industrial facilities was a key development of the Industrial Revolution, and many of the heavy industries so assembled have produced waste products both as artificial deposits and also as air- and waterborne contamination (see Waters et al. 1996 for a UK perspective). Iron works, foundries and other metal smelters, for instance, have produced large quantities of slag (the solidified glass of silicates and oxides left following removal of the molten metal), foundry sand (used as a moulding material) and spent refractory materials (used to line furnaces) with high concentrations of heavy metals (such as chromium, lead, nickel and zinc), sulphates, sulphides, asbestos, phenols and hydrocarbons. Prior to large-scale extraction of natural gas, gasworks in the 19th and early 20th centuries carbonised coal to produce town gas, producing a range of contaminants as by-products: sulphates, sulphides, phenols, coal tars, polyaromatic hydrocarbons, cyanides and heavy metals such as chromium, copper, lead, nickel and zinc.

The chemical industry produces diverse acids, adhesives, cleaning agents, cosmetics, dyes, explosives, fertilisers, food additives, industrial gases, paints, pharmaceuticals, pesticides, petrochemicals, plastics, textiles and so on, each with distinct waste products. Amongst these, increased reactive nitrogen, plastics and pesticides in particular provide conspicuous proxy indicators of Anthropocene strata (see Sections 4.3, 5.4 and 5.7).

A significant recent development is of large underground storage facilities. These have been developed for liquefied petroleum gas and compressed-air energy storage, while the subsurface is increasingly used as a source of geothermal energy and to provide long-term burial of nuclear and other hazardous wastes (Evans et al. 2009; Zalasiewicz et al. 2014c; Waters et al. 2018, in press).

2.5.2.7 Water Supplies and Sewerage

Modern urban conurbations need extensive water supplies and sewerage systems, while much agriculture is now wholly dependent on water from surface-water bodies or groundwater.

Reservoir dams and barrages: Cultivatable fields in arid regions of the Near East and North Africa were made possible by successive low check dams across seasonal watercourses in order to trap fertile sediment carried by floodwater (Barker 2002, Edgeworth 2011, 2013). This technique, still in practice, dates back to Neolithic times (Wilkinson 2003). In medieval Europe significant damming of watercourses to supply water power took place as early as the 12th century CE (Walter & Merritts 2008). In the eastern USA during the late 17th to the early 20th century, such milldams provided the primary control on the flux of sediments through river systems, causing burial of presettlement floodplain wetlands under thick layers of fine sediment (Walter & Merritts 2008).

Modern reservoirs cover about 200,000 km^2 across the globe (Hooke et al. 2012), with an estimated sediment mass trapped by large dams of in excess of three trillion tons (see Section 2.8.3). Although large dams were being built in the 19th and early 20th centuries, their numbers increase greatly after 1945, and most large dams were built over the last 70 years at a rate of more than two per day; more than half of these are in China (Syvitski & Kettner 2011). Many early dams were placed in highlands for the generation of hydroelectricity, but reservoirs are increasingly being sited within floodplains to provide flood control, irrigation and recreation (Syvitski & Kettner 2011).

Drainage channels/tunnels/aqueducts: In many agricultural areas, more than 80% of catchment areas can be drained by surface ditches or subsurface pipes that direct water to agricultural crops (Tarolli & Sofia 2016). Irrigation channels typically comprise small-scale surface ditches taking water from rivers into fields (e.g., Nile Valley). The Egyptians developed basin irrigation by at least 3000 BCE, with the construction of a network of earthen banks, parallel and perpendicular to the river, with floodwater diverted into basins by sluices.

Drainage ditches have long provided an important means for draining areas of high water table, e.g., Egypt and Babylonia. These have modified into subsurface clay tile drains in agricultural land, which date back to 5000 BCE in Crete and were certainly used by the Romans by 200 BCE. The process has led to extensive drainage and transformation of wetland environments (see Section 2.5.2.8 below), which in the United Kingdom became more widespread from the 1750s CE during the start of the Agricultural Revolution. Modern land drains use flexible perforated plastic piping made from PVC, less prone to fracturing than clay pipes and distinctive as an Anthropocene indicator.

Many of the world's longest tunnels transport water from source areas into urban centres. A Roman aqueduct supplied water to the ancient city of Gadara (modern Jordan), the longest section being 94 km (Döring 2007). The pipeline was built in the ancient 'qanat' style, with a series of vertical shafts connected underground by gently sloping tunnels. In comparison, more modern lined tunnels include the Thirlmere Aqueduct, England, which is 154 km long, built between 1890 and 1925 to supply water to Manchester. Aqueducts for water supply may also be constructed above ground. As with most other utilities, there is now a trend of using plastic piping.

Sewerage: In the past, raw sewage was spread on fields adjacent to cities to act as a fertiliser. Animal dung is still collected in many parts of the world to be used as plaster for mud buildings or floors and, when dried, to be burned to supply heat for cooking (e.g., in rural India).

The development of the integrated sewerage system in London was vital for maintaining the health of the rapidly expanding city. The 160 km of intercepting sewers, constructed between 1859 and 1865, was fed by 720 km of main sewers and 21,000 km of smaller local sewers. Construction of the interceptor system required 318 million bricks, 2.7 million m^3 of excavated earth, and 670,000 m^3 of concrete (Goodman & Chant 1999).

To combat the pumping of raw sewage into rivers, sewerage systems were developed in the 18th and early 19th centuries in Britain to store, but not to treat, the sewage. The sludges and solid wastes can contain high concentrations of cadmium, chromium, copper, lead, nickel and zinc and organic matter that may decompose to nitrogen-rich leachates, CH_4, CO_2 and H_2S.

Modern treatment works, with the distinctive Anthropocene signal of plastic piping, are now a common feature of many floodplains adjacent to urban areas. They screen solids from the sewage and aerate it in gravel beds to allow bacterial degradation of organic matter.

2.5.2.8 Flood Defences and Coastal Reclamation Schemes

Flood defences can include waterway replumbing, diversions, channel deepening, embankments and discharge focusing (Vörösmarty et al. 2003; Merritts et al. 2011; Pearson et al. 2015). Coastal engineering defences include groynes, jetties, seawalls, breakwaters and harbours (Syvitski et al. 2005b). Larger coastal infrastructure deposits have now become globally ubiquitous (Syvitski & Kettner 2011).

Protecting from riverine flooding: Along the Yellow River in China, as for many river systems, the levees bounding the river channel have been artificially raised to prevent flooding of farmland and settlements on the floodplain (see Section 2.8.4). As more sediment is deposited on the riverbed, superelevated channels form, which further increase the danger of flooding (Edgeworth 2013). The modern course of the Yellow River has been fixed by human intervention since the last major shift of the river and the formation of a new delta in the mid-19th century (Syvitski & Kettner 2011). Human modification of the Yellow River, though, has taken place for millennia, possibly since the Late Neolithic (Xu 2003; Edgeworth 2011), while catastrophic flooding has repeatedly buried settlements over the years. Beneath the modern city of Kaifeng, 9 km south of the present river, 8 m of river silt covers the remains of the medieval city of Dongjing. This in turn is underlain by 2 m of river silt, which overlies heavily compacted remains of the city of Daliang, dating from the fifth to the third century BCE (Johnson 1995; Edgeworth 2014). There the river is now elevated 10 m above the floodplain by successive rebuilding of the levees since 1950 (Johnson 1995).

Most levees are constructed rapidly through earth banks, with no distinctive Anthropocene signal other than the isolated presence of artefacts. In contrast, flood-control channels, which rapidly transport flood waters away from urban areas, are typically concrete-lined dry channels, occurring either at the surface or below city streets.

Protecting from coastal flooding and erosion: Engineered structures have long been used to provide coastal defences, control erosion and protect settlements from tides and waves. They include seawalls and barriers, as well as groynes, breakwaters and artificial headlands to inhibit lateral sediment transport and so maintain beaches. In the first century BCE, for instance, the Romans built a seawall/breakwater at Caesarea Maritima by sinking concrete-filled barges into position (Boyce et al. 2009).

Since the mid-20th century the scale of such construction has become enormous. Currently, in Japan about 56% of the coastline of Honshu Island is partly or fully artificial, and at least 43% of Japan's entire 29,751 km coastline is lined with concrete seawalls or other structures. A 370 km long concrete seawall in northeast Japan has been lengthened and raised to limit destruction from tsunamis, following the 2011 Tōhoku earthquake and tsunami. For example, the Hamaoka nuclear power plant is being protected by a seawall up to 21 m high. In China, a new 'Great Wall', >11,000 km long, already extends along more than 60% of the coastline (Ma et al. 2014). Such constructions profoundly modify coastal environments, such as tidal flats, that now lie inland of the walls. More localised structures include flood barriers in estuaries, which operate only during tidal floods and storm surges, like the Thames Barrier, which itself is part of an integrated system of flood defences designed to protect London from North Sea storm surges.

Extending land out to sea: Early examples of land reclamation developed mainly to increase the extent of agricultural land. The enclosure of The Wash embayment and associated Fenland in eastern England by a <100 km seawall transformed ~4,000 km^2 from wetlands into prime farmland, beginning in Roman times but more significantly from the 17th century. One consequence of drainage has been the complete removal from much of this area, by deflation and oxidation, of a layer of surface peat that was commonly >4 m thick, exposing the formerly buried landscape beneath (Smith et al. 2010).

In the Netherlands, land reclamation from the North Sea started over 2,000 years ago with construction of terpen (early dykes). In 1287 a catastrophic flood inundated former farmland, creating a new bay called the Zuiderzee. The subsequent history has been one of gradual reclamation of this inlet of the North Sea through construction of polders: low-lying tracts of land enclosed by embankments or dykes, creating an artificial hydrological disconnect from the sea. Windmills originally pumped water from the drained lands, which subsided, particularly through loss of peat (see previous paragraph), often to levels below sea level. To maintain this reclaimed farmland, continual pumping of water out of the polders is required. Water-table control is now done on industrial scales at pumping stations. One of the earliest large-scale projects was the Beemster Polder, completed in 1612 and adding 70 km^2 of reclaimed land (Figure 2.5.7). From 1927 to 1932, a 30.5 km long dyke (Afsluitdijk) was built, turning the Zuiderzee into the IJsselmeer freshwater lake. Significant modification of the dykes followed the 1953 flood disaster, and the Flevoland polder, reclaimed from 1955 to 1968, at 970 km^2 is the largest reclaimed island in the world. The country now has about 3,000 polders covering ~7,000 km^2, about 20% of the entire land surface. Despite vast terraforming resulting from the polderisation of the Netherlands, there has been relatively little anthropogenic movement of materials, limited mainly to that needed for the construction of the dykes.

Increasingly, such land reclamation is aimed towards creating more living space and helps construction of industrial sites or location of major airports. It now involves the transport and deposition of vast quantities of rock, sediment and soil. Approximately 3.15 billion tons of sand, crushed rock and engineering soils will eventually be used in the four land-reclamation projects to construct the Palm Islands Complex near Dubai in the United Arab Emirates (Syvitski & Kettner 2011). Construction of the 12.5 km^2 Hong Kong Chek Lap Kok Airport involved dredging 0.6 billion tons of marine clays and sands (Syvitski & Kettner 2011). This represents just one such reclamation of land from the sea, totalling around 60 km^2 by 1996, with about 25% of the surface area of Hong Kong Island reclaimed. The 11,000 km seawall of China, from 1950 to 2000, has seen enclosure of wetland areas increase by an average of 24,000 ha/yr, increasing dramatically to 40,000 ha/yr during rapid urbanisation in 2006–2010 (Ma et al. 2014).

2.5.2.9 Warfare

Defensive structures: The earliest known city walls are at Uruk in Mesopotamia, built before 3000 BCE (Lay 1992). Since then, such protective structures have commonly constrained the geometry of urban development (see Section 2.5.2.1); even cities such as Rome and Babylon, with populations exceeding half a million, were contained within small areas of 14 km^2 (Lay 1992).

The largest defensive structure is the Great Wall of China, 21,000 km long and comprising 16 separate walls and 44,000 forts and towers. The earliest constructions from the seventh century BCE are of layered mud and reed. During the Qin Dynasty (221–207 BCE), the walls were joined into a defence 6 m high and 10,000 km long, though now mostly degraded. The Ming construction of the 16th century is about 8,000 km long, 9 m wide and 7.5 m high, built of billions of bricks, each about 10 kg in weight, secured by lime mortar. The wall has been estimated to weigh 0.4 billion tons (Syvitski & Kettner 2011) and hence is modest in mass by comparison with the modern structures described above.

Modern defensive structures are commonly built of concrete. An early example is the ~600,000 m^3 concrete development built on Guernsey from 1941 to 1944 during Nazi occupation. It formed part of the Festung Europa (Fortress Europe), which on the mainland provided ~6 million m^3 of Atlantic defences against Allied invasion (www.festungguernsey.supanet.com/about_us.htm).

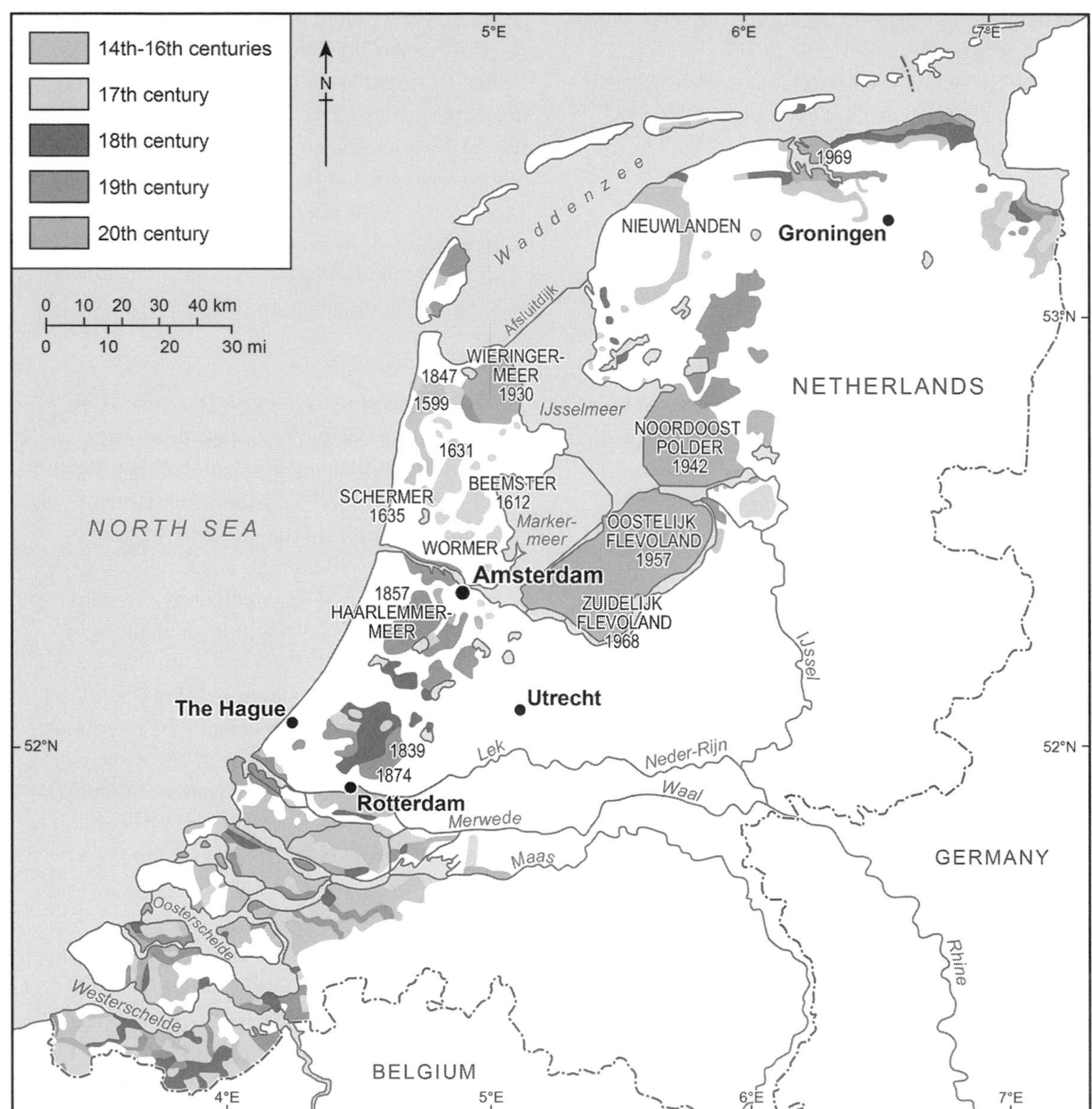

Figure 2.5.7 Polderisation of the Netherlands since the 14th century. (redrawn from the *Diercke Three Universal Atlas Diercke Niederlande* – p.113, figure 2). ©2009 Westermann Gruppe (A black-and-white version of this figure appears in some formats. For the colour version, please refer to the plate section.)

Bomb craters: Warfare impacts on geology, as well as on archaeology (Zalasiewicz & Zalasiewicz 2015). The nature of warfare changed following the invention of gunpowder, initially in China over 1,000 years ago, and subsequently as the technology spread across Asia and Europe during medieval times. The effect of gunpowder, used to fire shot and cannonballs, is seen in the archaeological record as distinct damage to skeletons and buildings, through the extraction of saltpetre as part of gunpowder production, and through the construction of powder mills in isolated locations.

The invention of the high explosive trinitrotoluene (TNT), used first in artillery shells by Germany in 1902, led to the industrialisation of warfare during WWI. High explosives changed the landscape through extensive cratering along the Western Front. British mines totalling >455 tons of ammonal explosive detonated beneath German lines during the battle of Messines on 7 June 1917, created 19 large craters, and were likely the largest planned explosion in history until the 1945 Trinity atomic-weapon test. On 18 April 1947 a British attempt to destroy the entire North Sea island of Heligoland and its extensive fortifications with a detonation of ~3.2 kilotons of TNT equivalent (Willmore 1949), considered the largest single non-nuclear intentional explosive detonation not for test purposes, showed that large-scale terraforming of landscapes had become reality after WWII. Between 1965 and 1971, Vietnam was pulverised by some 26 million explosions (Zalasiewicz & Zalasiewicz 2015). Such extensive cratering leaves little or no original soil surface undisturbed, with churned soil several metres thick mixed with human bones, which Zalasiewicz and Zalasiewicz (2015) compared with fossil bone beds and which Hupy & Schaetzl (2006) compared with bioturbation in coining the term 'bombturbation'.

Underground nuclear test explosions produce a mass of intensely radioactive, strongly shock-brecciated rock surrounding a melt core, in turn surrounded by roughly circular fault systems, outlining surface crater systems that, in the Yucca Flats test site, reach several hundred metres across (Grasso 2000). Uncontained subsurface explosions at shallower depths penetrate to the ground surface, producing conical craters, tens to hundreds of metres in diameter and depth, surrounded by ejecta (Hawkins & Wohletz 1996). The central cavity may subsequently collapse to form a chimney of debris, which, if it reaches the ground surface, forms a bowl-shaped subsidence crater that may be nearly a kilometre wide and several tens of metres deep (Hawkins & Wohletz 1996). In deep constrained detonations, there are four commonly recognised zones (Hawkins & Wohletz 1996): (1) the cavity, potentially floored by molten rock; (2) the crushed zone surrounding the cavity in which the rock mass has lost all of its former integrity; (3) the cracked zone in which the rock mass contains radial and concentric fractures; and (4) the zone of irreversible strain with deformation modifying porosity/permeability and material strength. These are argued to be similar to but distinguishable from natural impact craters (Waters et al. 2018, in press).

Nuclear testing has caused the localised conversion of sand into a glasslike substance known as 'atomsite' or as 'trinitite' in the United States (Eby et al. 2010) and as 'kharitonchik' in Kazakhstan (see Section 5.8). Trinitite from surface explosions generates glass beads, the product of airfall of molten glass, that appear similar to tektites produced during meteorite-impact explosions (Eby et al. 2010), like that seen as a result of the Chicxulub impact at the end of the Cretaceous.

Collapsed buildings/war debris: Warfare can lead to the ultimate destruction and abandonment of urban centres, such as the presumed city of Troy razed to the ground in 1180 BCE. Destruction can take place through rivers being diverted or levees being breached in order to flood towns. Abandonment can result from water being diverted away from fortified centres. Commonly, direct attacks on cities result in a layer of demolished buildings and charred remains that accumulated extremely rapidly and then acted as the foundations for subsequent regeneration. In bomb damage associated with the Blitz in London, 116,483 buildings were destroyed or damaged beyond repair from 1940 to 1945 (Ward 2015). Regeneration of the city occurred atop the debris of the damage, some of the rubble being recycled, much of it being transported to the Lea Valley to fill Hackney and Leyton Marshes and aid expansion of development there. In contrast, the contemporary bomb damage in Berlin was cleared from bomb-damaged neighbourhood and used to bury the remains of Albert Speer's unfinished military college to form the

Teufelsberg (the Devil's Mountain), 80 m in height and 1 km long, comprising some 75 million m^3 of rubble (Cocroft & Schofield 2012) that in effect forms a substantial lithostratigraphic unit of Anthropocene depositional age.

In summary, anthropogenic impacts of such landscape activities include not only the accumulation of artificial deposits but also soil erosion, slope failure, downstream and/or impounded sedimentation, gully development, creep, subsidence, scouring, coastline erosion, wetland/mangrove and dune destruction, and coastal inundation. Most of these processes were initiated in the Holocene, when significant perturbations began and developed, while the Anthropocene is marked by significant changes in scale and pattern of human impact that produce a distinctive and widespread stratigraphic signature.

2.6 Magnetostratigraphy

Colin N. Waters

The direction and intensity of the Earth's magnetic field have changed throughout Earth's history, and this variation provides a valuable palaeomagnetic dating method. Magnetic reversals, in particular, have been used in chronostratigraphy as a marker for chronostratigraphic boundaries because of their global reach and in geological terms are near-isochronous, e.g., for the base of the Quaternary System (see Section 1.3.1.5). Although there is no magnetic polarity change that can be used to help mark the start of the Anthropocene, variations in the magnetic-field strength can be considered. Most usefully, mineral-magnetic records from lake and marine sediments provide a tool for identifying major deforestation and soil-erosion events caused by significant anthropogenic changes in land use. Mineral-magnetic records in peats and lake sediments can also reflect particulate pollution from fossil-fuel burning.

2.6.1 Geomagnetic Variation

The Earth is a giant dipole magnet, but the direction and intensity of its magnetic field change on millennial scales. This geomagnetism is entirely independent of human impact or control. A particularly useful tool in stratigraphy is the record of the irregular polarity changes or palaeomagnetic reversals (or pole 'flips') in the Earth's magnetic field. These have occurred with an average frequency of four to five reversals per million years over the last 10 million years, but at some other geological times they were much less frequent, as in the Cretaceous, in which the C34 Normal Chron lasted about 42 million years (Ogg & Hinnov 2014). Many igneous rocks, including: volcanic lavas and tuffs; sediments, including some mudrocks and loess; and soils contain iron oxides. Some of these Fe minerals, such as magnetite, can preserve a residual magnetism or 'natural magnetic remanence (NRM)', which records the Earth's magnetic-field direction and intensity at the time of their formation.

The reversals can be recorded globally as magnetic chrons with effectively synchronous boundaries. The last substantial reversal occurred 786,000 years ago, the transition of polarities representing the boundary between the Matuyama Reverse Chron and the current Brunhes Normal Chron (Figure 2.6.1), and is estimated to have taken less than 100 years (Sagnotti et al. 2014). But there was a significant weakening in the magnetic field, the Laschamp event, that peaked 41,400 ± 2,000 years ago (Bonhommet & Zahringer 1969). It is associated with a short-duration polarity reversal that lasted about 440 years, with the transition from the normal field lasting about 250 years. The weakening of the geomagnetic-field strength allowed more cosmic rays to reach the Earth then, causing greater production of the cosmogenic isotopes ^{10}B and ^{14}C (e.g., Steinhilber et al. 2012). Since that time, the magnetic field has been strengthening (Muscheler et al. 2004), and it continued to do so until about 1800 CE, since when it has been decreasing (Steinhilber et al. 2012). Nevertheless, the rate of decrease and the current strength are within the normal range of variation in

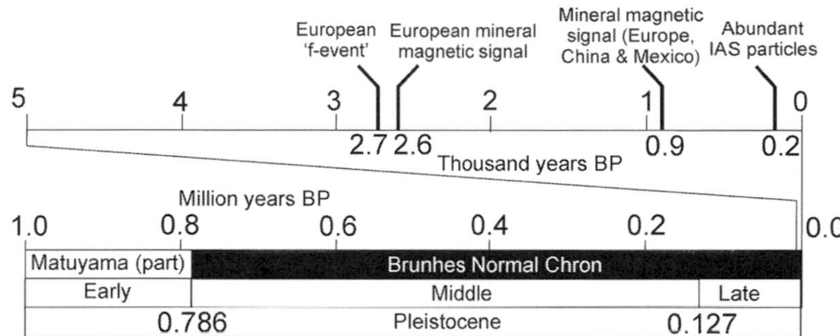

Figure 2.6.1 Key palaeomagnetic and mineral-magnetic events over the last one million years.

the field. The strength of the Earth's magnetic field has been measured continuously since about 1840, and its trend in strength (or 'dipole moment') has been one of constant decrease since then (Snowball et al. 2014). It is unclear whether this is part of a gradual polarity switch that may occur in the next thousand years or so, but at present there is no global palaeomagnetic event that could be used to define the base of the Anthropocene as currently understood (Snowball et al. 2014).

The magnetic field can also undergo 'excursions' when the magnetic field has a significant decrease in strength but does not undergo a reversal – these are so-called 'archaeomagnetic jerks' (Gallet et al. 2003). There was a global event, most strongly developed in mid- to high latitudes, coincident with a low in dipole latitude (direction) and peak in dipole moment (intensity) at ~ 2,700 cal/yr BP (the European 'f-event') that may be a potential chronostratigraphic marker (Gallet et al. 2003; Snowball et al. 2014). Other sharp cusps in geomagnetic-field direction and intensity occurred at 200 CE, 800 CE and 1400 CE (Gallet et al. 2003), but none are suitable as a marker for the Anthropocene Epoch as it is currently understood (Figure 2.6.1).

Human-manufactured materials that have been heated to high temperatures, such as ceramics, tiles, kilns, hearths, slag and brick funeral pyres, can also record the Earth's magnetic field at the time of cooling, producing a 'thermal remanent magnetisation' (TRM) (Snowball et al. 2014). These materials lose magnetism during intense heating, but the magnetism is restored when the material cools below a certain temperature (580°C for magnetite). Archaeomagnetic dating to decadal resolution can be used to date such objects heated in the last few thousand years using known past local alignments of the geomagnetic field (Eighmy & Sternberg 1990).

2.6.2 Mineral Magnetism and Land-Use Change

Mineral magnetism responds directly to anthropogenic impacts on the environment and so does provide a potential marker for the Anthropocene. The first demonstration of linkage between high concentrations of detrital ferrimagnetic minerals and catchment disturbance was provided by magnetic-susceptibility profiles in recent sediments of Lough Neagh, Northern Ireland (Thompson et al. 1975) and Loch Lomond, Scotland (Thompson & Morton 1978). Different types of soil have different magnetic properties, with certain well-drained, seasonally moist soils developing new Fe oxides that include magnetite (Maher 1986). Natural and human-induced fires (potentially related to deforestation) can result in combustion of organic compounds in soils, with Fe oxides converted in reducing conditions to magnetite (Maher 1986). Deforestation events, changes in agricultural practice and increased urban development commonly result in increased soil erosion, with the transfer of these Fe-oxide-bearing sediments to lakes and estuaries (Dearing & Jones 2003; Evans & Heller 2003). Mineral-magnetic studies are particularly suitable for recognising these concentrations of Fe oxides and hence of the soil-erosional events and have helped resolve a complex and diachronous history of deforestation (Snowball et al. 2014). Mineral-magnetic signals in Europe associated with catchment disturbance during expansion of agriculture include clusters of events around 2,600 cal/yr BP, broadly time equivalent to (though not causally linked with) the geomagnetic European 'f-event' (Snowball et al. 2014; Figure 2.6.1 herein). The largest signatures began around 1100 CE ± 100 yr (Figure 2.6.1), with similar signatures evident in China and Mexico at broadly the same time, though these are not precisely isochronous (Snowball et al. 2014). Erosion of soils has undoubtedly increased as a result of deforestation since the mid-20th century (e.g., Dearing & Jones 2003 and references therein), with increased sediment input to fluvial systems, but resultant increased mineral-magnetic signatures will still be forming.

2.6.3 Mineral Magnetism and Pollution

Mineral-magnetic signatures can also result from atmospheric pollution (Oldfield & Richardson 1990). This can include both natural and anthropogenic biomass burning, via trace quantities of iron within organic matter and soil components being converted to new magnetic Fe oxides (Snowball et al. 2014). These magnetic signatures may be able to record significant vegetation-clearance events associated with introduction of agriculture to a region, though such events are typically highly diachronous. The most prominent recent increase in mineral-magnetic abundance in sediments that can be used as a potential marker for the Anthropocene is associated with residual mineral particles generated from combustion of hydrocarbons (Hounslow 2018). Such fly ash includes spheroidal carbonaceous particles (SCPs) that may contain Fe oxides (Snowball et al. 2014) associated with oil burning; and magnetic inorganic ash spheres (IASs) that are produced through solid fossil-fuel burning, iron and steel manufacture and metal smelting (Oldfield et al. 2015; see Section 2.3.1 herein). Both types of particulates can be dispersed widely through atmospheric pollution, but the IASs provide the most distinct mineral-magnetic signature.

The first appearance and relative abundance of IAS particulates in sediments can be analysed through chemical extraction from the sediments, but it can be more rapidly studied through magnetic susceptibility (χ), saturation isothermal remanent magnetisation (SIRM), or hard isothermal remanent magnetisation (HIRM) signals. This type of record is particularly well preserved in peat bogs (Oldfield et al. 1978), soils (Maher 1986) and lake (McLean 1991), coastal (Plater et al. 1998) and offshore sediments (Horng et al. 2009). Isothermal remanent magnetisation is the remanence left in a sample after a steady field has been applied for a short time, SIRM being the maximum remanence, and is a useful technique for discriminating between magnetite and hematite. Elevated values of χ, SIRM and HIRM coincide with maximum abundance of magnetic spherules, in the case of offshore Korea represented by a sharp increase in 1987 and a peak in 1993 (Horng et al. 2009; Figure 2.6.2a, b and d herein).

Raised or 'ombrotrophic' peat bogs are especially suitable environments, as they only record direct input from the air, whereas other environments may also contain reworked particulates (Oldfield et al. 2015). However, the record may be complicated by the propensity of magnetite to dissolve in the waterlogged and acidic conditions present in peat mires (Oldfield et al. 2015). The earliest records of IASs are from the early 16th century of northwest England, associated with early iron manufacture. These particles first became abundant in western Europe and the eastern seaboard of North America in the early 19th century, and as the effects of the Industrial Revolution spread, industrial sources of these particulates expanded during the 19th century. Most sites show initial increase of magnetic pollution particles forming a '1900 CE–event' (Figure 2.6.1), representing an expression of major fuel burning in these industrialised areas (Snowball et al. 2014). However, in other parts of

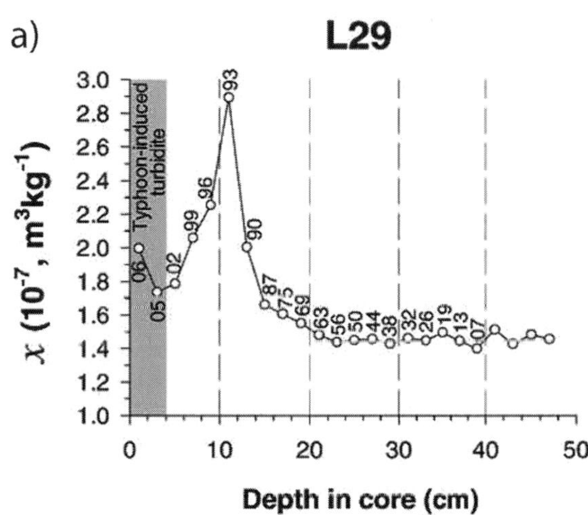

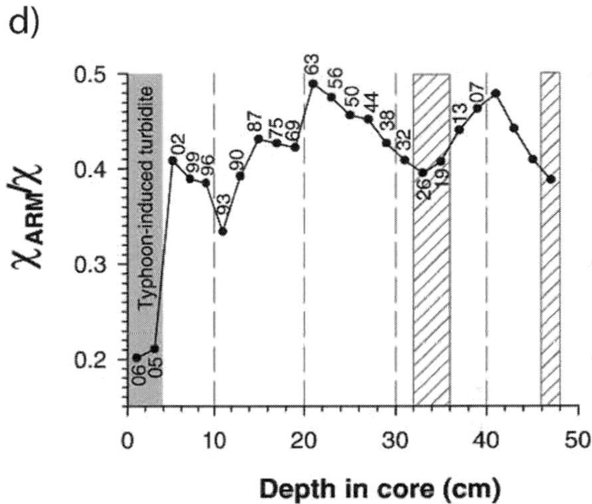

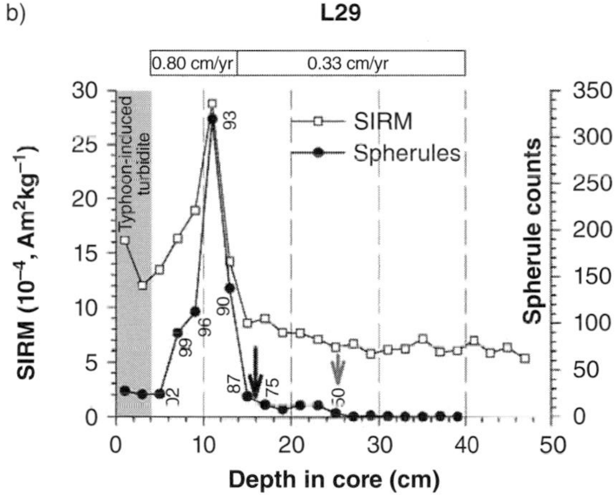

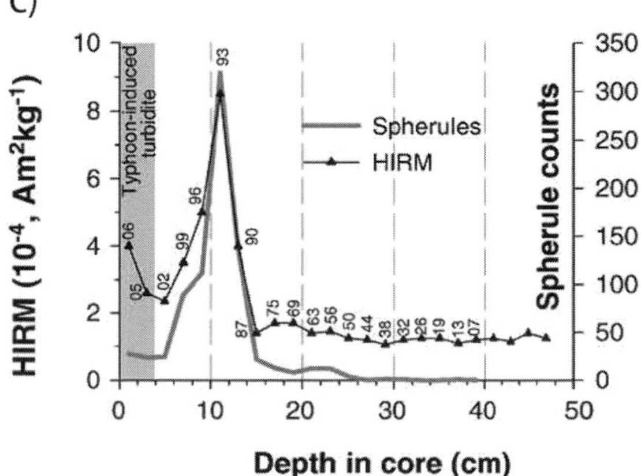

Figure 2.6.2 Marine-core profiles from Taiwan with ages shown along profiles, e.g., 02–2002 and 50–1950 (from Horng et al. 2009), showing (a) specific magnetic susceptibility (χ); (b) SIRM, numbers of magnetic spherules and sedimentation rates; (c) HIRM and magnetic spherule counts; and (d) χ_{arm}/χ showing variations in magnetite grain size, with possible turbidite layers that occurred prior to industrialisation marked with hatched bars. ©2009, with permission from Elsevier.

the world, signatures appear later, e.g., 1950s in eastern Asia. Offshore of Taiwan, the IAS signal relates to gradual industrialisation in the 1950s–1980s and subsequent rapid increase in magnetic spherules (Horng et al. 2009). Coal-fired power stations have declined both in numbers and in the magnitude of coal combustion over recent decades in Western countries, which, along with the introduction of scrubbing methodologies to clean the emitted air, has reduced the production of IASs (Oldfield et al. 2015). There appears to be a ~100-year lag in the first appearance of these magnetic pollution signatures in peats and subsequently in lake sediments, likely due to catchment and soil storage of the magnetic particulates, though differences in sediment age model constructions between peats and lake records is also possible (Snowball et al. 2014). The diachroneity of this signal reflects the gradual spread of industrialisation and for correlation is less suitable than the appearance of SCPs, which have a more pronounced and global mid-20th-century signal in response to increased oil combustion (Rose 2015; Section 2.3.1 herein).

The magnetic signal is commonly associated with increased heavy-metal concentrations. For example, in the Tees estuary of England, combined geochemical, magnetic and radionuclide signals linked to the historical record of contamination from mining and industrial activity have been resolved in sediment cores (Plater et al. 1998). The concentration of magnetic minerals (recorded as magnetic susceptibility) was used to characterise near-field industrial contamination, which began in ~1925, with a peak level in the mid-1950s, falling after the early 1980s (Plater et al. 1998). Analysis of low-frequency magnetic susceptibility (χ) provides a proxy for Zn, Pb and, to a lesser extent, Cu concentrations (Plater et al. 1998). Furthermore, records of χ_{arm} / χ and χ_{arm} / SIRM, reflecting saturation isothermal remanent magnetisation (SIRM) and the susceptibility of anhysteretic remanent magnetisation (χ_{arm}), provide information concerning magnetic grain size, a proxy for physical grain-size variations (Thompson & Oldfield 1986). Relatively low χ_{arm} / χ values are associated with coarser magnetic grains (Figure 2.6.2d); offshore from Korea, this equates with introduction of coarser IAS particles starting about 1963 during initial industrialisation (Horng et al. 2009).

2.7 A Pedology and Pedostratigraphy for the Anthropocene

Daniel deB. Richter, Sharon A. Billings and Colin N. Waters

2.7.1 Introduction

Agriculture, forestry, construction and mining are transforming more than half of Earth's approximately 13 billion hectares of soil. Farmers add amendments and mix by cultivation about 3,500 billion tons of soil for the production of food crops each year. About 35 billion tons of soil and rock are excavated for construction and mining (Hooke 2000), and rates of wind and water erosion have been accelerated to three- to fourfold background geological rates (Wilkinson & McElroy 2007). In addition, natural erosion mainly occurs in high, steep mountainsides, in landscapes completely different from those where land use has accelerated erosion due mainly to agriculture (Figure 2.7.1). Natural geomorphologic processes have hence been equalled or surpassed by human action (Figure 2.7.1). These erosional impacts on the Earth's surface are widely studied (Garcia-Ruiz et al. 2015), as are the resulting sediment problems of streams and rivers that drain human-disturbed landscapes. What is not well appreciated is that once detached and mobilised, most eroded particles do not travel far in fluvial systems but are deposited as sediments on hillslopes and floodplains. Where human-accelerated erosion occurs, former surfaces can be buried by metres or more of legacy sediment

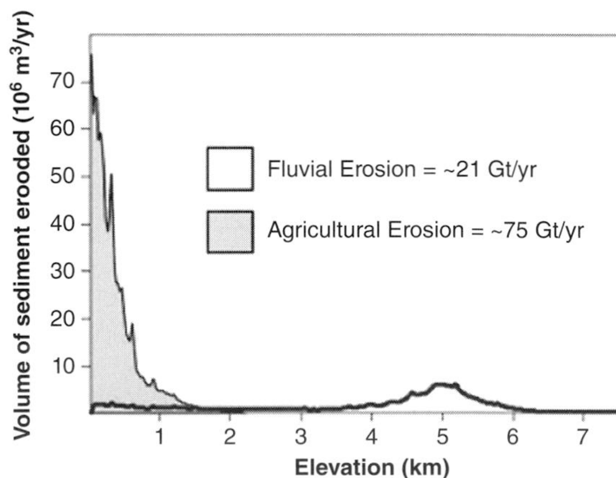

Figure 2.7.1 Ongoing fluvial erosion on Earth as a function of elevation and source, illustrating that most agriculture is practiced at below 1,000 m elevation on relatively low-slope landscapes when compared with high-slope natural erosion that occurs at relatively high elevation. The distributions emphasise the high potential for legacy-sediment deposition from agricultural sources (James 2013) and the new human signature for accelerated erosion and deposition. The y-axis is per metre of elevation. (after figure 10 of Wilkinson & McElroy 2007). © GSA

(James 2013). Indeed, Merritts et al. (2011) argue that the morphology of valleys has been so transformed as a result of legacy sediments that these new valley formations are important features of the Anthropocene (Figure 2.7.2). Human beings, says Hooke (2000), are 'the premier geomorphic agent sculpting the landscape'. Our actions result in an erasure of natural soil systems, both in eroded uplands and in sediment-inundated bottomlands (Figure 2.7.2).

Coincident with this acceleration of human alteration of soils, i.e., over the last decades and century, has been the remarkable international development of the sciences of pedology and pedostratigraphy, respectively the sciences of natural soil formation and natural preservation as palaeosols in the stratigraphic record. Pedology and pedostratigraphy have traditionally held human activities to be interruptions or disturbances to ongoing soil formation, as humans have not kept apart from the concept of soil and palaeosols as 'natural bodies'.

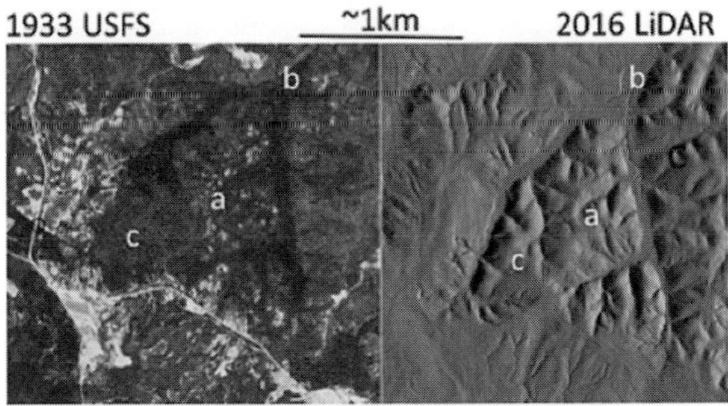

Figure 2.7.2 The landscape's surface roughness is profoundly increased with human occupation of the land (Tarolli & Sofia 2016). This 1933 aerial photo from the US Forest Service and a 2016 LiDAR image of the USA's Southern Piedmont (at the Calhoun Critical Zone Observatory in South Carolina) illustrate impacts from 19th- and early 20th-century farming in the form of (a) gullies, (b) legacy sediments, (c) terraces and incised stream channels, all features absent in an unusual area of lightly used land with strikingly low surface roughness that was generally forested with hardwoods in 1933. US Forest Service Photograph.

The growing human imprint on Earth's soil has created a need for new sciences (Richter et al. 2011). Two that we mention here, anthropedology and anthropedostratigraphy, both aim to better understand and quantify effects of human activities on soils and on palaeosols. Over the last few decades, pedology has assembled a literature that seeks to integrate human forcings fully within the model of soil formation (Bidwell & Hole 1965; Yaalon & Yaron 1966; Amundson & Jenny 1991; Arnold et al. 1990; Bryant & Galbraith 2002; Richter 2007; Richter & Yaalon 2012). In contrast, pedostratigraphy remains a science of natural systems (Salvador 1994; Remane et al. 1996; NACSN 2005) and has yet to open itself to human influences on sediments and strata.

Here, we briefly review the global transformation of Earth's soils from natural to human-natural bodies through the lens of the growth of agriculture, one of the most extensive of all human activities. We then consider the ongoing and important transition of the venerable science of pedology to anthropedology and conclude by considering the current state of pedostratigraphy and by arguing for the creation of a new science of anthropedostratigraphy.

2.7.2 A Brief History of Agricultural Soils

Agriculture is responsible for a wide variety of significant soil modifications over vast areas. As of ~2007, croplands extended across about 16.7 ± 2.4 million km^2 and pastures about 33.5 ± 5.7 million km^2, or about $12.8\% \pm 1.8\%$ and $25.8\% \pm 4.3\%$ of the Earth's current ice-free land surface (Hooke et al. 2012). In 1800, estimates for cropland and pasture were about 4% and 5% of the land surface, respectively (Hooke et al. 2012), the marked increase reflecting the approximately sevenfold growth in human population.

The Neolithic agricultural revolution represents the transition from hunter-gatherer lifestyles to those of agriculture and settlement that occurred in the Fertile Crescent of the Tigris and Euphrates and along the Nile, Indus and Yangtze River valleys as early as 10,000 to 13,000 years ago (Lal et al. 2007).

Meanwhile, agriculture was initiated in other distinct centres (Diamond & Bellwood 2003), including the Yangtze and Yellow River Basins (9,000 years ago), New Guinea (9,000–6,000 years ago) and Central and South America and sub-Saharan Africa (5,000–4,000 years ago). This Neolithic revolution spread westward across most of Europe by 7,700 years ago. Early cultivated crops included wheat, barley, millet, flax, rice, potatoes, beans, peas, lentils, squash and maize. Sheep, goats, cows, oxen and pigs were domesticated for food (Vigney 2011; Ruddiman 2013) and physical labour (Ellis 2011).

The transformation of many cultures of hunter-gatherers to settled agriculture brought with it changes in soils due to land management that included organic-matter amendments, flooding and ploughing. For example, the organic matter and char-enriched *terra preta de indio* of the Amazon basin are soils with such surprising features that they help us more fully appreciate the early development of agriculture. Originally created by pre-Columbian human populations potentially up to 2,500 years ago, *terra preta* soils still persist, though they have long been abandoned (Woods 2008). It is now known that those responsible for producing *terra preta* soils had initiated horticulture in the region ~5,000 years ago, cutting and burning fields, disseminating and domesticating food crops, and planting trees (Roosevelt 2013).

In northern Europe and elsewhere, agriculturally productive plaggen soils were formed by farmers as early as 6,000 years ago and continued to be cultivated through medieval and post-medieval times (Simpson 1997). Peat, heather sod and other organics were collected from outfield areas and first used as bedding for cattle and sheep, and then when saturated with slurry, it was spread as fertilising manure. Organic matter of plaggen soils can accumulate to over a metre deep and, like *terra preta* soils, contains abraded pottery sherds and large amounts of black-carbonised organic matter (Holliday 2004).

Early rice-production sites in China, such as Ciaoxieshan and Chuo-dun-shan in the Yangtze Delta, date to 6,000–5,000 years ago (Cao et al. 2006) and contain soils managed by flooding. Modification of hillslopes with agricultural terraces (Tarolli et al. 2014) may date to the Neolithic (~5,500 years ago) in Yemen (Wilkinson 2005). Terracing associated with dry stone-wall construction has a much later origin, from about the sixth century CE, with construction of the Andenes of the Wari culture of Peru. The process reduces soil erosion and surface-water runoff (Figure 2.7.3a) and is particularly useful for crops that require irrigation, such as rice in paddy fields.

Plough-soils are an extensive feature of most cultivated lands (Figure 2.7.3b) and represent a product of anthroturbation, i.e., soil mixing by humans. This practice was initiated through the use of the wooden ard (Lal et al. 2007), and traditional methods of ploughing by hand or with animals are

(a)

(b)

Figure 2.7.3 Two distinct types of agricultural soils: (a) terraces in Colca Canyon in Peru; (b) deep ploughed soil in Cornwall, England.

still widely practiced. Buried plough-soils have been found beneath Neolithic long barrows, together with patterns of ard marks from plough cuts into the natural subsoil (Ashbee et al. 1979). A significant advance in cultivation came with the modification of ploughs with iron shares from about 5,000 years ago (Ruddiman 2005). Known as the Roman plough, it was commonly used in Europe until the fifth century CE (Lal et al. 2007). Contemporary plough-soils contain a variety of artefacts, tools, plastics, chemical compounds and organic debris and have also been described as the world's 'greatest depository of archaeological material' (Slowikowski 1995).

In retrospect, technological changes happened slowly over most of this period, and even the Industrial Revolution, initially, had only limited influence on agricultural practice. Plough-soils from the 19th century in the United Kingdom contain occasional pieces of coal used by steam-powered ploughs, as well as coprolites that were quarried and spread over the fields as a phosphate fertiliser. Networks of clay pipes to stimulate drainage were laid in trenches cut into subsoils where hydraulic conductivity was low (Edgeworth et al. 2015).

Dramatic and extensive new techniques were introduced in the second half of the 20th century as part of the Green Revolution, which brought intensive farming of the land as an industrial operation (Barnett et al. 1995). The changes in mechanised agricultural practices, soil management and genetics of crop plants have significantly increased yields and, in many regions, greatly increased field sizes with removal of old field boundaries. The soils that support such food production now receive a wide variety of fertilisers, organic amendments, pesticides and organic wastes that include plastics and pharmaceutical compounds. Irrigation continues to expand in farmland area, occupying about 20% of all cultivated lands (FAOSTAT 2017). In poorly drained soils, drainage is facilitated by extensive plastic pipelines arranged with ever-increasing precision to control water availability, all with increased effectiveness and economy compared with traditional tile drainage (Rhoads et al. 2016). New cultivars of maize, wheat, rice and other grains have produced high-yielding varieties, which are genetically recognisable in plant pollen accumulating in modern plough-soils and sediments. New hybridised seeds and the widespread use of genetically modified crops also provide records of contemporary agriculture. But perhaps the clearest signature is the reduction in agricultural biodiversity associated with modern industrial farming, in the species and cultivars which dominate today's global food production (see Section 3.3).

2.7.3 The Bridge from Pedology to Anthropedology

Traditionally, pedology is the science of soil formation – soils formed via the interaction of climate, organisms, relief, parent material and time. If human activities were included at all, they were treated as organisms. Soils that were influenced by human activities were considered 'disturbed', 'artificial', 'not natural' or 'managed' and in this way have been set aside from pedology investigations of 'virgin soils', as was done by Hilgard (1860) when he wrote his essay 'What Is Soil?' In one of the most widely used soil textbooks in history, Lyon and Buckman (1946) defined pedology as 'soil science in its most restricted form', a science that aims to 'consider the soil purely as a natural body with little regard for practical utilisation'.

Since the 1960s, however, pedologists have created a remarkable literature that integrates human forcings in a model of soil formation that reflects the realities of soil transitioning from natural to human-natural systems (Bidwell & Hole 1965; Yaalon & Yaron 1966). Yaalon and Yaron (1966) articulately cautioned that the natural soil body was increasingly functioning as the parent material upon which human-affected changes operated. A number of pedologists addressed the fact that contemporary human beings do not fit well within the organism factor of soil formation (Amundson & Jenny 1991; Bryant and Galbraith 2002; Richter 2007; Richter et al. 2011), as the earlier brief agricultural history demonstrates. The importance of goals, cultural diversity and reasoning, combined with the pace at which humans are accelerating soil change and the technological power with which we impact the soil, all set humans apart from the organism factor (Figure 2.7.4). In fact, at least a few pedologists (Yaalon & Yaron 1966; Arnold et al. 1990; Dudal 2005; Richter & Yaalon 2012) have essentially argued that the concept of soil as a human-natural body is as important to pedology today as were the 19th-century articulations of the natural-body concept, which set the foundations for pedology by Hilgard (1860), Darwin (1881) and Dokuchaev (1883). Anthropedology has come of age, complete with a state-factor equation with six variables that include human beings (Dudal 2005).

This advance coincides with the relatively new but powerful development in pedology that most soils are inherently polygenetic (Buol et al. 2011). This means that soils accumulate and erase properties over time in response to the ebb and flow of dynamic soil-forming factors. Fundamental is that nearly all soils have palaeosolic properties, as most have long enough residence times to accumulate properties formed during past environments (Richter & Yaalon 2012). The high-order interactions of climate, biota, geomorphology, lithology and human action exert effects that range from ephemeral to long lived, sometimes for many million years (Bacon et al. 2012). As features in soil derive from past and present

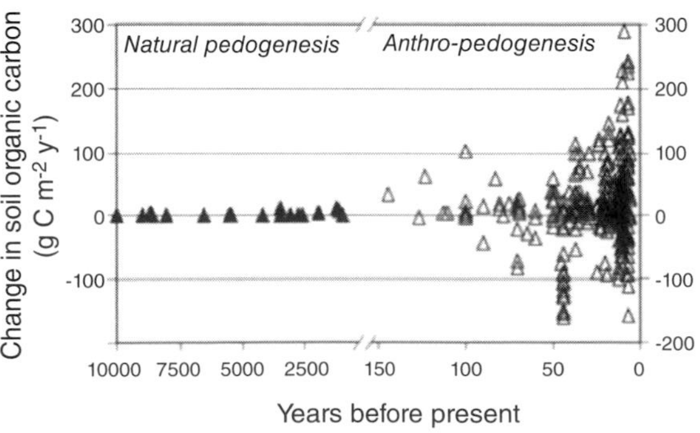

Figure 2.7.4 Rates of change in soil organic carbon (SOC) following natural soil formation, such as following volcanic deposits, floodplain deposits, landslides and beach deposits, all compared with rates of SOC change in response to cultivation, grazing, reforestation, manure applications and crop rotations. (reproduced from Richter & Yaalon 2012, figure 1)

processes, the metaphor of palimpsest, the ancient Greek word meaning overprint, describes the soil well (Targulian & Goryachkin 2008). Soils typically move through time in ebbs and flows (Butler 1959), not continuously following any neat chronological order. Soils are more like ancient palimpsests made from animal skins that were written upon, erased and overwritten. The skins were used and reused by different people and languages through time (Netz & Noel 2007); soils are similarly time transgressive, with prominent periodic erasures and overlays (Schaetzl & Thompson 2015).

An entirely fresh perspective on this situation can be obtained from Gilbert (1877), whose insights into the Earth sciences have only recently been rediscovered (Humphreys & Wilkinson 2007). In 1877, Gilbert stated with wonder, 'Over nearly the whole of the earth's surface, there is a soil, and wherever this exists we know that conditions are more favourable to weathering than to transportation.' Gilbert had realised a fundamental attribute of the planet, that across nearly all landscapes from the tundra to the tropics, weathering's production of soil particles and solutes (W) outpaces transport-related losses via erosion and dissolution (T).[2] Even in the most naturally erosive environments, W has kept pace with T, and soil profiles may be thin, but they accumulate. Given liquid water and biogeochemical weathering agents, W liberates mineral particles and inorganic solutes; T removes a fraction of those products, and soils accumulate the remainder.

A major problem for the Anthropocene, therefore, is that human activities are accelerating T relative to W, a shift that has enormous consequences for soils, sediments and stratigraphy, i.e., Earth's critical zone (Brantley et al. 2017). In the Anthropocene, recognising humanity as 'a fully fledged factor of soil formation' (Dudal 2005) not only enriches pedology but reinforces the vital role to be played by pedology

[2] Note that what Gilbert called 'soil' was later called 'regolith' (Merrill 1897), although in this book and others (Richter & Markewitz 1995), we follow the Gilbert tradition.

in science and society in the 21st century (Grunwald et al. 2011).

In 1961 the North American Commission on Stratigraphic Nomenclature formally recognised palaeosols as soil-stratigraphic units. These were not often used in subsequent years, and in 1983 and subsequently in 2005, the stratigraphic code was revised in part to expand the application of soil-stratigraphic units (NACSN 1983, 2005; Figure 2.7.5 herein). Pedostratigraphy has since become an important tool in interpreting long terrestrial (continental) records, specifically in reconstructing palaeoclimates via their correlation to the marine oxygen isotope records in deep ocean basins (Maher & Thompson 1992). Loessal palaeosols are widely used in Quaternary stratigraphy as stratigraphic markers, for example, in the long records of loess sequences in Western China.

A challenge in using soils in pedostratigraphy is that soils are fundamentally time transgressive, i.e., soils evolve during their formation. In addition, soils have horizons, layers that are generally horizontal with textural and mineralogical properties that often differ from horizons above and beneath. Soils usually have from two to more than five basic horizons, all of which develop in concert over time (Figure 2.7.5). Therefore, when Global Boundary Stratigraphic Sections and Points (GSSP) are proposed to define the bases for chronostratigraphic units (see Section 7.8), the need for an isochronous marker and complete depositional successions results in sections containing palaeosols being intentionally avoided. For example, it has been suggested that the base of an extensive anthropogenic soil horizon that has been modified by ploughing and fertilisation and may contain human artefacts could make a suitable GSSP at ~2,000 years ago (Certini & Scalenghe 2011). However, this proposal has not gained wide support for the reason that the age of onset of significant development of such a soil is highly diachronous (see Section 7.3.6).

2.7.4 Summary

Human forcings represent a new wave of polygenesis of soils. Given Gilbert's 1877 ideas on weathering and transportation, since human forcings have so substantially accelerated rates of soil transportation, the effects are being actively recorded in Earth's stratigraphy. Yet in the latest versions of the North

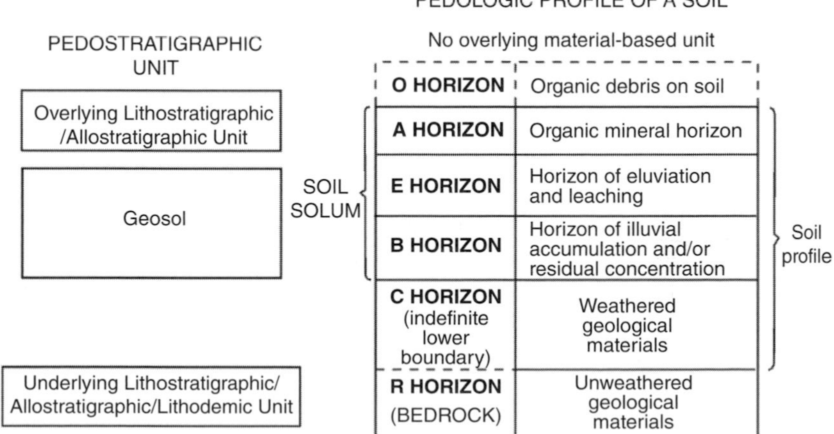

Figure 2.7.5 Relationships between pedostratigraphic units, geosols and pedologic profiles (NACSN 2005). The diagram illustrates the importance of the upper and lower boundaries of the geosol and that the indefiniteness of the lower boundary of the C horizon will cause the geosol to be bounded by the lower boundary of the B horizon. Not stated in the 2005 code is that the lower boundary of B horizons and C horizons are commonly gradients and typically indefinite. ©2005 AAPG. Reprinted by permission of the AAPG whose permission is required for further use.

American and international stratigraphic codes or guides (Remane et al. 1996; NACSN 2005), anthropic signals have yet to be described. These human-natural records, however, can inform us about the basis for identifying anthropic strata and boundaries and about how the Earth is transitioning during the Anthropocene from a natural to human-natural planet.

Human alteration of naturally formed soils and palaeosols thus includes soils eroded and buried under new layers of human-derived deposits. These human-generated event beds (composed of both underlying intact soil profiles and superimposed mixed soil materials, sometimes with evidence of renewed soil genesis) are lithostratigraphic signals of human-natural history. They are an inadvertent but common outcome of human land use. Revisiting Figure 2.7.1 emphasises the importance of accelerated erosion loss and legacy-sediment deposition when considering geological and human-affected pedo- and lithostratigraphies.

Over the Holocene, humans have increasingly transformed soil biologically, chemically and physically. Erosion from uplands (Gilbert's T), due mainly to agriculture, has been greatly increased, and as a result, legacy sediments inundate many footslopes and bottomlands. Palaeosol formation has been accelerated, as both fluvial and aeolian deposition have increased relative to long-term natural rates. As soils are transformed into human-natural systems, soil fluxes of energy, gases, water, solutes and solids with the atmosphere, hydrosphere, lithosphere and biosphere are altered in ways difficult to predict. The human imprint on Earth's soil has created the need for new sciences, those of anthropedogenesis and anthropedostratigraphy, to better understand human-altered soils and to track the effects of human-altered soils on palaeosols and stratigraphic records. For nearly 50 years, pedologists have been working to integrate human forcings into their genetic models. It is time for pedostratigraphy to follow suit. The scientific transformations of pedology to anthropedology and pedostratigraphy to anthropedostratigraphy signal clearly that a boundary is being crossed: the Earth is transitioning to a human-natural system associated with the Anthropocene.

2.8 Changes to Holocene/Anthropocene Patterns of Sedimentation from Terrestrial to Marine

James Syvitski, Jan Zalasiewicz and Colin P. Summerhayes

2.8.1 Introduction to Anthropocene River and Lake Deposits

2.8.1.1 River Deposits

Following the launching of the Anthropocene by Paul Crutzen in 2000, rivers were one of the first phenomena to be analysed in this context (Meybeck 2003; Syvitski 2003), and there have been a number of important subsequent studies (e.g., Syvitski & Kettner 2011; Merritts et al. 2011; Brown et al. 2013; Kelly et al. 2017). The history of human management and perturbation of rivers is long, with impacts to sedimentation patterns spreading in a time-transgressive pattern across landscapes over millennia (e.g., Lewin 2012; Williams et al. 2015a). Apparently subtle changes could have far-reaching effects, such as when, in the 19th century, fluvial sedimentation patterns changed across North America, as numbers of beavers and beaver dams fell sharply as a result of hunting (Kramer et al. 2011).

By comparison with lacustrine deposits (see Section 2.8.1.2), fluvial deposits show considerable internal complexity. The watercourse itself is only one part of the system, although it is the one most commonly featured in general discourse. It transports the sediment (most of which is en route to other sedimentary systems downstream), depositing some part of it in particular patterns across the floodplain. Considered sedimentologically (e.g., Reading 1996), there are three main such patterns. Most modern temperate-latitude rivers are of *meandering* type, where a single river channel actively meanders back and forth across a floodplain, leaving in any one place a coarse channel-floor deposit, overlain by a channel-side (point-bar) deposit, usually sandy, overlain in turn by muddy floodplain top sediments. This resulting fining-upward sedimentary succession is arranged in packets that reflect the pattern of river migration, the resultant erosion of earlier-formed deposits, and any overall aggradation of the whole fluvial succession (resulting, say, from tectonic subsidence). On very low-gradient floodplains and/or where floodplains are highly vegetated, the river channel may be hindered from actively migrating and so may stay in place in an *anastomosing* system, where channel and overbank deposits remain laterally segregated. Conversely, where gradients are steep, vegetation is sparse, and sedimentary input is high, then the river may form multiple unstable, rapidly shifting channels of a *braided* system, typically dominated by coarse deposits as fine sediment is rapidly swept farther downstream, as is common in, for instance, proglacial regions.

The resultant range in sediment body geometries and the complexities of their relation to small (annual to centennial) intervals of time may hinder the consistent recognition of a Holocene/Anthropocene boundary in modern fluvial deposits, and this factor was regarded by Autin and Holbrook (2012) as one reason to reject the concept of a formalised Anthropocene.

Nevertheless, Anthropocene river deposits may be locally distinguished from underlying and adjacent Holocene deposits by such components as artificial radionuclides, plastic debris and other incorporated technofossils, and the subfossil remains of invasive species. On a wider scale, such factors as the enormous expansion of dam building have caused large perturbations in fluvial-sediment supply and distribution, and this is discussed in Section 2.8.3.

2.8.1.2 Lake Deposits

Lakes are an important part of the Earth System: there are approximately 100 million on Earth greater than a hectare in area and about a million that cover more than a square kilometre (Wetzel 2001). Intricately linked to their geographical, hydrological and biological context, they cannot be regarded as isolated systems, and so they have been strongly affected by the changes to the terrestrial water cycle and sediment-mass fluxes described in Section 2.8.2, by changes to the ambient chemical environment, both local and global (Chapter 5), and by a variety of biological changes, including the global dissemination of neobiota (see Section 3.3). Crucial to providing that necessity of human existence, water, many lakes are directly adjacent to major conurbations, increasing their vulnerability to direct human impact. However, even remote lakes are subject to human influence via atmospheric deposition of particulates, e.g., fly ash (see Section 2.3.1) and chemically altered precipitation such as acid rain.

In terms of practical stratigraphy, Anthropocene lacustrine deposits are generally difficult to access for sampling, as they are mostly underwater – except when natural or anthropogenic hydrological changes have caused them to dry up in recent times. The classic example of such desiccation, as both environmental catastrophe and Anthropocene phenomenon, is Lake Aral on the Kazakh/Uzbek border. Formerly the fourth-largest lake in the world as regards area, it has now mostly disappeared due to large-scale Soviet-era diversion of formerly inflowing river water in the 1960s for irrigation of cropland. Similarly, Lake Chad in North Africa is a shallow lake (less than 7 m) present in a closed drainage basin that has reduced from an area of open water covering ~25,000 km^2 in 1963 to one covering 350 km^2 by the end of the 20th century (Coe & Foley 2001). This has resulted from significant reductions in rainfall and increased extraction for irrigation.

However, though (largely) still underwater, lake deposits compensate by being, once sampled by boat- or raft-borne coring devices, relatively straightforward as regards their stratigraphic pattern. Their deposits commonly form ordered strata that demonstrate good superposition and, especially in those lakes with low-oxygen bottom waters, tend not to be seriously disrupted by bioturbation, with some showing annual lamination. Lacustrine sediments include markedly diachronous patterns associated with anthropogenic impact, such as sediment influxes reflecting local and regional land-use changes (Edwards & Whittington 2001). However, they also include high-fidelity archives of a wide range of globally distributed and more or less synchronous human influence (e.g., Smol 2008). Therefore, they include important proving grounds for Holocene/ Anthropocene stratigraphy (Wolfe et al. 2013; Waters et al. 2018; discussed further in Section 7.8.4.4 herein).

2.8.2 Quaternary vs. Anthropocene Sediment Fluxes

2.8.2.1 Pleistocene

Of the estimated 43,000 trillion metric tons (43 Pt) of Earth's Quaternary sediment mass, 74% is marine-based and largely off the continental shelves, and 26% is continental, of which 85.4% is siliciclastic with the remainder being carbonate material (Hay 1994). For the last 2.6 million years, this Quaternary sediment was produced, transported and distributed across the Earth's surface, during multiple glaciations and sea-level fluctuations. Geophysical and sedimentological evidence points to glacial-phase deposition rates as being many times greater than interglacial (e.g., Holocene) depositional rates (Syvitski et al. 1987; Syvitski 1993; Forbes & Syvitski 1995). During the Pleistocene, Earth's land area expanded by up to 16 million km^2 as ocean levels fell, and many rivers discharged their loads directly into the deep ocean (Mulder & Syvitski 1996). Waxing ice sheets also delivered their sediment more directly to the deeper ocean.

Given the Quaternary mass estimates of 43 Pt (Hay 1994), divided by the duration of the Quaternary (2.6 Myr), the global Quaternary flux rate off of the landscape is ~16.5 billion tons per year (Gt/yr). The flux rate would be higher (23.9 Gt/yr) given a start to the Quaternary of 1.8 Myr, as was the case when Hay published his study, but only if the volume of sediment transported off the landscape between 1.8 and 2.6 Myr is excluded. The actual Quaternary flux rates must lie between 16.5 and 23.9 Gt/yr, with ~20 Gt/yr as a best estimate. How these volume-calculated Quaternary transport rates compare with modern observations on global sediment transport is central to the Anthropocene thesis.

2.8.2.2 Holocene

There are few to no historical observations of a river's sediment load during most of the Holocene Epoch, except for the last few hundred years. Earth scientists have used two approaches to ascertain the impact of humans on a river's sediment flux. The first approach is through the judicious use of observations from pristine rivers, located in remote, low-population areas with little human footprint (Meybeck et al. 2003); only 12.5% of the world's population lives in the Southern Hemisphere, and only 5.7% of the world's population lives above latitude 50°N (Kummu & Varis 2011). Observations that predate and postdate the installation of river infrastructure (e.g., dams, diversions) are also useful (Syvitski et al. 2005c). Exploration of the impact of specific environmental disturbances on a river's sediment load has also helped in quantifying the magnitude of a particular anthropogenic disturbance (e.g., Walling & Fan 2003; Yang et al. 2006). Predictive models are tested against such observations (Syvitski & Milliman 2007) and subsequently applied to the remaining global rivers (Morehead et al. 2003; Syvitski & Kettner 2011). These climate-sensitive models are used to understand how rivers behaved during the Holocene and transitioned into the Anthropocene (e.g., Syvitski et al. 2005c; Overeem et al. 2005; Kubo et al. 2006; Kettner et al. 2007; Kettner & Syvitski 2009; Cohen et al. 2013).

A second approach for determining sediment fluxes across the Holocene uses standard stratigraphic and mass-balance studies for understanding sedimentary volumes associated with a particular time interval and unique point source (Syvitski 1993). These source-to-sink methods often include chronological evaluation of cores along with deposit mapping to evaluate the depositional history (Ta et al. 2002; Hori et al. 2004; Correggiari et al. 2005; Stefani & Vincenzi 2005; Tesson et al. 2005; Vella et al. 2005; Tanabe et al. 2006). Such stratigraphic studies often highlight any human role in shaping the deposits.

Based on hundreds of studies, human action is capable of increasing the sediment flux of a river by orders of magnitude, particularly for smaller rivers (e.g., Syvitski & Milliman 2007).

The Holocene is an epoch when sea levels rose and then stabilised (see Section 6.3.1), vegetation expanded, and river systems organised and adjusted to the warmer Holocene climate. The pattern of Holocene to Anthropocene change may be illustrated by five Asian rivers: the Yellow, Mekong, Yangtze, Pearl and Red (Wang et al. 2011). In the Early to Mid-Holocene, these delivered some ~0.6 Gt/yr of sediment to the sea. In an interval from ~2,000 years ago up to the mid-20th century, the rivers increased their sediment load to ~2 Gt/yr, sediment supply increasing due to deforestation and agriculture. Hence, over that time, these five rivers delivered an extra 2,000 Gt of sediment to their coastal ocean due to human activities: a sediment mass equivalent to the natural load of all global rivers in 125 years. Then, from the mid-20th century on, these five rivers saw their sediment flux rapidly decrease back to ~0.6 Gt/yr due to sediment impounded (as discussed in Section 2.8.3, Figure 2.8.4) behind a series of major dams constructed over the last several decades. Such human manipulation of the landscape can therefore explain why the Anthropocene (~1950 to the present) has global sediment-flux rates and/or changes much larger than for either the Holocene or the Quaternary (Hay 1994).

2.8.2.3 Anthropocene

Observations of sediment transport began in the mid-19th century, with the first global flux estimates made in the mid-20th century (Syvitski 2003b). Sediment delivery to the modern global ocean is now considered at a rate of 26 Gt/yr (Syvitski et al. 2017), mostly in particulate form: 16 Gt/yr as fluvial particulates (94% via suspended load, 6% as bedload); 4 Gt/yr in dissolved form; ~1.5 Gt/yr as aeolian transported particulates; ~0.5 Gt/yr from coastal erosion; and ~4 Gt/yr from glacial contributions (25% by meltwater; 75% via the calving and melting of icebergs).[3] Compared to the Quaternary flux rates of ~20 Gt/yr, the modern or Anthropocene flux rate of 26 Gt/yr appears unreasonably high, particularly given the normally low interstadial depositional rates observed across most of the Holocene. The growth in the human footprint on Earth's landscape, through the Holocene and into the Anthropocene, is central to understanding the disparity in these flux values.

2.8.3 Dam and Reservoir Sedimentation

Human action can decrease the flux of fluvial sediment to the sea in a number of ways, such as riverbank hardening, water-diversion schemes and river-channel sand mining. In one example, embankments and levees along the Yellow River led to 50 Gt of river sediment being stored as a 20 km wide fluvial deposit, superelevated 5–20 m above the floodplain and running a river length of 1,000 km.

However, the impoundment of sediment behind dams has by far the largest impact on decreasing the flux of sediment to coastal areas (Syvitski & Kettner 2011). As of February 2017, there were 58,519 large dams registered in ICOLD (http://icold-cigb.org/GB/world_register/general_synthesis.asp), a subset of all world dams that now number in the millions. A large dam has a height >15 m and impounds more than 3 million m^3 of water. Total reservoir storage capacity is >7,000 km^3 (Palmieri et al. 2003). Most large dams are for irrigation purposes (49%), followed by hydroelectric generation (20%), water supply (11%), flood control (9%) and recreation (5%), with 6% of dams built for other reasons (e.g., navigation, fish breeding) (ICOLD 2017). An ever-growing number

[3] It is normal to use the Earth's shorelines as the measurement gateway for flux determinations. This minimises accounting for portions of a moving sediment mass more than once.

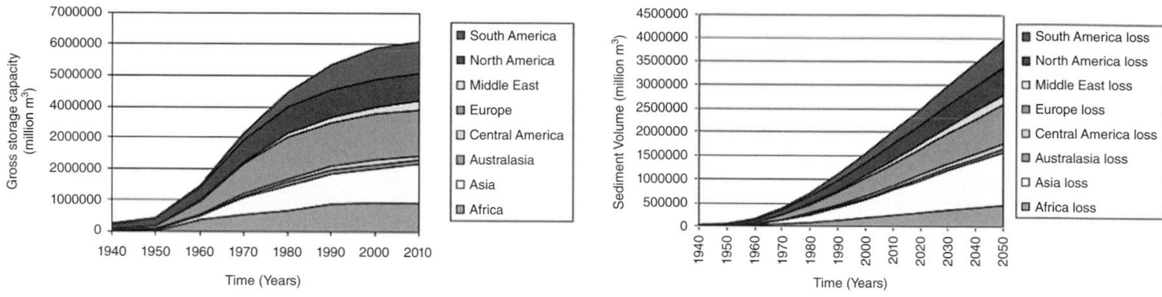

Figure 2.8.1 Estimated global impoundment of water (left) and sediment (right) behind dams. From figure 1.1. of Basson (2009).

(42%) of large dams serve multiple functions, and a majority of them are earth dams (63%). China has many (40%) of these large dams, although their storage capacity is just 8% of the global volume, and most Chinese dams were built after 1950. North America has 19% of world dams. Millions of people have been displaced through reservoirs drowning landscape. While the oldest large dam (Proserpina, Spain) was built in 130 CE, and Japan has 14 large dams that predate 1500 CE, the great majority of large dams, and hence the bulk of impoundment, are from since the mid-20th century (Figure 2.8.1).

Mahmood (1987) estimated that 50 km^3/yr (~65 Gt/yr) of sediment is trapped behind the world's large dams, and by 1986 ~1,100 km^3 (~1,400 Gt) of sediment had accumulated in the world reservoirs, similar to the estimate of Basson (2009). Palmieri et al. (2003) found the global trapping rate since 1986 to be 45 km^3/yr (58.5 Gt/yr), a rate considerably greater than either the 'average Quaternary' sediment flux of 20 Gt/yr or the modern sediment flux to the oceans of ~26 Gt/yr. The rate also indicates that 1,755 Gt of sediment has been sequestered behind dams since 1986 and, with the earlier estimate of Mahmood (1987), that there has been a total impoundment of 3,155 Gt since the mid-20th century (Figure 2.8.1). Figure 2.8.1 also indicates that the impoundment is truly global, with all continents except Antarctica affected.

In summary, humans through their interaction with landscapes both increase river sediment loads and concurrently (and sometimes sequentially) decrease fluvial loads through reservoir impoundments (Syvitski et al. 2005d). Syvitski & Kettner (2011) calculated pre-Anthropocene and Anthropocene flux rates and note that this addition to and reduction of global sediment loads have partially cancelled each other, with global suspended-sediment delivery to the coastal ocean reduced by just 15% (Table 2.8.1). The 3,155 Gt of impounded sediment is equivalent to the discharge of all (6,500) global rivers over a 208-year interval. Stated another way, the resulting deposit stored within reservoirs is equivalent of a 10 m thick deposit spread over 242,700 km^2, roughly the area of the United Kingdom, or a 5 m thick deposit covering California or Spain.

At regional scales, changes in river sediment loads are more dramatic (Table 2.8.1), with individual rivers offering examples of order-of-magnitude impacts. Dams along the Colorado, Nile, Indus and Yellow Rivers now have very low levels of sediment discharge (<0.2 Gt/yr), whereas these four rivers previously discharged 1.5 Gt/yr of sediment to the global ocean (Syvitski & Milliman 2007).

Individual examples include the Mississippi River reservoirs that trap 20 Mt/yr (UNESCO 2011), China's Three Gorges Dam that traps 178 Mt/yr (Hu et al. 2009), Volga River reservoirs that trap 16 Mt/yr (UNESCO 2011), and the 740 Yellow River reservoirs (to 1990) that trap 374 Mt/yr, mostly within the larger reservoirs (UNESCO 2011).

There have been few stratigraphic studies of the sediments impounded behind dams. One such study is of the sediments accumulating within Englebright

Table 2.8.1: **Flux of river water (*Q*) and suspended sediment (*Qs*) for pre-Anthropocene and Anthropocene conditions circa late 20th century (Syvitski & Kettner 2011)**

Landmass	*Q* (km^3/yr)	*Qs* with Humans (Gt/yr)	*Qs* Pre-Humans (Gt/yr)	(% change)
Africa	3,797	1.1	1.6	-31%
Asia	9,806	4.8	5.3	-10%
Australasia	608	0.28	0.24	16%
Europe	2,680	0.4	0.6	-33%
Indonesia	4,251	2.4	2.4	0%
North America	5,819	1.5	1.7	-12%
Oceans	20	0.004	0.003	33%
South America	11,529	2.4	3.3	-27%
Global	38,510	12.8	15.1	-15%

Note: Values represent continental coastline fluxes.

Lake, formed by the damming in 1940 of the Yuba River, California (Snyder et al. 2004; Figure 2.8.2 herein). The ~10 km long lake includes a prograding delta, the front of which is now roughly halfway along the lake. The deposits can exceed 30 m in thickness and are dominated by gravity-flow sand and gravels, while in front and extending to the dam are thinner muddy prodelta deposits of the order of 5 m thick. The succession spans the Holocene-Anthropocene boundary as understood here and might be identified by criteria including those applied to the succession that has accumulated behind the Hollenbeck Dam, described below.

Hollenbeck Dam in Connecticut is a small industrial dam dating from the mid-19th century, holding back a lake a few hundred metres long, which is now largely sediment filled. Kelleher (2016; Figure 2.8.3 herein) studied its ~5 m thick fill using cored samples and used measurements of 'bomb spike' ^{137}Cs activity to separate pre-1950 and post-1950 stratigraphic units (see Section 5.8), using this boundary as a reference to chart the distribution of rates and styles of sedimentation and of industrial-pollution levels of the succession. Reservoir deposits in general, like natural lake deposits, should include a variety of Anthropocene proxy indicators, including plastics (Section 4.3), pesticides (Section 5.7.2) and fly ash (Section 2.3.1), as well as artificial radionuclides (Section 5.8).

The succession in post-1950 reservoirs is effectively wholly Anthropocene in age, though nevertheless it can be substantial in scale. The Three Gorges Dam began to trap water and sediment in 2003, with rates of sedimentation in the reservoir estimated at 172 Mt/yr (Hu et al. 2009), suggesting that more than 2,000 million tons of sediment are already sequestered there. If spread over the ~1,000 square kilometres of the lake floor behind the dam, this would form a layer averaging 1 m thick, though it can be inferred to vary in thickness and facies in a manner akin to that in Englebright Lake.

2.8.4 Delta Deposition

2.8.4.1 Holocene Deltas

Many coastal deltas formed 6,500 to 8,500 years ago, as global Holocene sea-level rise slowed and stabilised following the melt of the Pleistocene ice sheets (see Section 6.3.1). The appearance of these deltas (their size, shape and number of distributary channels) is controlled by the interaction between boundary conditions and forcing factors (Syvitski & Saito 2007),

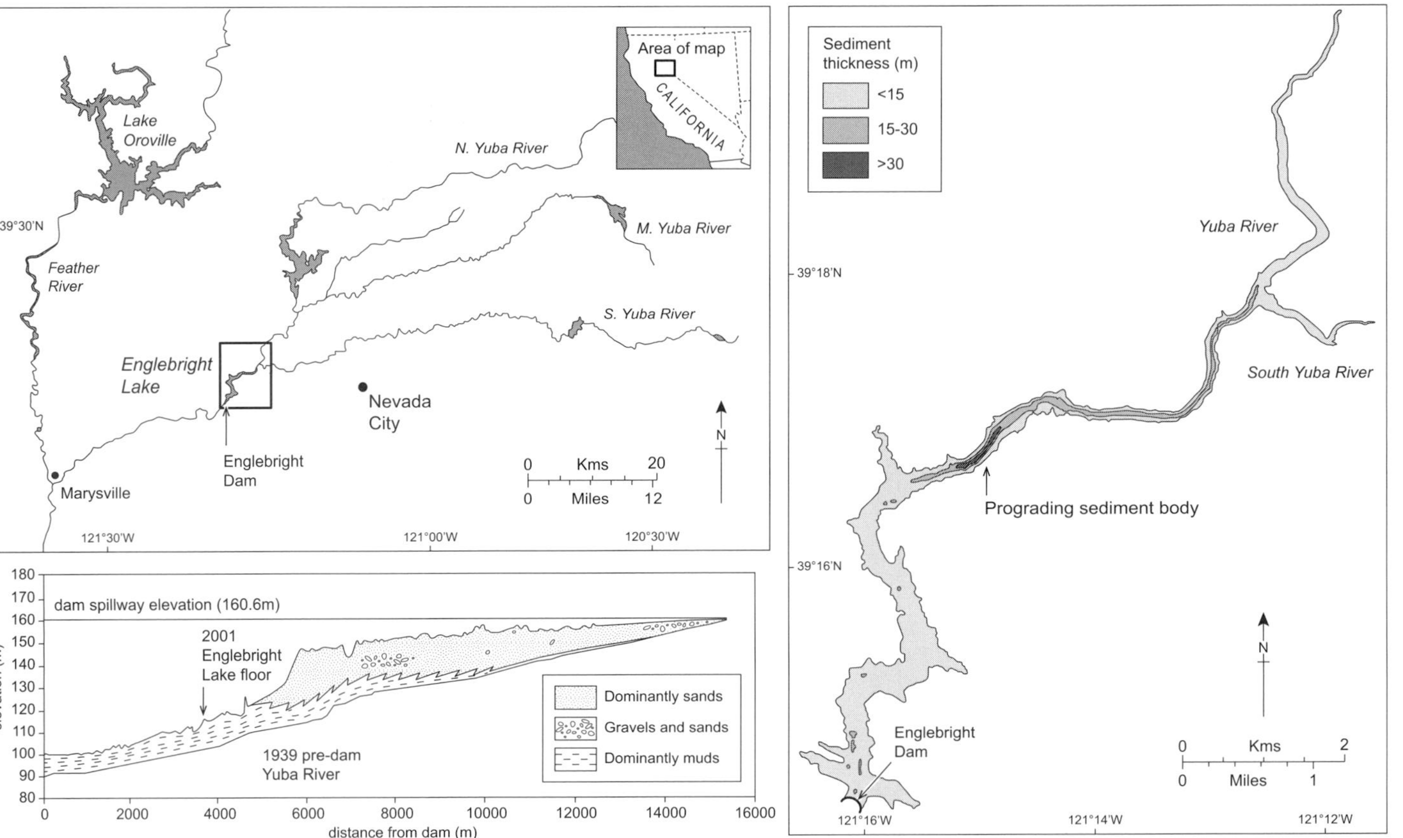

Figure 2.8.2 Location, morphology, and sediment succession in Englebright Lake, California. (adapted from Snyder et al. 2004). (A black-and-white version of this figure appears in some formats. For the colour version, please refer to the plate section.)

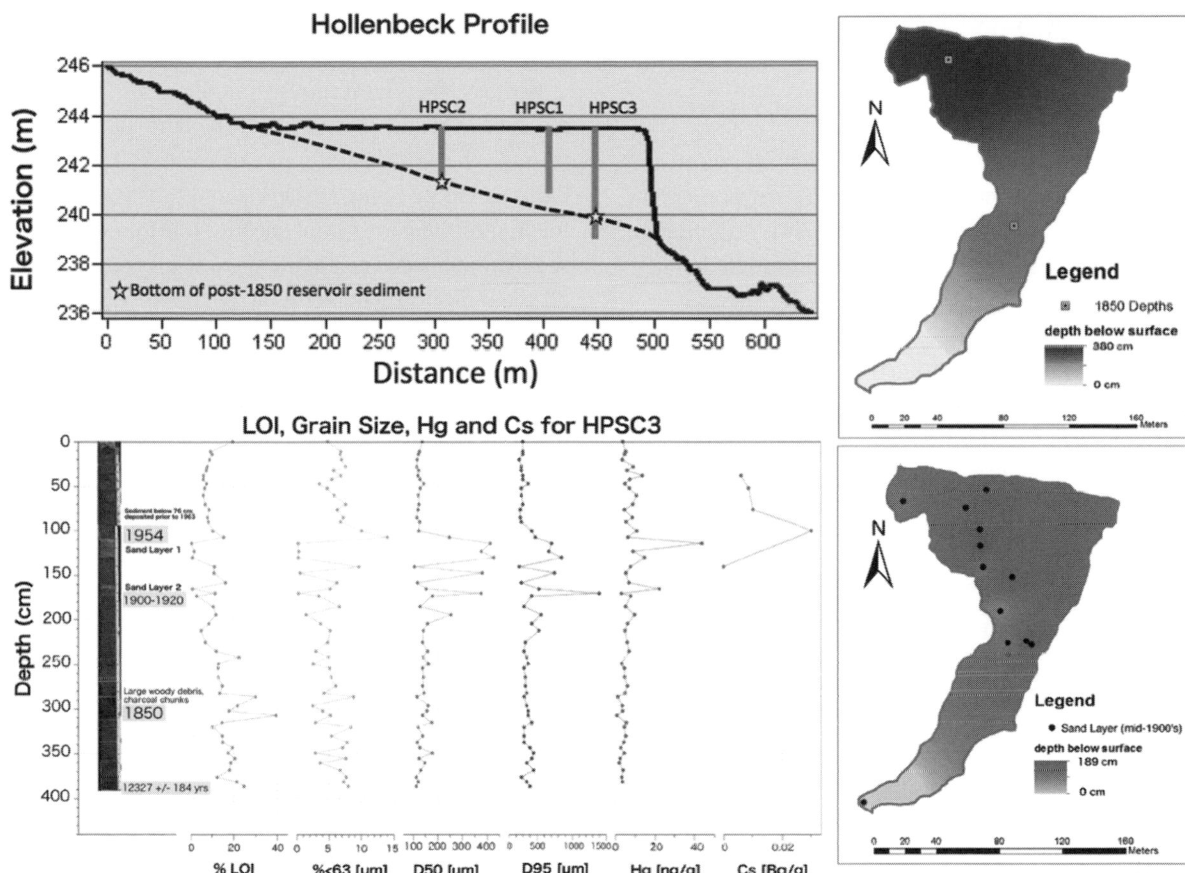

Figure 2.8.3 The succession behind Hollenbeck Dam, Connecticut, where 'bomb spike' ^{137}Cs is used to separate the succession into pre-1950 and post-1950 parts. (combined from figures 4, 5 and 6 of Kelleher 2016) (A black-and-white version of this figure appears in some formats. For the colour version, please refer to the plate section.)

including (1) sediment supply from drainage-basin rivers; (2) accommodation space for the incoming sediment and as controlled by sea-level fluctuations, offshore bathymetry, tectonics, subsidence, compaction and crustal isostasy and (3) coastal energy as delivered by waves, tides and ocean currents that work to disperse the incoming sediment. Climate factors, such as between polar vs. tropical influences, can alter the morphology of a coastal delta and define the character of not only river discharge (e.g., intra- and interannual variability) but also a delta's ecology (Syvitski et al. 2017). A delta's fine-grained, organic-rich soils get nourished every few years with flood inundation (Syvitski et al. 2009). The incoming fresh water flows over extremely flat surfaces (0.005 m/m for steep mountainous deltas to 0.00001 m/m for large deltaic systems; Syvitski & Saito 2007). As a consequence, river channels split into distributary channels that frequently shift their location. In essence, delta geomorphology reflects the fragile nature of these coastal systems – a quasi-equilibrium state to reflect the volume of sediment supply, the dispersal energy that distributes the sediment, and the nature of the local vegetation that covers and stabilises the sediment (Syvitski & Saito 2007).

The transient nature of a delta depends on perturbations to the major parameters that influence its surface. Vertical change in a delta's surface relative to local mean sea level, $\partial\eta/\partial xy$, is determined by five factors:

$$\partial\eta/\partial xy = (Ag - \mathrm{Er}) - \Delta E - C_n - C_A \pm M$$

Aggradation rate (*Ag*) is the volume of sediment delivered to and retained on a delta's surface as new sedimentary layers (Giosan et al. 2014). Aggradation is reduced by surface erosion (*Er*), although deltaic erosion rates are typically small. Under natural conditions and under the influence of the seasonal flood wave, a delta's aggradation rate varies from 1 to 10+ mm/yr when areally averaged. Flooding from ocean surges may also contribute sediment to the coastal portions of a delta.

The oceans' eustatic sea-level rate (ΔE) is influenced by (1) fluctuations in the global storage of terrestrial water (e.g., glaciers, ice sheets, groundwater, lakes and in modern times even reservoir volumes sequestered behind dams); and (2) changes in ocean water volume as impacted by changes in ocean water temperature (the steric effect). Today ΔE contributes 3+ mm/yr under the influence of global warming, a rate much accelerated compared to pre-Anthropocene times, except for the period just after the Pleistocene. Sea-level rise is a global process but with local manifestations (see Section 6.3.1). Asian deltas have recently been experiencing higher rates of sea-level rise due to the steric impact on dynamic (ocean) topography – western geostrophic (wind-driven) ocean boundary currents are the first areas of an ocean to expand under rising ocean temperatures.

Natural compaction (C_n) reduces the volume of deltaic deposits through dewatering, grain-packing realignment and organic-matter oxidation (typically $\leq$3 mm/yr; Syvitski 2008). Human-induced subsidence (C_A) reduces the volume of deltaic deposits through subsurface mining (oil, gas, or groundwater) and through human-influenced soil drainage and peat oxidation. Rates of subsidence can vary widely between deltas, depending on the magnitude of the anthropogenic driver, from a few mm/yr to hundreds of mm/yr, with groundwater withdrawal being the dominant reason behind much of the world's coastal subsidence (Tessler et al. 2015).

Isostasy, flexural response and tectonics (*M*) all affect the vertical movement of the land surface through the redistribution of earth masses from (1) Early Holocene rising sea levels and thus water loading of the continental shelves, (2) Holocene sedimentary deposits loading the world's coastal zone, (3) the impact of passive vs. active plate-boundary dynamics and (4) crustal rebound from melting ice sheets. In many cases the response to these shifting loads brings about a crustal–upper mantle flexural response that can be long lasting, e.g., the impact of Early Holocene rising sea levels can still be felt along the world's coastlines today. The *M* term is thus geographically variable, but rates are typically $<\pm 5$ mm/yr (Syvitski 2008).

2.8.4.2 Human Occupation of Deltas

A delta's low gradient is both attractive to and dangerous for human occupation and utilisation. A large flat delta with rich organic soil has the potential for agricultural development. These same flat terrains are easy to flood, both from their rivers and from the sea. Some cultures adapted to this episodic inundation by keeping their growing population centres upstream and off the delta (e.g., Ganges Delta). While annual river floods were needed for nourishing agricultural soils, the Nile Delta being the type example, floodwater inundation could also destroy habitats and livelihoods.

As Holocene deltas rapidly grew and shorelines shifted seaward, human occupants followed. Inhabitants were initially concerned with the ephemeral nature of water routes distributed across the delta. Distributary channels would episodically shift location. So while deltas offered lush vegetation, diverse forests, wildlife and fisheries, and rich muddy soil for agriculture, movement through these expansive and ultra-flat wetlands remained difficult. Additionally, tropical infections, particularly from disease-carrying insects, kept human populations low.

Cultures have now adopted engineering practices to control floods and reduce risk by constraining the floodwaters using levees, embankments and seawalls (see Section 2.5.2.8). This forces the sediment-laden water to exit the delta largely intact, with sediment being deposited at mouth bars, as shallow marine deposits, within the leveed channels, and along shorelines (Syvitski et al. 2005a, b, d). As delta swamplands were drained for agricultural purposes, soil and subsoil peats oxidised, with the side impact of subsiding the land surface (Renaud et al. 2013).

There are presently ~600 million people living on or near major deltas (using a 25 km wide envelope around each delta), primarily in megacities (Table 2.8.2). As societies urbanised, much of the population growth was a result of migration into delta cities. By 2050 the populations of many of these deltaic megacities will exceed 20 million and in some cases 30 million (Table 2.8.2).

2.8.4.3 Anthropocene Deltas

Once home to vast areas of wetlands, deltas now host extensive agriculture and aquaculture infrastructure, megacities, ports and other transportation hubs. There are two vectors for the creation or conversion of a Holocene delta into an Anthropocene delta. The first vector is defined by a few smaller deltas that would not exist except for the direct intervention of human actions (Syvitski 2008; Maselli & Trincardi 2013). Type examples of this direct vector include the formation of both the Huanghe or Yellow River Delta (China) and the Po River Delta (Italy). In both of these cases, before major human intervention a prograding delta could not form, as the main river channel(s) would change the location of the main river mouth as the river avulsed across the vast flood plain, and/or the coastal ocean energy could

Table 2.8.2: **Population growth of 12 delta cities (in millions), from 1975 when their metropolitan areas held 62 million people to 2010 when their population expanded to 140 million, a growth rate of 4%/yr**

City	Delta	1975	2010	2050
Karachi (Pakistan)	Indus	4.0	12.0	31.7
Kolkata (India)	GBM	7.9	15.1	33.0
Dhaka (Bangladesh)	GBM	2.2	13.0	35.2
Yangon (Myanmar)	Irrawaddy	1.8	6.3	10.8
Bangkok (Thailand)	Chao Phraya	3.8	8.2	11.9
Ho Chi Minh (Vietnam)	Mekong	2.8	10.0	26.0
Hanoi (Vietnam)	Red	1.9	4.2	10.0
Guangzhou (China)	Pearl	3.1	9.6	13.0
Shanghai (China)	Changjiang	11.4	20.0	21.3
Tianjin (China)	Huanghe	6.2	9.8	10.1
Cairo (Egypt)	Nile	6.0	16.9	24.0
Buenos Aires (Argentina)	Parana	10.9	14.2	15.5

Note: By 2050 (UN projections) these same cities will continue their growth to 242 million, but with exceptions at a reduced rate (2.6%/yr). GBM: Ganges-Brahmaputra-Meghna Delta. Data sourced from Wikipedia 2017 and UN DESA, Population Divisions 2017. Note some secondary data sources offer larger numbers, as they include metropolitan areas of undefined area.

effectively disperse the river's natural sediment load (Syvitski et al. 2005b).

In the case of the Po, two important alterations to the natural system set the stage for Anthropocene delta formation: (1) deforestation within the drainage basin greatly increased sediment supply to the coast; and (2) citizens of Bologna captured the Po's path via an engineered canal (~1150 CE) that was designed to become a direct commercial route to Venice through its lagoon. Unfortunately, the Po's discharge of unwanted sediment threatened to infill the Venice Lagoon; this forced the Venetians to construct a 7 km long canal (~1600 CE) so as to redirect and fix the Po River outlet from migrating back into the lagoon. Today the Po Delta remains highly engineered by river levees and seawalls to protect farmlands from being inundated given accelerated 20th-century subsidence from gas production.

During the 15th to early 19th centuries, the Yellow River was spatially fixed by artificial levees, discharging into the Yellow Sea where, similar to the situation for the early Po, local ocean energy was capable of distributing the incoming sediment along the coastline. In 1855, a levee breach near Kaifeng led to a full avulsion that altered the course of the Huanghe so that the river flowed into the lower-energy Bohai Sea. The new delta formed at a rate of ~22 km^2/yr. Human perturbation of the Loess Plateau had by then already increased the natural sediment load of the Yellow River by an order of magnitude (Syvitski & Milliman 2007; Syvitski & Kettner 2011). Extraction of groundwater for aquaculture and domestic needs has subsequently led to accelerated land subsidence – the Yellow River Delta has the highest observed rates of subsidence (1 m every four years; Higgins et al. 2013). Massive seawalls now offer coastal protection to the delta, particularly the large deltaic oil fields that have been put at risk.

The second vector for conversion of a Holocene delta into an Anthropocene delta is through pressures of an exploding delta population, along with concomitant exploitation of the natural services that deltas offer (Renaud et al. 2013). Dam interception of upstream river-borne sediment presently leaves many modern rivers entering a delta with a reduced sediment load (Vörösmarty et al. 2003; Syvitski et al. 2005d). In some rivers (e.g., Colorado, Ebro, Nile), little to no sediment now reaches the deltaic environment. Where distributary channels are embanked, floodwaters can no longer aggrade the land surface.

Delta subsidence has been consequent on the growing populations of deltas, attributed primarily to (1) groundwater extraction for potable water and aquaculture and aquaculture needs (e.g., Yellow, Mekong, Yangtze, Chao Phraya Deltas); (2) petroleum extraction (e.g., Po, Mississippi, Niger Deltas); and (3) peat oxidation (e.g., Rhine-Meuse Delta). Deltas are sinking at rates of tens to hundreds of mm/yr; on passive margins this is geologically unusual. Even without subsidence, eustatic sea-level rise caused by anthropogenic climate warming puts most deltas at risk of coastal drowning (Giosan et al. 2014). Combining sea-level rise and land subsidence, delta shorelines are rapidly receding, and great swaths of land are being given up to the marine environment (Saito et al. 2007). To accommodate inland surfaces below sea level, coastal barriers encircle shorelines (e.g., Rhine-Meuse Delta), and soils are pumped to limit saturation. In essence, deltas have become at-risk Anthropocene landscapes.

2.8.4.4 Anthropocene Deltaic Deposits

Deltas during the Holocene and the Anthropocene have received a pulse of sediment, lasting for hundreds of years and in some cases millennia, from upstream drainage-basin disturbances. Deforestation, changes in agricultural practices, road and city building, and mining are amongst the main reasons for this sediment pulse. Wang et al. (2011) estimate that five Asian deltas (Yellow, Mekong, Yangtze, Pearl, Red) received an extra 2,000 Gt of sediment (Figure 2.8.4). Given their combined area of 1.1 million km^2 area, this translates to an area-averaged deposit thickness of ~3 m. However, the actual

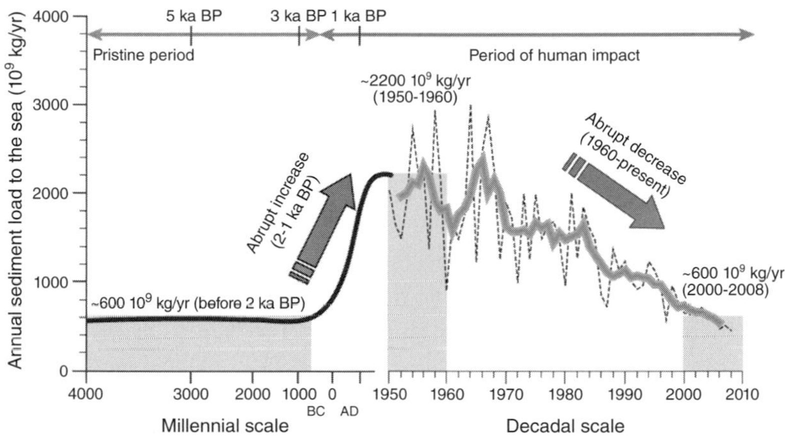

Figure 2.8.4 Mid-Holocene to Anthropocene history of sedimentation of five major Asian rivers. (from figure 14 of Wang et al. 2011). ©2011, with permission from Elsevier

deposits are distributed over a much smaller area, and thus the deposits are much thicker.

This initial burst of anthropogenic-contributed sediment was followed by an even more rapid reduction in sediment delivery as a result of dam building and reservoir sequestration of sediment, hardening of river-channel embankments, improved agricultural practices and even reforestation. In most cases the rivers carry much less sediment than under ambient Holocene conditions, of which little sediment gets deposited on the delta surface. Rather, the sediment is shot through the deltaic depositional environment to the marine coastal environment. During the stage of sediment reduction, accelerated subsidence has become the norm. This gross compaction of delta deposits impacts not just Holocene and Anthropocene deposits but pore spaces of deposits much older, both in the case of gas and oil mining and in the case of groundwater extraction in deeper aquifers. Oxidation of peats results in a greatly reduced organic content of the surficial layers.

The surface layers of Anthropocene deltas are commonly crisscrossed with drainage canals, with the Rhine-Meuse Delta being an exemplar. Sometimes these drainage canals are primarily built for geophysical surveys for oil exploration, the Mississippi and Niger Deltas being the type examples. Irrigation canals dissect many deltas (e.g., San Joaquin, Nile, Mekong and Pearl Deltas). The surface soils have become more saline (with Indus and Colorado Deltas being the type examples). The soils contain the residues of pesticides and commercial fertiliser and as a result have elevated metals (e.g., mercury, lead, cadmium, arsenic) and complex and persistent organic compounds (e.g., DDT, DDE) (see Sections 5.6 and 5.7). In addition, these sedimentary deltaic deposits often underlie the rapidly growing megacities, with their cement, metal, glass, plastic and other urban material (see Sections 2.2 and 2.5.2).

2.8.5 Marine Settings

2.8.5.1 Marine Settings

As Carl Sagan pointed out, referring to a photograph taken from the Voyager 1 spacecraft as it left the solar system on 14 February 1991, seen from space 'our planet is a lonely speck in the great enveloping cosmic dark . . . a pale blue dot', 72% of whose surface is covered by ocean (Sagan 1994).

While humans have been modifying the land surface in various ways for thousands of years, that vast expanse of ocean long remained largely immune to these changes, a reflection of the fact that much of it is far distant from our inhabited shores. Significant impacts began with fishing, this growing in significance over several centuries. In the past 100 years, there have been major changes to the balance of marine organisms, with precipitous population declines in particular of the larger creatures like whales, seals and sharks, as well as the larger food fish – much in the same way that early hunters managed to kill off most of the large land mammals over the past 30,000 years. As marine hunters became technologically smarter, they reduced many fisheries to a state close to collapse (Pauly et al. 2002; Roberts 2007, 2013).

We can extract the history of the oceans from their sediments, which comprise elements from river runoff, windblown dust, the remains of marine organisms, debris and chemicals from volcanic eruptions, and deposition from floating ice, as well as – since the early days of sail – human rubbish from ships and shore, including shipwrecks. Marine sedimentation takes place in a range of settings, from shallow to abyssal waters. We briefly describe their main characteristics, particularly as regards the preservation potential of Anthropocene deposits, before discussing the main types of human impact that allow their stratigraphic definition.

Coastal systems: Deltas, discussed in detail in Section 2.8.4, are the main sources of sedimentary input to the coast – or they were, at least, until much of the sediment they used to supply was held back behind dams. There are other kinds of coastal sediment bodies.

Estuaries, as well as coastal embayments without major riverine inputs (for instance, The Wash region of eastern England), are often associated with enhanced tidal action and reduced wave action. In Holocene times, these embayments have tended to be progressively infilled, with much fine sediment being derived not only from landward but also from seaward, brought by tidal currents. These sediments are often followed by peat accumulation as intertidal areas convert to freshwater reed or mangrove swamps and coastal salt marshes, depending on latitude. Between these embayments, more or less continuous beach systems commonly form on open coastlines in systems marked by considerable dynamism of the sedimentary environment, where both aggradation and erosion can be rapid, and coastal sediment bodies can accumulate or be destroyed even by single major storms. These deposits tend to be frequently reworked and coarse grained. Beaches may also retain coastal lagoons that gradually silt up. In subtropical regions where there is not too much sedimentary input from land, calcium carbonate can accumulate both biologically (as coral/algal reefs or as the remains of marine organisms such as bryozoa) and inorganically (as carbonate mud precipitated from warm waters supersaturated in calcium carbonate). Such carbonate platforms may be land attached or, like the Bahamas islands, isolated from land. Coral reefs, described in Section 3.4, provide a unique type of coastal system that may be close to land in the form of fringing reefs (e.g., the Great Barrier Reef) or divorced from land in the form of atolls, the drowned bases for most of which are slowly sinking extinct volcanoes.

Shelf systems: Coastal systems grade into continental-shelf systems via near-shore zones, which, like beaches, are more or less permanently reworked by storm waves and tidal currents and which are therefore dominated by gravel and sand. These range into deeper settings where tidal currents slacken and storm waves rarely or never affect the seafloor. Here, gravitational forces dominate, and fine-grained sediments (mostly winnowed out from shallower settings) settle, often via ebb surges from storms and tsunami (which can be effective transport mechanisms of debris, including anthropogenic debris, from land to the seafloor; Dawson & Stewart 2007). Many continental shelves are in fact relict Pleistocene erosion surfaces covered by coarse sediments that are being reworked by waves, which may locally be covered by an inshore mud belt of more recent origin. Algal reefs may occupy the shelf edge, as off the Nile and Amazon Deltas. In the case of the Nile, the input of sediment from land was channelled across the shelf to the Nile Fan, a deep-sea structure mantling the continental slope – at least that was true before the Nile was dammed at its mouth. In the Amazon case, ocean currents sweep the Amazon plume of river-derived mud north along the coast towards the Caribbean, leaving the shelf edge clear for the formation of an algal reef.

In general, preservation of an Anthropocene signature in marine shelf systems may be patchy, as these regions are commonly sediment starved following the postglacial sea-level rise, with many rivers still dumping their sediment load nearshore in estuaries that operate as sediment traps. And over the many shallow seafloors that are pervasively wave and/or current swept, long-term winnowing of finer sediment takes place towards the continental slope and rise regions. Hence, Anthropocene signatures are likely to be more persistently preserved in deep water, as we describe below, or possibly in shallow water inshore mud belts like those off Morocco and southwest Africa.

Deep sea: Deepwater systems include those of the continental slope (below the storm wave base and off the continental-shelf edge) and the continental rise below. Given a typical rate of accumulation of 20–30 cm/1,000 yr (Kennett 1982), Anthropocene sediment deposited from 1950 onwards could comprise a layer 1.3–2 cm thick on the continental slope. Taking the western margin of the United Kingdom as an example, Holocene sedimentation rates range from 2.1 to 6.5 cm/ 1,000 yr (Thomson et al. 2000), representing a layer about 0.4 cm thick for the Anthropocene, in cores from water 1,100 to 3,570 m deep. Bioturbation extends down to about 15 cm in cores from this environment, indicating the very high potential for smearing the Anthropocene signal down beyond its basal boundary. The most recent material in the cores (in the top 5 cm) tends to be organic carbon (C_{org}), representing the deposition of recent planktonic remains from the ocean surface. The rapid decrease in C_{org} below the sediment-water interface shows that these remains are rapidly used for food by organisms living on or in the seabed – and causing the bioturbation.

Submarine canyons that traverse the continental slope funnel material down to the deep-ocean floor. The canyons terminate in extensive clastic wedges of turbidite fans on the continental slope, which can reach over a thousand kilometres across the ocean floor. Further seaward, sediments deposited by turbidity currents flowing down canyons and across fans spread out across the deep-sea floor as 'turbidites' that fill hollows in the topography to form extensive abyssal plains covering some 36% of the deep-ocean floor (Kennett 1982; Weaver et al. 2000). Contourite drifts form another type of clastic wedge that fringes the continental margin, with sediment supplied by the deepwater contour (thermohaline) currents that, due to the Coriolis effect, hug certain of the lower continental slopes. These currents tend to redistribute fine terrigenous sediment brought to the deep sea down-canyon in turbidity currents. Sediments in contourites tend to accumulate at about 2-4 cm/1,000 yr (Stow 2001). Much of the continental slope is made up of largely relict hemipelagic mud deposited by rivers and streams debouching at the shelf edge when the sea level was lower. These muds comprise terrigenous material (silt and clay) from land mixed with the remains of marine plankton, to varying amounts depending on the development of the local hydrological system and the productivity of the local surface water. More runoff means more terrigenous components.

Beyond the continental slope and rise lie the regions of slowly accumulating deep-sea oozes, where there are both local highs (seamounts) and lows (submarine trenches). Away from abyssal plains covered by largely terrigenous turbidites, the deep-sea floor is mantled largely by three kinds of sediment (although between turbidite episodes these same kinds of sediment can cover abyssal plains too):

- Calcareous oozes, made up of the minute remains of plankton with calcium carbonate skeletons. These include the coccolithophorids (phytoplankton with calcite skeletons), the zooplankton in the form of the foraminifera (with calcite skeletons) and the pteropods, or sea butterflies (with aragonite skeletons). These oozes cover some 48% of the deep-sea floor above the depth at which calcium carbonate minerals dissolve (roughly 4,500 m depending on the ocean). They accumulate at rates of 0.3–5.0 cm/1,000 yr, the higher rates and deeper oozes being under centres of high productivity, for example, along the equator in the Pacific Ocean.

- Siliceous oozes, made up of the minute remains of diatoms or radiolaria – phytoplankton with silica skeletons. These cover another 15% of the deep-sea floor, usually below the depth at which carbonate minerals dissolve. They accumulate at rates of 0.2–1.0 cm/1,000 yr.
- Red clays, made up mostly of windblown desert dust. These also occur below the depth at which carbonate minerals dissolve. They accumulate at rates of 0.1–0.5 cm/1,000 yr. Red clay forms a minor constituent of both the calcareous and siliceous oozes, in some instances rising to about 70% of their composition.

At rates of accumulation of 0.5 cm/1,000 yr, common amongst all three of these sediment types, the period since 1950 would comprise a sediment layer a mere 0.033 cm thick, easily smeared by bioturbation. That is because all three environments tend to be highly oxidising, with burrowing organisms mixing the upper layers of the sediment down to depths of 5–10 cm. Anthropocene signals postdating 1950 (see Section 2.8.5.2) would be mixed and diluted with earlier signals in such sediments.

The greatest prospect for finding coherent Anthropocene marine successions are therefore where there is a high rate of sedimentation combined with conditions anoxic enough to prevent bioturbation. Good examples include the Black Sea, the Santa Barbara Basin (California), the Cariaco Basin (Venezuela), the Saanich Inlet (Canada) and other comparable environments (see Section 7.8.4.1). Here, annual layers of sediment may be preserved as varves (e.g., Hendy et al. 2015): Schimmelmann et al. (2016) noted 52 such marine sites.

2.8.5.2 Stratigraphically Significant Human Impacts

Anthropogenic coastal reworking: Within the last couple of centuries and particularly the last several decades, many coastal systems have seen large-scale change associated with such activities as the replacement of mangrove swamps for shrimp farms in low-latitude coastal regions and, in temperate climate zones, the draining of coastal wetlands for farmland. In the latter case, surface peat deposits have commonly disappeared through desiccation, deflation and oxidation in the production of 'negative stratigraphy', An example is the physical removal by these processes of up to 4 m of peat across ~2,000 km^2 of the English Fenland since the 17th century (Smith et al. 2010), to leave much of this area up to 2 m below sea level, with the area currently prevented from inundation by a combination of seawalls and continuous pumping. An Anthropocene boundary here approximates to a regional sedimentary hiatus, likely to be buried beneath marine transgressive deposits in coming decades and centuries.

Other sedimentary disturbances, many dating back to the mid-20th century or later, include the effects of sand and gravel extraction, which is increasing on shallow continental shelves worldwide as land supplies for both concrete manufacture and beach replenishment have become exhausted (https://na.unep.net/geas/archive/pdfs/GEAS_Mar2014_Sand_Mining.pdf). At the sites of extraction, both sediment fluxes and biotic community structures have changed (e.g., Krause et al. 2010). Other coastal or near-coastal mining takes place for minerals such as cassiterite (a source of tin), diamonds, quartz (for glass), and ilmenite, zircon, or rutile (for titanium). Mining on land, too, may lead to mine tailings washing into coastal seas, leaving a prominent sedimentary signature – as in the case of Cornish tin mining, for instance (see Section 2.5.2.5).

Such coastal modification is additional to the more obvious changes associated with urbanisation and the building of coastal defences, which now locally affect the majority of some extensive stretches of coastline (see Section 2.5.2.8).

Trawling effects: In general, human impacts on these open marine systems were later than those on land, with fishing-related declines in fish stocks dating locally from medieval times (Roberts 2007). Resulting changes to the structure of marine ecosystems and to organisms lower in the food chain were then likely small, even with the dramatic reported declines in fish catch (e.g., Myers & Worm 2003).

Beam trawling from fishing boats has been active in the North Sea since the 14th century, typically to water depths of up to 100 m (Martin et al. 2015). More profound impacts came with the industrial age, from ~1800 CE onwards, with the physical and biological disruption caused by greatly enhanced open-sea bottom trawling – literally a ploughing of the seafloor – associated with the advent of steam-powered ships and the development of ships with greater range and power. Otter trawling, developed in the 1880s and 1890s, has facilitated fishing to greater depths (Martin et al. 2015), and the industry expanded markedly from the early to mid-20th century, now reaching continental-slope settings in waters approaching a kilometre deep.

By the end of the 20th century, sea-bottom trawling was taking place across an area of some 15 million km^2 each year (Watling & Norse 1998). This now includes most of the world's continental shelves and significant areas of the upper continental slope (Puig et al. 2012), along with the upper surfaces of seamounts. The trawl-net doors ('otter boards') that hold the trawl net in place make gouges up to tens of centimetres deep in the sediments along the trawl path. The otter-door spread may be as wide as 80 m (Oberle et al. 2016). The trawl net itself drags along the seabed behind a 'footrope', churning up everything in its path and in effect destroying both the habitats and the ecology of the seafloor, especially in areas below wave base, to a thickness of some 35 cm (Oberle et al. 2015). The ploughing process winnows out the finer sediment fractions, which are commonly swept away by bottom currents, leaving behind coarsening-upward sedimentary signatures (Palanques et al. 2001; M. Coughlan, pers. comm.). Benthic animal and plant assemblages are profoundly altered (Watling & Norse 1998; Malakoff 2002). Some sensitive assemblages (e.g., deepwater coral systems) are decimated or destroyed (Sheppard 2006), while the process can smooth and reshape what was formerly rough submarine topography (Puig et al. 2012).

The resuspended mud forms sediment plumes that extend much more widely (Dounas et al. 2007; Ferré et al. 2008), reaching the continental slope. The mass of sediment resuspended by trawling is about the same as that entering the shelves through rivers (prior to damming; see Section 2.8.2), some ~22 Gt/yr (Oberle et al. 2016). Where these muds contain pollutants from land – for example, in the form of heavy metals or persistent organic pollutants (as described in Sections 5.6 and 5.7) – bottom trawling may assist in the process of displacing fine-grained pollutants onto the nearby continental slope, helping to enhance the Anthropocene signal there. Nevertheless, as already mentioned, Anthropocene sedimentary materials on the upper continental slope are likely to be smeared vertically by bioturbation.

Hydrocarbon exploitation: Offshore engineering projects have modified marine sediments locally, with many examples. For example, until recently it was standard practice to discard drill cuttings contaminated with oil onto the seabed around offshore oil and gas wells. In the North Sea, contamination tended to be greatest close to the wellhead and to decrease exponentially seaward in a zone around 10 km wide. Since the late 1990s, tougher regulations have changed the North Sea practice to 'skip and ship'; that is, to capture the oil-soaked cuttings and ship them away to some approved disposal site, though that may not happen in all drilling provinces. Decommissioning of offshore oil and gas installations involves the removal of the engineered structures but not the contaminated sediments, which will remain as an Anthropocene signal. The polluted nature of the contaminated sediments leads to their biodiversity being highly reduced, their main component being polychaete worms. In some places the remains of rigs have been left in situ as 'reefs' to encourage the subsequent growth of local fish and other species.

Oil and gas wells may discharge from seabed-processing facilities directly into floating offshore platforms – in which case there is no conventional drilling platform above the sea's surface once the wells have been drilled. In other instances, wellheads may discharge oil or gas directly into seabed pipelines that may lie on the seabed or in specially dug trenches. These too have to be removed on decommissioning.

Most of the growth in offshore oil and gas production took place away from the shoreline on the continental shelf from the late 1940s onwards and grew slowly. The first offshore well out of sight of land was drilled in the Gulf of Mexico off Louisiana in 1947. Offshore drilling and production expanded when supplies from conventional onshore sources began to plateau in the early 1970s. Continental-shelf production also eventually reached a plateau, forcing new production to move onto the continental slope, where it now takes place at depths of up to 2,500 m. Exploration drilling takes place even deeper.

The first offshore drillship was the *CUSS 1* developed in 1955 for the Mohole project to drill into the Earth's crust. An offshore oil-drilling ship, *Glomar Challenger*, was adapted in 1968 to provide the basis for the Deep Sea Drilling Project (DSDP), where the drilling took place for scientific rather than commercial purposes. The first purpose-built drilling semi-submersible, the *Ocean Driller*, was launched in 1963. Since then, many semi-submersibles have been purpose-designed for the drilling industry's mobile offshore fleet. As of June 2010, there were over 620 mobile offshore drilling rigs available for service (https://en.wikipedia.org/wiki/Offshore_drilling).

Sedimentary signals derived from these offshore oil and gas activities and dating from the 1950s onwards can now be expected from the continental shelves and slopes of the major offshore oil provinces in the North Sea, the Gulf of Mexico, the Santa Barbara Channel (California), the Grand Banks (Canada), Venezuela (Lake Maracaibo), the Caspian Sea, Sakhalin (Russia), the Campos Basin (Brazil), offshore Nigeria and Angola (Africa), Southeast Asia, the Persian Gulf, Western India, the Taranaki Basin (New Zealand) and Australia's Gippsland Basin and North West Shelf.

The contamination of coastal sediments by oil began early in the 20th century in response to spills from or the wrecking of oil tankers as tanker traffic grew to feed growing international needs. But the incidence of this kind of contamination was rather low while oil production remained largely onshore and at a total volume of less than 10 million barrels per day, which was the case up to about 1950. After that, tanker traffic and associated oil spills grew apace as world production grew exponentially between 1950 and 1972. Coastal contamination from tanker traffic has declined since 1992, when the International Maritime Organization mandated the transport of oil in tankers with double hulls. Evidently, while the contamination of coastal sediments from tanker accidents began before 1950, the great rise in oil production onshore and offshore from 1950 onwards means that most oil-contaminated coastal sediments are a product of the Anthropocene.

Oil spills can leave long-lasting signatures in coastal salt marshes, as in Buzzards Bay, Massachusetts, for instance. Buzzards Bay is a major transit route for small tanker and barge traffic transporting heating and industrial oil and gasoline north through the Cape Cod Canal into Sandwich, greater Boston and northern New England markets. Nearly 1.6 billion US gallons of oil pass through the canal annually (early 2000s), with additional deliveries made to New Bedford within Buzzards Bay. Reflecting that volume of traffic, the bay has been the site of several substantial oil spills (http://buzzardsbay.org/pastspills.htm). Spilled oil is still found buried in Buzzards Bay's salt marshes. Similar areas of funnelled tanker traffic have experienced similar incidences, for example, in the Persian Gulf.

One of the largest coastal spills from an oil tanker was that of the Exxon Valdez, in Prince William Sound, Alaska, in March 1989 (10.8 million US gallons of crude oil). Its residues remain in the coastal sediments. Oil can also be spilled from accidents on offshore rigs, for example, the IXTOC-1 well in the Mexican sector of the Gulf of Mexico in 1979 (130 million US gallons) and the Deep Water Horizon spill in the US sector in 2010 (210 million US gallons). Much of an offshore spill may be dispersed before reaching the coast, though some of the less easily degraded tarry compounds may end up in deepwater sediments. More noticeably, some of the spill may drift ashore into coastal areas of salt marsh, where it will tend to be buried, as in the case of Buzzards Bay.

Shipborne litter: Much of the early human signature in this marine realm comprises the wrecks and lost trade goods littering the seafloor, with much sea transport occurring close to coastal zones until the expansion of trade in the 1400s. Early trading ships used ballast to maintain stability for return journeys where cargo was not being transported, with the ballast commonly jettisoned as the ships approached harbour; an early example of this is in the first century BCE at the Roman harbour of Caesarea Maritima, Israel, where a laterally extensive ballast layer, 20–60 cm thick, comprises pottery and boulders, many of which are exotic (Boyce et al. 2009).

Coastal zones show a number of kinds of anthropogenic input or modification. One notable input comprises an accumulation of litter via material dropped overboard (jetsam) that is now seen in most surveys of the seafloor, where it is commonly easily distinguished from the surrounding (mostly very fine-grained) sediment (Figure 2.8.5). The incidence of jetsam has reached a scale rivalling that of ice-rafted debris (Ramirez-Llodra et al. 2011). Spreads of ash and clinker (combustion products from the coal that powered steamships) formed 'trackways' across the seafloor along the main shipping routes in immediately pre-Anthropocene time, ~1850 CE to ~1950 CE, while plastics, aluminium and other such more modern materials that clutter the seabed along shipping routes largely date from after 1950 CE (Ramirez-Llodra et al. 2011). In the distal, naturally slow-accumulating parts of the seafloor, it will take thousands of years to bury this debris.

The International Maritime Organization banned the practice of throwing litter overboard with the introduction of the International Convention for the Prevention of Pollution from Ships (MARPOL) in

Figure 2.8.5 An Anthropocene technofossil photographed on the seafloor of the Marianas Trench. Image courtesy of NOAA Office of Ocean Exploration and Research, 2016 Deepwater Exploration of the Marianas.

1973 and the introduction of its Annex V Protocol in 1978 (prevention of pollution by garbage from ships), which entered into force in 1988. A revised Annex V entered into force in 2013; this should reduce future anthropogenic input of this kind. However, much debris of human origin still finds its way to the coastal zone and the open sea as flotsam (essentially floating materials, which may include anything swept downriver by storms). The plastic garbage patches of the Pacific Ocean and the Sargasso Sea comprise both flotsam and jetsam, part of the former eventually converting to the latter by being weighted down by biofouling.

Telecommunication cables: Cable laying for telecommunications is a further cause of disturbance of ocean floors, broadly in the same locations as the main commercial ocean traffic routes. The first successful transatlantic telegraph cable in 1856 was followed a century later by the start of the global telephone network initiated with the first transatlantic cable in 1956 and a fibre-optic network across the Atlantic in 1988.

Anthropogenic chemical signals: A major perturbation of near-coastal marine environments arises out of the spreading 'dead zones' affected by (mostly seasonal) anoxia due to eutrophication from excess N and P derived from fertilisers being washed out to sea (Diaz & Rosenberg 2008; see Figure 5.4.3

herein). These chemical changes (see Section 5.4.1) profoundly modify benthic biota, both macrofauna and microfauna (Wilkinson et al. 2014; see Section 3.4 herein), towards those species that can quickly recolonise the seafloor following repeated kill events.

Persistent organic pollutants (POPs) have been used to define a post-mid-20th-century sedimentary unit in the marine realm (Paull et al. 2006), while artificial radionuclides are widespread and detectable contaminants; remarkably, they have even been found in a unit of glacimarine deposit (which is hence Anthropocene in depositional age) beneath the Pine Island glacier of Antarctica (Smith et al. 2016).

The deep sea is also a chemical sink with, for instance, remote regions such as the Marianas and Kermadec Trenches in the Pacific showing higher POP levels than coastal industrialised zones through bioaccumulation (Jamieson et al. 2017). On abyssal ocean floors, additional chemical signals such as those from artificial radionuclides are rapidly (e.g., Robison et al. 2005) if unevenly (Buesseler et al. 2007) transported to the sea via marine snow and the 'faecal express' (the rain of faecal matter beneath concentrations of organisms that live in the photic zone). The very slow accumulation rate over most of this realm means that this material is commonly thoroughly intermixed by bioturbation with pre-Anthropocene sediment so that the Holocene/Anthropocene boundary is blurred. However, the slow accumulation rate also means that anthropogenic input is concentrated through lack of dilution by 'natural' clastic and carbonate sediment input.

Plastics in the marine realm: Enclosed plastics in marine sediments are an unambiguous indicator of Anthropocene age, representing a new source of input dating from the 1960s, disseminated from shorelines, rivers and ships (see Section 4.3). They represent a low-density material that may be funnelled through submarine canyons in turbidity currents, undergo long-distance drifting before the material eventually sinks through biofouling, or be ingested by fish or plankton with subsequent transport to the seafloor as marine snow. By such means, marine environments from coastlines to deep remote seafloors now have a detectable signal of plastics (Zalasiewicz et al. 2016a and references therein). In Hawaii, accumulations of plastic debris have formed 'plastiglomerates' (Corcoran et al. 2014) in which melted plastic has bonded beach pebbles and sand to form a rock with good potential for long-term preservation.

2.8.5.3 Summary

The marine Anthropocene sedimentary succession, although it is only now beginning to be studied in detail per se, appears distinctive and widely traceable as a chronostratigraphic unit by a range of proxies. In some locations, where the rate of sedimentation is rapid or where the water conditions are anoxic, its lower boundary is recognisable, but in others, where the rate of sedimentation is low and conditions are oxidising, the boundary signal is blurred by bioturbation. Nevertheless, even there, using models of the bioturbation process, the thickness of the Anthropocene layer can be estimated.

3 The Biostratigraphic Signature of the Anthropocene

CONTENTS

In this chapter we examine how fossils are used to establish biostratigraphic boundaries, describe the biostratigraphic signature of the 'Late Quaternary Extinction' and evaluate possible biostratigraphic markers for a boundary for the Anthropocene Series, the material stratal counterpart of an Anthropocene Epoch. Examples are provided of how biostratigraphic markers have been used to define the boundaries of geological time units throughout the geological past. Emphasis is given to the global time-transgressive Late Pleistocene megafaunal extinctions, and how human-mediated extinctions have subsequently continued throughout the Holocene into the Anthropocene. We also demonstrate the record of invasive species as the basis for developing a biostratigraphic approach to help to constrain the choice of a potential Anthropocene Series boundary, looking at aquatic and terrestrial plants and animals, with illustrations from San Francisco Bay and Kaua'i (Hawaii). The chapter also describes the modification of the tropical reef ecosystem during the Anthropocene, and its trajectory under increasing human pressure.

3.1 Fossils as Markers of Geological Boundaries

Mark Williams, Anthony Barnosky, Jan Zalasiewicz, Martin J. Head and Ian Wilkinson

Fossils are important markers for chronostratigraphic and geochronological subdivisions of Phanerozoic time, the last ~541 million years of Earth history recorded in the rock record. Indeed, their usefulness extends further back in time into the Proterozoic. Although the first or last appearance datum of a particular fossil taxon is now used as the primary boundary marker for most units of the Geological Time Scale at system/period rank (e.g., Jurassic, Carboniferous; see Figure 1.3.1 and Section 1.3.1), this distinction historically has been provided by changes in fossil assemblages overall. These historical boundaries often reflect substantial perturbations in oceanography, climate or volcanic activity. Modern definitions of these boundaries, while reducing ambiguity by using single fossil datums, usually attempt to follow historical boundaries as far as is practical. Fossils are also key boundary markers at lower hierarchical ranks of the Geological Time Scale (epoch/series, age/stage, etc.). It is important, therefore, to consider recent changes in biota, their extinctions, originations and changing patterns of biogeographical distribution, as these may be important markers for an Anthropocene Series/Epoch boundary (Barnosky 2014).

Most fossils are *body fossils*, such as arthropods preserved with their dorsal exoskeletons (Figure 3.1.1) and more rarely with their soft anatomy (e.g., Hou et al. 2017). They can also include *trace fossils* of organism activity in sediments, such as burrows and tracks. Chemical signatures of life ('biomarkers') can yield evidence of quite specific biological presence. Body fossils include those from the macroscopic remains of organisms, such as mollusc shells, mammal bones or plant leaves, and from the microscopic remains of organisms, such as pollen and foraminifera. From an Anthropocene perspective, it is important to consider what a fossil is and indeed what distinguishes it from a merely dead organism, given the seamless transition from death to geological preservation. There are various interpretations of the term 'fossil', but its etymology – the Latin *fossilis* meaning 'obtained by digging' – continues to be generally relevant. Perhaps the most flexible and useful definition of a fossil is offered by the International Code for the Nomenclature of Algae, Fungi and Plants (McNeill et al. 2012), which distinguishes fossil from non-fossil material by 'stratigraphic relations at the site of original occurrence'. In other words, if the dead organism occurs in a sedimentary context where normal stratigraphic principles apply, then it can be treated as a fossil.

Figure 3.1.1 A body fossil, 3.4 cm long, of the early Cambrian trilobitomorph *Acanthomeridion serratum*, from Maotianshan, Chengjiang, Yunnan Province, China. (Courtesy of Professor Hou Xianguang; registration number YKLP 10290. Yunnan Key Laboratory for Palaeobiology.)

Fossils are used in geology for two main reasons. Firstly, abundant, diverse and well-preserved fossil successions are sought, in unbroken sedimentary successions, to provide abundant evidence not only directly for ecological and evolutionary change but also for use as proxies for climatic, oceanographic and other types of planetary change. Secondly, fossils are used routinely to date natural exposures or borehole successions of sedimentary deposits and to allow those to be correlated with each other by being placed in the chronological framework of the Geological Time Scale.

Organisms have evolved different anatomies and ecological relationships over geological time, and these changes are preserved in the fossil record. As a result, a rock succession from 400 million years ago, accumulated during the Devonian Period, has a very different array of fossil organisms from one that formed 100 million years ago during the Cretaceous Period, and this in turn differs from assemblages of organisms living today. This principle of the evolutionary succession of life can be used to recognise biostratigraphic intervals called biozones (or biostratigraphic zones), essentially rock successions and their contained fossils that are characteristic of a particular interval of geological time. The biozones most widely applied in the rock record are those characterised by the appearance of a particular species (or set of species) that is common, morphologically distinctive and geographically widespread, crosses environmental gradients, is geologically short lived and is close to being isochronous in its first appearance (time equivalent everywhere it first appears). Stratigraphic control here is provided by the lowest (evolutionary origin) or highest (extinction) occurrence of key species, which are recognised through morphology, on the assumption that a distinct morphology represents a distinct biological species. Although this assumption is debatable (and indeed although individual palaeontologists may argue over the validity of particular recognised species), this approach has enabled the construction of a highly effective geological time scale. Generally, although not exclusively, it is body fossils, or rather the hard skeletons of these, recognised by their shape (i.e., as 'morphospecies'), which provide high-resolution biostratigraphic markers in deposits of Phanerozoic age, with a resolution that may subdivide time into episodes as brief as half a million years.

Different types of biozones can be defined (Figure 3.1.2 herein; Salvador 1994; see also Barnosky 2014 for a comprehensive explanation). A *range zone* is the body of strata (or in some cases unconsolidated sedimentary deposits) representing the known stratigraphic and geographical range of occurrence of a specific taxon (a group of one or more populations of an organism or organisms considered by taxonomists to form a unit such as a species, family or group), a taxon-range zone, or a combination of two taxa (a concurrent-range zone). In this context too, the local range of a taxon may be of particular importance in the Anthropocene, where the local range of a neobiotic species (e.g., an invasive and so 'new' species in a region) may be significant (see Section 3.3 below). A *lineage zone* is the body of strata containing specimens representing a specific segment of an evolutionary lineage. It may represent the entire range of a taxon within a lineage or only that part of the range of the taxon below the appearance of a descendent taxon. An *assemblage zone* is the body of strata characterised by an assemblage of three or more fossil taxa that distinguishes it in biostratigraphic character from adjacent strata. An *abundance zone* is a body of strata in which the abundance of a particular taxon or specified group of taxa is significantly greater than is usual for adjacent parts of the succession. Finally, an *interval zone* (see also Section 4.2) is a body of fossiliferous strata between two specified biohorizons. A biohorizon is typically either the highest or lowest stratigraphic occurrence of a given taxon. Interval zones differ from range zones in that they are defined on the lowest or highest occurrences of different taxa, whereas taxon-range zones are defined by the uppermost and lowermost occurrence of a single taxon, and concurrent-range zones by the co-occurrence of

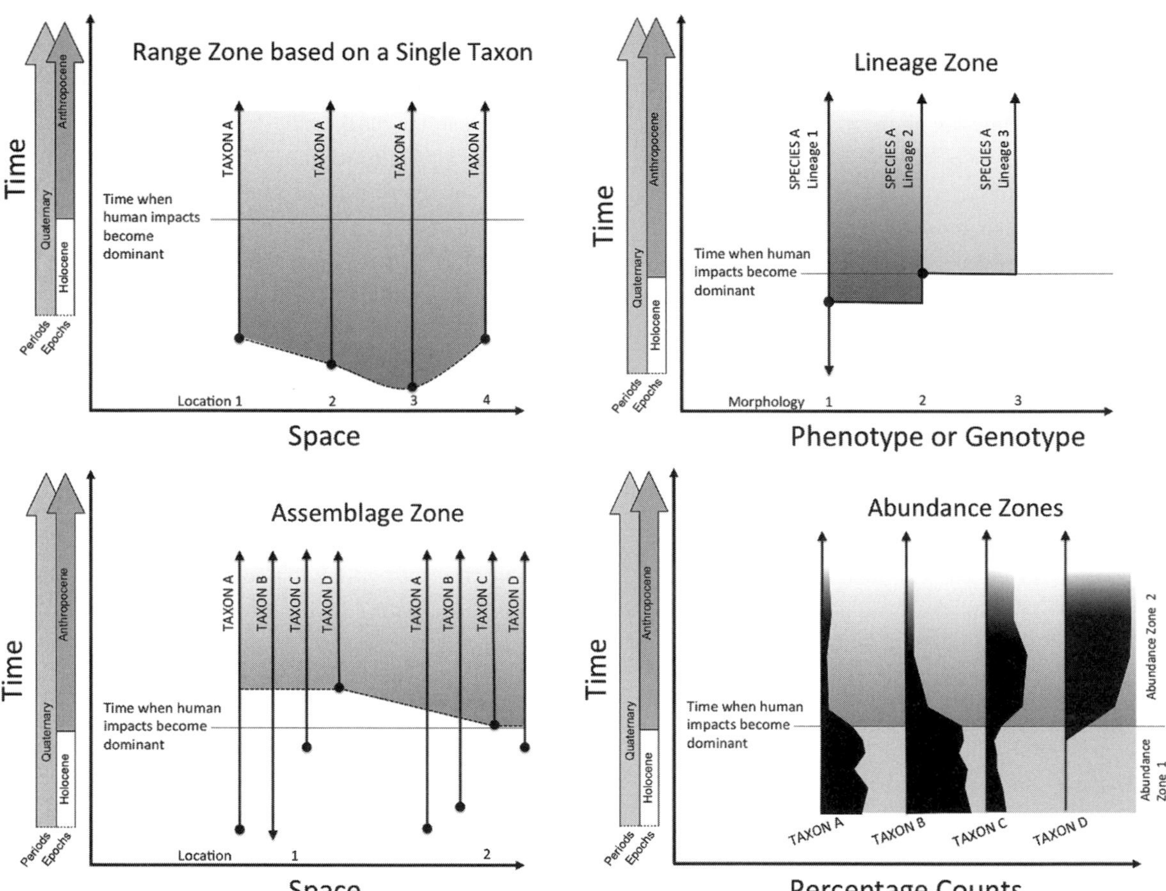

Figure 3.1.2 Four types of biozones and their utility for defining the Anthropocene (see Section 3.1 herein and Barnosky 2014). Vertical lines represent ranges of taxa (e.g., a specific plant or animal species). Dots at the ends of lines indicate first (lowermost) and last (topmost) appearance in a region. Arrows (downward facing) indicate that a taxon (such as a mollusc species in a marine basin) was already present in a region – or continues to persist (upward-facing arrows). Top left, the taxon-range biozone probably has limited utility for the Anthropocene, because few, widespread and easily fossilisable morphospecies are known to have originated in the past 300 years. Top right, a lineage biozone may be useful for characterising the Anthropocene, for example, where it is based on the evolution, hybridisation and genetic modification of crop plants such as maize and rice. Bottom left, an assemblage biozone may be readily identifiable for the Anthropocene based on combinations of indigenous and neobiotic species. Bottom right, the abundance biozone (showing four different taxa) may also have great utility in characterising the Anthropocene, particularly where humans have greatly increased the numbers of individual domesticated species, such as the chicken (Bennett et al. in press), and conversely decimated populations of wild species.

two taxa. Assemblage zones differ from interval zones by the co-occurrence of three or more taxa. Ecobiostratigraphy is a particular utilisation of the assemblage-biozone approach (see Section 3.4) where distinctive fossil assemblages are linked to specific palaeoenvironments that have changed over time. Using fossil assemblages to track these changes in the sedimentary record allows correlation, in some cases, with a precision of just hundreds of years or less.

Although the fossil record extends back more than three billion years (e.g., Allwood et al. 2009; Wacey et al. 2011; Dodd et al. 2017), the early record is sporadic and, prior to 0.58 billion years ago, is mostly characterised by microorganisms or macroscopic layered structures such as stromatolites made by prokaryotic cells. These generally have limited use in biostratigraphy, and they do not contribute to any internationally accepted chronostratigraphic subdivision of time or rock based on their first appearance. Fossils are of paramount importance, though, for subdividing the rock succession of the past ~541 million years, and the base of the Cambrian System and thus of the Phanerozoic Eonothem is recognised globally on the identification of a suite of trace fossils – especially *Treptichnus pedum* – in the rock succession (Landing et al. 2013; see Section 1.3.1.1 and Figure 1.3.3 herein). Although body fossils interpreted to be animals are present in rocks of latest Precambrian (Ediacaran) age (e.g., Yin et al. 2015), animals first become globally widespread and abundant in the Cambrian System and, moreover, can be referred to organism groups that have modern representatives, including sponges, various worms, arthropods and vertebrates (Hou et al. 2017). The fossil record of the Cambrian System is also replete with organisms that showed rapid evolution, such as the trilobites, and which therefore demarcate a series of biozones within Cambrian strata (e.g., Rushton et al. 2011). Over time, the skeletons of many different organisms have provided high-resolution biozonations of the rock record, including conodonts (microfossil remains of early vertebrates) in the lower Paleozoic, the ammonoids in the upper Paleozoic and Mesozoic, and coccolithophorid algae in the Cenozoic. The extension of a complex multicellular biosphere (plants, fungi and animals) to the land, beginning during the Ordovician Period (Wellman et al. 2003), allows a similar resolution for the subdivision of terrestrial rock successions based on fossils. In this case, microscopic fossils (pollen, spores) are particularly important for biostratigraphy.

Whilst the appearance of new species can define a biozone, wholesale changes to the biosphere are in some cases fundamental to recognising major chronostratigraphic subdivisions of time as reflected by fossil content in rocks, such as at the Proterozoic–Phanerozoic boundary, established at the point where the trace-fossil record of animal activity becomes diverse (Landing et al. 2013; see Section 1.3.1.1 and Figure 1.3.3 herein), and the Cretaceous–Paleogene boundary, which is characterised by an abrupt mass extinction event. Some major chronostratigraphic boundaries, such as the Ordovician–Silurian boundary, are recognised by a small biostratigraphic change, in that case on the lowest occurrence of the graptolite *Akidograptus ascensus* (Melchin et al. 2012; see Section 1.3.1.2 herein).

For the more recent past of the Quaternary Period (the last 2.6 million years), and even more particularly for the Holocene Epoch (the last 11,700 years), a somewhat different approach needs to be used than is the case for most of the Phanerozoic. The Quaternary Period is too brief for many evolutionary appearances or extinctions, although there have been some, and these are indeed exploited for biostratigraphy (see below). However, the dominant biostratigraphic signal here, both on land and in the sea, is one that is climatically driven, where alternations between warm and cold or wet and dry conditions have, in any one place, seen the immigration and emigration of different biotic communities that have left a stratigraphic record of this biological succession. It is often the same species that repetitively appear and disappear across several climate cycles, but the order of their appearance or their relative abundance may differ between successive cycles, and indeed some species may be absent from a particular cycle (for instance, in the last interglacial phase, both humans and horses were absent from the British Isles, even though they were present there in both the preceding and succeeding interglacials).

For the Pleistocene and Holocene, typical fossil groups used for biostratigraphy in the marine realm are foraminifera, coccoliths and dinoflagellates, while on land a wide variety of taxa are used, including vertebrates (both large and small); plants (macroscopic remains, pollen and spores); aquatic algae such as diatoms; and invertebrates including gastropods, bivalves, beetles and even the preserved head capsules of chironomids (midges). Increasingly throughout this interval, particularly in the Holocene, human remains and artefacts are used, although their study (archaeology) is typically more directed towards understanding human history than for the correlation of geological strata. While the culturally controlled artefacts assemblages recognised by archaeologists are used to define time units (Mesolithic, Iron Age, etc. – see Section 7.2), they are known to be diachronous, often markedly so.

The main planks of Pleistocene and Holocene chronostratigraphy (Pillans & Gibbard 2012) are based not so much on biostratigraphy as on other forms of stratigraphy, notably chemostratigraphy, which responds to the many climate oscillations of this time, expressed particularly in the marine oxygen isotope record, but also magnetostratigraphy (see Sections 1.3.1.5 and 1.3.1.6). Biostratigraphic events help to characterise the main boundaries, but they do not act as primary markers per se. For example, the base of the Quaternary, chosen to mark a time when Northern Hemisphere glaciation was intensifying, actually corresponds to an interglacial (Marine Isotope Stage 103; Figure 1.3.7). Its precise position was chosen because (1) it corresponds to an important lithological marker (the Nicola bed in Sicily; Rio et al. 1998) that allows calibration to the astrochronological timescale, and (2) its position is just ~1 m above a major palaeomagnetic boundary (between the normal Gauss and the reversed Matuyama magnetochrons), which allows the recognition of this boundary on a global scale in both marine and terrestrial deposits (Gibbard & Head 2010; Head & Gibbard 2015). This is not to say that biostratigraphy has no role to play in the characterisation and recognition of the Neogene–Quaternary boundary. A succession of recognised origins and extinctions of biostratigraphically important calcareous microplankton species both precedes and post-dates this level (Pillans & Gibbard 2012, figure 30.3), although none are coincident with it. Using an ecobiostratigraphic approach, a detailed picture emerges. A study of dinoflagellate cysts from deep cores in the North Atlantic, conducted at the scale of individual climate cycles, provides a highly resolved biostratigraphic characterisation of the Neogene–Quaternary boundary interval and indeed reveals a significant shift in North Atlantic Current circulation and enhanced seasonality at ~2.6 Ma (Hennissen et al. 2014, 2015, 2017). Nonetheless, although the Quaternary is marked by a geologically rapid succession of marked climate oscillations (described in detail in Section 6.1.1), the overall rate of biological evolution, as seen in species extinction and origination rates, seems not to have been geologically unusual (Barnosky 2005).

Within the Late Pleistocene and Holocene, fossils continue to be used both as a proxy for climate change, reflecting such events as regional to global climate changes, including the Younger Dryas event and the 8.2 and 4.2 ka events used to subdivide the Holocene (Walker et al. 2012). Fossils are also used to track the strongly diachronous human disturbance associated with population change, agriculture and urbanisation (e.g., Dearing et al. 2012).

The biostratigraphic pattern of the Anthropocene differs. In part, it is beginning to reflect the anthropogenic global warming of the past few decades (Letcher 2016 and papers therein). Although these climate-driven geographical-range shifts are in their early stages (they will likely become more pronounced over coming decades), the patterns of local to regional human disturbance that developed through the Holocene are intensifying (Steffen et al. 2015; and see Chapter 7 herein). The combination of these factors has recently been joined by globalisation and the associated mass transit of goods, people and organisms, which has led to what has been called the Homogenocene (Samways 1999), in some ways synonymous with the Anthropocene but emphasising the role of invasive species.

A further factor is accelerated or directed evolution driven by various human actions, both deliberately and inadvertently. This includes pathogens becoming antibiotic resistant (Gillings & Paulsen 2014) and invertebrates and plants becoming pesticide and herbicide resistant, almost wholly a post-mid-20th-century phenomenon (Borel 2017). There is also the modification of food crops and animals by selective breeding and more recently by genetic engineering; this produces physiological modifications and also – of importance to biostratigraphy – distinct and numerically abundant new morphospecies like the modern broiler chicken (Bennett et al. in press).

These phenomena include an array of rapid and ongoing biological changes that may be used to provide an ultra-high-resolution biostratigraphy, where the rate of change is orders of magnitude faster than the geological norm and in some ways is comparable to that of technological evolution (see Section 4.2). These changes are only just beginning to be documented in a biostratigraphic context.

In the following sections we investigate potential biostratigraphic markers that might bracket an Anthropocene Series boundary, following an approximate chronological order from the megafaunal extinctions of the Late Pleistocene to the neobiota of the 21st century.

3.2 Late Quaternary Extinctions

Anthony Barnosky, Ian Wilkinson, Jan Zalasiewicz and Mark Williams

3.2.1 The Megafauna

By the Late Pleistocene, more sophisticated human toolkits and growing human population (Barnosky 2008) combined to precipitate a prominent extinction event, known as the Late Quaternary Extinction (LQE). This provides a clear biostratigraphic marker in many parts of the world, albeit one that is time transgressive over some 50,000 years globally, progressing from Australia to Eurasia and the Americas to isolated islands and, since the 20th century, becoming increasingly apparent in Africa. The time transgression of the signal follows the migration of humans to various parts of the globe. Thus, the LQE seems to be the first clear sign that humans had emerged from their long but ecologically less disruptive history to play a greater role on the world stage. As a consequence, about half the world's megafaunal mammal species (usually regarded as species with average body mass exceeding 44 kg), as well as many large bird and reptile species, went extinct. While the details of extinction intensity, geographical patterning and chronology may well have been influenced by interaction with Late Pleistocene climate changes that transformed many local ecological settings (Koch & Barnosky 2006; Brook & Barnosky 2012; Wroe et al. 2013; Villavicencio et al. 2015), human involvement seems clear because most extinctions occurred only after (and often relatively soon after) people first colonised a region, and the extinct taxa had survived many previous glacial–interglacial transitions (Koch & Barnosky 2006). This inference is upheld by a recent global analysis of extinct mammal species whose body size was larger than 10 kg, which identified a link with human environmental pressure and recognised a heightened extinction rate in those regions where *Homo sapiens* arrived as the first hominin (Sandom et al. 2013). Therefore, the LQE event is the first strong signal of human-driven dramatic alterations to the global biota.

The strong anthropogenic signal begins in the Late Pleistocene of Australia, where the LQE took place against a background of decreasing megafaunal diversity that may be linked to long-term trends of increasing aridity (Wroe et al. 2013). Prior to the arrival of humans, the Australian megafauna is interpreted to have been relatively diverse, but a wave of extinctions swept through between 51 ka and 40 ka, likely within about 10,000 years of the arrival of humans, which is thought to have been near 65 ka (Clarkson et al. 2017). It is likely that en route to Australia humans caused similar disruptions, but these are not (yet) recorded in the known fossil record.

The extinction of several large mammals has been interpreted as a combined result of hunting, of the disruption of Australasian ecologies – for example, through burning of vegetation – and of climate change. Amongst the animals that disappeared, the large flightless bird *Genyornis newtoni* became extinct at ~47 ka and suffered from human predation (Miller et al. 2016). Conclusions linking megafaunal extinctions to human activities in Australia are contested, and long-term climate change may also have been an important driver (Wroe & Field 2006; Wroe et al. 2013; Cohen et al. 2015).

In North America, the megafauna extinction was focused between 13.5 and 11.5 ka (Barnosky 2008; Koch & Barnosky 2006) at a time of rapid climate change and of the migration of Clovis hunters into the region, although an earlier pulse of extinction took place in the far north of the continent during a pronounced cooling between 35 and 25 ka (Koch & Barnosky 2006). In South America, the LQE spreads over several thousands of years, and the timing of extinctions varies from region to region; the major extinction pulse seems to begin after human arrival at 14.6 ka, but some taxa may have persisted for another ~7,000 years (Barnosky & Lindsey 2010; Prado et al. 2015; Villavicencio et al. 2015). Megafauna extinctions in the Americas have been variously attributed to primarily human impacts, climate change or combinations of these (Koch & Barnosky 2006), but most recent work highlights a major role for humans (Koch & Barnosky 2006; Barnosky & Lindsey 2010; Brook & Barnosky 2012; Villavicencio et al. 2015). The contention that a comet explosion helped precipitate the LQE (Firestone et al. 2007) has not gained much empirical support (Surovell et al. 2009). The coincidence of human arrival with end-Pleistocene climate change, notably the Younger Dryas in North America and analogous rapid warm-cold-warm fluctuations in South America, may have produced the much more severe extinction of megafauna in this region relative to areas with long-lived human habitation such as Eurasia (Koch & Barnosky 2006; Barnosky & Lindsey 2010; Villavicencio et al. 2015).

In Eurasia the LQE was characterised by relatively minor losses, but these were notable nonetheless, including such Pleistocene icons as mammoths and giant Irish elk. Here too, the combination of end-Pleistocene climate change and expanding *H. sapiens* populations appears important (Nogues-Bravo et al. 2008). An early phase of extinction of species that were common during interglacial times occurred between 49 ka and 25 ka, a period of cooling leading towards the last glacial maximum and coinciding with expansion of *H. sapiens* through that part of the world (Koch & Barnosky 2006). The later phase of extinction saw the loss of such species as mammoths and Irish elk beginning about 14 ka, coincident with warming (the Bølling-Allerød warm interval, discussed in Section 6.1.2.1) that heralded the end of the Pleistocene Epoch and further growth of human populations (Koch & Barnosky 2006). However, these animals lingered in refugia of favourable climate and low or absent human populations until ~8 ka (e.g., Irish elk in Siberia) to ~4 ka (e.g., mammoths on Wrangel Island; Stuart et al. 2002, 2004).

As a potential guide to the base of the Anthropocene, the LQE suffers from being time transgressive over some 50,000 years, and even setting a boundary at the youngest pulse of extinction would mean that the Anthropocene largely overlaps the Holocene. However, regional biostratigraphic units that illustrate the stepwise transition into a human-dominated Earth System may be defined, and such an approach is illustrated by the recent definition of the Santarosean and Saintaugustinean North American Land Mammal Ages of ~14 ka ago and the mid-16th century, respectively (Barnosky et al. 2014).

3.2.2 More Recent Patterns of Extinction and Extirpation in Seas, in Lakes, on Land and on Islands

Focusing on the later Holocene record, investigations of marine environments show a long period of low-level and spatially localised ecological degradation beginning about 5,000 years ago and extending to ~1800 CE (Wilkinson et al. 2014). This process began in the Mediterranean, but by the mid-20th century, marine ecological degradation was not only global but accelerating rapidly, reaching its apogee about 1950–1952. Over the same time and continuing to the present, the over-exploitation of many marine species has led to severe reduction of megafauna in the sea, echoing what took place on the land during the LQE (McCauley et al. 2015). This acceleration was caused principally by rapid industrialisation during and after World War II (see Section 7.5) and an intensification of agriculture with increased fertiliser use to sustain rapid population growth. By the late 20th and early 21st centuries, national and international laws and regulations concerning the environment locally resulted in a phase of recovery within the marine microbiota (Yasuhara et al. 2012).

This global pattern of ecological change, encapsulated in the model of ecological degradation by Yasuhara et al. (2012), is applicable in both marine and terrestrial contexts. In England – the first modern industrial country, with a long history of environmental change – terrestrial ecological degradation as indicated by species loss (involving fungi, plants, birds, mammals, insects, etc.) can be demonstrated using Species Recovery Trust data. Of the 421 species that have disappeared from England since 1800 CE, the date of extirpation is known accurately for 333 (Figure 3.2.1). In the phase of inception, prior to about 1900 CE, there were as many years with extirpation as without. Then, with rare exceptions, extirpations occurred annually between 1900 and 1974 CE, peaking at 1951–1953 CE, whereas the late 20th century saw a return to the pre-1900 situation. Using this model, it is possible to suggest that the acceleration phase in the early to mid-1940s might mark the Holocene/Anthropocene transition, although equally the apogee phase in 1951–1953 CE might mark a regional (i.e., for England) biostratigraphic boundary. Fossilisable elements of this process include plant spores, pollen and beetle elytra (hardened forewings).

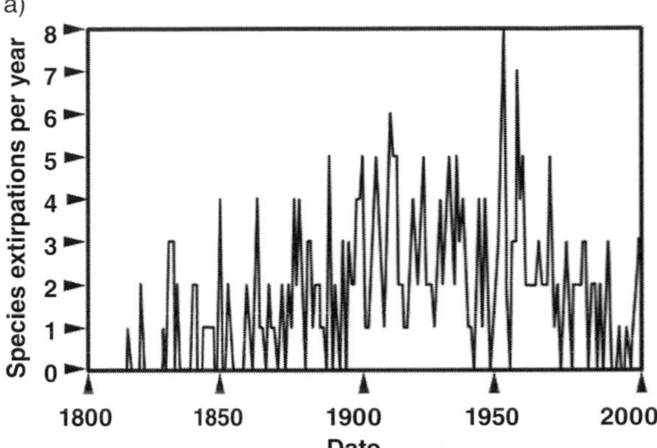

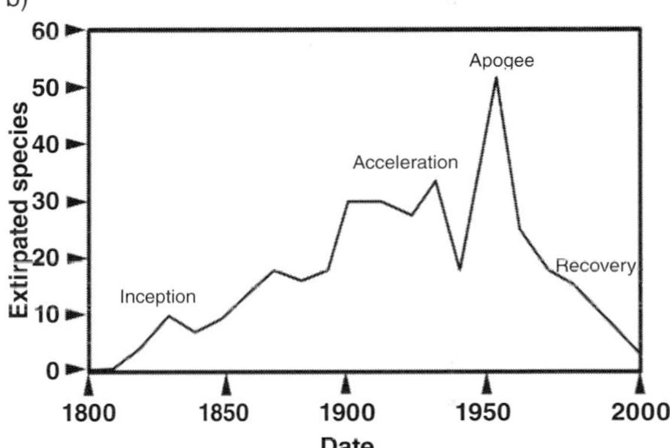

Figure 3.2.1 (a) Annual extirpation of terrestrial organisms in England since 1800 (of a total of 421 extirpated species, 333 have well documented extirpation dates). (b) Graphical representation of ecological degradation in England, as represented by the number of extirpated species with time generalised on a decadal scale, to show that the phase of inception was followed by a phase of acceleration culminating in apogee. Based on data from the Species Recovery Trust (www.speciesrecoverytrust.org.uk/Images/ll/England'slostspecies.pdf).

Freshwater lakes are particularly susceptible to anthropogenic ecological degradation, affecting microfloral and microfaunal groups at the base of the food chain. Lake Sevan in Armenia is a classic example of how local, human-induced stress radically changed the ecological trajectory of the lake biota. Water abstraction from Lake Sevan for irrigation and hydroelectric power, particularly after 1950 CE, lowered water level by more than 20 m, profoundly influencing the ecosystem, including temperature (Legovich 1979; Hovhannissian 1994), sedimentation, substrate and light penetration (e.g., Latti 1932; Reznikov 1984; Hovhannissian 1994) and oxygen and nitrogen concentration (Hovhannissian 1994, 1996), with concomitant changes to the trophic structure (Friedman 1950; Markosyan 1959; Meshkova 1976; Gambaryan 1979; Legovich 1979; Ostrovsky 1983; Kurashov & Belyakov 1987; Simonyan 1988, 1991; Hovhannissian 1994; Wilkinson & Gulakyan 2010).

The various publications show that between 1948 CE and the mid-1970s, the previously rich diatom population was replaced by filamentous green algae and cyanobacteria. Macrophytes – the food source of many organisms in the benthic community – decreased from more than 20 abundant species to 1 (although 13 rarer species occur), with *Chara*, *Potamogeton* and *Zannichelia* almost disappearing. The gammarid biomass sharply declined, and the diversity of deepwater cladocerans and ostracods collapsed. Between 1972 and 1980 CE, the population of zooplankton changed from being dominated by copepods and cladocerans to one dominated by smaller zooplankton such as rotifers. The zoobenthos that was dominated by gammarids, chironomids and oligochaetes – prior to 1948 CE – declined, mirroring the reduction of macrophytes. Lake-margin ostracod communities saw increase in some taxa (especially environmentally tolerant taxa) and the disappearance of others. Higher in the food chain, there was near extinction of the four endemic subspecies of the Sevan trout (*Salmo ischchan*), while favouring competitors such as carp. Such dramatic biological changes have undoubtedly left a high-resolution biostratigraphic record reflecting anthropogenic impact since the 1950s, especially since many of the microorganism groups involved have hard skeletons known to preserve over geological time scales.

Island biotas have shown marked extinction and extirpation patterns following colonisation by humans. Thus, in New Zealand, moa populations were large and viable immediately prior to the arrival of the first Polynesian colonists around the late 13th century (Allentoft et al. 2014). After people arrived, nine species of moa became extinct within two centuries. In the notorious case of the dodo (*Raphus cucullatus*) on Mauritius, the last reported sighting was in 1662 CE, only a few decades after the species was first documented. It has an extensive fossil record in the Mare aux Songes swamp area of Mauritius (Rijsdijk et al. 2009), which also preserves many other recently extinct birds, mammals and reptiles (this swamp also has a distinctive anthropogenic signal, in that it was backfilled with hard core during late British colonial rule [in 1943 CE] to alleviate the spread of malaria; Rijsdijk et al. 2009).

Biotic extinctions are often coupled with a fossil record of neobiota – species that have extended beyond their pre-anthropogenic geographical range as a result of deliberate or accidental human introduction; for example, the widespread introduction of the Pacific rat across the islands of Polynesia (e.g., Matisoo-Smith et al. 1998). The signature of the neobiota is a key component of the Anthropocene biosphere (McNeely 2001). It has been unfolding ever since humans began to translocate animals and plants in the Late Pleistocene and has been accelerating since the increasing globalisation of trade from the 15th century onwards.

For biostratigraphy, neobiota have many advantages over extinctions and extirpations of indigenous biota. Many extinctions are of species that possessed only a localised distribution, and though evocative of human impact, as in the case of the Yangtze River dolphin (*Lipotes vexillifer*; see Turvey et al. 2007), they do not leave a widespread signal approximating an isochronous surface (see Section 3.1). By contrast, many neobiota are globally widespread and geologically recent, meaning that they might be used to define an approximately isochronous surface for chronostratigraphic correlation. Some are species with hard skeletons (such as bivalves or foraminifera) that invade sedimentary deposits en masse, giving them a high preservation potential.

In the next section we examine the record of neobiotic species as regards developing a biostratigraphy to help constrain the choice of a potential Anthropocene Series boundary.

3.3 The Biostratigraphic Signal of the Neobiota

Mark Williams, Jan Zalasiewicz, David Aldridge, Colin N. Waters, Valentin Bault, Martin J. Head and Anthony Barnosky

Many organisms have become widespread as a result of human interaction, including animals that are feral (e.g., black, brown and Pacific rats), organisms that have been domesticated (e.g., maize, rice, tulips, Atlantic salmon, rainbow trout, chickens, pigs) and those that have been domesticated but subsequently became feral (e.g., pigeons, cats, pigs) at least locally. The extent of neobiotic inundations into new regions introduces the possibility of biozonations based on species incomings. Although this process has been unfolding for millennia, many of the neobiota were introduced during the 19th and 20th centuries. In a neobiotic biostratigraphic signal, the most useful organisms are those that possess skeletons, that have a good fossil record prior to their translocation and that are widespread. Invasive neobiota may represent very good Anthropocene markers because they can quickly reach remarkably high abundances in new environments as a result of escape from predators, parasites and disease and through efficient exploitation of novel resources.

3.3.1 Aquatic Neobiotic Species

Aquatic neobiota may live in either fresh water or the sea, with some species occupying both, such as salmon. Marine species in particular have become widely translocated via canals, shipping (in ballast water or from hull fouling) and aquaculture. Approximately 10,000 species have been transported around the globe in the ballast tanks of ships (Bax et al. 2003).

Aquatic microbiota, here defined as micro-invertebrates (e.g., ostracods, nematodes) and unicellular eukaryotes (e.g., diatoms, dinoflagellates), are particularly susceptible to transportation, being small and abundant. Many translocations occurred in the 20th century and dovetail with other potential stratigraphic markers of the Anthropocene, especially radiogenic signatures (Waters et al. 2015, 2016; Section 5.8 herein). For example, the benthic agglutinating foraminifer *Trochammina hadai* is a Holocene species originating in the estuaries and coastal areas of Japan (McGann et al. 2012). It was translocated across the Pacific Ocean into the coastal western United States, possibly in the 19th century, but more likely in the early/mid-20th century (McGann et al. 2012), and since then has become dominant, often comprising 50% of the foraminiferal fauna, to the detriment of local foraminifer species (McGann et al. 2000). *T. hadai* therefore provides a potential widespread biostratigraphic marker in the Americas (McGann et al. 2012) associated with markedly reduced abundances of indigenous benthic foraminifera.

Many invasive species of dinoflagellates have contributed to the marine neobiota, although only those producing a resistant resting cyst within their life cycle will have left a biostratigraphic signal. Translocations of such cyst-producing species during the 20th century have been mediated by shipping and aquaculture or in some cases may represent geographical range expansion in response to changes in global climate. *Gymnodinium catenatum* is an important warm-water toxic marine planktonic dinoflagellate species found today in many parts of the world. It was introduced to Tasmania in the early 1970s, most likely via the ballast water of cargo ships from Japan (McMinn 1997). It is considered an invasive species in the northeast Atlantic, and the sedimentary record of its cysts along the Atlantic coast of the Iberian Peninsula indicates an arrival in the southern part of the area in 1889 ± 10 years. It expanded and became established progressively northward along the Iberian coast during the 20th century, culminating with the first of many toxic blooms in 1976 CE. The cause of its arrival on the Iberian Peninsula is uncertain, but its northward expansion appears to reflect warming and long-term hydrographic changes (Ribeiro et al. 2012, 2016). The global spread of *Gymnodinium catenatum* during the 20th century has therefore left a permanent if complex biostratigraphic signal, reflecting both translocation via international shipping and geographical range expansion in response to climate change.

Skeleton-bearing species have been widely dispersed by humans in the 20th century. A classic example is the 'ship worm' bivalve mollusc *Teredo navalis*, which likely originated along Atlantic European coasts but has subsequently spread to temperate and tropical seas worldwide. This animal can withstand a range of salinities and temperatures, facilitating its dispersal in the wooden hulls of ships. The dispersal of this species over time needs proper documentation (its fossil record is unstudied) and likely occurred over centuries. However, its progression along the Eastern Seaboard of North America is well documented in the 19th and 20th centuries, and on the West Coast the species invaded San Francisco Bay in the early 20th century (Cohen 2004; Fofonoff et al. 2017; Figure 3.3.3 herein) and therefore predates the invading benthic foraminifera *T. hadai* later in that century (McGann et al. 2012).

A more recent and well-dated example amongst marine neobiota is the Amur River clam, *Potamocorbula amurensis*. This species is native to coastal regions of East Asia between latitudes 53°N and 22°N. It was transferred to San Francisco Bay in 1986 CE, likely via ballast water (Carlton et al. 1990), where it rapidly became abundant. A similarly important biostratigraphic signature in marine systems is through the recent proliferation of the Pacific oyster, *Crassostrea gigas* (Figure 3.3.1). This species, native to Japan and northeast Asia, is the most widely grown bivalve in aquaculture and has subsequently spread through natural spatfall (the settling and attachment of young bivalves) into large parts of Europe, North America, Australia and New Zealand. The species can create reefs in the intertidal and shallow subtidal zone by cementing shells to each other (Padilla 2010). Pacific oysters were introduced into the Dutch Wadden Sea in 1983 CE and showed dramatic increases in abundance in the 1990s. On dense reefs each square metre may contain more than 500 adults, weighing more than 100 kg/m^2 (Fey et al. 2010). Oysters occur commonly as fossils in the sedimentary record (e.g., Marie et al. 2011), and so we may reasonably expect an Anthropocene signature from *C. gigas* reefs.

Figure 3.3.1 Reef of the neobiotic Pacific oyster, *Crassostrea gigas*, in the intertidal zone of the Dutch Wadden Sea. Individual animals can reach 30 cm in length and partially cement to other live oysters or empty shells, growing the reef upward and outward. (Photograph taken by Tom Ysebaert, Wageningen Marine Research.)

The barnacle *Balanus amphitrite* may also be useful for biostratigraphy. Likely originating either in the tropical Indian Ocean or the Pacific, and having a fossil record extending back at least into the Pleistocene (Cohen 2011 and references therein), *B. amphitrite* now is widely distributed from New

Zealand to California and has well-documented first appearances as a neobiotic species in San Francisco Bay during 1938–1939 CE, whereas its first occurrence in UK waters, at Shoreham Harbour, Sussex, was in 1937 CE (JNNC 2016). A potentially widespread global marker amongst shallow-marine biota around the mid-20th century, its record in sedimentary deposits has not yet been systematically documented.

Non-marine aquatic invaders include the zebra mussel, *Dreissena polymorpha*. Native to the Ponto-Caspian region of Eastern Europe, the species first spread into western Europe during the late 18th and early 19th centuries, reaching the United Kingdom in 1824 CE (Aldridge et al. 2004). A secondary expansion in the late 20th century saw the species become a widespread and prolific biofouling nuisance in North America and parts of Europe, including Ireland, Italy and northern Spain (van der Velde & Rajagopal 2010). Zebra mussels can displace indigenous European mollusc species, particularly native unionids (Sousa et al. 2011), which are showing dramatic declines across the continent (Lopes-Lima et al. 2016). Since both invader and displaced species are shell bearers, there is a strong possibility that these molluscs will leave a biostratigraphic marker of the Anthropocene, not least because zebra mussels can occur in densities of 11,000 individuals/m^2 (Aldridge et al. 2004).

Other widespread neobiotic species in rivers and lakes include the American signal crayfish (*Pacifastacus leniusculus*), which could leave a fossil record via its hard arthropod exoskeleton. It also creates extensive burrow systems in riverbanks, providing a potential trace-fossil record of its presence. The signal crayfish was introduced from North America to Europe (Sweden) in 1960 CE for aquaculture. It is widely distributed from Scandinavia to Greece, occurring in rivers and lakes from coastal regions to sub-Alpine settings (Johnsen & Taugbøl 2010). Signal crayfish have displaced native European crayfish, in part through the transmission of fungal 'crayfish plague', to which European species are especially susceptible. European and signal crayfish share habitats (Gallardo & Aldridge 2013), and so their changing fortunes will likely be recorded in stratigraphic successions.

The spiny water flea, *Bythotrephes longimanus*, was introduced from its native Baltic Sea to the Great Lakes of North America in the early 1980s in ballast water from transatlantic ships (Johannsson et al. 1991). It is now established in all the Great Lakes except Lake Ontario. Cladocerans have chitinous exoskeletons that are preserved in lake sediments. *B. longimanus* is a large predatory planktonic cladoceran that preys on smaller cladocerans. Its introduction has led to a decline in abundance and species richness amongst indigenous cladocerans, these changes having a cascading effect on other biota within the lake ecosystem (Strecker & Arnott 2010).

3.3.2 Terrestrial Neobiota

3.3.2.1 Plants

Crop cultivation, like the domestication of animals, has multiple origins (Zeder 2011), with the earliest fossil evidence of sustained agriculture preserved in grains and chaff of emmer and einkorn in the pre-pottery Neolithic of the Euphrates Valley from about 10,500 years ago. Records from Israel suggest that humans may have been experimenting with crop cultivation considerably earlier, 23,000 years ago (Snir et al. 2015). Early agriculture arose independently in, for example, China, the Americas, Southeast Asia and Africa (Martin & Sauerborn 2013). These early origins are often associated with particular crop types: maize in Central America, wheat in Southwest Asia, rice in China. To some degree these crops remain dominant foodstuffs in their indigenous regions, although China is now the second-largest producer of maize after the United States.

Maize, first domesticated in Mexico more than 7,000 years ago, spread through South America over several thousand years (Piperno 2011) and reached North America much later, becoming widespread from about 900 CE (Emerson et al. 2005). Maize spread to the 'Old World' after the Iberian colonisation of the Americas in the 16th century. This interval of time, which extends into the early 17th century, is known as the Columbian Exchange and has been mooted as a possible contender for an Anthropocene Series boundary (Lewis & Maslin 2015). It is associated with the large-scale exchange of plants, animals and diseases that have left some significant biostratigraphic signatures, although these are complex and diachronous over centuries.

Accelerated globalisation in the 20th century included the spread of major food crops, and this will be reflected in the sedimentary record. Maize production in the 21st century, now on all major landmasses bar Greenland and Antarctica, exceeds 1 billion metric tons per year, higher than that of both rice (forecast 511 million tons for 2018) and wheat (forecast 754 million tons for 2018; FAO, accessed June 2018). Maize leaves a fossil record of pollen and phytoliths (e.g., Pohl et al. 2007), but its record of translocation across the world is highly time transgressive (see figure 2 of Zalasiewicz et al. 2015a), reflecting human cultural development in different regions prior to connectivity via globalisation. There has been no detailed evaluation of the overall sedimentary record of maize pollen and phytoliths, but it seems clear that this domesticated crop does not provide an isochronous signal that can be used to locate an Anthropocene Series boundary worldwide. However, along with wheat and rice as globally distributed cultivars, detailed constraints on the chronology of its spread could help establish regionally useful biostratigraphic correlative levels in Holocene–Anthropocene successions.

Similarly, the extension of certain crops through Polynesia, such as the sweet potato (*Ipomoea batatas*) in the 13th century CE (Wilmshurst et al. 2011), tracks the spread of humans and establishes a connection with Central or South America, where this plant was first domesticated more than 5,000 years ago. The geographical expansion of such crops is important for tracking the timing of human colonisation events in remote or previously uninhabited regions, but it does not provide a globally synchronous marker for the Anthropocene.

More recently and widely translocated plants, such as the invasive Japanese knotweed (*Fallopia japonica*) and *Rhododendron poniticum*, are also likely to have been time transgressive in their distribution, though over a narrower time interval.

3.3.2.2 Animals

Like plants, many widely dispersed neobiotic animal species have a record of translocation that is time transgressive and that has been developing since the latest Pleistocene. For example, dogs have left a biostratigraphic signal from a complex multicentre domestication process, leaving a fossil record in Europe, the Levant, the Middle East, Northern China and Kamchatka from about 12,000 years ago (Larson et al. 2012). The initial domestication of dogs overlaps the geographical range of wolves in the Old and New World, whilst the spread of dogs as neobiota into sub-Saharan Africa and South America follows the spread of agriculture (Larson et al. 2012), leaving a time-transgressive stratigraphic signature that is also characteristic of other domesticated animals, including pigs, sheep, cows and chickens. In general, with domesticated animals, there are two biostratigraphically significant factors: the geographical spread of the species from its place of origin, regardless of whether or not it is morphologically modified from the wild form, and then selective breeding that alters fossilisable parts of their morphology, eventually to result in what might be regarded as new fossil morphospecies (regardless of whether they can still interbreed with the wild form or with other morphospecies grading back to the wild type). Either of these processes can, at any given place, provide a biostratigraphic first-appearance datum.

Rats are amongst the most widespread non-domesticated animal species, having been translocated across all continents except Antarctica. Three species, *Rattus rattus* (black rat), *R. exulans*

(Pacific or Polynesian rat) and *R. norvegicus* (brown rat) are particularly widespread. All three are thought to have originated in Asia, the black and Pacific rat in Southeast Asia and the brown rat in continental Asia. The appearance of these species beyond their native range provides good stratigraphic indices of human migration into uninhabited terranes. For example, in New Zealand, human colonisation is linked with the arrival of *R. exulans* in the period 1230–1282 CE (Wilmshurst et al. 2011). Skeletons of *R. exulans* find their way into sedimentary deposits elsewhere in Polynesia; for example, *R. exulans* is present in the sinkhole succession of Kaua'i, Hawaii, in sediments dated with a minimum age of 1241 CE (see Section 3.3.3.2). Polynesian colonists transferred many neobiota, including pigs (*Sus scrofa*), dogs (*Canis familiaris*) and the sweet potato (*Ipomoea batatus*), as they extended across the Pacific in the 13th century.

There are stratigraphic signals from neobiotic terrestrial invertebrates, too. The giant African snail (*Lissachatina fulica*) has invaded Polynesia, Southeast Asia, India, China and North and South America. In South America, the snail may have invaded ecosystems from the early 1800s onwards, and it is particularly common in urban areas (Thiengo et al. 2007). However, as with vertebrate neobiota, few studies have attempted to recover the temporal record of such species from sedimentary deposits. The giant African snail with its robust shell – some 20 cm long in adults – is fossilisable, having been recorded in the recent sedimentary deposits of the Maha'ulepu sinkhole succession of Kaua'i, Hawaii, where its introduction to that region can be precisely dated to 1936 CE (Burney et al. 2001). Because the giant African snail has also been introduced into Florida but is now being deliberately eradicated in that region, the theoretical possibility of defining brief local range zones is introduced (see Section 3.1).

3.3.2.3 Microbiota as Disease Vectors

Many vertebrate and invertebrate neobiota are vectors of disease. The giant African snail, for example, acts as an intermediate host for the nematode *Angiostrongylus cantonensis*, which can cause meningoencephalitis in humans (Thiengo et al. 2007). Such pathogenic neobiota leave no direct fossil record, just as ancient pathogens were presumably an important but invisible control on the ancient fossil record. These diseases, though important in documenting the impact of neobiota on human ecologies, are generally sporadic, regionally specific and time transgressive in their occurrence and therefore unlikely to provide precise biostratigraphic control for the Anthropocene. However, as an important influence on the abundance and distribution of humans, they are an influence on human-driven stratigraphic change.

The extinction of the smallpox virus (in December 1979 CE) is a profound example of the biological impact of humans in the 20th century. Smallpox strongly impacted human populations through classical and historical times, and when transferred to the Americas in the 16th century by European colonists (for example, in 1520 CE to Mexico, and in the same decade to Peru), it caused widespread epidemics. This human catastrophe contributed to a wholesale cultural change in which Spanish-era buildings supplanted the earlier pre-Colombian buildings, a change reflected in the accumulating urban stratigraphy. Although the virus is not fossilisable, the effects of the disease on human bones, including the infection *osteomyelitis variolosa*, will be recognisable in archaeological and geological strata.

Some diseases, such as the Black Death (*Yersinia pestis*), transmitted by fleas carried on black rats, have left distinctive signals of their impact in mass human-burial sites within major cities, and these can be correlated with historical accounts of pandemic (plague) spread of these diseases from classical to modern times, in the case of the Black Death. In London, for example, there are widespread plague pits with skeletal remains that date to a major outbreak in 1665–1666 CE (Johnson 2018).

In more recent times tree populations have been decimated by introduced pathogens against which they may have little resistance. A particularly virulent strain of Dutch elm disease, an ascomycete fungus carried by the elm bark beetle, was introduced to the UK in the early 1960s from logs imported from eastern North America (Harwood et al. 2011). The disease resulted in the death of almost 30 million trees between 1970 and 1990 (Potter & Urquhart 2017), and this will have left a geologically lasting signature in the record of pollen in sediments across many parts of the UK. A stratigraphic pollen record from southern England indeed shows this elm decline very distinctly (Perry & Moore 1987).

3.3.3 Neobiota-Based Biostratigraphies

Here we explore how biostratigraphic markers of the Anthropocene can be widely recognised in successions that are geographically remote from each other, either for making local correlations within a region or for making wider correlations between different parts of the world. We use the shallow-marine succession in San Francisco Bay and a 'sinkhole' succession from Kaua'i, Hawaii, as examples. Each has its own distinctive indigenous fauna and flora, but by the 19th century both regions show strong signals of neobiota. The presence of such distinctive patterns, the result of various forms of accelerating human impact on Earth, means that the biostratigraphic potential of the Anthropocene, despite its very short timescale to date, may be at least as great as that of earlier intervals within the Quaternary.

3.3.3.1 San Francisco Bay

The San Francisco Bay Delta area is one of the most invaded aquatic ecosystems in the world. Cohen et al. (1998) identified some 234 neobiotic species, including invertebrates, vertebrates and vascular plants (see Figure 3.3.2). They noted an accelerating trend of invasions over 145 years, about half occurring since 1960. Recent studies of shallow subtidal settings in the bay have identified soft sediment communities numerically dominated by these invasive species (Jimenez & Ruiz 2016). The presence of so many neobiotic species in the bay suggests that a high-resolution biostratigraphy might be developed from their fossil/sedimentary record, not least because many of the species bear hard skeletal structures (Figure 3.3.2). Sediment cores with Pleistocene and Holocene microfossils are well known from the bay (e.g., Sloan 1992; McGann et al. 2002) and provide a comparison with the natural, pre-human-influenced coastal assemblages.

Here we focus on some of the molluscs with hard, mineralised shells that have a high potential to be preserved in the sedimentary record of the bay (Figure 3.3.2) and that are widespread species, although we note that foraminifera, such as *T. hadai* mentioned above, are also biostratigraphically significant. We suggest that a biozonation of the bay succession could identify a pre-neobiota interval characterised by indigenous species that continued into the mid-19th century. Succeeding this is an interval of early neobiota invasions (see also Cohen et al. 1998, figure 1) that could be characterised by the first appearance of the oyster *Crassostrea virginica* – potentially defining (the base of) a *Crassostrea virginica* Biozone – a species that was introduced to the bay in 1869 CE for aquaculture. This bivalve is a native to the Atlantic coast of North America, and its introduction here and elsewhere is associated with the development of oyster fisheries. *C. virginica* was also introduced into the estuaries of the Hawaiian Islands at about the same time (1866 CE), hence providing a possible biostratigraphic connection with a remote island succession in the mid-Pacific (Figure 3.3.3). Nevertheless, the introduction of *C. virginica* to the San Francisco Bay was ultimately unsuccessful, and so its palaeontological record – if any – would likely define a narrow interval only into the early 20th century.

This early phase of neobiota invasions into the bay is followed by pan-Pacific connections between North America and East Asia, signalled by, for example, the arrival of the Pacific oyster *Crassostrea gigas* – which

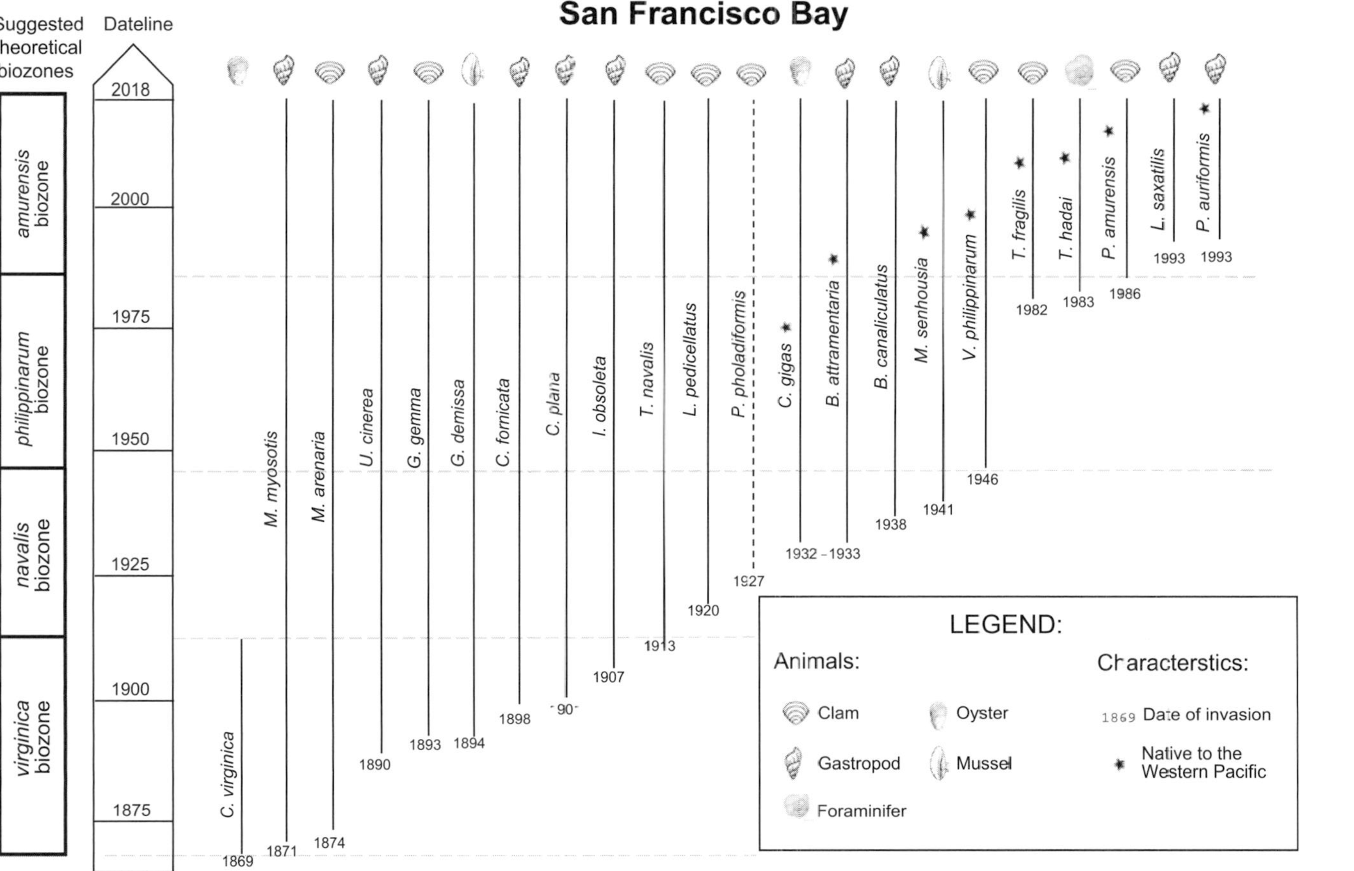

Figure 3.3.2 A theoretical biostratigraphy of introduced 'shell-bearing' species in San Francisco Bay, based on the chronology of their appearance. Species with recalcitrant skeletons, such as bivalves that occur in high densities, are likely to record the best biostratigraphic signal. Patterns depicted here require ground truthing in the sedimentary succession of the bay itself to identify which taxa may leave a lasting geological signal of anthropogenic impact. Ranges compiled from a variety of sources, including Barrett (1963), Carlton (1979, 1992), Carlton et al. (1990), Carlton and Cohen (1998) and Fofonoff et al. (2017). The introduction of *C. virginica* for aquaculture was ultimately unsuccessful. Ranges of other taxa depicted through to present are sometimes based on time-intermittent observation of the species in the bay. The timing of first introduction for several of the species figured here differs in other embayments along the California coast.

is native to Japan – into the bay about 1932 CE, and by the introduction of the East Asian bivalve *Venerupis philippinarum*, first recognised in the San Francisco Bay in 1946 CE (Cohen 2011), though there are earlier records elsewhere on the Pacific coast of North America. Locally in the bay this species occurs in abundances of 2,000 individuals/m^2. It is a widely introduced species, occurring from Hawaii (introduced in the 1880s, established by 1918 CE) to Europe, and might be used to designate a *Venerupis philippinarum* Biozone in the bay that also signals growing globalisation and pan-Pacific trade links. A later stage of colonisation from East Asia is typified by the Amur River clam *Potamocorbula amurensis* (Figure 3.3.2), which was introduced into the bay in 1986 CE. Depending on its sedimentary record, this might also be used as basis for a biozone. *P. amurensis* originated in the northern Pacific, where it is a native to the waters of China, Siberia, Korea and Japan. Although a relatively small species about 2.5 cm across, it occurs in huge numbers in muddy substrates within the bay: up to 48,000 individuals have been recorded per m^2 of sediment. It was first discovered in Suisun Bay (Northern California) in 1986 CE but has spread widely and rapidly. It was likely introduced through ballast water discharged from a ship (Cohen 2011) and has caused wholesale change to ecosystems.

3.3.3.2 Kaua'i, Hawaii

The fauna and flora of remote islands are usually highly distinctive, being subject to long-lived geographical isolation that has fostered allopatric (isolated geographical) speciation. Species on small islands often have narrow geographical ranges, and sedimentary successions on such islands record an indigenous fossil fauna and flora. As a result, remote island successions are particularly instructive in showing a diversity of local biostratigraphic signatures that, latterly, records the profound influence of humans introducing a direct and indirect cargo of non-indigenous organisms. Understanding island biostratigraphies is therefore instructive for quantifying the geographical extension of humans to remote geographical areas. These changes can potentially be correlated more widely, via the introduction of widespread terrestrial neobiota such as rats or edible marine molluscs such as oysters.

The sedimentary successions of the Maha'ulepu sinkhole and the caves of Kaua'i, Hawaii (Burney et al. 2001), provide a good example of where the extirpation of geographically local taxa can be combined with signals of global change, in this case the introduction of rats and other fauna. The sinkhole succession shows a floral and faunal record over the past 9.5 yrs that accumulated in a freshwater to brackish lake, with periodic marine incursions, which includes diatoms, land snails and bivalves. The diatoms, coupled with multiple ^{14}C dates, provide a detailed stratigraphy for the sinkhole and cave succession that records step changes in anthropogenic influence through Polynesian and European ingressions to the island.

The first evidence of human impact in the Maha'ulepu sinkhole is from an introduced Pacific rat dated to between 1039 and 1241 CE (Burney et al. 2001). Significant changes in land-snail species recorded through the succession indicate the impact of firstly Polynesian and latterly European colonisers and their associated neobiota. Declines in several species coincide with Polynesian activity, whilst all indigenous species of snails, bar *Cookeconcha* cf. *psaucicostrata*, became extinct during the 19th and early 20th centuries. In the 20th century, *Lissachatina fulica* (in 1936 CE) – the giant African snail – and *Euglandina rosea* (in 1957 CE) – the cannibal snail of Central America – were introduced. Anecdotal evidence suggests the extinction of the last indigenous snail population of *Cookeconcha* cf. *psaucicostrata* during the 20th century. These changes provide a detailed and unfolding biostratigraphic record of human impact on the local biota that is supported by diatom, bird, pollen and other fossil data (Burnley et al. 2001). The presence of bivalves such as *C. virginica* in the marine waters around Oahu and of the land-based giant African snail suggest the possibility of intercontinental correlation (Figure 3.3.3).

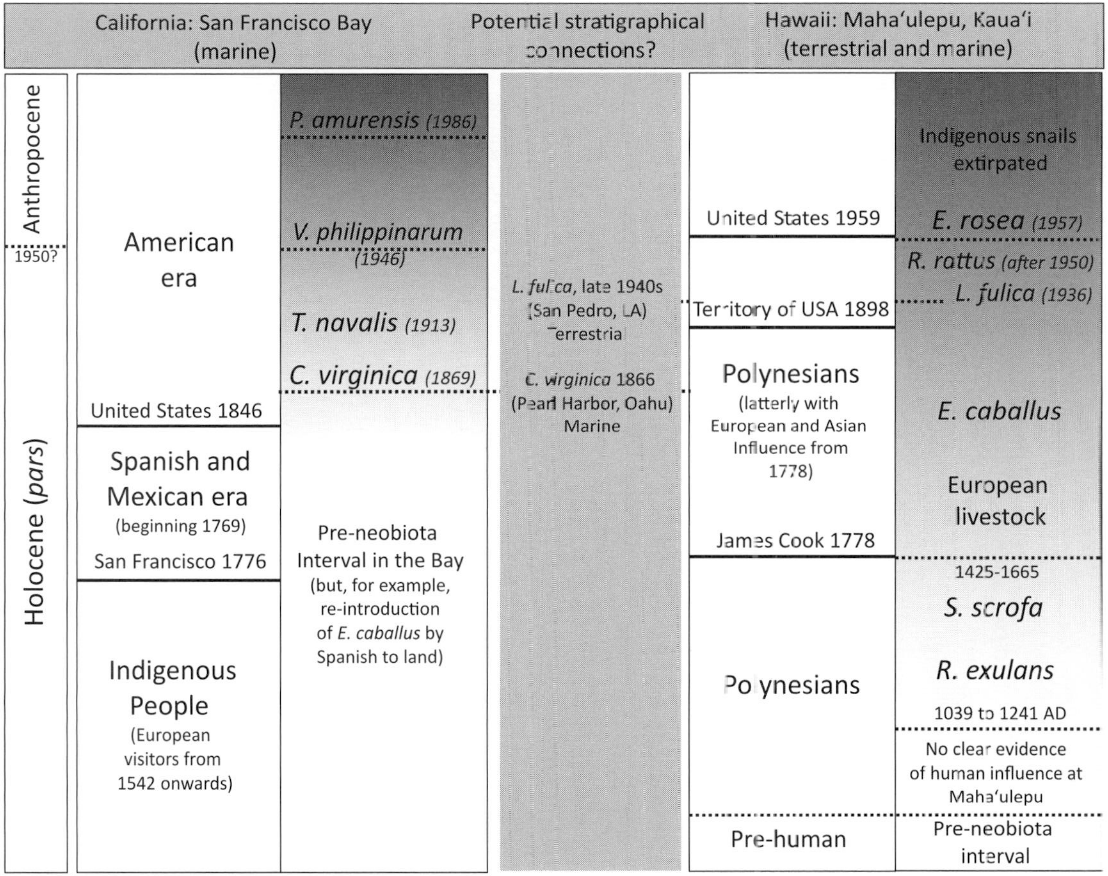

California: San Francisco Bay (marine)
Potential stratigraphical connections?
Hawaii: Maha'ulepu, Kaua'i (terrestrial and marine)

Anthropocene
1950?
Holocene (*pars*)

American era
United States 1846
Spanish and Mexican era (beginning 1769)
San Francisco 1776
Indigenous People (European visitors from 1542 onwards)

P. amurensis (1986)
V. philippinarum (1946)
T. navalis (1913)
C. virginica (1869)
Pre-neobiota Interval in the Bay (but, for example, re-introduction of *E. caballus* by Spanish to land)

L. fulica, late 1940s (San Pedro, LA) Terrestrial
C. virginica 1866 (Pearl Harbor, Oahu) Marine

United States 1959
Territory of USA 1898
Polynesians (latterly with European and Asian Influence from 1778)
James Cook 1778
Polynesians
Pre-human

Indigenous snails extirpated
E. rosea (1957)
R. rattus (after 1950)
L. fulica (1936)
E. caballus
European livestock
1425-1665
S. scrofa
R. exulans
1039 to 1241 AD
No clear evidence of human influence at Maha'ulepu
Pre-neobiota interval

Figure 3.3.3 Selected mollusc species introductions into San Francisco Bay (compiled from Carlton et al. 1990; Cohen 2004, 2011; Fofonoff et al. 2017; Committee on Nonnative Oysters in the Chesapeake Bay, National Research Council 2004) and terrestrial invasive species in the Maha'ulepu sinkhole succession of Kaua'i, Hawaii (extracted from Burney et al. 2001). Cultural human changes are indicated in the left-hand column for both regions and neobiota in the right-hand column. Theoretical biostratigraphic connections between San Francisco Bay and Hawaii, some 4,000 km away, are suggested.

3.4 Using the State of Reefs for Anthropocene Stratigraphy: An Ecostratigraphic Approach

Reinhold Leinfelder

3.4.1 General Considerations

Throughout Earth history, reef organisms have had only minor potential for biostratigraphic use, with exceptions such as some Paleozoic rugose corals and several Cretaceous reef-building rudist bivalves. However, reefs as ecosystems have shown pronounced evolutionary changes as well as marked adaptation to distinct but changing ecological reef settings, and they hence were prone to reflect large-scale synchronous global impacts as well as regional ecological changes.

Therefore, extensive episodes of global reef decline during major Earth-history global extinction events, followed by millions of years of reef recovery times, provide excellent global correlation markers. Although the causes for these global extinctions are not all fully understood, elevated sea surface temperatures (SSTs) combined with a strong lowering of marine seawater alkalinity and sometimes turnover of oceans into an oxygen-deficient state appear to be associated with most, if not all, extinction events. In most cases extensive volcanism (at times perhaps triggering additional methane clathrate degassing and blowouts from ocean sediments) was related to the extinction events, with additionally an asteroid impact being substantiated for the Cretaceous/Paleogene boundary. Reef recovery rates were very slow, of a minimum duration of a few million years and a maximum of 140 million years in coral-rich, open-water tropical reefs, extending from the mid–Late Devonian extinction to the Late Triassic onset of shallow-water scleractinian reefs (Figure 3.4.1). Other reef types, such as richthofeniid reefs, muddy reefs with a low-diversity fauna of corals or deeper-water mounds rich in phylloid algae, siliceous sponges, calcisponges and bryozoans continued to grow during this extensive gap. This might also be important for reassessing reef growth during the recent episode of Anthropocene reef crisis (Stanley 2001a, b).[1]

Less often used for stratigraphic purposes are shorter-term changes in the growth episodes, architecture and community structure of reefs, as controlled by sea-level change. Reef expansion episodes are most typically associated with relative sea-level rises. During times of high sea level, reefs may be covered by other sediments, while when sea levels are low, reefs are often exposed subaerially and cease to grow. Such dynamic patterns can be highly significant for deciphering sea-level changes lasting ~0.5–3 Ma or less, governing available space for reef growth, wave energy level and the influx of siliciclastic sediments and nutrients from terrestrial areas. The last of these may result in regional shallow-water oxygen depletion and cessation of reef growth. This relationship has made reefs an important ecostratigraphic tool for constraining sea-level history in ancient tropical marine environments, especially during the Mesozoic, where the new scleractinian coral reefs thrived in a great variety of settings next to other reef types such as siliceous sponge reefs and microbial reefs, each having its distinct environmental framing. The dependence of reef organisms and reef types on particular environmental settings does not preclude using them for stratigraphic correlation, so long as there is a good appreciation of controls on reef parameters and how to model them in an ecostratigraphic context (Hofmann 1981; Sokolov 1986; Oloriz et al. 1995). Consequently, ancient reefs have been used to constrain global sea-level histories with generally good results. Their geometric growth patterns,

[1] For overviews of reef evolution and extinctions through Earth history, see Wood, (1999), Leinfelder and Nose (1999), Stanley (2001a, b) and Veron (1995, 2008); for global extinction events, see also Buggisch (1991), Copper (2001), Flügel and Senowbari-Daryan (2001), Hautmann (2012) and Clarkson et al. (2015).

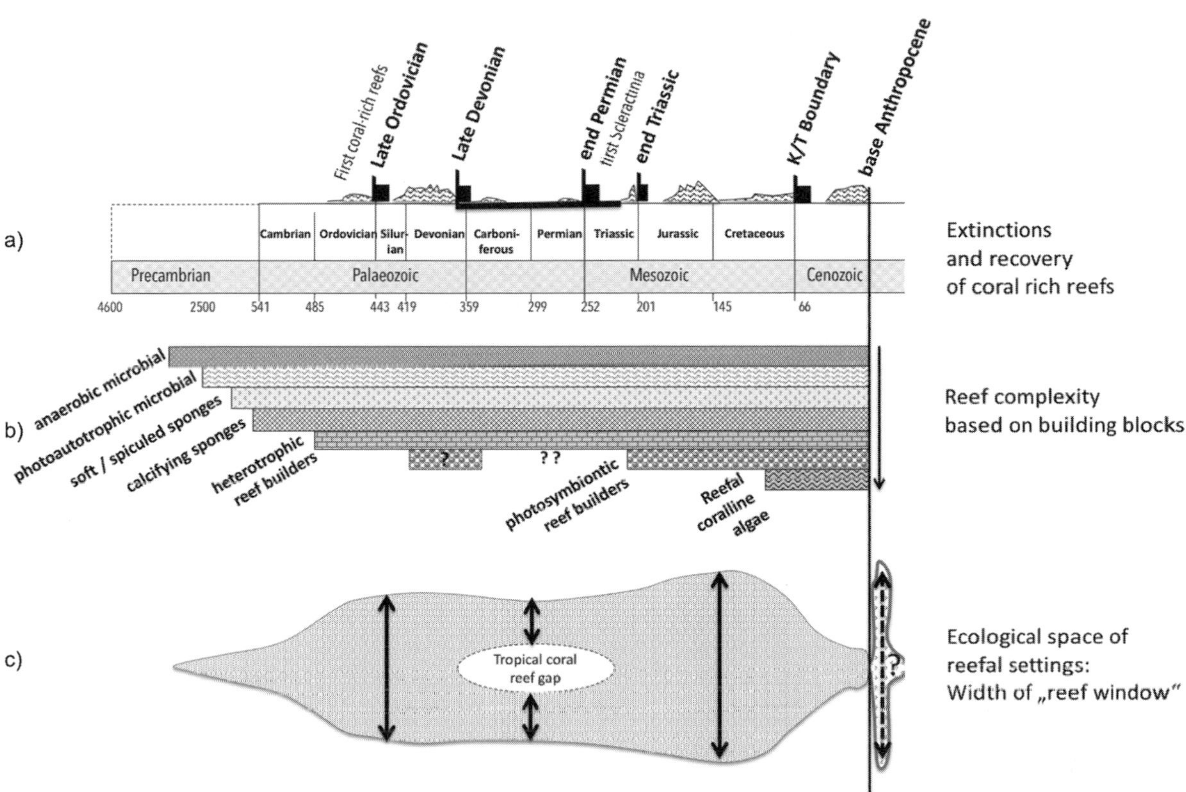

Figure 3.4.1 The past, present and possible future of reefs.
(a) Coral-rich reefs through time. Growth episodes (highs), major Earth-history extinction events (vertical labelled lines), reef recovery times (black rectangles); bold horizontal line is the 140-million-year-long gap of tropical, coral rich reefs. Numbers are millions of years. Precambrian not to scale. Modified after Veron (2008).
(b) The concept of increasing overall reef complexity, based on subsequent addition of reefal building blocks, which persisted throughout Earth history. Note that modern tropical coral reefs still contain microbial crusts within reef caves and include demosponges as well as calcifying sclerosponges (based on Leinfelder & Nose 1999).
(c) After a number of changes from the Proterozoic (stromatolite reefs) to the Mid-Paleozoic (possibly already including photosymbiontic reef organisms, chiefly stromatoporoids and some tabulate corals) and the long gap of tropical coral-rich reefs from the latest Devonian to the Late Triassic, the width of reef windows, as defined as the maximum available space of reef settings, became probably largest during the Jurassic, with microbialite reefs and a great variety of different siliceous sponge reefs and tropical coral reefs co-occurring. After optimising the photosymbiontic system and with the onset of coralline algae conquering high-energy settings, modern tropical coral reefs are in a much narrower, more superoligotrophic window than, e.g., Jurassic coral reefs. For the Anthropocene, a best-case scenario is shown, with reef compositions changing and relic reef types that are adapted to higher nutrients, more runoff, elevated temperatures and reduced alkalinity reoccurring in a volatile fashion. The widening of the Anthropocene reef window (which is not reflecting size and frequency of reefs) corresponds to the ecostratigraphic reef episodes (1–5b) as outlined in Figure 3.4.3. Modified from Leinfelder et al. (2012).

composition, indicators for shifts in terrigenous runoff or even oxygen concentration, can be used to establish different patterns of reef growth. This in turn allows the detection of sea-level fluctuations and enables their use for chronostratigraphic purposes (e.g., Sarg 1988; Schlager 1992; Leinfelder 1997, 2001; Leinfelder & Wilson 1998; Leinfelder & Schmid 2000; Leinfelder et al. 2002).

It is only in Cenozoic reefs that direct proxies for sea surface temperature (SST) can be derived from oxygen and Ca/Sr isotopes (Figure 3.4.2) and seawater alkalinity from boron isotopes (for pH) (see Section 5.3). Such proxies are recorded in the annual skeletal growth bands, which can be either directly dated by ^{14}C measurements, dated by backward counting of growth rings, or correlated via characteristic waxing-waning sets of growth rings, as done with tree rings. SSTs appear to be recorded with a mostly negligible vital effect on isotopic composition within the aragonitic skeleton. Nevertheless, there are restrictions: oxygen isotopes are particularly dependent on salinity and might also need other corrections, such as the amount of isotopically light water being bound in polar ice caps, while the possible effects of diagenesis also need to be taken into account. For coral skeletons, stable isotopes are the key instruments for identifying global or regional shifts in seawater temperature for the Cenozoic, chiefly in the Neogene, and may also be used as stratigraphic tools (e.g., Fairbanks & Matthews 1978; Chappell & Shackleton 1986; Ahmad et al. 2011; Zinke et al. 2014; Tierney et al. 2015; DeLong et al. 2016).

Although most present tropical reefs are dependent on photosymbiosis, letting them flourish within a narrow range of warm, superoligotrophic (with low nutrients and high dissolved oxygen) waters, there are still some modern reefs that, from their organic composition and environmental adaptations, resemble geologically much older reefs and hence are termed 'atavistic' by Leinfelder et al. (2012). Such

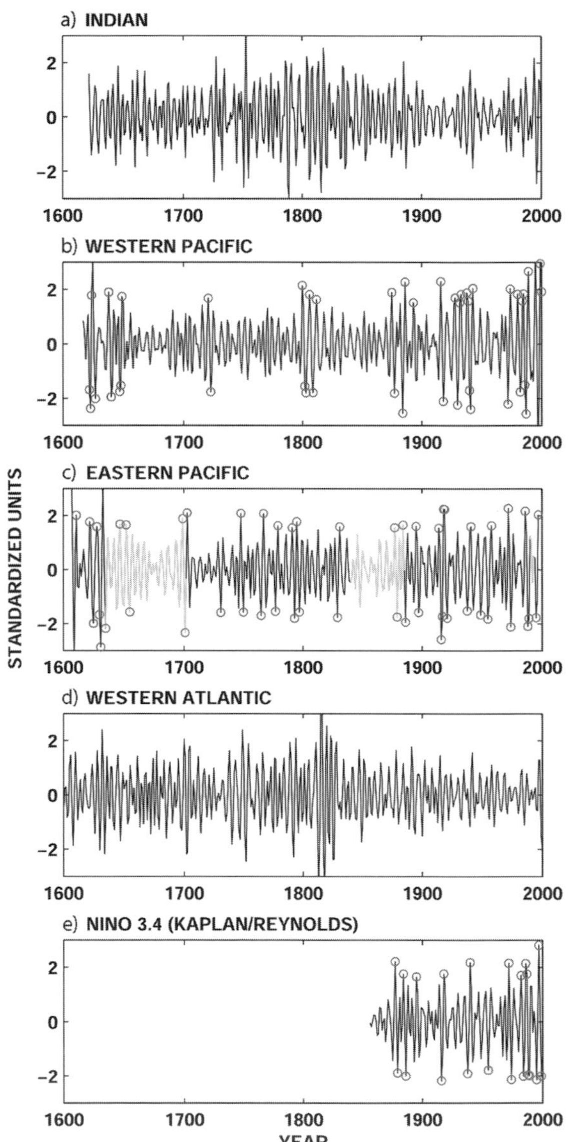

Figure 3.4.2 Comparison of SST records between the years 1600 and 2000 CE as reconstructed from coral skeletons (mostly using $^{16}O/^{18}O$ isotope and Ca/Sr ratios) from different tropical reefal settings, allowing correlation of SSTs across oceans. Shown are averages within the ENSO (three to seven) year band, (a)–(d); (e) shows a comparison with instrumental data. (from Tierney et al. 2015)

reefs are composed of more robust, more resilient, more adaptable or even 'Lazarus'-type relic taxa, which allow them to thrive in high-sediment and -nutrient settings, deeper waters and generally unstable environmental regimes. Atavistic reefs and Lazarus-reef taxa can be expected to become more common as the Anthropocene progresses (see Section 3.4.3).

3.4.2 The Present Situation of Tropical Reefs with Regard to Stratigraphic Correlation

Despite today's narrow superoligotrophic reef window, Quaternary tropical coral reefs are basically resilient ecosystems that can tolerate or easily recover from occasional temporary disturbances. Damage caused by a tropical cyclone or coral bleaching due to temperature peaks is reparable if sufficient regeneration time is available before a new disturbance occurs, so long as the overall ecological conditions for reef growth are in order. As in the grazing of savannahs, an occasional storm may even foster biodiversity by providing fresh substrata to larvae of slower-growing corals. Other corals, such as some acroporoids, even use fracturing for vegetative reproduction and more extensive dissemination, with coral branch fragments becoming cemented to the ground by surviving polyps and restarting colonial growth. Such natural resilience is rapidly diminishing in most recent coral reefs. Overfishing, water pollution (owing to runoff of soil, fertilisers, other chemicals, plastic particles, etc.), global warming and the decreasing pH of higher-latitude tropical marine waters have dramatically increased the vulnerability of reefs (e.g., Wells & Hanna 1992; Hughes et al. 2003, 2017a; Hoegh-Guldberg 2007; Heiss & Leinfelder 2008; Burke et al. 2011; Schoepf et al. 2015). Hence, natural perturbations, such as El Niño–type temperature peaks (discussed in Section 6.1), have stronger and much more widespread ecological effects. It is highly probable that the increase in the frequency and intensity of both high-temperature and storm events is caused by the anthropogenic rise of atmospheric CO_2 (e.g., Reed et al. 2015; GFDL 2017). In addition, rainfall associated with tropical cyclones may result in pulses of strongly increased influx of soil particles, nutrients and pesticides from agriculture in the hinterland, as frequently happens on the Great Barrier Reef (Brodie et al. 2013). This results in eutrophication peaks in the oligotrophic tropical reef ecosystem, leading to blooms of planktic algae and severe overgrowth of corals by soft filamentous benthic algae. Owing to the overall weakening of the reef system by these factors, corals are more vulnerable to natural diseases. Stressed coral reefs have difficulty in recovering, especially if individual events such as hurricanes or bleaching events coupled with temperature peaks occur too often. In a stratigraphic context, ecological perturbation events in coral reefs can affect vast areas and occur penecontemporaneously in different regions or even globally across the entire tropical zone, and hence they should be stratigraphically correlatable.

After severe bleaching events in 1997/1998, 2002 and partly in 2010, 2016 was one of the strongest bleaching events in the Great Barrier Reef (GBR). About 90% of the surveyed 1,156 individual reefs of the GBR complex were affected, and >60% of all corals were bleached. There was no recovery time, because another devastating bleaching event took place in 2017, possibly representing the most pronounced coral bleaching in the history of the GBR. It is reported that about 70% of the shallow-water corals around the tourist town of Port Douglas died, with similar values around Cairns and Townsville (GBRMPA 2017; Hughes et al. 2017b). In addition, the devastating effects of Cyclone Debbie, which swept across Australia in March 2017, not only smashed many reef regions into rubble but once again swept mud, fertilisers and pollutants into the reef regions (Robertson 2017). Hence, as with earlier bleaching events, especially the 1997/1998 event, the 2016 bleaching event can be correlated not only across the major part of the GBR (e.g., Cantin & Lough

2014) but also across the Indian Ocean, other parts of the Pacific and the Caribbean. There are many studies on how bleaching events are recorded not only in the SST-recording isotopic oxygen signal but also in growth patterns or isotopic proxies of reduced calcification (e.g., Pereira et al. 2015; DeLong et al. 2016; DeCarlo & Cohen 2017), but using this for interregional to global time-slice correlations based on coral skeleton characteristics and proxies is still in its infancy (e.g., Neukom et al. 2014; Tierney et al. 2015; Abram et al. 2016; Figure 3.4.2 herein).

3.4.3 The Application of Reef Stratigraphy: Ecostratigraphic Scenarios for the Anthropocene

3.4.3.1 Ecostratigraphic Scenario 1: Correlating Reefs in Deteriorating Ecological Settings (Figure 3.4.3a)

Since stone corals are good recorders of SSTs through possessing annual skeletal growth bands and mirroring SSTs in equilibrium conditions via oxygen or Sr/Ca-isotopes, they can monitor rising SSTs as long as they are not killed by the heat (e.g., Ahmad et al. 2011; Tierney et al. 2015; DeLong et al. 2016). Despite recording only local temperatures, such measurements can be correlated using growth-ring counts or other characteristics, such as waxing/waning patterns of sets of growth rings or even ^{14}C ages. SST-peak-related bleaching events, as discussed above, can be correlated across the entire globe via combining SST proxies and micro-erosion events across reefs. Storms, if pronounced, especially as recovery times now are slow, should also be correlatable across wide areas of reefs via disruptive surfaces associated with a dominance of reef rubble within the reefs and lagoonal settings. Overfishing and eutrophication of reefs are resulting in a strong reduction of reef coral diversity, overgrowth by algal turfs (which can be fossilised as microbored surfaces), or macroborings by organisms such as sponges and bivalves. Interruptions of reef growth, caused by coastal runoff, may also be discernible and probably correlatable via cessation of reef growth in association with biological and physical erosion, as well as via coatings of terrigenous material.

If anthropogenic greenhouse gases continue to be emitted at the current rates, even strong attempts at local or regional reef management or reef protection will not help prevent highly diverse and structured coral reefs from disappearing, possibly as early as the mid-21st century. In such scenarios, ocean acidification spreading from higher latitudes towards the equator will prevent coral reefs 'escaping' poleward from rapidly rising tropical SSTs (e.g., Hoegh-Goldberg et al. 1999, 2007, 2009; Hughes et al. 2017a, b). 'Escape' to deeper, somewhat cooler waters would also be largely hindered by these being too turbid because of the increasing runoff of nutrients creating more plankton. Nevertheless, it is important to recognise that coral reefs have survived frequent warmings and coolings within the Pleistocene, some of which were rapid while others were not. Furthermore, it is clear that the same corals that populate the Great Barrier Reef also exist on the much warmer reefs around New Guinea. Hence we cannot predict with confidence that present-day reefs will die out completely as the oceans continue to warm. Given sufficient time, reef organisms may be able to adapt, although the centres and patterns of reef growth may well shift in time as the climate changes.

Relating this reactivity of reefs to an Anthropocene under 'business as usual' conditions (in terms of climate change and environmental pollution) would mean that ecological change of coral reefs relative to the earlier Holocene should, at least to some extent, be detectable from the 15th century onwards. Holocene reefs likely began to transform since Columbian times (Jackson 1997), especially due to the onset of intensive fishing, often resulting in local to regional overfishing, including that of sea turtles and sea mammals. These early changes may characterise the initial anthropogenic imprint on reefs, expressed in the form of reduced diversity and coral coverage and subsequently increasing in response to the effects of growth in population, world trade and industrialisation during the mid-20th century (see Section 7.5).

The suggested base of the Anthropocene around the mid-20th century is expected to be correlatable across coral reefs (see Section 7.8.4.2 for use of corals as a potential medium for the placement of a GSSP) by the following:

(1) Spikes of radioactive fallout from nuclear bomb tests preserved in coral skeletons, beginning in the early 1950s, peaking in the 1960s and continuing until the 1980s (Waters et al. 2016)
(2) Possibly also enrichment of Pb from industrial activity, since corals are able to accumulate heavy metals (e.g., Berry et al. 2013)
(3) An increase of plastic and other anthropogenic particles trapped in interstitial reef voids and cavities from the 1950s to present

Younger ecostratigraphic correlation within reefs might be possible by the following:

(1) The nearly complete disappearance of sea urchins owing to a pervasive infection in the Caribbean in the 1980s (Knowlton 2001)
(2) A strong decrease in coral-reef diversities since the late 1980s/early 1990s, especially with the strong reduction to near disappearance of Caribbean acroporoid corals, paralleled by an increase of filter feeders and boring organisms (e.g., Seemann et al. 2012; Lirmann et al. 2014)
(3) Correlation of the 1997/1998, 2002 and 2016 global bleaching events, using coral skeletons (see Figure 3.4.2)
(4) Correlation of the onset of invasive species, such as occurrences of teeth and other skeletal parts of the lion fish in the Caribbean in the early 21st century (Schofield 2009), possibly preservable in muddy reef lagoons

Despite many efforts to support coral reefs, there is general scientific agreement that under 'business as usual' conditions there will be a global demise or severe reduction of coral reefs somewhere between the middle and end of the 21st century (Figure 3.4.3a), providing another important biostratigraphic marker horizon. As pointed out above, Earth history is no comfort in this respect, because recovery following global shallow-water reef extinctions during the Phanerozoic took place over many millions of years (see Section 3.4.1).

3.4.3.2 Ecostratigraphic Scenario 2: Using 'Assisted' Coral-Reef Episodes of the Anthropocene

Some reef researchers see the evolutionary and epigenetic adaptivity of corals and other reef organisms as much larger than previously thought, believing that especially in a combined effort of (1) keeping atmospheric CO_2 below 450 ppm, (2) extensive management of reefs (in relation to overfishing, eutrophication, sediment runoff and other pollution as well as tourist or shipping damage) via marine parks, and (3) 'assisted' approaches to enlarge resilience of coral reefs, there might be progress towards enhancing coral-reef development during the future Anthropocene (see also Section 3.3). There are presently many studies in the new field of 'assisted adaptability' aiding the evolution of coral reefs. For example, recovery after mechanical reef damage, such as tropical storms or the collision of a ship on a reef, could be assisted by replanting cultured coral (for an overview see Ferse 2008), although costs would be extremely high on a larger scale, and long-term success has not yet been demonstrated. Some coral species are potentially better adapted to higher water temperatures (Schoepf et al. 2015) or minor acidification (Shamberger et al. 2014), but these cannot be transplanted into other regions so far, a restriction which is not fully understood yet. Some working groups experiment on 'assisted evolution' in order to breed resistant species (e.g., van Oppen et al. 2015), but despite some local success (e.g., Zayasu & Shinzato 2016), so far there have been several setbacks (Hughes et al. 2017a). Hence the exact temporal onset of a recovery phase through allowing and enhancing adaptation and dissemination of substitute corals is difficult to predict, especially since many factors have to be taken into account. Here, three aspects are considered:

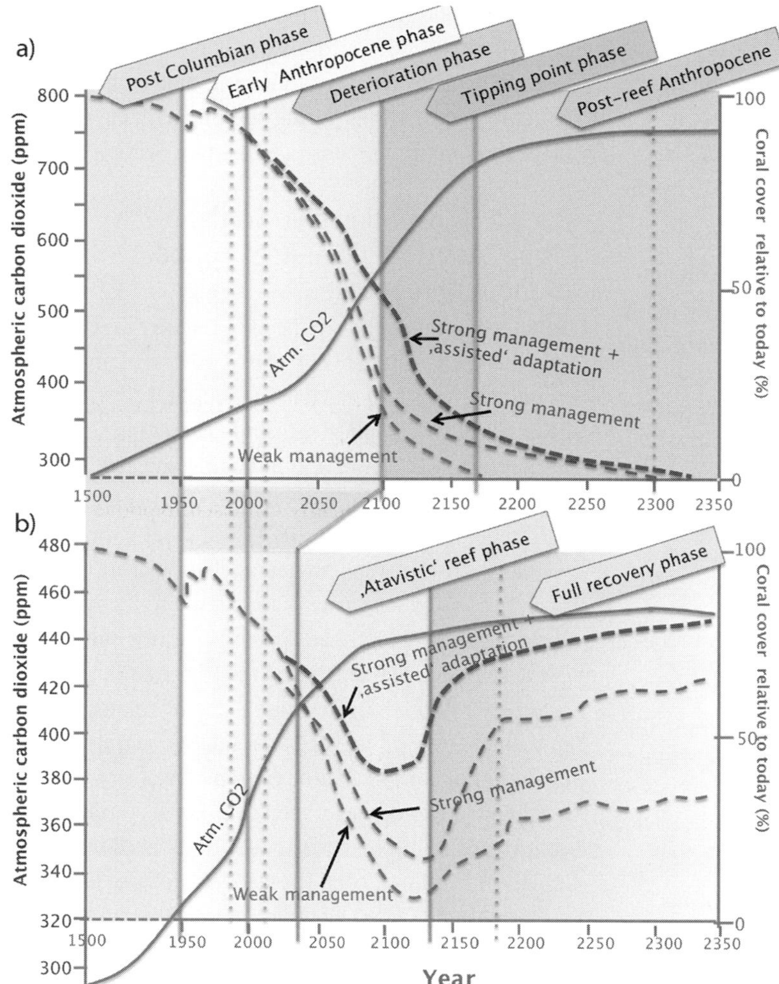

Figure 3.4.3 Use of Anthropocene coral reefs for ecostratigraphic purposes. Shown are two conceptual scenarios. (a) 'Business as usual' (BAU) scenario (relative to anthropogenic atmospheric CO_2 emissions); (b) an integrated CO_2-mitigation/reef-management/'assisted'-adaptation scenario. Based on two scenarios taken from Hoegh-Guldberg et al. (2009), as developed for the Coral Reef Triangle, adapted and extended. In the BAU scenario (a), where atmospheric CO_2 will keep rising, not even strong management (with or without 'assisted' adaptation) will help coral reefs. The combined mitigation/management/adaptation scenario (b) will also not see a rapid reversal of the negative trend, but after going through a 'vale of tears', it might eventually see the return of fully developed coral reefs.
Ecostratigraphic episodes:
a+b:
(1) Post-Colombian coral-reef phase ('pre-Anthropocene', ~1492–1950 CE)
(2) Early Anthropocene coral-reef phase (~1950 to 2000 CE), showing increasing, often punctuated decline of coral reefs. Further subdividable by ecocrisis events such as the near disappearance of *Diadema* sea urchins in the Caribbean or severe global bleaching events (dotted line).
(3) Deterioration phase (from 2000 CE onwards): accelerated decline of coral reef, with more frequent punctuation events (e.g., 2016 bleaching event, dotted line) and rapid deterioration of coral coverage and coral reef occurrences.

(1) **Assisting partial adaptation towards elevated SSTs and reduced alkalinity:** Hughes et al. (2017a) suggest that whereas scenarios for 'business as usual' on SSTs and shallow-water ocean acidification threaten living coral reefs with no chance of survival, the prognosis would be very different under a mitigation scenario. SSTs would still continue to rise slightly in the short term (2010–2039 CE), even if global emissions began to fall, but this would be only in the range of 0.32°C to 0.48°C. From 2039 to 2099 CE, SSTs would begin to stabilise, with temperatures changing, depending on different reef provinces, from +0.20°C to -0.05°C. Consequently, under this low-emission scenario, coral reefs would experience an increase of SSTs from 0.30°C to 0.68°C within the period from 2010 to the end of the 21st century. The entire range of temperature increase, including the past century, would still be near or above 1°C, which is considered to be critical to coral reefs, but at least some of them could continue to exist. There is evidence that the geographical range of tropical species does not contract towards the equator, owing to acidification, but rather some species expand towards the subtropics, fleeing increases in SST despite having to handle slight decreases in aragonite concentration (Pandolfi 2015; Poloczanska et al. 2016). After the mass bleaching event of 1997/1998 in the Maldives, the reefs changed their composition to some extent. The more temperature-resistant and robust scleractinian coral *Pavona varians* survived better than *Montipora* and spread rapidly at the cost of other species (Loch et al. 2002, 2007). Other corals, such as *Acropora hyacinthus* from the Pacific volcanic island Ofu, can withstand temperatures of up to 38°C by activating special genes (Barshis et al. 2013). However, heat-tolerant corals cannot simply be transferred from one site to another, as shown, e.g., by Polato et al. (2010) for larvae of the Caribbean star coral *Montastrea faveolata*. Yet the possible exchange of photosymbionts to more heat-tolerant types (e.g., Rohwer & Youle 2010), such as the recently discovered *Symbiodinium* C-type *S. thermophilium*, which can withstand up to 36°C as a symbiont (Hume et al. 2015), together with a much better understanding of comparative genomics of reef corals (e.g., Bhattacharya et al. 2016), gives some hope that the original, frequently criticised hypothesis of 'adaptive coral bleaching' might have at least partial applicability (Buddemeier & Fautin 1993; Buddemeier et al. 2004). Hughes et al. (2017a) suggest that provided a full integrated management of other stressors, such as overfishing and pollution, is introduced by using new integrated heuristic models, which include socioeconomic drivers, and provided coral reefs are allowed and assisted to change their composition, there may be a chance that functional reefs, albeit partly with a different set of corals, might persist.

(2) **Allowing and supporting an episode of atavistic reef growth:** The future of Anthropocene coral reefs relies on reducing emissions of CO_2 from fossil fuels and on new forms of governance, management and educational concepts (Leinfelder 2017) to limit the human impact on coral reefs.

Figure 3.4.3 (*cont.*) BAU scenario (a):
(4a) Tipping-point phase (~2100–2130 CE): episode of catastrophic tipping-point-type collapse of all coral reefs, characterised by dying reefs.
(5a) Post-reef Anthropocene (from about mid-22nd century onwards: largely coral-free Anthropocene oceans, characterised by biogenic and physical reef reworking.
Mitigation/management/adaptation scenario (b):
(4b) 'Atavistic' reef phase (~2030–2130 CE): this is the 'vale of tears'-adaptive episode of 'atavistic' and other atypical coral reefs (see text for explanation).
(5b) Full recovery phase (from ~mid-21st century onwards): renewed episode of large-scale oligotrophic Anthropocene coral reefs, latest from ~2200 CE onwards (dotted line). (A black-and-white version of this figure appears in some formats. For a colour version, please refer to the plate section.)

However, it may also rely on protecting and supporting exceptional, in part atavistic reef types that so far thrive only in places different from the classic modern oligotrophic reef settings and can grow under elevated sediment and nutrient influx, as do the Brazilian Abrolhos reefs (Leão 1982; Leinfelder & Leão 2000; Leão & Kikuchi 2001), the Iraqi reefs off the Shatt al Arab (Pohl et al. 2013) and the recently discovered muddy-water reefs off the Amazon mouth (Moura et al. 2016). In addition, many classical reefs in the Caribbean also transform into more nutrient- and sediment-tolerant low-diversity reef ecosystems, which also resemble earlier reef types from Earth history by being less rigid, more meadowlike and more adapted to elevated coastal-runoff nutrients and heat. This is shown by the disappearance of acroporoid, as well as massive corals, in favour of *Porites* and *Siderastrea* corals, soft sponges, large numbers of brittle stars and many other changes. In the Almirante Bay of the Caribbean of Panama, such a transition is recorded from the 1990s (Greb et al. 1996; Berry et al. 2013; Seemann 2013; Seemann et al. 2012. Even more unexpectedly, reef thickets and bioherms composed of glass sponges, adapted to elevated nutrients, which were thought to be extinct at least since the Cretaceous, have been rediscovered alive in the Pacific off British Columbia (Conway et al. 2001). Having been threatened by fisheries, they recently became fully protected (Johnson 2017), giving such truly atavistic reefs from the age of the dinosaurs a chance to thrive into the future Anthropocene.

(3) **Living with the 'vale of tears':** The well-known study on the Indonesian coral-reef triangle (Hoegh-Guldberg et al. 2009) also concludes that reefs will only have a chance if (a) strong management (including new protection and social-management schemes, as also outlined by Hughes et al. 2017a) is implemented and (b) emissions from fossil fuels stop around 2050. But even in this positive case, reefs would have to cross a critical 'vale of tears' phase (Figure 3.4.3b). Such specialised reefs would only have a chance to survive if allowed to adapt to the new conditions in a natural way or helped via human-assisted adaptation. Such reefs would be of lower diversity, having fewer ecological niches, and would probably be short-lived and patchy. They would most likely not play an important role in coastal protection and possibly only a reduced role in supporting fish stocks. However, after such a phase of volatility and adaptation, reefs might hopefully recover in the late 22nd or 23rd century, possibly forming large, structured reefs again, though different in taxon structure relative to Holocene ones.

Combining all these aspects and fusing them into an ecostratigraphic context, one can assume, under a mitigation/management/assisted-adaptation scenario, the following reef-growth-based, correlatable ecostratigraphic episodes.

(1) The present decline phase might persist till about 2030–2040 CE, to be followed by a transition stage to lower diversity, mixotrophic, managed reefs, which frequently change their characteristics. This means that coral reefs would continue to grow, albeit in a completely different form, with the high-diversity, stable communities retreating in favour of more volatile, new, short-lived types, with some just exchanging their key constructive elements, while many others would show atavistic features, such as being adapted to higher nutrient levels, sediment runoff, deeper settings, warmer SSTs or lower pH. All these reefs would have to be assisted and redesigned in various ways to assist them to go through a 'vale of tears' possibly for the next 100, if not 200, years (Hoegh-Guldberg et al. 2009; Leinfelder et al. 2012; Hughes et al. 2017a).
(2) With some hope, another extended recovery phase would set in not earlier than 2100 or 2200 CE, with coral reefs stabilising and re-diversifying again in clear, oligotrophic, tropical waters, returning to robust, resilient and long-lasting behaviour, but nevertheless with a new set of coral-reef ecologies. Such a change would provide another easily recognisable ecostratigraphic boundary (Figures 3.4.1 and 3.4.3b).

4 The Technosphere and Its Physical Stratigraphic Record

CONTENTS

We examine here the concept of the technosphere and its relationship to the Anthropocene, with emphasis on the human artefacts – especially plastics – that are a physical record of both the technosphere and Anthropocene. The technosphere is an autonomous global system, the Earth's newest 'sphere', which drives the Anthropocene through interplay of technological and human systems and structures. A science-based, physically-grounded approach to investigating the technosphere is outlined, consistent with human, social and technological forces. As examples, two scenarios exploring human prospects in the Anthropocene are developed, focusing on the global problem of technological acceleration, and the societal response to global warming. We show, too, how a stratigraphic signal of humans extending back in excess of 2 million years is evident from their artefacts, or technofossils, modern examples of which can help define an Anthropocene Series. Principal amongst these are objects made of plastic, the evolution, spread and preservation potential of which are significant to their providing high-resolution stratigraphic markers for the Anthropocene.

4.1 The Technosphere and Its Relation to the Anthropocene

Peter Haff

The technosphere is the autonomous global system that drives the Anthropocene. It comprises the world's humans and its technological systems, including transportation, communication, power transmission and financial networks; governments and their associated bureaucracies; military, educational and scientific establishments; religious institutions and political parties; and artistic, political, environmental, cultural and other social movements (Haff 2012, 2014a). The technosphere, a geologically young global system, is Earth's newest 'sphere', joining the four classical spheres of air (atmosphere), water (hydrosphere), rock (lithosphere) and life (biosphere) (Haff 2014a). A science-based, physically grounded approach to investigating the technosphere that is consistent with human, social and technological forces is outlined below. The aim is a non-anthropocentric description of the technosphere, built upon properties common to any dynamic system (taken here as a set of parts that dissipate energy as they collectively act to define the actions by which the system is recognised). Amongst those system properties is agency – the capacity to pursue a purpose. The language of purpose, when the latter is interpreted physically, provides a useful way to describe certain aspects of the dynamics of the technosphere (Haff 2012), including channelisation of the intentional behaviour of humans towards support of technospheric functionality. As examples of this approach, two scenarios exploring human prospects in the Anthropocene are developed. These focus on dynamical aspects, respectively, of the global problem of technological acceleration and of the social response to global warming. Dealing successfully with these challenges requires resolution of the contest between human intentionality and technospheric agency.

4.1.1 The Geological and Social Anthropocenes

There is more than one Anthropocene. Natural scientists have described an Anthropocene associated with multiple impacts inflicted by humans and technology on the Earth's land surface, its water bodies, its atmosphere and its biology (e.g., Crutzen & Stoermer 2000; Zalasiewicz et al. 2011b). These impacts and their consequences for the material Earth are a main point of interest for scientists who study phenomena of the 'geological Anthropocene'. The effort to make a case for the Anthropocene as a new, formal unit of geological time is directed mostly at technical questions regarding the nature of potential stratigraphic markers, such as their age, distribution, composition, texture, structure, preservation potential and uniqueness (Waters et al. 2016). The evidence and arguments needed to make such a determination deal largely with characteristics of the rock record and thus lie squarely within the geological Anthropocene.

Social scientists and humanists, on the other hand, have implicitly defined a 'social Anthropocene' that engages the conditions, motivations and histories of the world's people, including the role of politics, creative works and agency. A principal emphasis is on human actions and interactions and, in a time of rapid planetary change, a reconceptualisation of the human place in the world (e.g., Emmett & Lekan 2016). In the social Anthropocene, human intentions are treated as one of the important determinants of how the world unfolds, in contrast to the geological Anthropocene, where intentionality, when it is at issue, is usually relegated to boundary or initial conditions.

The presence of different foci for social and physical analyses sometimes leads to an implicit assumption that these two approaches to understanding the world can be carried out independently from one another. Often they can. However, such an assumption can lead to misinterpretations of Anthropocene processes. Thus,

Earth scientists working on the Anthropocene generally do not invoke intentional cause as an element of scientific explanation, regarding human behaviour as lying beyond their professional purview. Nonetheless, humans themselves are products of the Earth, so there is a case to be made for an interpretation of Earth System dynamics that incorporates from the outset the effects of human purposive activity (Haff 2014a, 2014b, 2016; Garrett 2014). Similarly, although an anthropocentric outlook often serves well for investigations of the social Anthropocene, the Earth origin of humans stands as a warning against too wide a separation between human-centred and physical conceptions of the world.

4.1.2 A New Earth Sphere: The Technosphere

The 'sphere' suffix calls attention to similarities between the technosphere and the classical Earth spheres – atmosphere, hydrosphere, lithosphere and biosphere (Haff 2014a). Like the other spheres, the technosphere is global and ubiquitous; its technological and human components carpet the Earth. A network of roads, paths, pipes, railways, airline routes and shipping lanes supports long-distance transport of mass and energy, reminiscent of the circulation of atmospheric and ocean currents. In emerging as a global phenomenon, the technosphere has joined the classical spheres to become an autonomous Earth system, operating without direct human control. Every sphere, including the technosphere, depends on resources and services supplied by sister spheres to maintain its mode of operation. However, the technosphere is currently overwhelming the ability of other spheres (Steffen et al. 2007) to meet its demand for raw materials and essential services such as waste recycling. This condition of overdependence puts a question mark on the technosphere – an indication of an uncertain future, not least for its human components. It also bears noting that the newest sphere shares with the older spheres a feature of critical importance to humans; namely, that without the resources that the technosphere makes available for their sustenance, a large fraction of the Anthropocene human population could not survive. For example, Erisman et al. (2008) estimate that about half the present world population depends for its existence on industrial-scale artificial nitrogen fixation (Haber-Bosch process). Technology – like air, water, soil and food – has become a necessity of human life.

4.1.3 Agency in the Anthropocene

Agency, or the expression of purpose, characterises the behaviour of the technosphere (Haff 2016). This is not only because its human parts are purposive entities but also, as discussed below, because the technosphere itself is endowed with intrinsic purposiveness. The physical basis of agency in living systems and its mechanisms of emergence have been studied from a scientific perspective by a number of authors, using concepts such as self-organisation and autocatalysis (e.g., Ulanowicz 1997; Juarrero 1999; Deacon 2012). The present chapter relies instead on a more primitive notion of agency deriving from generic or 'regulative' system rules that reflect the requirements of being organised. The resulting description of agency represents a snapshot in time; it does not address the processes by which agency emerges, only the status of regulative purpose in extant systems. Human intentionality remains of course an essential feature in the function of the technosphere and will be discussed further below, but it is not the teleological factor analysed here.

Looking at agency from a physical perspective forces attention on what might appear to be a fundamental difference between the social and the geological Anthropocenes. The contrast between the goal-directed nature of much of human behaviour – i.e., in its evident responsiveness to a 'final' cause (or purpose) – and the apparent lack of agency in inanimate objects like rocks seems to suggest that certain human actions have a distinguishing quality that lies beyond the usual 'efficient' cause-and-effect explanations common to science (e.g., as might be used in explaining the motion of a struck snooker ball; Falcon 2015). In order to move towards a more unified framing of causation, earlier work on the regulative properties of dynamical systems (Haff 2014b, 2016) is used here to construct a physical, non-anthropocentric picture of system behaviour that is general enough to include purposive processes.

A regulative property or effect is one that is characteristic of a system but is independent of system identity. It is distinguished from a constitutive property or effect that depends on the specifics of the system (i.e., that depends on its constitution). For example, the particular local changes in momentum that produce the forces that cause the parts of an isolated system to move in such-and-such a way are constitutive events, but the requirement that the system's total momentum remain constant (i.e., that it be conserved) is one of its regulative properties. In the same way, regulative agency is a generic system property but does not determine the specific ways in which the system expresses purposeful activity.

The physical basis of regulative agency in a dynamical system stems from the requirement that the system must behave in a way that allows it to survive, i.e., to maintain the collective or organised action of its parts (Haff 2016). If the system did not behave in this way, it would not be available for analysis. The expression of regulative agency by a system can be summarised by saying that the system acts as if it had a goal – in this case, acting as if it were trying to survive – notwithstanding that most systems, for example, a can opener or the technosphere itself, do not consciously take action or have intentions. The fundamental goal-oriented property common to every dynamical system can be called its intrinsic purposiveness or *intrinsic agency*. The corresponding *intrinsic purpose* of the technosphere is to behave in a way that allows it to survive. In terms of human interests, the intrinsic purpose of the technosphere is to maintain its own immediate welfare, not that of the world's human population.

Because a dynamical system expresses the collective action of its parts, the parts inherit their own type of regulative agency from the system, acting as if they were trying to support the system's intrinsic purpose. This corollary of a system's intrinsic agency is taken as the *functional agency* of its parts, on the basis of which they express functional purpose. The functional purpose of a hammer is not to hit the nail but to help the construction company of which the hammer is a part to survive. The functional purpose of the carpenter is not to use the hammer to hit the nail, although that may be his intention, but through his capacity as a component of the construction company to help the company achieve its intrinsic purpose.

A second corollary of intrinsic agency stems from the requirement that system components must actually pursue their functional purpose if the system is to survive. For a part to act in support of a system requires that it be physically able to do so. The system acts to ensure this ability by providing a suitable environment for the functional activity of its parts. It also provides constraints and, for its human components, incentives that help align the realisation of the functional purpose of a part with the system's own intrinsic purpose. The property of enabling and guiding the functional purpose of its parts is the system's *provisional agency*. For example, the construction company provides building materials for its employees, without which they could not do their job; it provides supervision for workers to ensure the work is done correctly and on schedule; and it provides pay sufficient to keep the workers on the job.

4.1.4 Aristotle in the Technosphere

Our own intentional actions and their consequences seem so clear to us (for example, in human contributions to development and use of technology) that it is easy to conclude that the technosphere is an expression of human intention. However, intention is not a regulative quantity but a constitutive mechanism through which (but not only through which) human functional purpose can be realised. Although the technosphere is in part a consequence of human intentionality, those intentional actions are highly conditioned; they tend to be channelled towards human functional purpose by provisional forces and incentives applied by the technosphere. Functional and provisional purpose work together to support a positive feedback loop that integrates the behaviours of parts, including intentional human parts, into the action of the whole; namely, the technosphere in pursuit of its intrinsic purpose. This loop can be called the *formal cause* of the behaviour of the parts of the system (e.g., Ulanowicz 1997), in analogy, through its form or pattern, to Aristotle's causal category of the same name (Falcon 2015). Pushing back the question of causation one step further to the origin of the formal feedback structure leads to the *final cause*, another of Aristotle's categories, represented here by the intrinsic purpose of the system. With respect to a construction company, for example, the final cause of the carpenter striking the nail is the company's intrinsic purpose, this purpose being to maintain its survival. From a larger-scale perspective, the final cause reflects the intrinsic agency of the technosphere. Neither formal nor final cause introduces new kinds of force into the technosphere. Rather, these terms serve as summaries of certain physical effects embodied in the collective behaviour of parts. Within this framework, humans, despite their special quality as intentional entities, are subject to the same regulative requirements stemming from technospheric agency as are technological artefacts and systems, underlining the lack of distinction in this analysis between the geological and the social Anthropocenes (and between artefacts and humans).

The next two sections offer examples of how adoption of a regulative approach to the purposive dynamics of the technosphere reframes the role of human intention from prime mover of technological change to a mechanism subject to technospheric agency. The first example considers how co-option of human intentions by the technosphere leads to the acceleration of technology, while the second looks at the prospects for human influence on the technosphere's future trajectory.

4.1.5 Co-option of Human Intention: Acceleration of the Technosphere

Human intentional activity is nudged by incentives towards actions that support the technosphere. Incentivisation can take many forms, for example as technology manifested in devices, methods and knowledge that make it easier, quicker and cheaper to accomplish with better results some desired or suggested goal. In other words, a principal incentive offered by the technosphere is efficiency, a product that today has become, in essence, a commodity. One result is a flood of physical products, like ballpoint pens and plastic bags. The subsequent surge in material waste constitutes a distinctive component of the stratigraphy of the Anthropocene (Zalasiewicz et al. 2014d).

When an increase in efficiency makes it possible to do once impossible things, like interact instantaneously with anyone in the world, whole new wants come into play, such as immersion in social media. Because the technosphere operates through provisional agency to turn human intention to its own uses, intention and efficiency reinforce each other in service to the technosphere. Efficiency is able to channel the direction of intentional activity because it optimises and eases the path towards what people want. For example, continuing increases in the efficiency of energy technology have made possible the realisation of more ways to use energy and of greater output per joule (e.g., Nordhaus 1996), trends that have in turn enabled the rising demand for power during the Anthropocene.

Similarly, when an increase in technological efficiency acts to make incentives for human intention to focus on creating and using products of even greater efficiency, notable in the exponential increase in computing power under Moore's law (e.g., Lundstrom 2003), then technology accelerates. Continued technological acceleration presages a fundamental problem for humans because human capacity to deal with events that unfold ever more rapidly is biologically limited. Under a continuously accelerating regime, the timescale for significant technological change will eventually become too short for humans to deal effectively with ensuing consequences, such as the emergence of unexpected and large-scale environmental and social disruptions. This effect is already in evidence. Global warming, discussed below, is an early example of the kind of disruptive phenomenon that can be expected in the future from an ever-faster technosphere. In the long run, neither humans nor a liveable environment can stand up to sustained technological acceleration. However, because the technosphere gives us what we think we want, its accelerated state is not yet high on the list of perceived challenges to human well-being (see, however, last chapter of West 2017).

4.1.6 Collective Intention in the Technosphere: The Social Thermostat

Other problems stemming from the behaviour of the technosphere are more widely recognised, such as its contribution to environmental pollution and global warming. One may imagine that, despite co-option, individual human intentions might somehow cohere to moderate such dangerous behaviour and that, if enough humans climb on-board, the collective force of human resolve might be able to turn the trajectory of the technosphere in a direction more favourable to future human well-being, perhaps even touching the brake on acceleration. But it is necessary to keep in mind that in this scenario there are two parties at the bargaining table, each with its own agenda – a collection of humans of similar outlook acting together as a system and the technosphere. Even granted the presence of a large number of humans intent on taming the technosphere, it is a question not only of what humans want but also of what the technosphere wants. However large the human movement may be, and especially if it *is* large, the resulting collective effort will have need of energy, materials, information and transportation, quantities that only the technosphere can provide at the required scale. These necessities will be supplied to the extent that the movement itself functions as a component of the technosphere by pursuing its functional purpose in support of the host system.

Still, being tethered to the technosphere does not imply that humans cannot influence its behaviour. Systems of sufficient complexity often contain internal negative feedback mechanisms that act to maintain critical dynamic variables at or near certain set-point values, as in the case of temperature in a thermostatically controlled office building. As part of the technosphere, the modern environmental movement, with its concerns about global warming, resembles an emerging social thermostat. However, control over the value of system variables like temperature is only part of the thermostat story. Conventional thermostatic-control mechanisms are parts of larger systems (for example, a company) with respect to which they have a functional purpose, realised in their support of system (corporate) metabolism. By providing a comfortable, temperature-controlled environment for office workers, a building's thermostat fulfils this purpose. The consequence of temperature control is increased total throughput by workers and thus greater use of energy and resources by the company that employs them. Enhancement of corporate energy and material flows that support its intrinsic purpose, not the comfort of its workers, is the primary reason that the company provides a temperature-controlled office environment.

The lesson of the office thermostat for the social thermostat is that regulating an environmental variable such as temperature is not the same thing as regulating a system's metabolic rate, i.e., its rate of energy dissipation. Where rates of metabolic processes are in fact closely controlled, as in many designed systems and in biological organisms, the controls act to ensure longevity sufficient to achieve utility of design or to allow organismal reproduction. The technosphere, however, was not designed, nor has it evolved under the selection pressure of Darwinian evolution. It has emerged possessing no global mechanism of metabolic regulation. Regulation of metabolism introduces the possibility of a new timescale into system dynamics – a lifetime – the time over which the system exists in a stable metabolic state. But without an intrinsic lifetime, i.e., lacking enforced set-point values for energy use, the technosphere acts only in the moment, without regard to the more distant future, necessarily biased towards increasing consumption of energy and materials. The system races ahead like a forest fire without much concern for its own longevity. With only one technosphere to work with, and thus lacking the corrective effect of multiple trials and errors such as organismal evolution provides in biological systems, imposing limitations on global energy use in a controlled (smooth) manner by a social movement is likely to be difficult.

However, energy use per se is not the problem. A trajectory of increasing energy use by the technosphere does not in itself imply that the Earth environment will become less habitable by humans: at a high-enough power level, any environmental outcome is physically possible. Piggybacking on increasing energy use by alignment of social action with the current momentum (and intrinsic purpose) of the technosphere may represent an easier path to environmental goals than working against that momentum by attempting to countervail the built-in tendency of the technosphere to increase its metabolic rate. If increased energy flows can accomplish environmental desiderata like recycling waste, including carbon, then, in return for a jacked-up metabolic rate, the technosphere might be able to provide, besides food, water, transportation and other necessities of civilisation, a more stable climate and a less polluted planet. A strategy of increased energy use by the technosphere to safeguard environmental resources is not a recommendation but an illustration of how a non-anthropocentric approach to technospheric dynamics can suggest otherwise counterintuitive possibilities for dealing with challenges of the Anthropocene.

4.1.7 The Two Anthropocenes Again

It seems evident that technology is shouldering itself into the world as the next new thing in a long line of planetary innovations, bringing along with it a powerful source of agency – the technosphere. The future to be experienced by humans depends on the degree to which they recognise and engage that agency. The standard cause-and-effect (efficient) framing of the geological Anthropocene is insufficient by itself to confront this future, because other forms of causation are in play. Similarly, despite its embrace of the role of human purpose, the social Anthropocene lacks the single-minded devotion to efficient cause necessary to penetrate the workings of the physical world or of technological systems. The formal-cause framework of a physically based teleological dynamics expands and merges these two perspectives, exposing the needs and demands of the technosphere to analytic and operative capabilities of both Anthropocenes. From this dynamical perspective, it seems clear that there are no purely technological answers to many of the challenges humans face in the Anthropocene; contributions from each culture are necessary in order for humans to engage the entire range of questions raised by the behaviour of the new world system. The first step along this road is the most critical – namely, for humans to recognise that the technosphere has agency, and that that agency is not the same as our own.

4.2 Technofossil Stratigraphy

Jan Zalasiewicz, Colin N. Waters, Mark Williams and Anthony Barnosky

In addition to using traditional palaeontological markers to help define an Anthropocene Series (for which see Chapter 3), a stratigraphic signal of humans is also evident from their artefacts, technofossils sensu Zalasiewicz et al. (2014d, 2016b). These also represent a physical record of the evolution of the technosphere. Technofossils have distinctive characteristics that mark them out as being different from trace fossils (see Section 3.1). For example, they are often fashioned from materials that either are rare in nature or do not occur naturally, like native aluminium or plastic. They are produced into a myriad array of forms that do not have parallels in nature, from a hand-worked stone implement, immediately discernible by its fracture patterns from a naturally eroded pebble, to a skyscraper (and its component parts), fashioned from steel, concrete, glass and plastic. Uniquely to nature, some technofossils may have no functionality beyond the symbolic. Technofossils are often crafted from materials that have long-term geological durability. For this reason it is possible to consider the development of a technostratigraphy, rather in the way that fossils are used to define biostratigraphy (Section 3.1), from the succession of technofossils in successively younger intervals of sedimentary strata (e.g., stone tools at the bottom, silicon chips and iPhones at the top). Although other hominins, great apes, birds and some molluscs are known to have used, or use tools, we here limit the term technofossil to structures made by species of *Homo*.

The earliest biostratigraphic marker of *Homo* is from Late Pliocene body fossils in Ethiopia that are dated to between 2.8 and 2.75 million years old (Villmoare et al. 2015; see Section 7.2 herein). A stratigraphic record of stone tools extends back prior to this, but such tools are likely the product of work by australopithecines. Stone was the principal medium of technofossils for more than two million years and continues to be a major component up to the present, used in everyday items from kitchen worktops to roads. Early *Homo* likely fashioned technofossils from a range of media, including wood, as suggested from the tool use of other great apes, but only the more recalcitrant stone structures and some bone have survived, being almost exclusively the technofossil record of *Homo* for the Late Pliocene Epoch, as well as the Gelasian and Calabrian ages of the Pleistocene Epoch (see Section 1.3 for the description of these chronostratigraphic epochs). This early stratigraphic record of technofossils is complex and has been thoroughly explored through archaeological analysis of material cultures. Specific artefact assemblages are associated with particular cultures and typically say more about societies or geographical context than about precise chronology. This is certainly the case for technofossils characteristic of cultural terms such as Palaeolithic, Oldowan, Acheulean, Mousterian, Aurignacean, or Bronze Age and Iron Age. Although they have a time-significance, such artefacts are often markedly time transgressive from one region to another.

Later in the Pleistocene, during the Ionian Stage (~781,000–126,000 years ago), new types of technofossil supplement the stone-tool record. These include artefacts associated with fire use that may extend back nearly half a million years, being more extensive from about 125,000 years ago (Section 7.3.2). Technofossils associated with fire include burnt bones, burnt stone tools, hearths and wood ash. These structures are important markers of the growing ability of humans to regulate the environment around them and, in the case of structures such as hearths, have been interpreted as indicating a sense of 'domestic' space (Karkanas et al. 2007 and references therein).

The diversity of technofossils increases during the terminal Tarantian Stage of the Pleistocene (~126,000–11,700 years ago) to include artefacts with an apparently abstract or aesthetic quality, such as pieces of ochre with carved patterns and shells used for beads, recovered in the Blombos Cave of southern Africa, dating from about 75,000 years ago (Henshilwood et al. 2002, 2009; d'Errico et al. 2005), engraved ostrich eggshells from Diepkloof, dated to over 55,000 years ago (Rigaud et al. 2006), the static artwork of cave paintings as far afield as Europe and Indonesia from about 40,000 years ago (Aubert et al. 2014), and the variety of portable art figurines made, for example, from mammoth ivory in Europe, before 30,000 years ago (Conrad 2003, 2009).

A widespread stratigraphic signature of technofossils evolves with the use of ceramics in the latest Pleistocene (see Section 2.2.3). The earliest ceramics are of figurines, dating to about 30,000 years ago in Europe (e.g., Farbstein et al. 2012 and references therein). These predate the earliest ceramic vessels, which are of East Asian occurrence and are associated with the processing of food. Pottery from Jiangxi Province, China, thought to have been used in cooking, is dated to 20,000 years ago (Wu et al. 2012), while in Japan, pottery of the incipient Jōmon Period dates to about 15,000 years ago (Craig et al. 2013). The use of ceramic pottery for storing and processing food evolved independently in many geographical regions across the world during the Late Pleistocene and Holocene, and these developments are highly time transgressive.

During the Holocene Epoch, the range of different media from which technofossils were constructed and the morphological diversity of these technofossils increased dramatically to include pure or alloyed metals that are rare or absent naturally (see Section 2.1), novel anthropogenic rocks including brick and concrete (see Sections 2.2.1 and 2.2.2), glass, and some early varieties of plastics (see Section 4.3; Figure 4.2.1). Brick technology, like stone-tool and ceramic technologies before it, developed in many places, with some of the earliest recognised uses being in the Near East and Indus Valley civilisations of the eighth and seventh millennia BCE. Concrete, too, has a long history of development from classical times to present, though its use in the 20th and 21st centuries dramatically accelerated.

Although initial developments of different technofossil types typically show time-transgressive stratigraphic patterns and therefore do not provide a degree of resolution for precise stratigraphic correlation worldwide, the types of technofossils that evolved in the 20th century may offer better utility for technostratigraphy and might be used to define interval biozones (see Section 3.1 and Figure 4.2.2). These include technofossil types that can be considered within a kind of taxonomy comparable to biological taxonomy. For instance, a generic type might be the ballpoint pen, first patented in 1888 but only developed for significant manufacture by the Biró brothers in 1943. Post-WWII, commercial production began for the public, and over the next few years, as designs rapidly evolved and prices dropped, true mass-market production began. A single ballpoint-pen species as Anthropocene indicator might be the Bic Crystal pen, first produced in 1950 and now with over a hundred billion individual pens sold (and subsequently discarded) (https://en.wikipedia.org/wiki/Bic_Cristal) – that is, averaging 200 for each square kilometre of Earth's surface, land and sea. Within the Bic Crystal 'species', subspecific variants might reflect the replacement of a stainless steel ball in 1961 by the more durable ball of tungsten carbide now used. Other Anthropocene indicator technofossils might be the plastic credit card and polyethylene bag, which both appeared commercially from ~1950. Slightly younger Anthropocene strata might be marked by the detachable ring-pull device on cans in 1963, which evolved into the attached ring pull in 1975, while compact discs might (from today's perspective) reflect an even later Anthropocene interval, first being released in 1982 (Zalasiewicz et al. 2016b). Such mass-market technofossils attained a global distribution almost instantaneously.

Given the capabilities of modern manufacture, the number of different 'families', 'genera', and 'species' of technofossil, although never properly tallied, certainly far outstrips current biological diversity and might now be comparable with the total

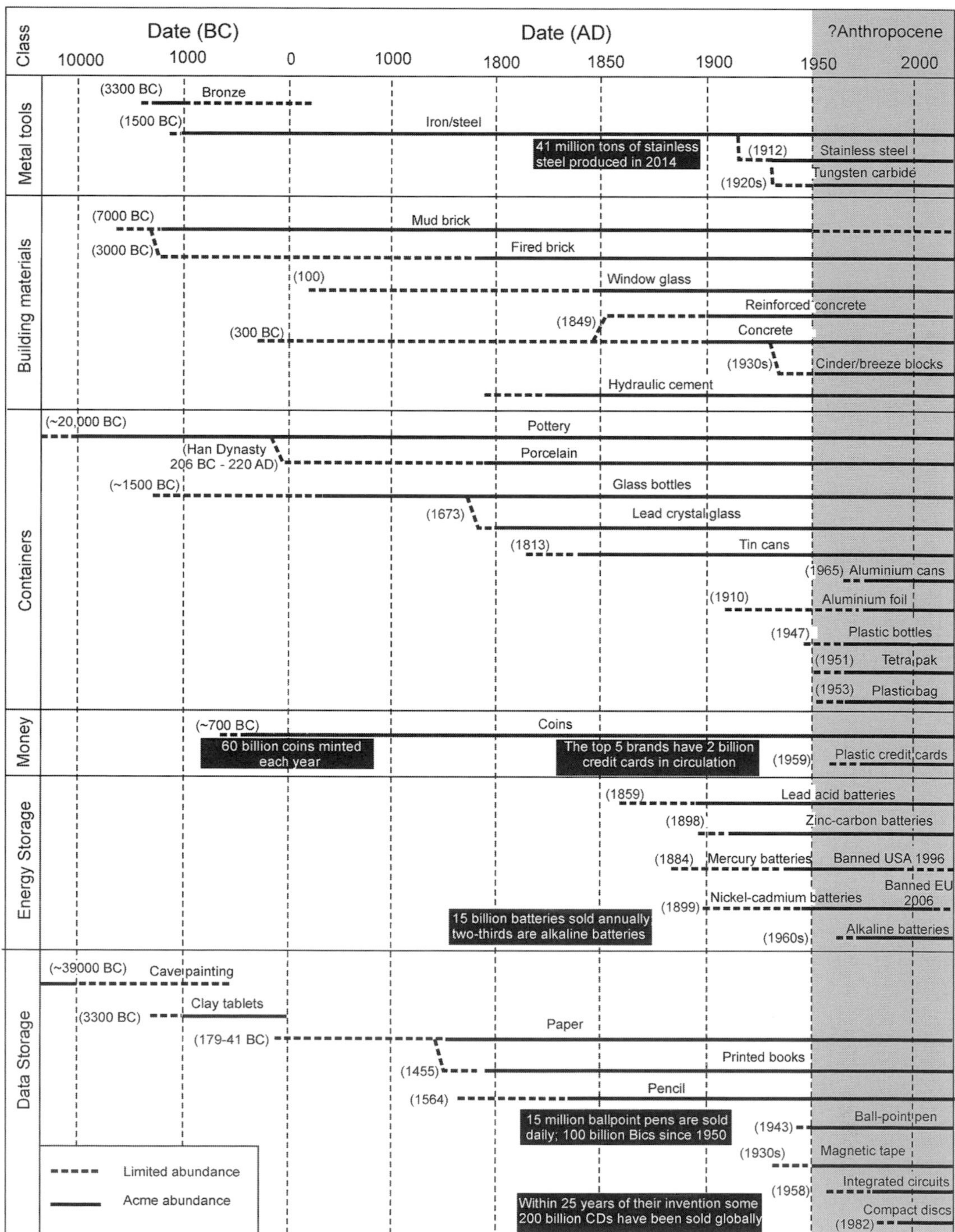

Figure 4.2.1 A stratigraphy of technofossil types from selected human cultural and technological innovations from Late Palaeolithic culture to present. (modified from figure 6 of Williams et al. 2016)

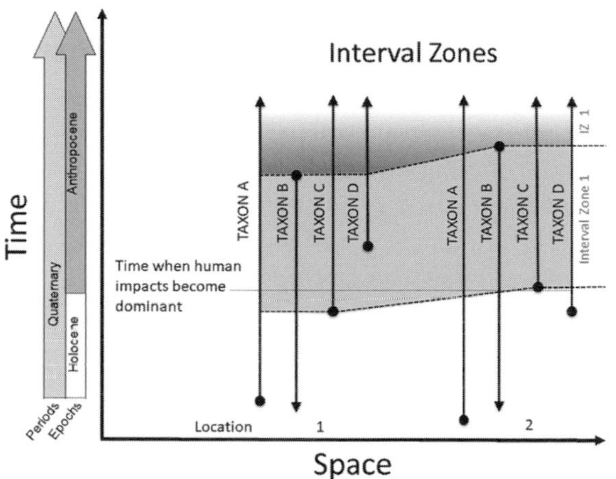

Figure 4.2.2 Interval zones (see Section 3.1) based on technofossils may be useful for defining the Anthropocene in the 20th century, based, for example, on various kinds of plastics (see Figure 4.3.2). In this figure 'taxon' would refer to a specific kind of technofossil rather than a biological species. Figure adapted from figure 7 of Barnosky (2014).

of biological diversity that has ever existed on Earth (Zalasiewicz et al. 2016b). Hence, the practical stratigraphic potential, even with just a small selection of some common 'taxa', is considerable (Figure 4.2.1) and has already been used in stratigraphic studies – for instance, in precise dating of deposits from extreme flood events by using entrained litter items, with, in this example, additional precision from preserved production dates (Hoffmann & Reicherter 2014).

The pace of technofossil evolution is currently quickening, as seen for example in smartphones. Thus, Apple's iPhone, first released in mid-2007, has gone through rapid evolution over a decade, with near-globally synchronous releases of the iPhone 5s and 5c in 2013 that included China. The eventual technostratigraphic signal of iPhones is more difficult to pin down, but many (millions) end up in landfills in China and Africa, leading to a widespread if discontinuous stratigraphic layer forming early in the 21st century. Attempts to improve the recyclability of these devices (for example, using Apple's recycling robot 'Liam') may reduce the stratigraphic impact of these devices but will postdate a widespread technostratigraphic signal from landfill sites dated to the earliest 21st century. One might therefore contemplate a technostratigraphic interval characterised by iPhones, with 2013 marking the globally synchronous occurrence of the iPhone 5.

4.3 The Stratigraphy of Plastics and Their Preservation in Geological Records

Reinhold Leinfelder and Juliana Assunção Ivar do Sul

Plastics are a striking invention, flourishing during the late 20th and early 21st centuries. They are light, robust and easy to produce and can be designed with bespoke properties: large, small, massive, super-thin, smooth, rough, flexible, stretchable, elastic, rigid, porous, dense, heat-resistant, durable, isolating, transparent, colourful and so on. We use them nearly everywhere: in the household, for recreation, during work, on travel, for tools and machines. This book would not have been written without plastic. Starting with the clock or smartphone waking us, the toothbrush, shower gel (often containing skin-cleansing microplastic particles), our shirt or blouse with microfibres, the coffee machine, the soles of our shoes and the stuff of our raincoats, the bus, train, or car that we take to work, the pens we make notes with, computers and their accessories, the credit cards we use to pay for all of these, and even the gum some of us might chew during the day – plastic is everywhere.

Most plastics we now use, at the end of their (often very brief) life cycles, accumulate in the environment, where they continue to be durable over long timescales. As plastics started to be produced and to accumulate at a globally significant scale after WWII, they became an important stratigraphic marker for characterising and even subdividing the Anthropocene. Their survival is enhanced by burial, as in landfills (cf. Tansel & Yildiz 2011), where they may become fossilised or reworked by future erosion.

4.3.1 Plastics as a Distinctive Technofossil Group

As noted in Section 4.2, technofossils show rapid technological evolution, and those made of plastics are no exception, contributing to their stratigraphic utility.

Organic technofossils: Plastics comprise organic matter engineered by humans: malleable solids made of high-molecular-weight organic polymers. They are durable, resembling naturally recalcitrant organic materials such as chitin or sporopollenin, and physically inert. But depending on their composition, especially with monomers such as softeners, they can be chemically active, releasing toxic gases and adsorbing other organic and inorganic molecules, including persistent organic pollutants (POPs) such as dioxins (e.g., Lithner et al. 2011) and metals such as cadmium and lead (Massos & Turner 2017).

Fossil fuel for manufacturing and energy: Many components of the physical technosphere (Haff 2014a; Section 4.1 herein) are built up by humans from inorganic resources, such as metal ores, rare earths, sand, clay, or limestone, which we mine, disaggregate, transport, refine and reassemble using high energy input, mostly from fossil fuels. By contrast, most plastics are entirely synthetic – primarily made from petrochemicals – although some are cellulose based, especially the filter fibres in cigarette ends. Fossil hydrocarbons, hence, provide the basic material for synthetic plastic and the energy to make it. Thus photosynthetic processes produce biomass, which is then converted to hydrocarbons by geological processes, which is subsequently reassembled into new organic matter in the form of plastic. Producing plastic to some extent thus copies natural processes and could be considered an early form of organic engineering. Today, about 8% of total annual oil and gas production is used for producing plastics (Thompson et al. 2009), showing the magnitude and omnipresence of plastics in our daily lives.

Ubiquitous presence: Plastics are a highly visible, globally ubiquitous indicator of the Anthropocene and provide us with a refined stratigraphic tool. They are an important component within the anthropogenic physical technosphere and are widely distributed in most other sediment types, being deposited by both natural (lakes, coastal sands, oceans) or engineered (soils, dammed lakes) processes. A modest proportion is burnt in incinerators (Geyer et al. 2017), adding to anthropogenic CO_2 production. Germany, sometimes labelled as 'recycling world champion',[1] reused 99% of all collected plastic waste in 2015; 46% was reused as working material, 53% was used for energy production. and only 1% was dumped (UBA 2016). In 2015, Europe overall used 39.5% of collected plastic waste for 'energy recovery', 29.7% was recycled, and 30.8% went into landfills (PlasticsEurope 2016). Globally, most plastic within the technosphere is deposited within waste dumps, often situated in elevated positions (for groundwater protection) or in other insecure settings. Such landfills are geologically relatively short lived and will be prone to future erosion and re-sedimentation (Zalasiewicz et al. 2016a; Geyer et al. 2017).

4.3.2 Production and Degradation of Plastic Materials: An Overview

The same features that make plastics indispensable to our daily lives (i.e., lightweight, cheap and resistant) are also responsible for their worldwide distribution in the environment. Plastic trajectories are extensive, with long pathways from the continent to the sea, from surface to deep waters, and from the sediment surface to deep sedimentary layers. All environmental compartments (i.e., sea water, sediments, biota and air) are affected and can potentially accumulate and export plastics, especially microplastics (<5 mm dimension), to the others (see Zalasiewicz et al. 2016a; Figure 4.3.1 herein).

---

[1] E.g., Deutsche Welle, 24.8.2015: Recycling-Weltmeister Deutschland (by Wolfgang Dick), http://dw.com/de/recycling-weltmeister-deutschland/a-18668262.

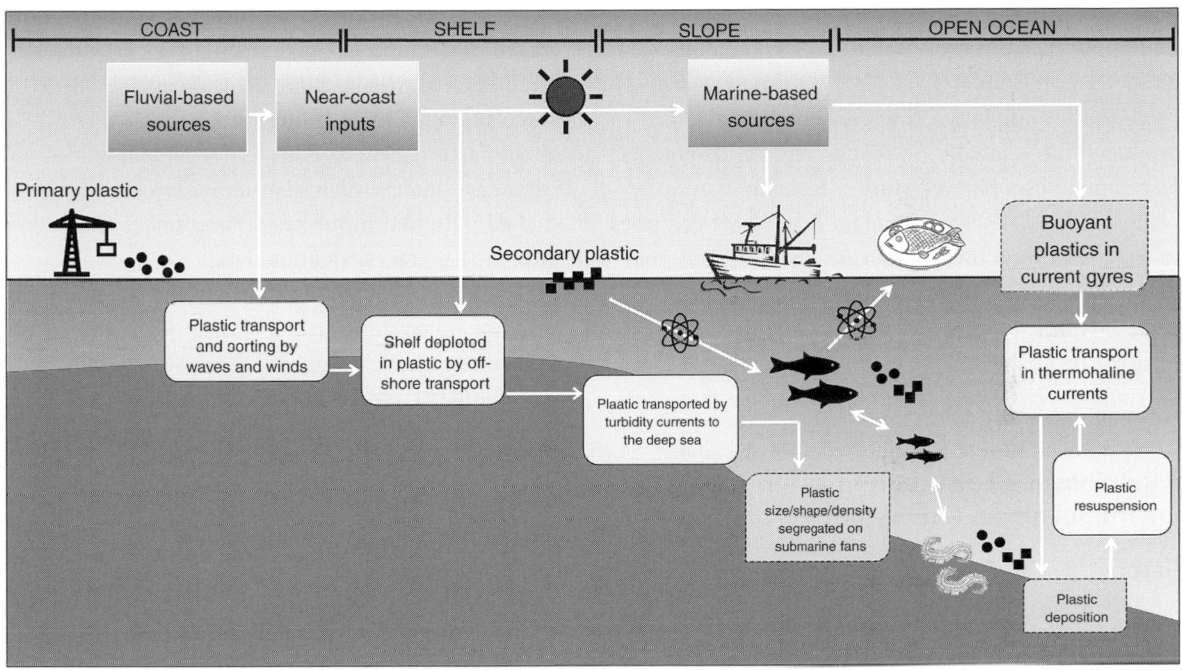

Figure 4.3.1 Conceptual model of plastic sources, transport and accumulation in the marine realm. Land-based sources are the main contributors of plastics to the ocean. Microplastics are ingested by organisms from all levels of marine food webs, together with adsorbed toxins that could reach humans through seafood. The final repository of microplastics is sedimentary deposits. Modified from figure 12 of Zalasiewicz et al. (2016a, figure 12). ©2016, with permission from Elsevier.

Production: Annual plastic production has increased more than 100 times, from about 2 million metric tons in the 1950s to 322 million tons in 2015, excluding fibres made from PET, polyamide and polyacrylate polymers. Geyer et al. (2017) calculated the total of plastic resins and polyester, polyamide and acrylic (PP+A) fibres produced from 1950 to the end of 2015 to be 7.8 billion tons (Gt), 50% having been produced in the last 13 years (2004–2017); including plastic additives brings this figure to 8.3 Gt, more than enough to wrap the entire globe in a layer of cling film. The current largest producer of plastics is China (with about 28% of the total in 2015), followed by the European Union plus Norway and Switzerland (18.5%) and the countries of the North American free-trade zone (18.5%). Within the EU, Germany is by far the largest producer (24.6% of total EU production). In the EU, nearly 40% is for packing, most used a single time and discarded; about 20% is in the construction sector and 9% in the car industry (Plastics Europe 2016).

Fragmentation: From macro- to microplastics: Information on plastic degradation and fragmentation in the environment is limited. Rates of weathering depend on polymer types (Shah et al. 2008; Andrady 2011) and variables such as incidence of sunlight, mechanical impact, seawater salinity, air and water temperature, hydrostatic pressure and biofouling (GESAMP 2016). UV radiation is by far the most significant factor controlling the residence times of plastics on beaches (Andrady 2011), where it makes plastic more brittle and liable to fragment into smaller pieces but seemingly does not cause 'complete' decay to primary components. Information on microplastic fragmentation in sedimentary strata is yet more limited. Where degradation and fragmentation continues, microplastics eventually produce nanoplastics, particles typically tens of nanometres in diameter (see Hanvey et al. [2017] for recent size proposed terminology). Nanoplastics may also be produced intentionally, for drug delivery, detergents, or cosmetic use. Limited studies indicate many plastics will persist in sediments for at least centuries to millennia, and their properties make it likely that they will resemble shells or other recalcitrant materials in leaving permanent casts and moulds in lithified rocks even if the plastic itself alters or decomposes over long timescales (see Section 4.3.5).

4.3.3 Plastic Redistribution in the Environment: An Overview

Geyer et al. (2017) estimate that by the end of 2015, all plastic waste ever generated from primary plastics will amount to 5.8 billion metric tons (Gt), of which 700 million metric tons (Mt) will be PP+A fibres. Only 30% of these plastics (2.5 Gt) are presently in use. Global recycling is small (12%), with about 600 Mt having been recycled between 1995 to the end of 2015, 90% of it recycled only once. About 800 Mt have been incinerated, whereas 60% (4.9 Gt) of all plastics ever produced have been discarded, now deposited in landfills or freely in the environment (Geyer et al. 2017). Thus, the lifespan of all plastics is short relative to, for example, iron and cement.

Plastic particles are lightweight and so are much more mobile than other human-made materials such as ceramics or glass. They are easily transported by wind (Gasperi et al. 2015) and water through the environment, where they accumulate. From being local 'litter' a few decades ago, plastics are increasingly recognised as a major environmental problem on land and in the sea (e.g., Ivar do Sul & Costa 2014; GESAMP 2016); even casual observation shows that macroplastic debris (>5 mm dimension) may be found in most inhabited environments. Microplastics are less easily visible to the naked eye, but methods for their analysis in sediment are being developed. They can be extracted from water by filtering and separated from sediment via sieving or density separation using centrifuge and high-density solutions (e.g., Nuelle et al. 2014; Woodall et al. 2014; Corcoran et al. 2015; Hanvey et al. 2017). Microplastics occur in both terrestrial and marine sediments, though their distribution on land is less studied than that in the sea (Thompson et al. 2009; Rillig 2012; Eerkes-Medrano et al. 2015).

Nanoplastics (spheroids, platelets and fibres with at least one dimension <100 nm), with their large surface-to-volume ratio, have increased capacity to adsorb organic compounds and potentially can penetrate organic cell walls. They have been shown to affect the growth and reproduction of at least some aquatic invertebrates (e.g., Besseling et al. 2014; Della Torre et al. 2014; Velzboer et al. 2014). The distribution of nanoplastic particles in the natural environment is poorly known because of the technical difficulty of isolating them from water or sediments, but they are clearly becoming increasingly commonly dispersed.

Plastic within non-aquatic terrestrial sediments: On land and away from shorelines, plastic litter is widely distributed in the surface environment, most clearly in and around urban areas via casual littering (Thompson et al. 2009; Rillig 2012). The distribution of plastic accumulations below ground is determined by landfill sites where, in the last few decades, plastics have come to make up ~10% by weight of the buried waste (Thompson et al. 2009). Where landfill sites have been mapped out and their operation dated, plastic-rich sedimentary deposits up to several tens of metres thick may be delineated. Increasing plastics production in the 1960s coincided with increased proportions of single-use goods, rapidly increasing the proportion of plastics in landfill from the 1970s (Ford et al. 2014). The use of plastics in agriculture has also grown since the 1960s, Hussain and Hamid (2003) noting that global agricultural consumption of plastics was ~2.5 million tons per year at that time. Plastics are being incorporated into many cultivated soils, becoming thoroughly mixed into the full depth of ploughing. Legislation across many parts of the world has stimulated increasing reuse and recycling of plastic goods, but so far this has restricted rather than overturned the growth of plastic disposal. The problem is greater in some developing countries, where the arrival of abundant packaged goods is associated with inefficient waste disposal (a developing country, Rwanda, though banned plastic bags in 2008).

The flux of plastics into sedimentary environments includes relevant but non-obvious pathways. Microplastic particles are released in the form of cigarette ends, from cosmetics, by washing synthetic textiles and from car-tyre abrasion. Of all plastic particles being released into the environment, an estimated 35% are due to washing textiles and 28% by car-tyre abrasion; road networks have increasingly become corridors of microplastic deposition, and much eventually reaches the seas, contributing to up to 30% of the ocean's plastic soup (Boucher & Friot 2017). Plastics are also widely used in the laying of underground cables and pipes for services and communications, often under or along roads.

Rivers as microplastic hotspots: Although the scientific community has only recently studied freshwater systems with respect to microplastics (Ecrkes-Medrano et al. 2015), they have already been found in lakes (e.g., Eriksen et al. 2013; Imhof et al. 2013; Free et al. 2014; Zbyszewski et al. 2014) and rivers, such as the Thames (Morritt et al. 2014), Danube (Lechner et al. 2014), and Yangtze (Zhao et al. 2014). Microplastics enter rivers via wind, storm sewers and wastewater-treatment plants (i.e., as primary-sourced microplastics). The low density of the most commonly produced plastics means that much stays within the water column or upon its surface and is transported farther downstream or out to lakes and seas (Sadri & Thompson 2014). Thus, rivers are major conduits for the transport of plastics from land to lakes or the sea. However, low-density plastics have also been found in lake-bottom sediments, having been deposited as a result of density increase by mineral fillers during production or by mineral adsorption while in the water column (Corcoran et al. 2015; Corcoran 2015).

In rivers, macroplastics (such as plastic bags) often act as baffles to trap sediment. Along lake shorelines and riverbanks, microplastics tend to become trapped in organic debris brought in by waves and currents (Zbyszewski et al. 2014; Corcoran et al. 2015). In addition, high-density macroplastics may accumulate within channel bedload, where mobile plastic elements in the traction carpet may be abraded rapidly (Williams & Simmons 1996) and eventually reduced to microplastic particles. Between rivers and the sea, mangroves can also trap plastics (Ivar do Sul et al. 2014), this phenomenon being called the Christmas-tree effect (Williams & Simmons 1996).

Plastic in the oceans: The sea is the final resting place for a range of different types of human litter, from glass to metals to building waste, though plastics form the greatest modern component. Making up some 10% of human refuse by weight (which corresponds to a much higher percentage by volume), plastics are selectively transported by wind and water to make up to 50%–90% of marine litter (Barnes et al. 2009; Jambeck et al. 2015). Current estimates indicate that significant amounts of plastic produced annually will ultimately reach coastlines and oceans as litter (Rochman et al. 2013).

Macro- and microplastics entering the sea have considerable environmental impact. Since the 1960s and especially in the last decade, both the phenomenon itself and study into it have grown markedly, particularly for microplastics (Ivar do Sul & Costa 2014; Leinfelder & Haum 2015; Avio et al. 2017). Before being deposited, and depending on their density and size, plastic particles may take long journeys through the water column (Figure 4.3.1), sometimes being overgrown (e.g., by small oysters, bryozoans, or corals; Gregory 2009), sinking slowly, being grazed on, and then floating up again. Even at the seafloor, plastics may be incorporated by filter- and sediment-feeding organisms, at times being expelled back into the water. Microplastics hence interact with the oceans at various scales, from the subcellular to the ecosystems level (Wright et al. 2013; Galloway et al. 2017). In addition, some unknown amount of marine plastics will eventually return to the human population through the marine food web (Dehaut et al. 2016), whereas the remainder becomes incorporated as sediment in the marine realm.

4.3.4 The Fate of Microplastics in Ocean Settings

Microplastic accumulations at the ocean surface: Once within the sea, low-density plastics such as polyethylene (PE) and polypropylene (PP) – which together comprise ~56% of output in Europe (PlasticsEurope 2016) – initially float in seawater. These low-density plastics are moved by wind stress and by surface currents and so can encircle the Earth, becoming concentrated, for instance, in mid-ocean gyres such as the Great Pacific Garbage Patch, some thousand kilometres in diameter (Moore et al. 2001; Ryan et al. 2009; Law et al. 2014). Beaches on remote ocean islands trap some of this shallow floating material.

Zettler et al. (2013) found that polyethylene and polypropylene in seawater are colonised by a complex microbial community, which they referred to as the 'plastisphere'. Consequently, with relatively higher densities, microplastics are then transported through the water column by various processes (see also Lobelle & Cunliffe 2011). Ocean gyres, indeed, show modelled concentrations of surface plastic debris within the midlatitudes of all oceans that mimic atmospheric circulation patterns of radiogenic fallout (e.g., Waters et al. 2015).

Plastics and microplastics stranded on beaches or deposited in marine sediments: Coastlines and beaches have understandably attracted much attention, given their sensitive status in human society and the high visibility of plastic litter deposited there. The monitoring of beach litter, mostly macroplastics, is typically done by counting items at the surface per unit length (e.g., per 100 m) of coastline and noting such aspects as type, composition, weight and volume. A recent study of Korean beaches (Hong et al. 2014) found 300–1,000 items/100 m, including polystyrene fishing buoys and plastic bags and bottles. Cigarette-filter ends are generally the single most common item found in such studies and in beach clean-ups. Of the approximately six trillion cigarettes smoked annually, the filter-bearing tips of over four trillion end up as litter each year (Carlozo 2008).

Plastics are virtually omnipresent in the coastal zone globally, not only in densely populated regions but also in remote areas. Barnes (2005) noted substantial amounts of macroplastics on remote islands. On some islands such as Diego Garcia, hermit crabs have taken to using plastic bottle tops as homes (see also Reed 2015). Barnes (2005) also noted a diminishing trend of plastics from equator to pole in the Southern Ocean, although noticeable amounts still reach Antarctic coasts. In Hawaii, accumulations of plastic debris have formed what Corcoran et al. (2014) referred to as 'plastiglomerates', in which melted plastic associated with campfires has bonded beach pebbles and sand to form a rock (wildfires and volcanic activity might also cause such melting); these dense hybrid plastic-sediment materials have good potential for burial and long-term preservation. Lavers and Bond (2017) recorded the highest density of plastic debris worldwide, with up to 671.6 items/m^2 (mean 239.4 and SD $\pm$ 347.3 items/m^2) on uninhabited Henderson Island beaches in the South Pacific. In addition, about 68% of debris (up to 4,496 pieces/m^2) was buried up to 10 cm deep.

In the dynamic beach environment, objects can be buried and exhumed many times (Smith & Markic 2013). Overall, the few studies involving depth profiles of beaches (e.g., Turra et al. 2014) suggest that plastic items may locally extend downward for as much as 2 m, with there being an order of magnitude more buried plastic than surface plastic. Hence, sediment bodies are forming in the coastal zones that, if seen in cross section, could contain sufficient macroplastic material to be recognisable to the field geologist as a post-mid-20th-century deposit. In some instances, these macroplastic fragments are already visible in beach-rock deposits, as along the Basque coast (Irabien et al. 2015).

In the deep sea, surveys by dredging or by remotely operated underwater vehicle (ROV) cameras have shown the spread of larger plastic fragments. Bottles, plastic bags and abandoned fishing nets are abundant (Watters et al. 2010; Richards & Beger 2011; Tubau et al. 2015; Corcoran 2015 and references therein) and are often concentrated by topography or currents into submarine lows, such as the bottoms of submarine canyons (Schlining et al. 2013; Tubau et al. 2015). Tubau et al. (2015) described ROV dive sites in the submarine canyons of the northwest Mediterranean at depths of 140–1,731 m that showed that plastics were the dominant component of litter (72%), most being observed on canyon floors at depths of more than 1,000 m, likely carried there by downslope flows originating near shore. Litter density ranged up to 11.8 items/100 m survey line and averaged between 8,000 and 15,000 items/km^2, reaching a maximum of 167,540 litter items/km^2 at one site (Tubau et al. 2015). Pham et al. (2014) considered that the relative scarcity of macroplastic objects on marine shelves was because they were being current swept into deep water, particularly via submarine canyons. Such deeper water and submarine-canyon environments, being less disturbed by bottom trawling than are shelf sediments, may provide a good record of the history of plastic influx associated with the Anthropocene. This new plastic-dominated debris layer overlies the debris of previous centuries. Overall, this earlier material is sparser, but a notable component is clinker from the old coal-fired steamships, thrown overboard en route and hence forming 'pavements' below the sailing routes (Ramirez-Llodra et al. 2011; see Section 2.8.5 herein).

Particles of microplastics are more abundant and more widely and evenly distributed than are macroplastics, and they can be recognised even in samples as small as 50 g of coastal sediment (Browne et al. 2010, 2011). Microplastics can include relatively large particles, several mm in diameter, such as resin pellets that are near-ubiquitous in some beach sediments.

Small (<1 mm) microplastics are particularly abundant. Largely composed of microfibres detached from machine-washed artificial fabrics (Browne et al. 2011) and transported via sewage outfalls to rivers and dumped sewage sludge, these have become very widely dispersed. Browne et al. (2011) suggested that fibres have become incorporated in and routinely extractable from shoreline sediments throughout the world, in quantities that range from tens to hundreds of fibres per litre of sediment. For example, Dekiff et al. (2014) reported ~5–25 microplastic particles (mostly microfibres) per kilogram of sediment for Norderney (North Sea), whereas Reis (2014) found an average of 66 fibres/kg on the Baltic island of Fehmarn. This potentially provides a near-ubiquitous signature of the Anthropocene in coastal settings.

Plastic fragments with densities >1 gm/cm^3, including PVC, sink in seawater. They can then be moved by tidal and storm-driven currents in shallow water, and by various gravity-driven currents (e.g., turbidity and contour currents) in deep water, before finally being deposited (see Zalasiewicz et al. 2016a, figure 12). Fischer et al. (2015) reported microplastics, mainly fibres, at depths of >5,000 m in the Kuril-Kamchatka Trench and adjacent abyssal plain. Even at these great depths, concentrations were as high as 2,000 pieces/m^2. Woodall et al. (2014) (see also Goldberg 1997 and Van Cauwenberghe et al. 2013 for earlier records) also reported microplastics, mainly as fibres, in deep-sea sediment-core samples from the Atlantic, the Mediterranean and seamounts of the southwest Indian Ocean. Abundances ranged from 1.4 to 40 fibres (average 13.4) per 50 ml of sediment. That was some four orders of magnitude more abundant than in the contaminated surface waters above. Even the Indian Ocean seamounts, which showed the lowest abundances, were conservatively calculated to have 4 billion fibres per km^2 (Woodall et al. 2014).

How did the plastics get to these ocean floors, far distant from land? There are a number of potential routes to the oceans' depths. Fibres are mostly composed of acrylic and polyester, which are denser than seawater. These behave like fine clay particles, slowly drifting in storm- or turbidity-current-generated nepheloid plumes or carried by thermohaline currents. Low-density microplastics, too, have sunk to the ocean floor. These could have been ingested by zooplankton and ejected as faecal pellets, or they could have sunk with the plankton when they died or travelled within the faeces or bodies of fish that ate the zooplankton (Boerger et al. 2010; Cole et al. 2013; Wright et al. 2013; Setälä et al. 2014). The microplastics could also have been caught up in gelatinous marine snow. In this respect, microplastics behave in a similar way to other microplanktonic taxa preserved in the geological record (e.g., coccoliths in deep-sea oozes) and represent a primary tool of biostratigraphic correlation in the geological record because of widespread distribution within strata that are likely preservable long into the future.

Microplastics may even reach marine ice. Significant amounts of microplastics (38–234 particles/m^3) have been found frozen in Arctic sea ice, having seemingly been derived from the Pacific Ocean (Obbard et al. 2014). The Arctic is thus a major global sink for these tiny plastic particles. However, melting at current rates could unlock over one trillion pieces of microplastics over the next decade. Rayon was the most common material, much of it from cigarette filters (one cigarette filter tip comprises ~10,000 rayon fibres) and hygiene products. Other materials included polyester, nylon, polypropylene, polystyrene (PS), acrylic and polyethylene.

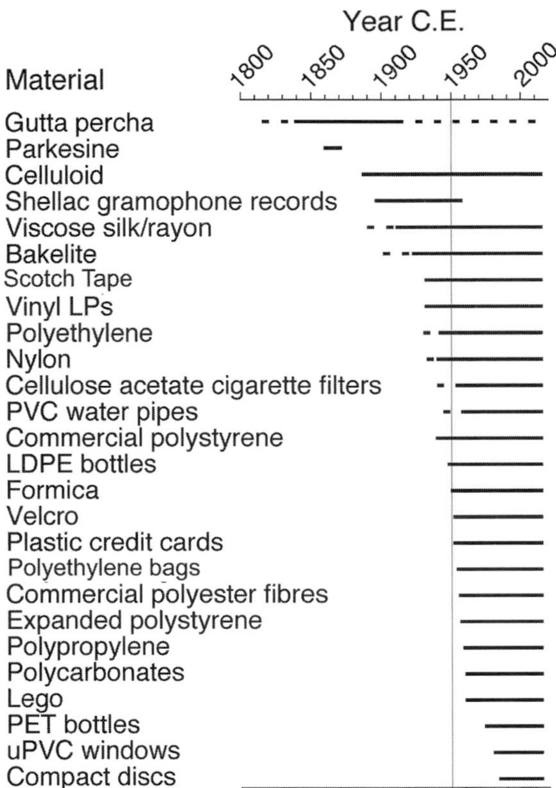

Figure 4.3.2 Stratigraphic appearance of some major types of plastics and plastic artefacts. Gutta-percha, the hardened sap of any of eight tree species from Southeast Asia, is not strictly a plastic. Nevertheless, it features in some early histories of this material. Between 1850 and 1899, some 27,000 tons were laid on the seafloor to serve as insulation for telegraph cables due to its resistance to saltwater corrosion (Tully 2009). Adapted from information mostly in http://bpf.co.uk/Plastipe-dia/Plastics_History/Default.aspx. Extracted from figure 1 of Zalasiewicz et al. (2016a). ©2016, with permission from Elsevier.

4.3.5 Plastics as a High-Resolution Stratigraphic Tool

Zalasiewicz et al. (2015b, 2016a) and Waters et al. (2016) suggested that plastics are a good indicator for identifying Anthropocene sedimentary deposits, since most plastics only occur (in increasing amounts) in post-WWII times (Figure 4.3.2). Plastics are clearly long lived on human timescales, especially when buried and beyond the reach of solar ultraviolet (UV) light. Plastics as a whole are resistant to microbial attack, and this underlies a good deal of their practical utility and their longevity in the environment. Nevertheless, some evidence of digestion by microbes has locally been observed (Harshvardhan & Jha 2013; Yang et al. 2014; Yoshida et al. 2016; see also Kasirajan & Ngouajio 2012), and plastics may host microbial communities different to the generally ambient ones (McCormick et al. 2014). The extent to which higher organisms are able to digest plastics has been rarely studied so far and is a matter of scientific debate. Bombelli et al. (2017a) suggested that the caterpillar of the wax moth *Galleria mellonella* can digest polyethylene; Weber et al. (2017) considered the methods used inadequate for proving true digestion rather than just shredding, ingesting and excreting the plastic unchanged, which is however rejected by Bombelli et al. (2017b). The sudden appearance of plastics as a widespread new addition to the surface environment, together with the rapid evolutionary rates observed in microbes subject to strong selective pressures, suggests that microbial degradation may become more common over time, not least because any microbes that can use plastics as a food source will be selectively advantaged. Nevertheless, this is currently a minor factor – and it must be noted that many eminently digestible and decomposable organic tissues (shell because of its organic matrix, bone, wood) may be commonly fossilised once buried.

Colder temperatures within the deep ocean and a lack of UV light make plastics on the seabed more likely to be preserved. In these conditions, they are said to last for 'centuries to millennia' (Gregory & Andrady 2003), mostly via inference from short-period laboratory studies. Over longer timescales, their diagenesis and fossilisation potential once buried in strata is a topic of considerable academic interest, although of no analytical study yet, as far as we are aware. The nearest comparison is with the long-chain polymers in recalcitrant organic fossils such as wood, spores, pollen and graptolites. Many plastics might behave similarly to recalcitrant non-carbonate shelled organisms, or perhaps to amber, over geological timescales. Some might change into bitumen, and depending on burial conditions, hydrocarbons released during diagenesis might contribute to future oil and gas deposits.

4.3.6 Outlook: The Role of Plastics for Anthropocene Studies

Plastics are now part of our everyday life. They have entered and for some time will still enter sedimentary deposits of many terrestrial and aquatic settings, eventually becoming incorporated into sedimentary deposits of the Anthropocene as technofossils, varying greatly in shape, size and chemical composition. Analysing these deposits allows us to monitor where and in what quantities plastics have been distributed. At the same time, the onset of plastic-containing deposits might provide a globally correlatable horizon that could help define the base of the Anthropocene. In addition, various types of plastics are expected to provide a refined subdivision and correlation tool within sedimentary deposits within the Anthropocene (Figure 4.3.2).

Plastics in the environment are, however, not just a useful stratigraphic tool but also a threat to organisms, including humans. Hence, an ultimate goal is to diminish the footprint of plastics within sediments (and environments as a whole) by better recycling and reuse. Whether this is successful or not, plastic will leave a long-duration and globally correlatable succession in sedimentary deposits worldwide that may help not only to define and subdivide the Anthropocene but also to monitor the potential success of reducing plastic impact on environments into the future. And should it be possible to achieve plastic-free sedimentary deposits in a future phase of the Anthropocene, this would again provide a clear stratigraphic subunit, the post-plastic zone of a future Anthropocene. For now, plastics provide a distinctive physical record of technosphere evolution during the late 20th and early 21st centuries.

5 Anthropocene Chemostratigraphy

CONTENTS

Chemostratigraphy, the combination of geochemistry with stratigraphy, can be used to define patterns of changing chemical composition through time that can provide proxies for geological age, which in turn may be useful in the classification of the Anthropocene as a chronostratigraphic unit. We discuss the patterns in terms of their geological causes, though their use in helping to develop a geological time framework is of prime concern here. Anthropocene archives considered here include modifications to the carbon cycle, such as changes in atmospheric carbon dioxide (CO_2) and methane (CH_4) concentrations, and stable carbon isotope records preserved in geological successions. Increased greenhouse gas emissions are resulting in ocean acidification, measurable in carbonates using a boron isotope proxy. Stratigraphic signals also result from the doubling of reactive nitrogen and phosphorus, mainly in agricultural fertiliser use, and from increases in atmospheric sulphur, mainly from hydrocarbon combustion. Environmental pollution more broadly includes distribution of diverse metal and organic compounds sourced from industrial activities, along with radioisotopes derived from fallout from nuclear device testing.

5.1 Capture of Geochemical Changes in Archives

Ian J. Fairchild, Jan Zalasiewicz, Colin P. Summerhayes and Colin N. Waters

Geochemistry is the study of the variation in chemistry of Earth materials, including both solid rocks and the fluids from which they form. Chemostratigraphy combines geochemistry with stratigraphy, defining the patterns of chemical composition of geological materials in an age sequence. Hence the geochemistry of these materials provides proxies for age to enable classification of the corresponding rock units into chronostratigraphic units (see Section 1.3). These patterns may further be used to help constrain geological processes operating within and between those time units, but their use in helping to build and use a geological time framework is of prime concern here.

Anthropocene archives form a subset of the geological rock record. Rocks are predominantly composed of minerals with a defined structure built of a systematic array of atoms and with a limited range of chemical composition (Section 2.1). Minerals can be present in a variety of combinations within rocks, including within the fossils that these rocks may contain. However, particularly for lower-temperature and younger geological materials, there will be a content of poorly crystalline mineraloids and organic material, whose bulk chemical composition may be obtainable but whose building blocks are difficult to define.

5.1.1 Chemostratigraphy in the Geological Record

Global chemostratigraphic markers may reflect very widely dispersed, chemically distinctive material produced by rare, large-scale events, such as the iridium-rich dust globally scattered from the bolide impact at the Cretaceous-Paleogene boundary (see Section 1.3.1.3) or changes to chemical cycling caused by climate/oceanographic change. Of value in the latter case are changes in relative proportions of isotopes of common elements, notably of oxygen and carbon, particularly where such changes are global in extent and synchronous or near-synchronous. The patterns of oxygen isotopes in marine microfossils provide a chemostratigraphy that reflects climate change particularly vividly over the Cenozoic, as in the now-iconic Zachos curve (Zachos et al. 2001; see also Zachos et al. 2008; Grossman 2012; Lisiecki & Raymo 2005; and Sections 1.2.1.5 and 1.2.1.6 herein), while patterns of carbon isotopes, exchanged between carbon reservoirs in the land and ocean/atmosphere system, have been used to define time boundaries such as the Paleocene-Eocene boundary (see Section 1.3.1.4). In the Neoproterozoic, the pronounced carbon isotope fluctuations associated with postglacial transgression at the end of the Marinoan ice age are an important aspect of the definition of the Cryogenian-Ediacaran system boundary (Knoll et al. 2006). Ongoing discussions on the ~720 Ma Tonian-Cryogenian boundary (Shields-Zhou et al. 2016) suggest that a combination of carbon and strontium ($^{87}Sr/^{86}Sr$) isotopes will likely be employed as primary parameters for correlation.

Figure 5.1.1a gives a diagrammatic representation of different genetic components that can be distinguished in a sedimentary rock, and Figure 5.1.1b provides examples of geological materials with different combinations of these components. The chemostratigraphic parameters described above belong to component 3, chemical/biological precipitates, most often marine $CaCO_3$, reflecting global chemical changes in the ocean. However, other chemical changes in rock strata vary through time and space, e.g., component 1 (terrigenous detritus), which reflects the provenance (sediment source areas). Input of this sort can result in local time markers, but these are rarely of wider correlatory value. Component 2 includes minerals such as clays that may be of climatic significance, but in practice, this component is not present with sufficient reliability to be used for chemostratigraphy. There are also phases (component 4) produced by post-depositional chemical changes (diagenesis) that

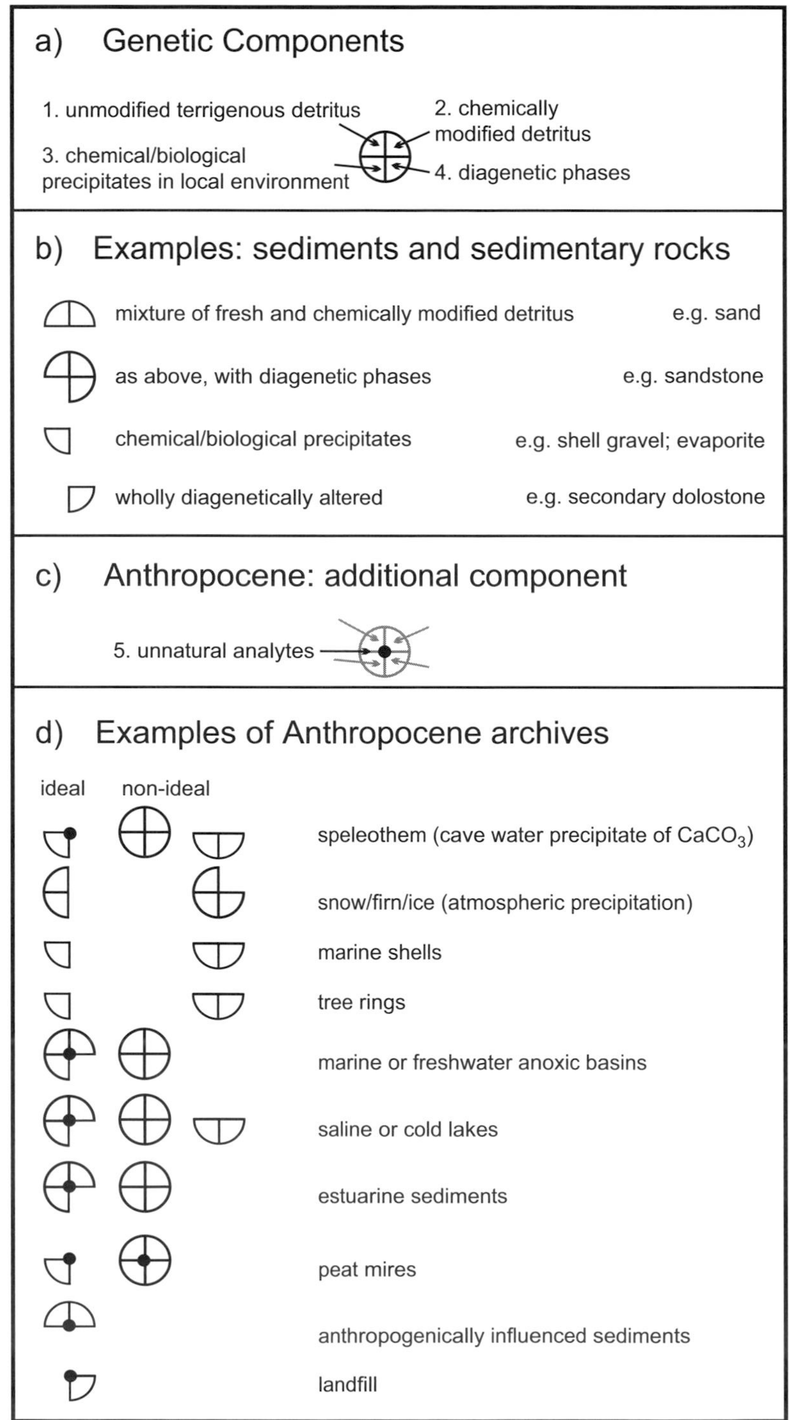

Figure 5.1.1 Components of sedimentary rocks and Anthropocene archives. (a) Genetic components. (b) examples of combinations of genetic components in sediments and sedimentary rocks. (c) Additional anthropogenic component. (d) Examples of mixtures of genetic components in ideal and non-ideal archives for the Anthropocene. Parts (a) and (b) are adapted from Fairchild et al. (1998) in the context of establishing a rationale for chemical analysis of sediments and sedimentary rocks. Parts (c) and (d) are new.

may be mediated by reactive components such as organic matter or the movement of subterranean fluids. Mixing by bioturbation can also be included in this category. The patterns of change over time and space, together with the establishment of the range of background chemical variation, must be assessed and resolved in order to extract the optimal time-specific chemical signals within strata.

5.1.2 Chemostratigraphy of the Anthropocene

The material flows of the Anthropocene (Figure 5.1.2) include both natural and anthropogenic fluxes. The anthropogenic addition to chemostratigraphic signatures is diverse and in some cases novel, strikingly marked or both. One aspect is the presence of unnatural materials (e.g., many persistent organic pollutants, plutonium from nuclear reactors, plastics, concrete, pulverised fuel ash) as shown diagrammatically in Figure 5.1.1c. In addition, the surface abundances of many elements in detritus are dominated or significantly perturbed (15%–50% mobilisation) by human activities such as deforestation, mining and construction (Figure 5.1.2 herein; Sen & Peuckner-Ehrenbrink 2012). The chemostratigraphic patterns produced include examples involving elemental and molecular abundance and variations in isotopic ratios. Our current knowledge of these chemical impacts is incomplete, especially regarding the distribution and preservation of complex new chemical forms, since

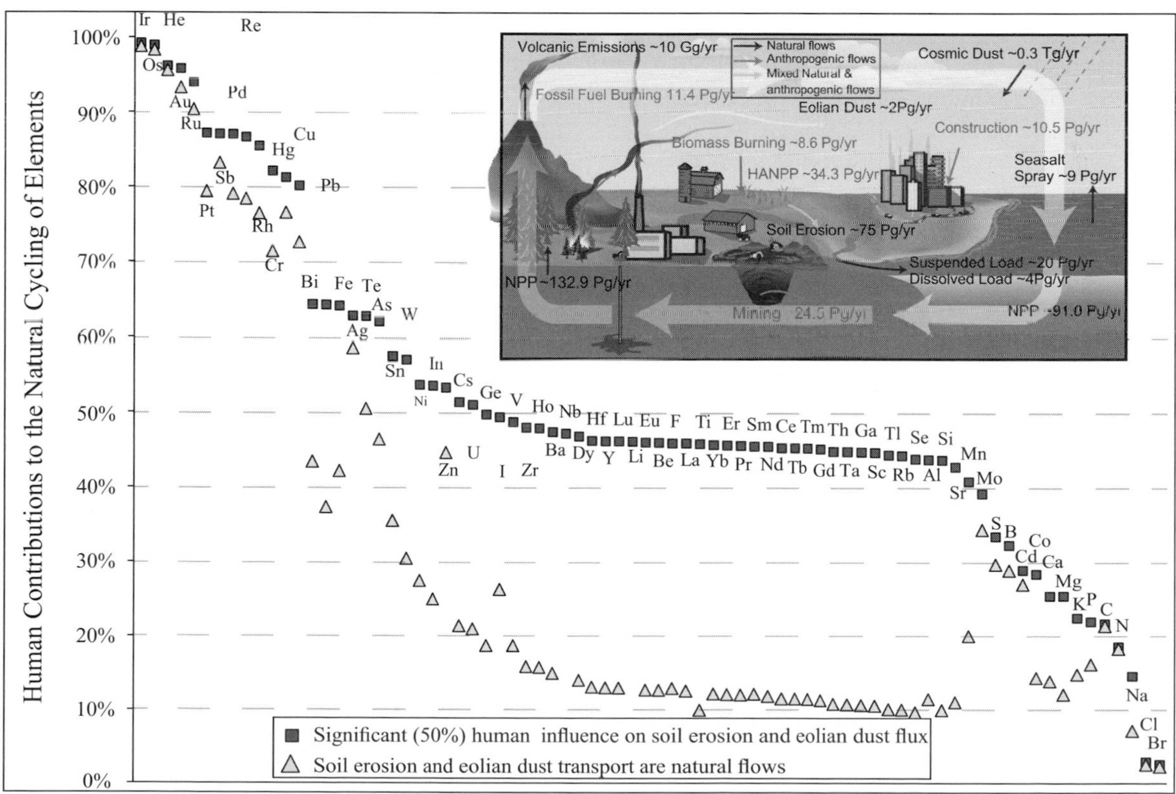

Figure 5.1.2 Anthropogenic domination and perturbation of elemental surface fluxes with a diagrammatic inset of the fluxes (reproduced with permission from Sen & Peuckner-Ehrenbrink 2012). The yellow datapoints do not allow for anthropogenic increases in aeolian sediment fluxes, whereas the red points allow for a 50% increase in these fluxes. ©2012 American Chemical Society. (A black-and-white version of this figure appears in some formats. For a colour version, please refer to the plate section.)

their stratigraphic potential has been studied only patchily. Nevertheless, the general chemostratigraphic component of the evidence base for the Anthropocene is considerable.

In the following text, we discuss some aspects of the chemostratigraphic potential of different archives in the Anthropocene by means of their genetic components (Figure 5.1.1d). In addition to these considerations, the ability to produce a precise (ideally annually resolved) chronology is important, as is the accessibility and preservability of the archives and their collective ability to record correlatable signals from different geographical domains.

Calcareous cave deposits (speleothems) ideally display a $CaCO_3$ chemistry characteristic of cave water, which in turn reflects that of the overlying environment and atmosphere, including sulphate and radiocarbon. In principle, unnatural or natural biomarkers may also be preserved, although they are not often studied. In non-ideal cases, sediment impurities may be present, and/or primary $CaCO_3$ (e.g., aragonite) may be diagenetically altered. On ice sheets, the chemistry is that of surface snow overlying firn and glacial ice successively. The precipitate corresponds to atmospheric moisture containing passively entrained impurities including both windblown dust and aerosol as well as trapped air. The solid phase records information on the S and N cycles, and air inclusions record information on C and N cycles, while dust patterns and sea salt reflect windiness. Although unnatural materials may be present, in practice they are not likely to be detectable. Post-depositional processes affecting the archives in non-ideal cases include wind reworking, reactions affecting the partial pressure of carbon dioxide (pCO_2) in cores containing carbonate dust and, conceivably, migration of brine along crystal boundaries in ice.

For the Anthropocene, marine shells and corals are normally well-preserved calcareous precipitates, although rarely there is some diagenetic change of metastable carbonate. Stable carbon isotopes here can reflect changing atmospheric composition. Trees likewise record this change in their cellulose, while sulphur is largely retained in proteins of the inner cell wall, but migration of other elements in sap is likely.

Natural sediments of lakes, estuaries and oceans may be suitable Anthropocene archives if annually laminated or, in the case of estuarine deposits, displaying rapid sedimentation rates. Inevitably they contain a wide range of detrital components 1 and 2 (Figure 5.1.1) mixed with organic matter created in the environment (component 3) and with the possibility of exotic unnatural materials (component 5) ranging from adsorbed plutonium to manufactured organic compounds. New diagenetic materials (component 4) are likely in response to organic matter diagenesis under the anoxic conditions that promote annual lamination. In saline lakes, chemical precipitates should retain a clear signal of changing lake chemistry.

Peats are distinctive in that, although dominated by in situ production (component 3), they are excellent archives for accumulation of pollutants, including unnatural materials. However, secondary changes such as sulphide precipitation can occur, and there can be a varying input of detritus. 'Anthropogenically influenced sediments' refers to mixtures of natural terrigenous materials (components 1 and 2) with unnatural components (component 5), such as may be found in beach or shoreface environments near contaminated sites, for example. Finally, landfill is dominated by unnatural materials (used in a loose sense here to include human-reworked organic materials), which undergo significant post-depositional change under reducing conditions.

The chemical systems are discussed below, and the archives are reviewed in more detail in Chapter 7.

5.2 Carbon

Jan Zalasiewicz and Colin N. Waters

The carbon signal of the Anthropocene is central, because much of the energy used to drive anthropogenic processes comes from the combustion of carbon-based fuels in geologically large amounts and at geologically extraordinary rates. This combustion gives rise to a variety of signals, some of which are particulate (such as fly ash, described in Section 2.3), while others affect the Earth's heat balance and therefore climate (Section 6.1); yet others, though, may be termed chemostratigraphic and are described below.

5.2.1 Atmospheric Carbon Dioxide (CO_2) Concentrations

About half of the CO_2 produced from fossil-fuel combustion (with lesser sources from vegetation combustion/decay, land clearance, cement manufacture and other processes) accumulates in the atmosphere, and a significant part of this persists for centuries to many millennia (Archer et al. 2009; Clark et al. 2016). Changing concentrations of CO_2 provide a direct stratigraphic signal where atmospheric gases are preserved in strata. This is something that only commonly happens where snow accumulates to form the ice of glaciers and ice caps, with bubbles of fossil air being trapped in the ice layers. The ice sheets of Greenland and Antarctica contain such records of fossil air going back ~125,000 years (e.g., NEEM 2013) and ~840,000 years (e.g., Lüthi et al. 2008; Section 6.1.1 herein) respectively, and this detailed and continuous record forms the context for the Anthropocene pattern.

The Anthropocene gas record itself is a product of the analysis of air samples, continuing the classic work of Charles Keeling (Keeling 1960). In ice cores there is a delay between the snowfall and the 'fossilisation' of the air, the latter only happening deep in the snowpack once it has compacted enough to convert continuous air passages (with more or less free exchange with the atmosphere) into isolated bubbles with trapped air. The offset between the age of snowfall and the age of trapped air depends upon the rate of snowfall but typically takes some decades; where accumulation rates are low, such as at the South Pole, the offset can be in excess of a millennium (cf. Rubino et al. 2013).

'Splicing' of the fossil-air CO_2 record obtained from different ice cores with the modern-air CO_2 record that began in 1957 has illuminated the course and context of the extraordinary recent rise in atmospheric CO_2 concentrations. Over the past 800,000 years (as far as the current record extends), CO_2 levels have oscillated in a regular, astronomically driven pattern between ~180 and ~280 parts per million (ppm) (Lüthi et al. 2008; Figure 5.2.1 herein); lower and higher values coincide with Quaternary glacial and interglacial phases, respectively (see Section 6.1.1 and Figure 6.1.3).

The Late Pleistocene to Holocene record shows a somewhat irregular rise from ~180 to ~260 ppm between 17 ka and 11 ka and then a very slow overall decline to ~255 ppm at ~7 ka (Figure 5.2.2a). This was followed by a very slow rise to ~280 ppm by ~1800 CE, which has been ascribed, though controversially, to a drip feed of carbon dioxide into the atmosphere from the effects of early farming (Ruddiman 2003, 2013; cf. Elsig et al. 2009). This slight increase may have been sufficient to delay a return to glacial conditions (Ganopolski et al. 2016). Superimposed upon this gradual two-phase trend are approximately centennial-scale oscillations of a few ppm, one of which, a short-duration dip of <10 ppm around 1610 CE (Figure 5.2.2b), has been suggested as a possible marker for an Anthropocene boundary (Lewis & Maslin 2015; cf. Zalasiewicz et al. 2015a; see Section 7.3.7 herein).

While early examinations of ice cores suggested that there was a delay of some hundreds of years or more between the rise in temperature and the rise in CO_2 at glacial terminations (Petit et al. 1999), careful re-examination of Antarctic ice cores from between the Last Glacial Maximum and the Holocene confirms that in fact the two rises were more or less synchronous (Parrenin et al. 2013), as expected from the basic physics of warming ocean water (for more detail, see Section 6.1.2.1).

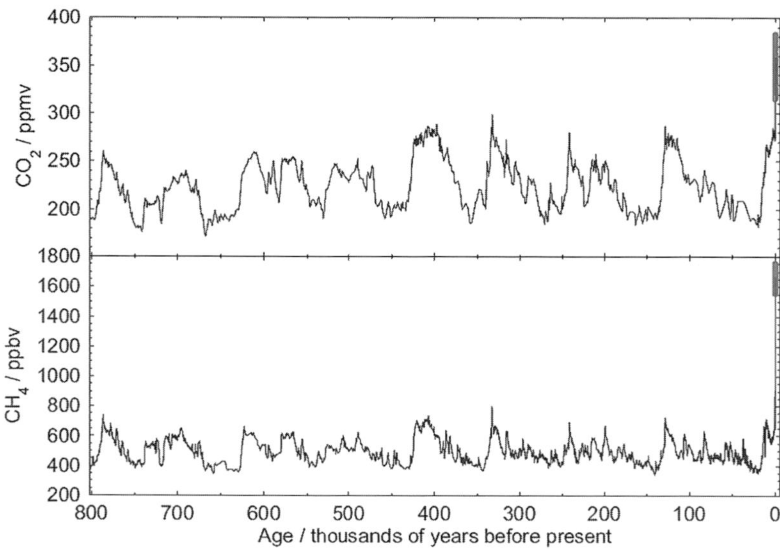

Figure 5.2.1 CO_2 and CH_4 concentrations over the last 800 ka from Antarctic ice cores (from figure 2 of Wolff 2014). The main curves are from Dome C and Vostok, the line marking an upturn in concentrations over the last 2,000 years is from Law Dome and the broad grey curve is from modern atmospheric measurements.

From ~1800 CE the trend changed markedly, with an accelerating rise to ~310 ppm by 1950 CE (Etheridge et al. 1996; MacFarling-Meure et al. 2006; Rubino et al. 2013; Figure 5.2.2c herein) and then further sharp acceleration at about 1965 leading to its current global average concentration of ~405 ppm, reached in 2017. The increases in CO_2 concentrations of >2 ppm/yr in 2012–2015, with 2015 showing a rise of 3.05 ppm, represent the highest annual increases yet recorded (http://noaa.gov/news/record-annual-increase-of-carbon-dioxide-observed-at-mauna-loa-for-2015). The rate of rise between 1950 and 2015 is ~100 times faster than the Late Pleistocene to Early Holocene rise (and the 2012–2015 rise is ~200 times faster), while the total amount of change since pre-industrial times, at 120 ppm, exceeds the change seen between glacial and interglacial phases of the Quaternary (Figure 5.2.1).

The total anthropogenic carbon dioxide input into the atmosphere (125 ppm as of 2017, nearly one-third of the total) has a mass of nearly 1 trillion metric tons (Tt = 10^{12}). Thought of as gas at atmospheric pressure, it is the equivalent of a pure CO_2 gas layer ~1 m thick mantling the entire Earth, and it is currently growing at about a millimetre a fortnight (Zalasiewicz et al. 2016b). This does not represent all anthropogenic CO_2 emission – approximately half has been dissolved into the oceans (see Section 5.3) or assimilated by ocean phytoplankton and terrestrial vegetation.

This substantial and geologically extraordinary rise is only directly captured, with the brief delay noted above, in ice, especially in the polar ice caps: a high-resolution but geologically short-lived archive. However, there are other, more permanently fixed proxies of this large and rapid ground-to-surface hydrocarbon transfer (described in Section 5.2.3).

5.2.2 Atmospheric Methane (CH_4) Concentrations

Methane is present in even smaller amounts in the atmosphere than is CO_2 and so is measured in parts per billion. However, it is yet more powerful as a greenhouse gas and shows a complex pattern of fluctuations in the polar ice record that reflects a balance between production/emission (from swamps, warming permafrost, ruminant stomachs and agriculture) and

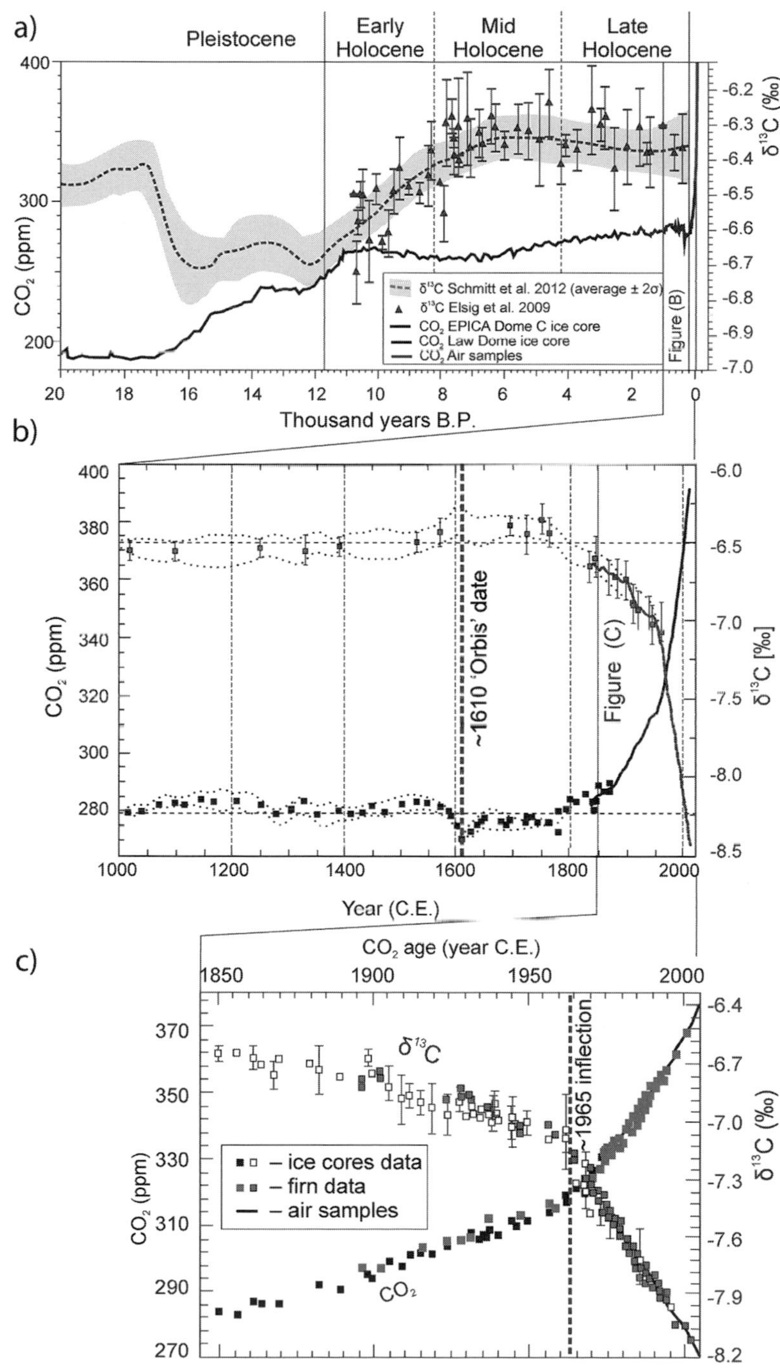

Figure 5.2.2 Glaciochemical CO_2 concentrations and carbon isotope ratios (from figure 5 of Waters et al. 2016 and references therein). (a) Atmospheric CO_2 from the Antarctic Law Dome and EPICA Dome C ice cores combined with data from observed measurements and $\delta^{13}C$ from atmospheric CO_2; (b) CO_2 concentration and $\delta^{13}C$ from atmospheric CO_2 from the Law Dome ice cores showing a 10 ppm dip in CO_2 recognised as the Orbis event of Lewis and Maslin (2015); (c) CO_2 concentration and $\delta^{13}C$ from atmospheric CO_2 from the Law Dome ice core, firn data and air samples showing inflections at ~1965 CE. Reprinted with permission from AAAS.

destruction (in the atmosphere, it is quickly oxidised to CO_2). These (global) fluctuations are stratigraphically useful in helping to correlate the Greenland and Antarctic ice-core records, which show differences in other proxies that reflect substantial contrasts in climate history between North and South hemispheres.

The Quaternary pattern of CH_4 seen in the ice-core record is similar to that of CO_2, showing rhythmic decreases (in glacial phases) and increases (in interglacial phases) between ~400 and ~800 parts per billion (ppb) (Loulergue et al. 2008; Wolff 2014). From this context, the positive excursion into Anthropocene levels appears yet more striking than that of carbon dioxide (Figure 5.2.1).

The Holocene methane record fluctuates between about 590 and 760 ppb (Figure 5.2.3b), and this may reflect fluctuations that are either natural, particularly in the extent of wetlands (Blunier et al. 1995), or anthropogenic, through biomass burning (Ferretti et al. 2005) or the effect of early farming (Ruddiman 2013). From about 1875 CE, though, CH_4 more than doubles to its present level of ~1,800 ppb (Nisbet et al. 2016). This rise is thought due to emissions from growing wetlands, but it may also include emissions from anthropogenic sources, including decaying vegetation in reservoirs, burning of vegetation, escapes of natural gas from pipelines and so on.

The CH_4 rise, as with that of CO_2, is, as a stratigraphic record, mainly written in the geologically temporary ice layers. As with CO_2, though, it has a wider importance both in signalling a significant change to the Earth System (Section 7.3.4) and in having wider effects on Earth processes such as climate (Section 6.1).

5.2.3 Stable Carbon Isotope Record and the Suess Effect

Living matter preferentially takes in a little more of the abundant stable isotopic form of carbon, ^{12}C, than of its less abundant and heavier stable isotope, ^{13}C. Hence, when a large amount of organically derived carbon (such as coal and petroleum) is buried, taking its excess of light carbon with it, then the surface carbon reservoir becomes correspondingly enriched in the heavy ^{13}C isotope, and this is reflected in the composition of plants and shells formed within the surface environment.

If the buried coal and petroleum are then released from the ground (e.g., by being burnt for energy), then the excess of ^{12}C floods back to the surface environment, changing ('lightening') the composition of the surface carbon reservoir. This change is then reflected in plants and shells, which take up this new isotopic pattern. Moreover, while modern plants also contain the rare and unstable radioactive isotope ^{14}C, there is none of it in fossil fuels. The addition of extra ^{14}C to the atmosphere from atomic testing changes the isotopic composition of the carbon in living things. This interference is called a 'bomb peak' ^{14}C and is often used in forensic as well as environmental studies. Another interference is known as the Suess effect, after the chemist Hans Suess who first noted it as a factor influencing the precision of radiocarbon dating. The Suess effect is also used to refer to the addition of petroleum-derived ^{12}C to the atmosphere, changing the $^{12}C/^{13}C$ ratio. This changing ratio is, like the perturbation of atmospheric CO_2, already marked, tracing out as a steepening increase in lighter carbon that is now about 2 permil (parts per thousand) above industrial levels (Elsig et al. 2009; Schmitt et al. 2012; Figure 5.2.2b herein). This change is quite large in geological terms and much more pronounced than any change in the Holocene (a slight tendency towards heavier carbon in the surface environment of less than 0.5 permil).

The Suess effect on stable carbon isotopes has been recognised in diverse environments, including tree rings (e.g., Loader et al. 2013), shells, bones, limestones, marine microfossil shells (e.g., Al-Rousan et al. 2004, corals (e.g., Sherwood et al. 2005a; Swart et al. 2010) and the atmospheric composition of air bubbles trapped in ice. It is not yet as large as the pronounced carbon isotope anomaly (again a lightening – of ~5 permil, associated with an ancient global warming event) that has been chosen to mark the Paleocene-Eocene boundary (Aubrey et al. 2007;

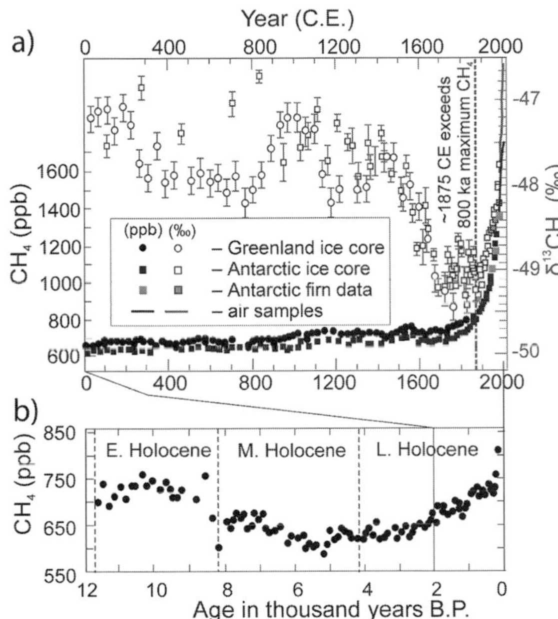

Figure 5.2.3 Glaciochemical CH_4 concentrations and carbon isotope ratios (from figure 5 of Waters et al. 2016 and references therein). (a) Antarctic and Greenland ice-core and firn records for CH_4 concentration and $\delta^{13}CH_4$ for the last two millennia; (b) Greenland ice-core records for CH_4 concentration throughout the Holocene. Reprinted with permission from AAAS.

see Section 1.3.1.4 herein), but it has been produced much more quickly and is correspondingly sharp as a stratigraphic signal. Carbon-cycle perturbations of the geological past, whether involving the surface carbon reservoir becoming overall lighter or heavier, have become key indicators associated with a number of major chronostratigraphic boundaries, including those at the base of the Ediacaran, Cambrian, Triassic and Jurassic periods (Gradstein et al. 2012), as well as the Eocene Epoch and Hirnantian Age.

The ratio of ^{12}C–^{13}C is expressed as $\delta^{13}C$. As well as being used as a stratigraphic marker for atmospheric CO_2, it can also be used to determine the likely cause of the methane increase. The $\delta^{13}C$ curve for CH_4 shows a marked increase of ~1.5 permil towards lighter carbon from ~1500 to 1700 CE (Figure 5.2.3a), perhaps a response to reduced biomass burning (Ferretti et al. 2005). An abrupt shift towards heavier $\delta^{13}C$ values in atmospheric CH_4 of ~2.5 permil since ~1875 CE (Figure 5.2.3a) is considered to be in response to increased pyrogenic emissions from fires – presumably arising from forest clearance for agriculture (Ferretti et al. 2005). That shift has been followed over the last 12 years by a negative shift of 0.17 permil of $\delta^{13}C$ in the increasing volume of atmospheric CH_4, attributed to biogenic sources such as an expansion of tropical wetlands or emissions from increased agricultural sources, like ruminants and rice paddies, and a decrease in biomass burning (Nisbet et al. 2016).

5.3 Boron Isotopes as a Proxy for Oceanic pH

Colin N. Waters, Jan Zalasiewicz, Reinhold Leinfelder and Colin P. Summerhayes

5.3.1 Ocean Acidification

About a third (27%) of the carbon dioxide released by burning hydrocarbons between 1959 and 2014 is dissolved in the sea (Canadell et al. 2007; Le Quéré et al. 2015). This helps slow the rate of increase of CO_2 in the atmosphere, but at the expense of changing the chemistry of ocean waters. Dissolution of CO_2 in seawater lowers its pH, so far in surface waters by 0.1 part of a pH unit, from ~8.2 during pre-industrial times to ~8.1 today (Tyrell 2011). That represents a ~30% increase in hydrogen ions, given that the pH scale is logarithmic.

Stratigraphically detectable effects of this 'acidification' (a relative change, given that ocean waters are still higher in pH (i.e., more alkaline) than the neutral point of 7.0) include its effect on the calcification of shells, especially of organisms that build their shells out of the more soluble calcium carbonate mineral, aragonite, such as corals (for future pH-based scenarios see, e.g., Gattuso et al. 2015; Hughes et al. 2017a, b). The picture is complicated, because the correlation of pH and chamber thicknesses in calcareous nannoplankton suggests that decreasing pH may possibly assist shell growth in those organisms (Meier et al. 2014; Bolton et al. 2016; see Section 3.4.3 herein for more). Ocean acidification also causes dissolution, rather than precipitation, of calcium carbonate sediment and shell remains. At present, most calcium carbonate present in shallow marine waters persists because surface waters are supersaturated with calcium carbonate. But as pressure increases and temperature decreases with water depth, and as sinking organic matter is oxidised, releasing CO_2, the carbonate saturation of seawater decreases with depth. That means that about half of the biogenic calcium carbonate 'snow' settling in deep oceanic waters dissolves (Sabine et al. 2004). At a certain depth, which varies with the extent of carbonate saturation, the rate of carbonate dissolution increases dramatically. Below that depth, known as the lysocline, the rate of dissolution increases to the point where it matches the rate of supply, and no carbonate can accumulate at greater depths. That point forms the carbonate compensation depth, or CCD. The CCD lies at an average depth of about 4,500 m but reaches 5,000–6,000 m in the Atlantic and only 4,000–5,000 m in the Pacific; it lies in greater depths in equatorial regions, where the rate of supply of carbonate sediment is high, and at shallower depths in polar regions, where extremely cold bottom waters are more corrosive because cold water contains more CO_2 and dissolves more carbonate than warm water does. The CCD was generally shallower in Cretaceous to Eocene times (Tyrell & Zeebe 2004), when CO_2 was more abundant in the atmosphere, in turn increasing the concentration of CO_2 in the ocean, which decreased the carbonate saturation of seawater, making carbonates dissolve at shallower levels. At times of enhanced CO_2 release, as during the Paleocene-Eocene Thermal Maximum, carbonates dissolved at yet shallower levels. In the future, as the CO_2 concentration in the atmosphere increases, its abundance in the ocean will increase further. Ocean circulation processes will then ensure that deep waters eventually become undersaturated in carbonate, which will make deep-ocean carbonate sediments dissolve and so raise the CCD and the lysocline.

Oceanic pH and CO_2 concentrations vary spatially due to deepwater upwelling, local variations in biological productivity and inflows of freshwater (Pearson & Palmer 2000). The highest concentrations of anthropogenic CO_2 tend to be in surface waters, with mixing to deep waters being on a centennial to millennial scale; the greatest mixing throughout the water column is in the North Atlantic, followed by the Southern Ocean north of the Polar Front (Sabine et al. 2004). Therefore, the contemporary effects of oceanic acidification due to modern anthropogenic CO_2 generation will likely be most prominent in those places but negligible elsewhere in deep ocean waters. Ultimately, as anthropogenic CO_2 penetrates the whole ocean depth, making deep waters undersaturated in carbonate, deep-sea oozes formed from calcareous phytoplankton and foraminiferal remains will dissolve and become increasingly rare, as occurred during the Paleocene-Eocene Thermal Maximum (PETM) described in Sections 1.2.1.4 and 6.1.1 (Tyrell 2011).

5.3.2 Use of Boron Isotopes as a Deep-Time Proxy for Ocean pH

Boron present in the oceans occurs as either boric acid $B(OH)_3$ or borate $B(OH)_4^-$ with the former enriched in the ^{11}B isotope and the latter (relatively enriched in the ^{10}B isotope) preferentially incorporated into marine carbonates. As this process of relative enrichment depends upon the pH of the water, it results in the isotopic fractionation of the two isotopes, which can be measured from the $^{11}B/^{10}B$ ratio in seawater and marine carbonates and recorded as $\delta^{11}B$ (Hemming & Hanson 1992; Pagani et al. 2005). The residence time of oceanic boron is some 14–20 Myrs, so variation in the $\delta^{11}B$ of seawater is unlikely over durations of millions of years (Pagani et al. 2005).

Boron isotope ratios of planktonic foraminifer shells have been used to estimate the pH of surface-layer seawater throughout the past 60 million years. The pH values so obtained can be used to reconstruct

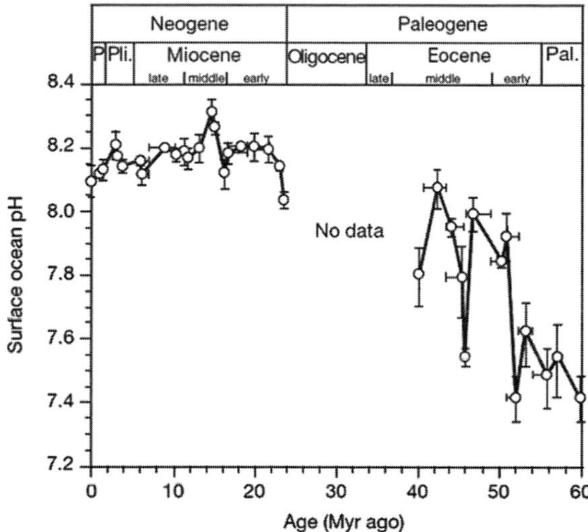

Figure 5.3.1 Sea surface pH for the past 60 Myr derived from $\delta^{11}B$ analysis of foraminifer shells. (reprinted with permission from figure 1 of Pearson & Palmer 2000). P., Pleistocene; Pli., Pliocene; Pal., Paleocene. ©2000 Nature

atmospheric CO_2 concentrations (Pearson & Palmer 2000). Care must be taken in applying this technique, because different foraminifer species calcify at different water depths, which may have distinctly different pH values (Pearson & Palmer 2000). The data show that pH steadily increased during the early Cenozoic, reaching peak values of 8.3 U (Units) in the Mid-Miocene, about 14 Myrs ago (Figure 5.3.1), and has subsequently declined to the modern values described earlier.

Boron isotopes extracted from the planktonic foraminifer *Globigerinoides sacculifer* provide a proxy of pH for the last 2.1 Myrs and suggest that the present-day atmospheric pCO_2 is the highest seen during that period (Hönisch et al. 2009). Prior to the Mid-Pleistocene Transition (MPT), at ~1.25–0.7 Myrs ago, $\delta^{11}B$ was characterised by glacial intervals with more acidic ocean chemistries (pH 8.24) relative to glacials during the last 700,000 years (pH 8.29), whereas pH during interglacials was comparable between the two intervals and overall notably more acidic (pH 8.14) (Hönisch et al. 2009). Correspondingly, CO_2 as measured in bubbles of fossil air from ice cores was also more abundant in interglacial than in glacial atmospheres (Wolff 2011; Section 5.2.1 herein).

5.3.3 Use of Boron Isotopes in Modern Carbonates as a Potential Anthropocene Marker

Modern oceanic pH values have only been routinely observed for less than a couple of decades (e.g., http://hahana.soest.hawaii.edu/hot/trends/trends.html), and although the trend of decreasing sea surface pH with rising temperature is evident in these observations, variation during the early and mid-20th century has not been determined directly. $\delta^{11}B$ isotopic records from foraminifera and corals have been used to provide a modern proxy for seawater pH and are described below.

Data from a *Porites* coral in the Great Barrier Reef covering an interval from 1800 to 2004 CE are used as evidence for recent acidification, which shows a strong positive correlation from 1940 to 2004 CE with the $\delta^{13}C$ Suess effect (Section 5.2.3), i.e., increasing dissolved CO_2, and with warming deduced from increasing Mg/Ca ratios (a proxy for sea surface temperature or SST) (Wei et al. 2009; Figure 5.3.2 herein). Pre-industrial boron isotope signals in the coral demonstrate 22- and 10-year cycles strongly controlled by the Pacific Decadal Oscillation (Wei et al. 2009). From 1940 to 2004, the $\delta^{11}B$ signal decreased from 25.1‰ to <22‰, representing a decrease in pH of about 0.2–0.3 U, but with marked annual oscillations of 0.5 U around 1940 and 1998 (Wei et al. 2009). Further analysis of *Porites* from Chichijima (Japan) and Tahiti shows no systematic variation in $\delta^{11}B$ between 1914 and 1954 CE, but after 1960 a clear decrease is observed, consistent with increased acidification (Kubota et al. 2015). This provides a potential signal that links ocean acidification and the Suess effect across the Anthropocene boundary, as described in this book. However, algal symbionts associated with corals may have a 'vital effect' on the $\delta^{11}B$ signal, according to

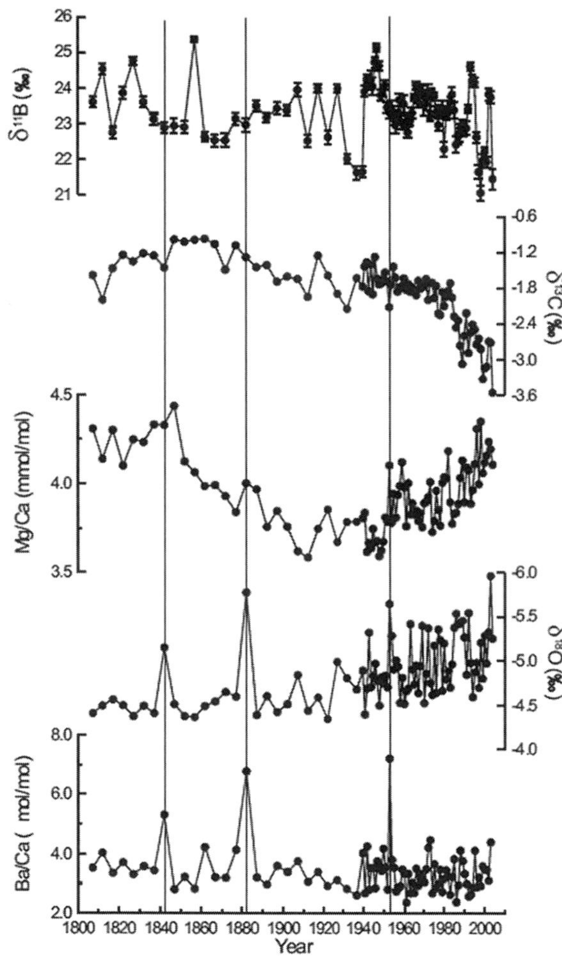

Figure 5.3.2 Temporal variations of $\delta^{11}B$, $\delta^{13}C$, Mg/Ca, $\delta^{18}O$ and Ba/Ca ratios of *Porites* coral from the Great Barrier Reef. Vertical lines indicate the pronounced oscillations of $\delta^{18}O$ and Ba/Ca ratios around 1840, 1880 and 1953 CE. (from figure 4 of Wei et al. 2009). ©2009, with permission from Elsevier

Hemming et al. (1998), who questioned the suitability of corals for studies of the boron isotope signal. Similar effects have been proposed for planktonic foraminifers (Pagani et al. 2005). Post-depositional diagenesis results in enrichment of ^{10}B, particularly within bulk carbonate samples and corals (Pagani et al. 2005), providing further concern about the viability of the technique. The $\delta^{11}B$ signal in *Porites* corals in the Great Barrier Reef shows seasonal cycles of amplitude typically greater than 1 permil and much greater than that observed in the seawater, with lower isotopic values and hence lower pH in the summer (D'Olivo & McCulloch 2017). In addition, the nature of the $\delta^{11}B$ signal is considered by D'Olivo and McCulloch (2017) to vary markedly in response to bleaching events (see Section 3.4). Nevertheless, it seems clear from comparisons across long timescales that the boron isotope signature does generally tie in with past estimates of seawater pCO_2 (Figure 5.3.1). With regard to using this signal in chronostratigraphic characterisation of the Anthropocene, at present the annual to decadal variability is obscuring the underlying signal attributable to changes in anthropogenic CO_2.

5.4 Nitrogen and Phosphorus

Jan Zalasiewicz

5.4.1 Nitrogen

By far the greater part of the Earth's surface nitrogen store is as the stable and generally unreactive dinitrogen molecule N_2 in the atmosphere. In nature, formation of the essential biologically available forms of nitrogen is a difficult and energetically demanding chemical process that only a few organisms are able to carry out. Nitrogen-fixing bacteria initially produce simple ammonia-based compounds that are then mostly oxidised to nitrates (the main biologically available form) by nitrifying bacteria. Other forms of bacteria (denitrifying bacteria) can use nitrates as an energy source to return them back to the dinitrogen form.

These changes are the basis of the nitrogen cycle (Lenton 2016), which was only minimally perturbed by humans until the 19th century, when mined nitrate deposits (notably of sodium nitrate from Chilean caliche deposits) began to be used to help boost agricultural production. This could not keep

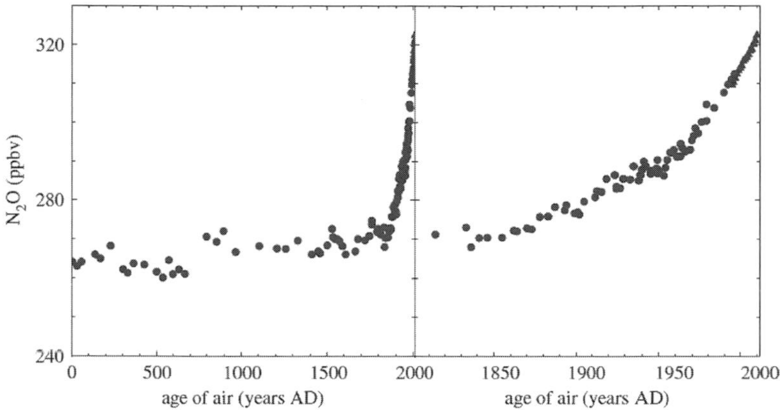

Figure 5.4.1 N_2O concentrations in the Law Dome ice core (and firn (blue circles) over the last 2,000 years and 200 years and annual averages for South Pole air (blue triangles). (from figure 1 of Wolff 2013) (A black-and-white version of this figure appears in some formats. For the colour version, please refer to the plate section.)

up with increasing demand. A critical breakthrough was made by Fritz Haber and Carl Bosch, who invented a process to synthesise ammonia from atmospheric nitrogen. That made the supply of nitrate potentially limitless, albeit calling for considerable energy input. Industrial production started in 1913 and climbed steeply from the mid-20th century to its present level of some 450 million tons of nitrate fertiliser per year (Steffen et al. 2015; see Section 7.5 and Figures 7.5.1 and 7.5.3 herein). This production keeps something like half of Earth's human population alive.

The resulting impact is, proportionately, greater than the coeval change to the carbon cycle (Section 5.2) and is likely to be the greatest perturbation to the Earth's nitrogen cycle for 2.5 billion years (Canfield et al. 2010). Stratigraphic impacts include increasing concentration of N_2O in polar ice, initiated ~1850 and with a further upturn from ~1950 (Figure 5.4.1), and changes in the ratios of stable isotopes of nitrogen (specifically, decreasing amounts of the heavy isotope ^{15}N relative to ^{14}N in far northern lakes (Holtgrieve et al. 2011; Wolfe et al. 2013) and polar ice (Hastings et al. 2009). These isotopic shifts started around the beginning of the 20th century (when nitrogen supply from fossil-fuel combustion was proportionately a more significant source) but became significantly more pronounced in the mid-20th century as fertiliser production soared and as unused nitrate was blown worldwide by the wind. Slight differences between the isotopic signatures from lake and ice cores mainly reflect fractionation during transport and deposition (Wolfe et al. 2013). Overall, nitrogen isotopes provide a robust and correlatable stratigraphic signal of anthropogenic change.

This record of anthropogenic nitrogen perturbation seen in isotope ratios is mirrored in the stratigraphic record of the Greenland ice sheet, which shows a marked 20th-century increase in total nitrogen oxides (NO_x) deposition, especially marked in the mid-20th century (Hastings et al. 2009; Figure 5.4.2 herein). This increase is much less pronounced or absent in Antarctic ice, which is far more distant from anthropogenic sources of nitrogen and (in this respect) is nearer to being pristine.

The increase in surface-reactive nitrogen has produced a number of secondary stratigraphic signals, including marked biological effects. The nitrogen isotope changes in the far northern records are accompanied by marked changes in the populations of diatoms (aquatic single-celled algae that produce a fossilisable silica skeleton) (Wolfe et al. 2013). Indeed, it was changes to lacustrine diatom populations resulting from industrialisation that prompted Eugene Stoermer to invent and start using the term 'Anthropocene' informally several years before Paul Crutzen first

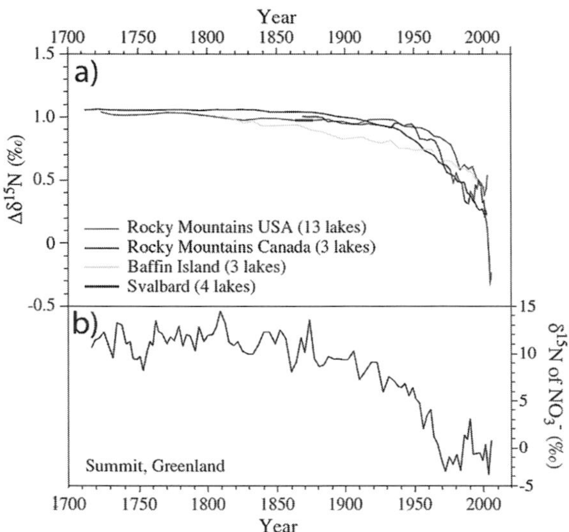

Figure 5.4.2 Changes in nitrogen-isotope ratios (a) in high-latitude northern lakes and (b) in nitrate deposition on the Greenland ice cap. Reprinted from figure 6 of Wolfe et al. (2013). ©2013, with permission from Elsevier. (A black-and-white version of this figure appears in some formats. For a colour version, please refer to the plate section.)

used it. And as nitrate fertiliser (with phosphorus as well, as discussed below) is washed off the land into rivers and coastal seas, it has over recent decades produced widespread oxygen-starved 'dead zones'.

The dead zones form through over-fertilisation of seawater, leading to the formation of large plankton blooms that then die, sink and decay, creating hypoxic conditions in which most or all macrobenthos suffocates (Diaz & Rosenberg 2008) and in which fish die. About 400 individual dead zones have now been identified, covering a total area of ~250,000 km^2; well-known ones are in the Gulf of Mexico, Chesapeake Bay on the US East Coast and the Baltic Sea (Figure 5.4.3).

Most dead zones are seasonal, with hypoxia and benthic kill-off in the summer and oxygen returning via autumn and winter storms. They can be identified stratigraphically through an increase in hypoxia-tolerant benthic foraminifera, for example, in the Gulf of Mexico (Osterman et al. 2008; Figure 5.4.4 herein). Stratigraphic analysis shows that near-shore hypoxia in this region occurred naturally at roughly centennial intervals but is now annual.

5.4.2 Phosphorus

Phosphorus, like nitrogen, is a key limiting nutrient for organic growth and hence for agricultural production. Farmers during the 19th century had by trial and error realised that the addition of animal bone to soils increased crop yields, and this source of fertiliser was developed through the mid- to late part of the century. Sources included animal bone from abattoirs; human bone from battlefields (at its height with some three million human skeletons, mostly scavenged from European battlefields, being recycled in this way in Britain; Perry 1992); dinosaur and marine-reptile bones and phosphatised faecal matter derived from the coprolite trade of the mid-19th century in England (O'Connor 2001) and mainly sourced from Mesozoic and Cenozoic 'bone beds'; and South American guano deposits (Mathew 1970).

Production in the 20th century continued to rise in the pattern of the Great Acceleration, mainly from geologically rare stratiform phosphate deposits located in a few countries (notably Morocco). Surface-reactive phosphorus has, like the stock of nitrogen, more than doubled (Filippelli 2002 – a trend that cannot be maintained like that of nitrogen, given the limited resources available in the Earth's crust (Cordell et al. 2009). Nevertheless, substantial amounts of sedimentary phosphate are known to cover the seabed (e.g., on New Zealand's Chatham Rise) and may be mineable if the economics are favourable (Cook & Shergold 1986).

Despite this increase, the stratigraphic signal from phosphate use is unlikely to be as clear as that of carbon or nitrogen, given the lack of isotopic signatures that may be used as tracers of this perturbation. However, it has contributed to indirect stratigraphic signals, such as the marine dead zones referred to above. And, soils and sediments impacted by phosphate fertiliser will contain raised concentrations of phosphate.

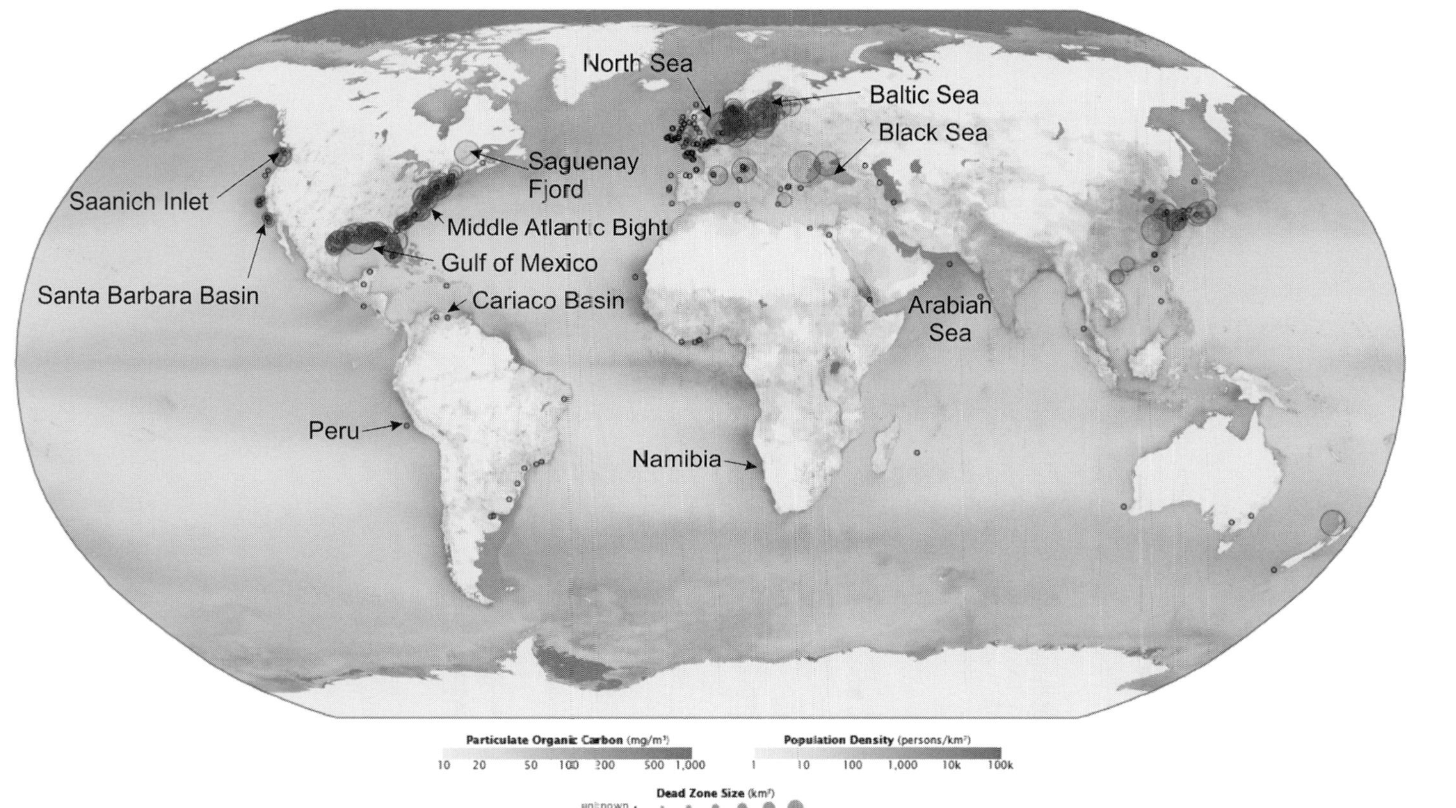

Figure 5.4.3 Distribution of marine dead zones globally. (from NASA Earth Observatory https://earthobservatory.nasa.gov/IOTD/view.php?id=44677); Aquatic Dead Zones generated 17th July 2010) (A black-and-white version of this figure appears in some formats. For a colour version, please refer to the plate section.)

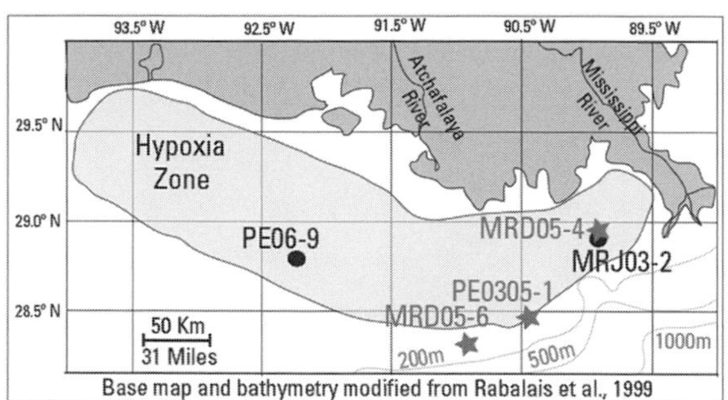

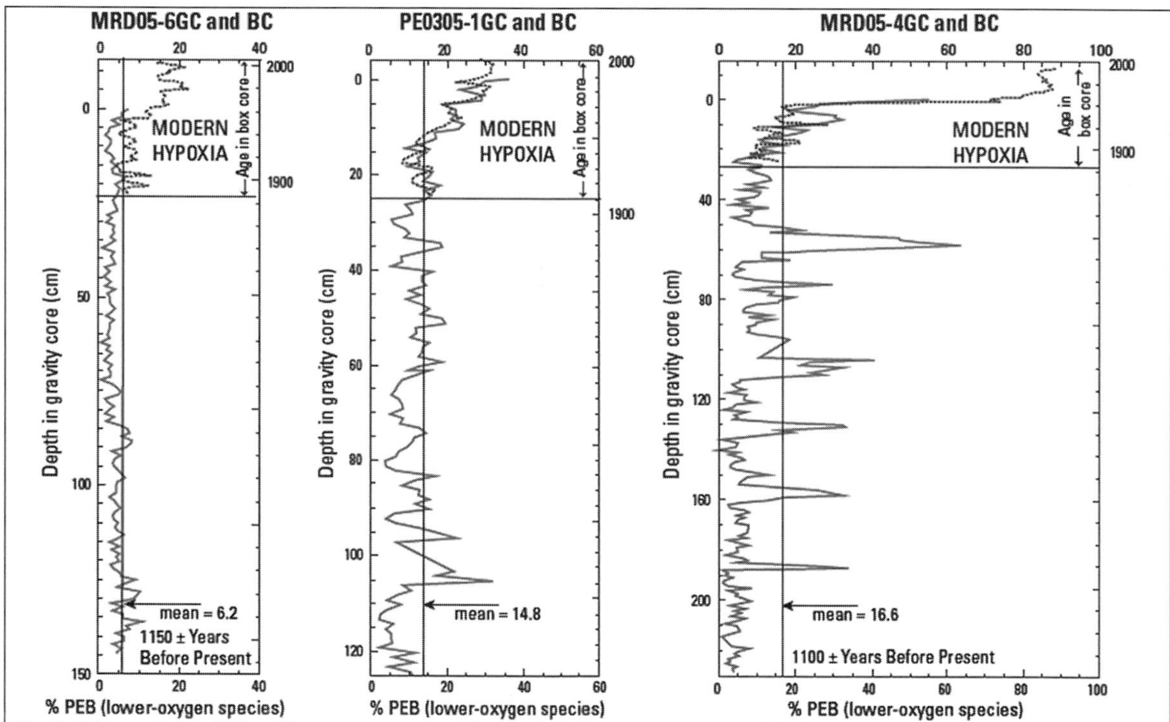

Figure 5.4.4 The dead zone on the Louisiana shelf in the Gulf of Mexico (top) and the response to changing bottom conditions as clearly seen by the temporal pattern of the PEB Index (the cumulative percentage of three foraminifer species, *Protononium atlanticum*, *Epistominella vitrea* and *Buliminella morgana*). From figures 1 and 3 of Osterman et al. (2008). ©2018, with permission from Elsevier.

5.5 Sulphur

Ian J. Fairchild

The sulphur signal of the Anthropocene is derived primarily from aerosols sourced from activities such as the burning of coal containing mineral sulphide impurities, and it is preserved in sedimentary archives as sulphate. Scattering of incoming radiation by sulphate aerosols in the stratosphere has mitigated greenhouse-gas warming of the troposphere during the Anthropocene with a forcing of -0.4 ± 0.2 W m^{-2} (Boucher et al. 2013). While these aerosols are commonly referred to as sulphate aerosols, they largely comprise droplets of sulphuric acid formed by the combination of sulphur dioxide gas with water vapour.

Sulphur provides a useful adjunct to the anthropogenic carbon, nitrogen and radiogenic isotope signals in continental environmental archives, having a mid-20th-century profile aligned with the Great Acceleration (Section 7.2.5) in the Northern Hemisphere, with local to regional variations. This is superimposed on a natural background that also has strong spatial variations and that can change dramatically after volcanic eruptions that provide periodic inputs of sulphur dioxide to the atmosphere.

5.5.1 The Global S Cycle: Phases, Fluxes, Lifetimes and Distributions

Although the anthropogenic interest centres on the atmosphere, this represents a small reservoir of sulphur (~5 million metric tons or Mt) compared with marine biota (30 Mt), lakes and rivers (300 Mt), soils and land biota ($3*10^5$ Mt) and ocean water ($1.3*10^9$ Mt) (Charlson et al. 1992). The residence time of sulphate (SO_4^{2-}) ions in seawater is several million years, so marine archives are not capable of recording the short-lived anthropogenic disturbances. Anthropogenic sulphate aerosols in the troposphere readily serve as cloud condensation nuclei and so have an atmospheric lifetime of only around one week (Boucher et al. 2013). For this reason, sulphur archives derived from such aerosols have strong spatial variations. In contrast, large explosive volcanic eruptions can transmit sulphur to the stratosphere, resulting in a global distribution of sulphate over a lifetime of typically one to three years.

Sulphate constitutes 7.7% of sea salt by mass. Bursting sea-spray bubbles produce fine sulphate aerosol particles (typically about 0.15 mg), totalling 1,400–6,800 Mt/yr (Boucher et al. 2013), 90% of which are deposited over the oceans (Brimblecome 2005). These, along with sulphates in aeolian dust and associated with some organic particles, are referred to as primary aerosols. The associated sea-salt content of terrestrial rainfall decreases rapidly within a few tens of kilometres inland. Workers seeking to emphasise other sources of sulphate in natural waters and ice cores often calculate 'non-sea-salt sulphate' by removing that fraction of total sulphate from an analysis that is in proportion to the chloride or sodium content seen in seawater.

Sulphur dioxide (SO_2) in the atmosphere arising from pollutants and volcanic emissions totals around 150 Mt S/yr. Although some sulphur dioxide is removed by 'dry' deposition onto leaves, most is oxidised to sulphuric acid, mainly within micron-sized (secondary) aerosol droplets, leading to deposition in the form of sulphate. The oxidants include the hydroxyl radical OH, O_3 and H_2O_2, and the process can be catalysed by transition metals (von Glasow et al. 2009). Hydrogen sulphide (H_2S) is also released naturally from biotic and pollution sources, probably in comparable amounts to sulphur dioxide, but is normally not discussed separately, as it is both reactive and analysed less often.

Marine algae release the volatile species dimethylsulphide (DMS) at rates of around 20 Mt S/yr (Brimblecombe 2005). This in part is oxidised to sulphate within aerosols in the atmosphere, forming mostly MSA (methane sulphonic acid), which can be broken down by bacteria to CO_2 and sulphate in soils. Other minor species present in the atmosphere that are largely derived from the oceans include carbon disulphide (CS_2) and carbonyl sulphide (COS), with emissions totalling 1 Mt S/yr (Brimblecombe 2005). COS has recently emerged as an indicator of gross primary productivity on land, as it is destroyed by hydrolysis when taken up by plant leaves, but also 20% of the emissions arise from anthropogenic manufacture of rayon, which largely became concentrated in China over the past 30 years (Campbell et al. 2015). Its atmospheric lifetime is 2 years.

Sulphate aerosols in the stratosphere can increase by up to two orders of magnitude a few months after volcanic eruptions. Starting with the work of Zielinski et al. (1994), ice cores have provided detailed records of past volcanic activity. However, mismatches between the abundance of volcanic ash particles and non-sea-salt sulphate in ice cores

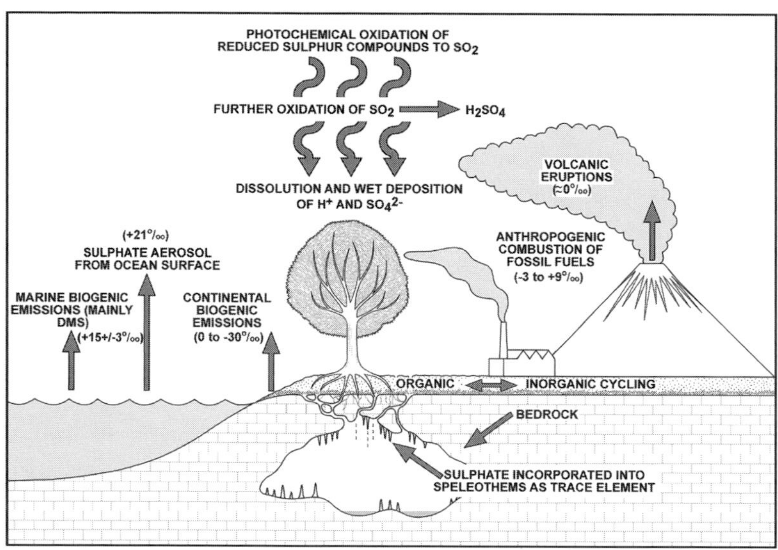

Figure 5.5.1 Sources and cycling of sulphur contributing to archives with characteristic $\delta^{34}S$ compositions (Reprinted from figure 1 of Wynn et al. 2008). The diagram also illustrates issues connected with soils and speleothem archives in caves. ©2008, with permission from Elsevier.

show that there is great variability in the capacity of individual volcanic eruptions to produce significant sulphur emissions. Relatively cool silica-rich magmas may yield little sulphur, while only highly explosive eruptions will result in significant output of ash and aerosols to the stratosphere, ensuring wide geographical distribution. Volcanic fumaroles provide more local and continuous outputs of sulphur species. Variations in subsurface fluid oxidation state in volcanic systems lead to differing proportions of oxidised and reduced forms of gaseous sulphur species (von Glasow et al. 2009). Although estimating the time-averaged volcanic output of sulphur species is difficult, it is likely much smaller than current anthropogenic emissions.

The stable isotope (δ ^{34}S) composition of sulphate sources provides a complementary parameter to assess the impact of anthropogenic S inputs (Figure 5.5.1 herein, Wynn et al. 2008). In particular, volcanic and pollution sources tend to have much lower δ ^{34}S signatures than do marine-derived S or rock sulphates.

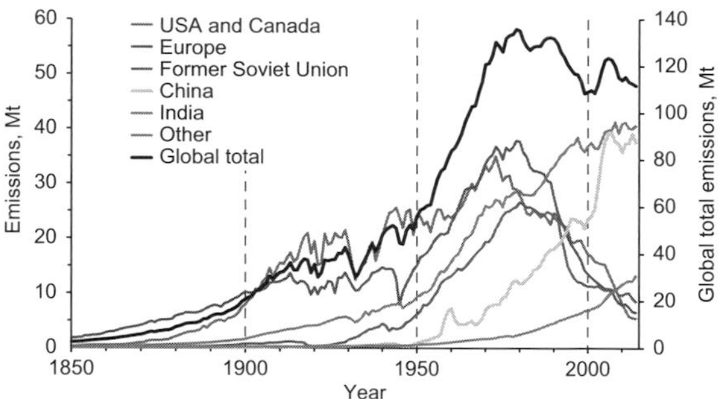

Figure 5.5.2 Emission inventories (in Mt SO_2) by region derived from bottom-up methodologies. Plotted from data in Hoesly et al. (2017). (A black-and-white version of this figure appears in some formats. For a colour version, please refer to the plate section.)

5.5.2 Anthropogenic Emissions

The primary source of pollutant atmospheric sulphur is the combustion of coal, with major contributions, particularly since 1950, also from industrial processes such as oil and gas refining and metal smelting (Smith et al. 2011; Figure 5.5.2a herein). In addition to pyrite, coal contains associated organic S species such as

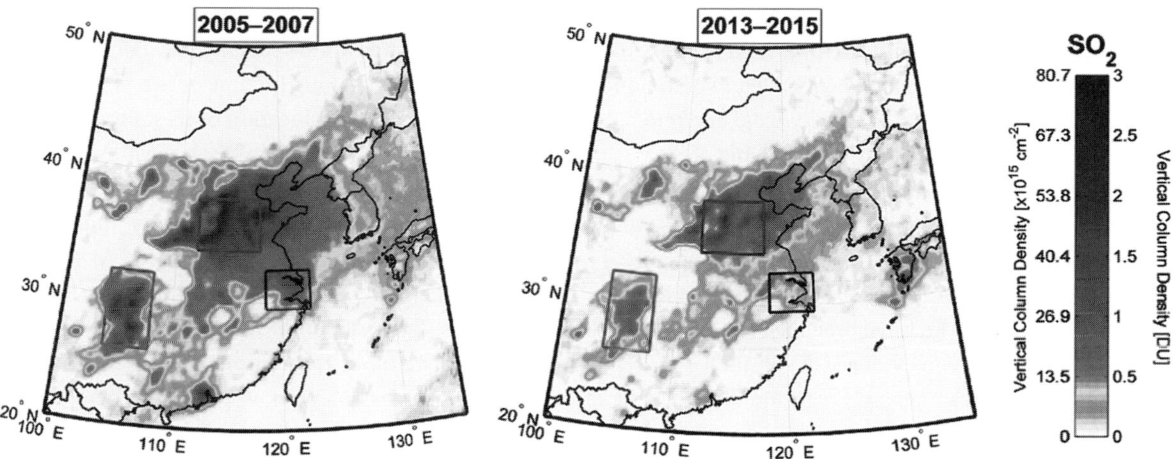

Figure 5.5.3 Satellite-derived OMI images of changing sulphur emissions over China. Emissions in hotspots are clearly falling between 2005–2007 and between 2013–2015, reversing the strong rise in total emissions seen in earlier years (Figure 5.5.2). From part of figure 5 of Krotkov et al. (2016). (A black-and-white version of this figure appears in some formats. For a colour version, please refer to the plate section.)

thiols, thiophenes and sulphides (Brimblecombe 2005). The 20th century saw dramatic increases in S emissions from burning fossil fuels, as determined by 'bottom-up' methods of summing national or regional totals. In particular, all industrialised areas saw significant rises in the decades following 1950 (Figure 5.5.2). The sulphuric acid generated made the largest contribution to the environmental problem of acid rain, but emissions levelled off globally in the 1970s as falling emissions in North America and western Europe were offset by developing economies (Hoesly et al. 2017; Figure 5.5.2 herein). Global coal production has actually increased to the present day, but the key to reduced emissions has been the introduction of desulphurisation technologies such as the treatment of exhaust gases with lime.

The geographical source of S emissions gives a good guide to the abundance of secondary sulphate aerosols because of their short atmospheric lifetime. Over the past 20 years, development of satellite remote-sensing techniques has led to an intimate knowledge of changing S loading by using advanced UV–visible light spectrometers such as the Ozone Monitoring Instrument (OMI). Satellite approaches have led to identification of 10% additional sources globally that were not identified in the bottom-up methods (McLinden et al. 2016). The evolving fine-scale pattern of industrial S emissions can be imaged (Figure 5.5.3 herein; Krotkov et al. 2016) and gives an indication of expected spatial variability in continental S archives. One disadvantage of the satellite technology is that the OMI instrument sees only around half the total emissions over the wider regions because of a high (40 kt/yr) detection limit per pixel (McLinden et al. 2016), so the spatial variability in archives would be less than seen in OMI images. Between 2005 and 2013, OMI-detected anthropogenic emissions declined from >50 to 40 Mt/yr compared with total ~20 Mt/yr from point volcanic sources.

5.5.3 Ice-Core Archives

Sulphate is passively entrained into snow, making snow an excellent tracer of its atmospheric composition. Seasonality of deposition is expected because of varying meteorological conditions, e.g., very high-resolution studies in northern Greenland reveal a spring-summer peak of total sulphate in both pre-industrial and modern parts of the record (Bigler et al. 2002). A high-accumulation site in coastal East Antarctica (E Wilkes Land) revealed summer-season enrichment in both non-sea-salt sulphate (regarded as entirely biogenic from DMS) and MSA, corresponding to the timing of phytoplankton blooms in the surrounding oceans (Caiazzo et al. 2017). The sulphate is physically present in snow and underlying firn columns as solid cryogenic salts, while in deeper Antarctic ice sulphuric acid brines along crystal boundaries have been identified (Sakurai et al. 2016), although no evidence of subsurface migration of such brines has yet been found. Hence the ice cores from East Antarctica and Greenland provide the definitive long-term hemispheric scale records of atmospheric sulphur loading, expressed as non-sea-salt abundance. The main variability relates to short-term loading from volcanically derived aerosols, and the relative preservation of major eruptions shows that while tropical eruptions are preserved in both polar regions, Icelandic activity is prominent in Greenland and absent from Antarctica (Sigl et al. 2015). There can still be ambiguity as to the nature of the sulphur loading where the location of the source volcano is unknown. It is helpful, therefore, that characteristic non-mass-dependent sulphur-isotope signatures in ice cores (revealed by analysis of the isotopes ^{32}S, ^{33}S and ^{34}S) have been found to be present only in the products of stratospheric eruptions, reflecting the distinct photochemical reactions that occur there (Baroni et al. 2008).

No discernible anthropogenic total S loading is detectable in Antarctica (Figure 5.5.4 herein; Sigl et al. 2015), consistent with the origin from DMS of non-sea-salt sulphate at the Wilkes Land site noted above. This lack of anthropogenic input facilitates regional variations in the interpretation of $\delta^{34}S$ within Antarctica, with lower values in West Antarctica implying greater volcanogenic/stratospheric input (Kunasek et al. 2010).

However, one minor S species does show anthropogenic impact in Antarctica. Firn air measurements show COS increasing from a steady pre-industrial level to a peak of ~550 permil in the late 20th century and dropping below 500 permil at the beginning of the 21st century (Montzka et al. 2004). The preservation of this anthropogenic signal in Antarctica reflects the relatively long lifetime of two to three years for COS rather than the location of its source.

In contrast to Antarctica, a distinct anthropogenic hump is seen in the Greenland non-sea-salt S. The levels rose during the early 20th century and peaked in the mid-1970s (Figure 5.5.4 herein; NEEM site; Sigl et al. 2015). Broadly similar patterns are seen at other Arctic sites (Goto-Azuma & Koerner 2001), depending on the relative input from European and North American sources, which have slightly different emission patterns. At the Col du Dôme, French Alps (Preunkert & Legrand 2013), sulphate (largely non-sea salt) peaks around 1980, in accord with emissions in nearby countries, and its rise is accompanied by rising levels of water-soluble organic species (aerosols). In contrast, a thousand-year Himalayan record displays, against a lower background level, a different pattern of non-sea-salt-sulphate change. There, values nearly double between 1970 and 1997 (Duan et al. 2007). This clearly reflects the rising S emission trends in East Asia. Sulphur isotope studies in ice cores are limited by available sample size, but Patris et al. (2002) confirmed the increasingly industrial origins of sulphur in Greenland from 1860 to 1970 by Greenland's progressively lower $\delta^{34}S$ values.

5.5.4 Terrestrial Archives

Sulphur assimilation in plant proteins is primarily in specific amino acids, which is also the form in which they are found in tree rings, mirroring Anthropocene atmospheric changes (Fairchild et al. 2009). Advances in analytical capability now permit direct analysis of wood samples to derive both concentration and isotopic data for sulphur (Wynn et al. 2014), opening the way to wider use of these archives in studies of the S cycle. In terrestrial peat archives, S will be present in various oxidation states in humic substances, with possibly additional pyrite. Jeker and Krahenbühl (2001) found a close match between declining S concentrations in the atmosphere and those in peat

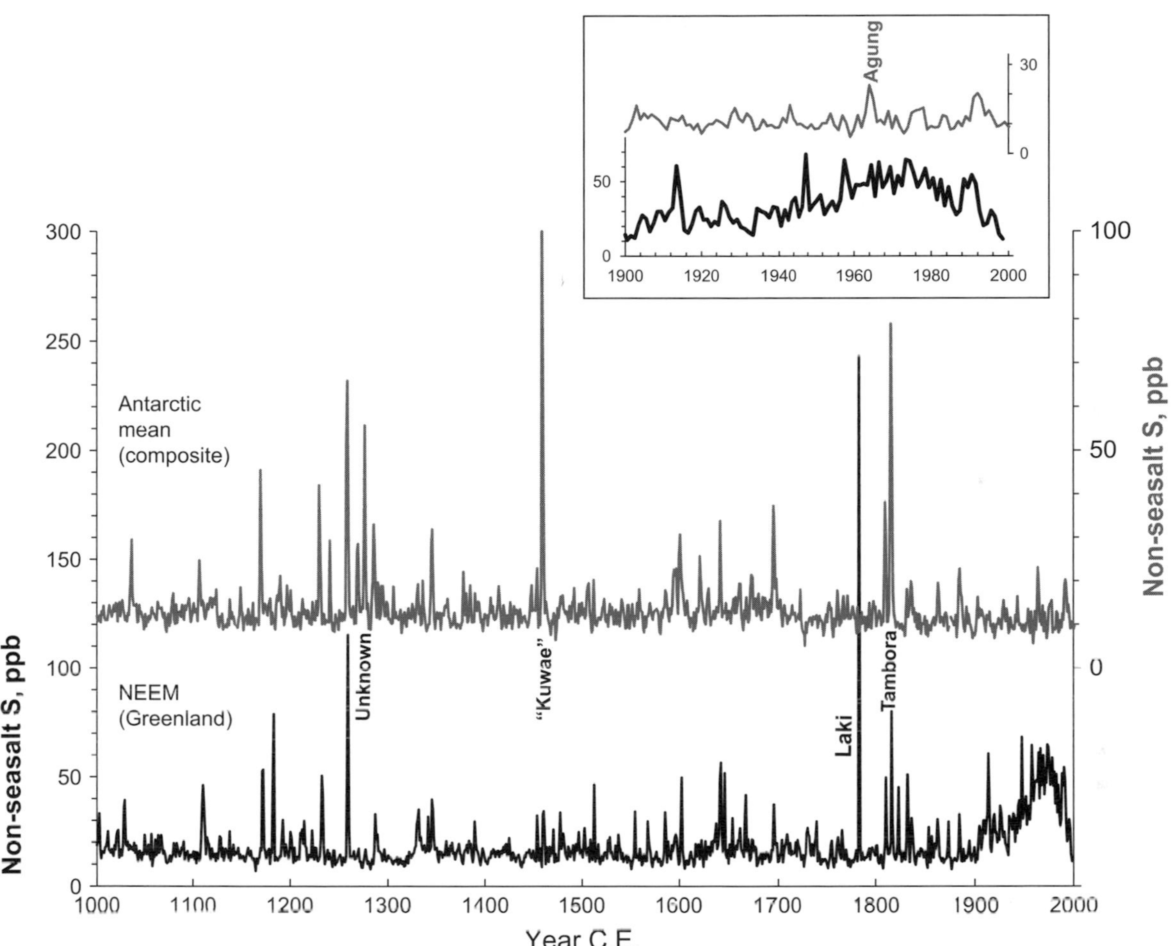

Figure 5.5.4 Non-sea-salt concentrations compared from Antarctica and Greenland ice cores with some of the most prominent volcanic eruptions labelled; inset shows detail of the 20th century plotted from data archived by Sigl et al. (2015). An anthropogenic late-20th-century peak is visible in the Greenland record but not in Antarctica.

profiles from two bogs in Switzerland. There are, however, additional difficulties in obtaining a precise age model in peat, which in this case was determined by ^{210}Pb dating. Secondary changes due to dissimilatory sulphate reduction have also been found in central European and British sites (Novak et al. 2005). Hence wood appears to be a more attractive option for chemostratigraphic archives.

Following the pioneering study of Frisia et al. (2005), the use of carbonate-associated sulphate in speleothems has been recognised as a tracer of changes in atmospheric sulphur composition, moderated by bedrock contributions and S cycling in soil and vegetation. In such karstic systems, studies of oxygen isotopes in aqueous sulphate (Wynn et al. 2013) confirmed that S has been cycled through plant matter, leading to a lag in speleothem archives compared with the atmosphere. In Section 7.8.4, speleothem and tree examples from central Europe are illustrated (Fairchild et al. 2009; Wynn et al. 2014), marked by characteristic variations in both abundance and isotopic composition of sulphur, reflecting changes in the sulphur loading of the regional atmospheric boundary layer.

5.6 Metals

Agnieszka Gałuszka and Michael Wagreich

The human-induced pollution of the environment with metals has a long and well-documented history. The first evidence of anthropogenic pollution from biomass burning was found in deposits 30,000–40,000 years old (Neanderthal hearths) from Gorham's Cave in Gibraltar (Monge et al. 2015). Metal-ore extraction and smelting started in prehistoric times. The earliest record of pollution from metal mining, dating back to 7,000–8,000 years, was found in the lake sediments of Keweenaw Peninsula, USA (Pompeani et al. 2013). A long history of the use of metals by humans caused changes in the metals' regional geochemical distribution pattern as they were extracted from ores in mineralised areas, where natural geochemical anomalies occurred, and were transported to the place of their final use elsewhere in the world. Many published papers have documented the time trends of metal levels from historic (or even prehistoric) times up to the present (e.g., Rosman et al. 1997; Shotyk et al. 2002; Allan et al. 2015; Cooke & Bindler 2015; Marx et al. 2016). Information about long-term emission trends can be obtained from studies of environmental-change archives, such as ice, peat or sediment cores and speleothems.

Metals are ubiquitous in the environment. Their wide distribution in all environmental compartments is caused by natural and anthropogenic factors. Metals are constituents of minerals that are released to the atmosphere from volcanic emissions. They are also components of organisms, either as essential micronutrients (e.g., copper, iron, manganese, zinc) or as non-essential xenobiotics (i.e., as foreign chemical substances, e.g., cadmium, mercury, lead). Extraction of metals from ores and industrial production of metals and alloys as well as disposal of metal-containing waste are important anthropogenic sources of metals in the environment. Emissions of metals from these sources are a natural consequence of their use in a vast number of applications. Perhaps surprisingly, a further source that has contributed substantially to global pollution with metals is fossil-fuel combustion and the resultant atmospheric transport of particulates so generated. The well-documented global-scale enrichment of different environmental systems in lead, copper and zinc, all of whose rates of emission into the atmosphere have increased dramatically in the 20th century (Callender 2007), makes them one of the most important stratigraphic markers of the Anthropocene.

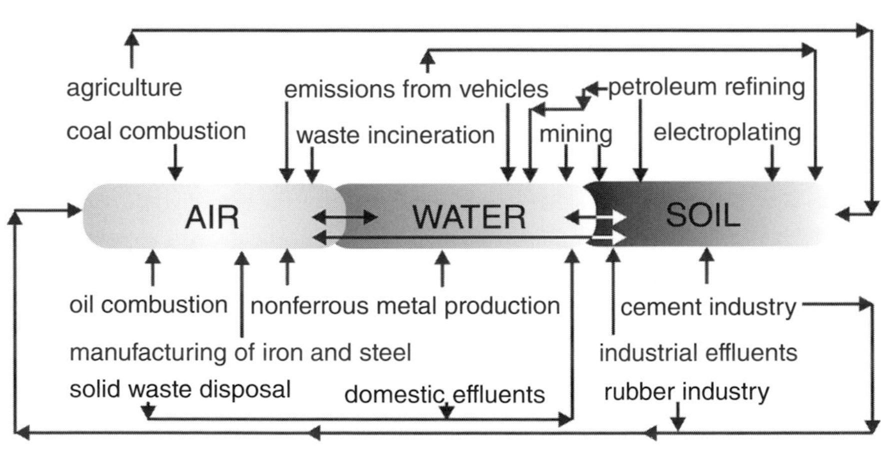

Figure 5.6.1 Major anthropogenic sources of metals in the environment.

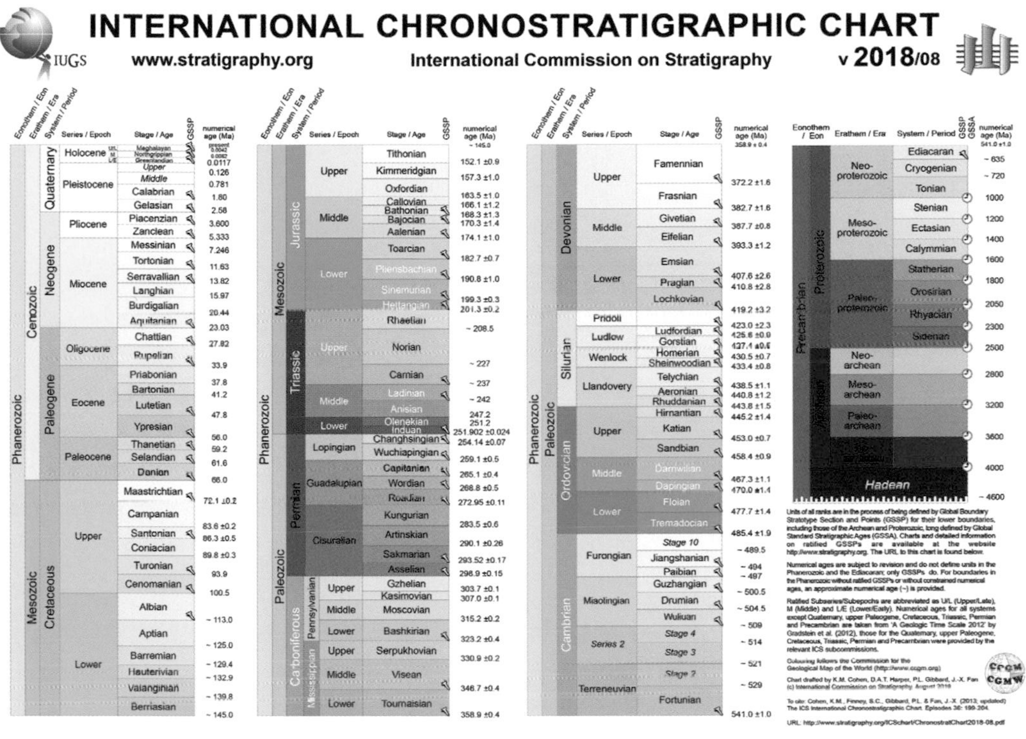

Figure 1.3.1 The Geological Time Scale of the International Commission on Stratigraphy (http://stratigraphy.org/index.php/ics-chart-timescale). Reproduced by permission © ICS International Commission on Stratigraphy 2018.

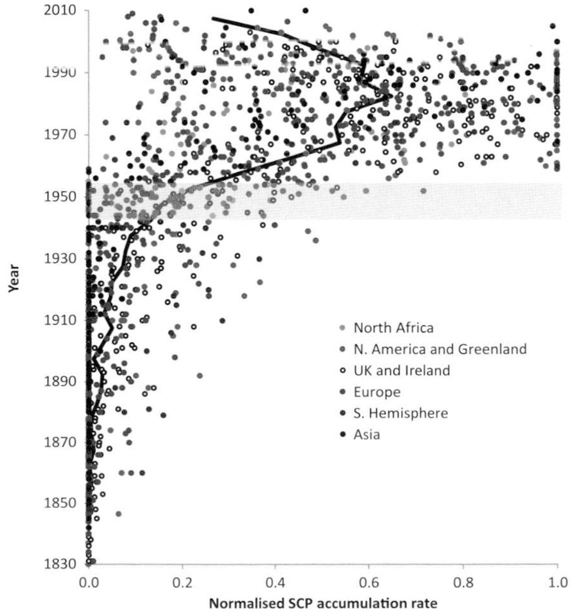

Figure 2.3.3 Global synthesis of SCP lake-sediment accumulation-rate data. Historical SCP data are normalised to peak accumulation rate (=1.0). Each region is coloured separately. Grey horizontal bar is 1950 ± 5 representing the Great Acceleration. Data from Rose (2015).

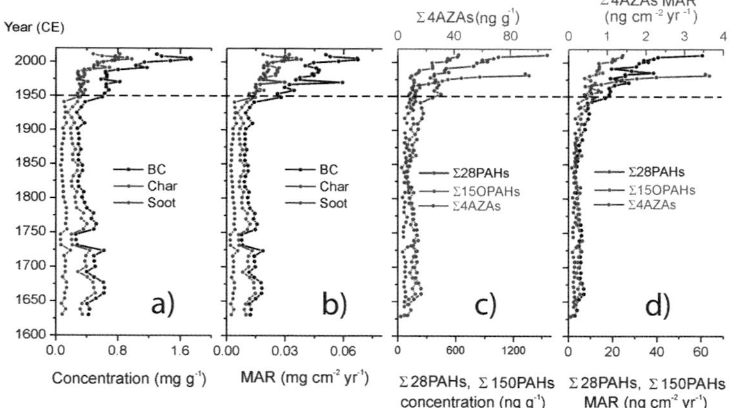

Figure 2.4.2 Historical variations of concentrations and mass-accumulation rates of black carbon (BC), char, soot, parent-PAHs, oxygenated PAHs (OPAHs), and azaarenes (AZAs) in the Huguangyan Maar Lake (adapted from figure 2 of Han et al. 2016).

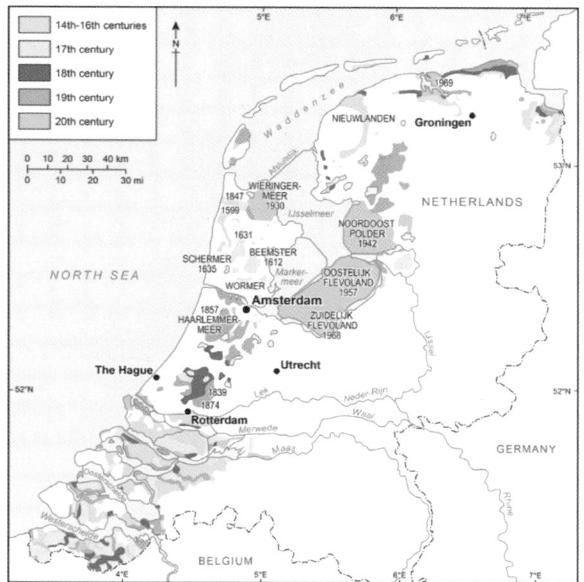

Figure 2.5.7 Polderisation of the Netherlands since the 14th century (redrawn from the *Diercke Three Universal Atlas Diercke Niederlande* – p.113, figure 2). ©2009 Westermann Gruppe.

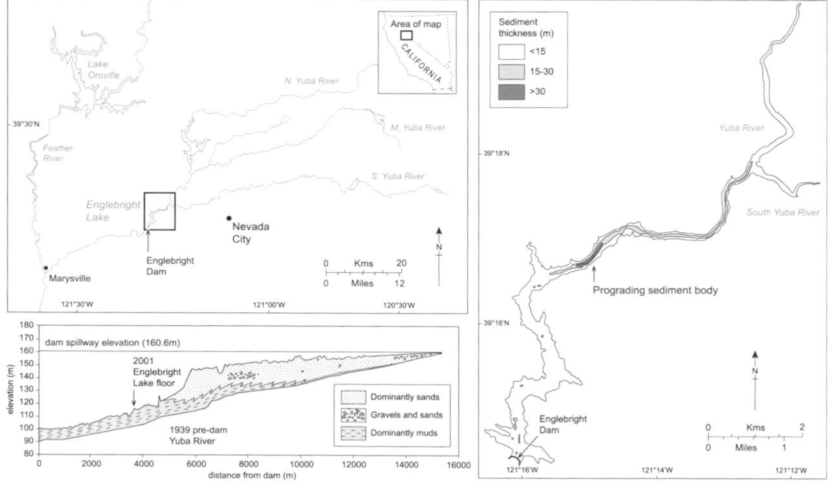

Figure 2.8.2 Location, morphology, and sediment succession in Englebright Lake, California (adapted from Snyder et al. 2004).

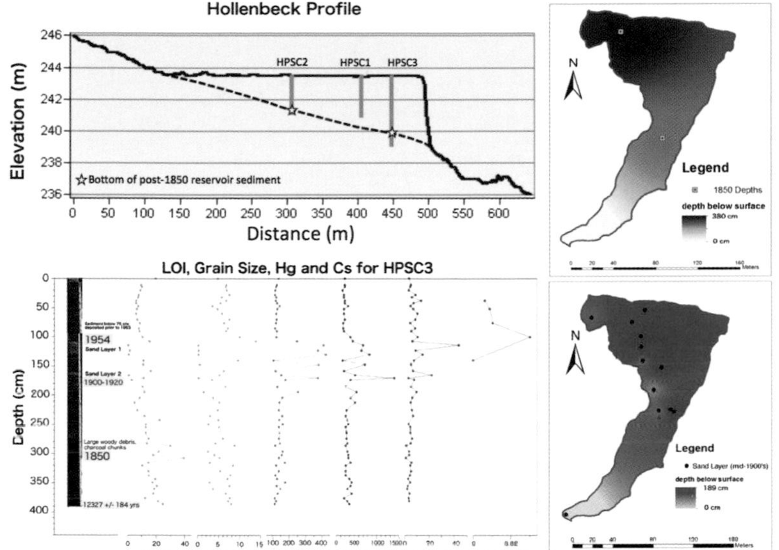

Figure 2.8.3 The succession behind Hollenbeck Dam, Connecticut, where "bomb spike" caesium-137 is used to separate the succession into pre-1950 and post-1950 parts (combined from figures 4, 5 and 6 of Kelleher 2016).

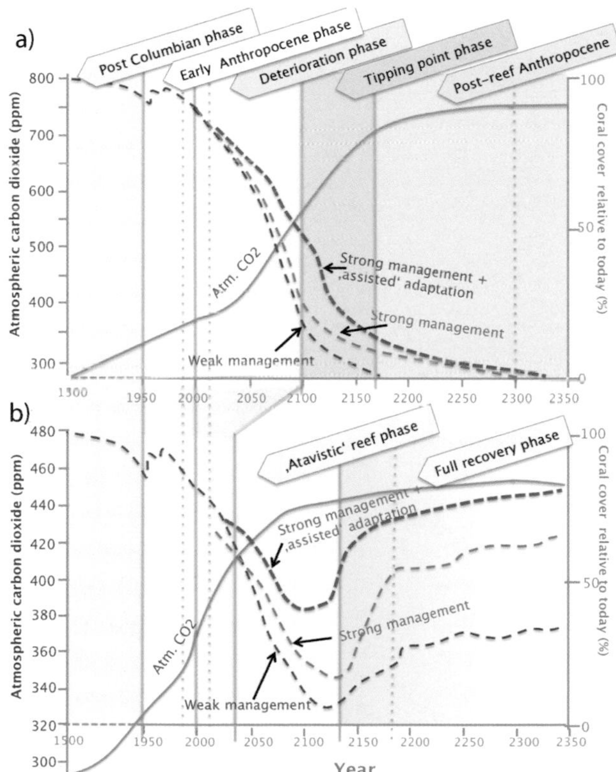

Figure 3.4.3 Use of Anthropocene coral reefs for ecostratigraphic purposes. Shown are two conceptual scenarios. (a) 'Business as usual' (BAU) scenario (relative to anthropogenic atmospheric CO_2 emissions); (b) an integrated CO_2-mitigation/reef-management/'assisted'-adaptation scenario. Based on two scenarios taken from Hoegh-Guldberg et al. (2009), as developed for the Coral Reef Triangle, adapted and extended. In the BAU scenario (a), where atmospheric CO_2 will keep rising, not even strong management (with or without 'assisted' adaptation) will help coral reefs. The combined mitigation/management/adaptation scenario (b) will also not see a rapid reversal of the negative trend, but after going through a 'vale of tears', it might eventually see the return of fully developed coral reefs.

Ecostratigraphic episodes:

a+b:

(1) Post-Colombian coral-reef phase ('pre-Anthropocene', ~1492–1950 CE)

(2) Early Anthropocene coral-reef phase (~1950 to 2000 CE), showing increasing, often punctuated decline of coral reefs. Further subdividable by ecocrisis events such as the near disappearance of *Diadema* sea urchins in the Caribbean or severe global bleaching events (dotted line).

(3) Deterioration phase (from 2000 CE onwards): accelerated decline of coral reef, with more frequent punctuation events (e.g., 2016 bleaching event, dotted line) and rapid deterioration of coral coverage and coral reef occurrences.

BAU scenario (a):

(4a) Tipping-point phase (~2100–2130 CE): episode of catastrophic tipping-point-type collapse of all coral reefs, characterised by dying reefs.

(5a) Post-reef Anthropocene (from about mid-22nd century onwards: largely coral-free Anthropocene oceans, characterised by biogenic and physical reef reworking.

Mitigation/management/adaptation scenario (b):

(4b) 'Atavistic' reef phase (~2030–2130 CE): this is the 'vale of tears'-adaptive episode of 'atavistic' and other atypical coral reefs (see text for explanation).

(5b) Full recovery phase (from ~mid-21st century onwards): renewed episode of large-scale oligotrophic Anthropocene coral reefs, latest from ~2200 CE onwards (dotted line).

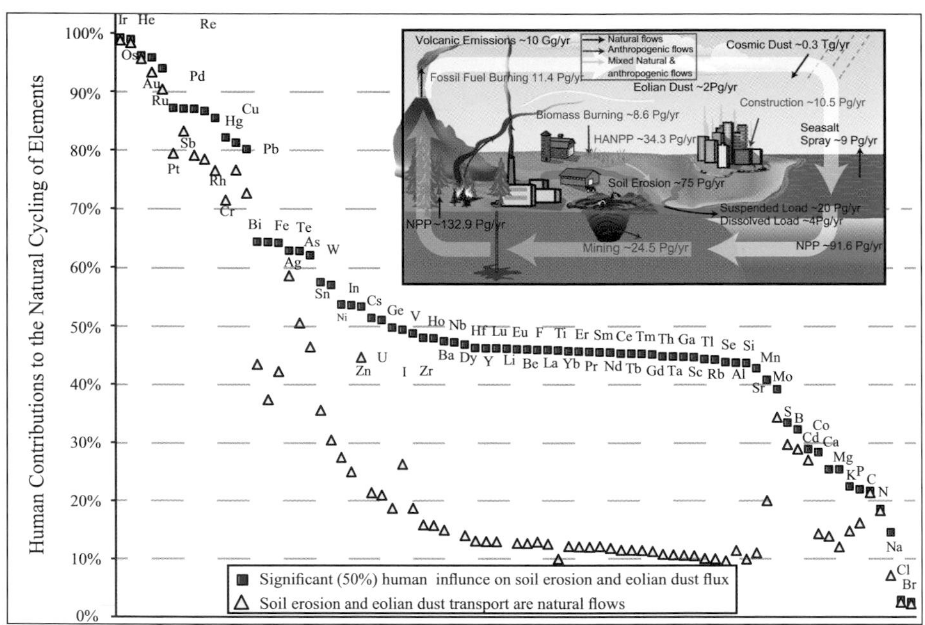

Figure 5.1.2 Anthropogenic domination and perturbation of elemental surface fluxes with a diagrammatic inset of the fluxes (reproduced with permission from Sen & Peuckner-Ehrenbrink 2012). The yellow datapoints do not allow for anthropogenic increases in aeolian sediment fluxes, whereas the red points allow for a 50% increase in these fluxes. ©2012 American Chemical Society.

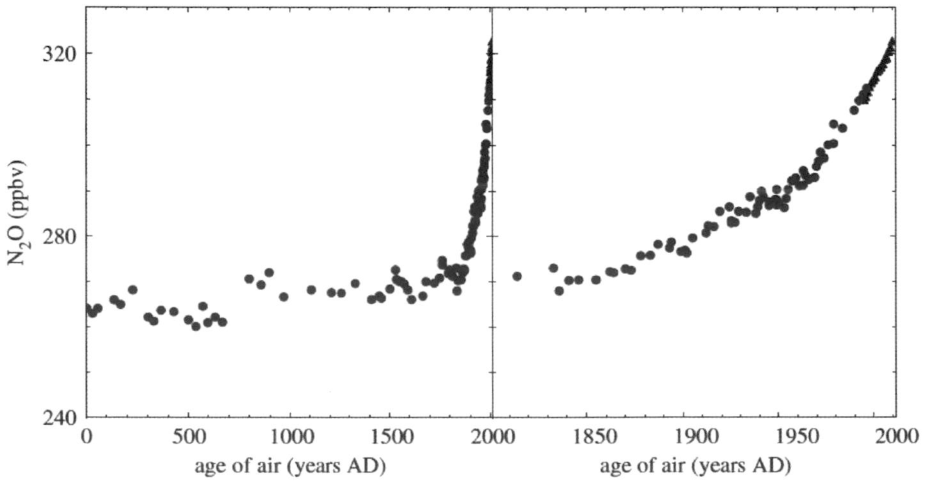

Figure 5.4.1 N_2O concentrations in the Law Dome ice core (and firn (blue circles) over the last 2,000 years and 200 years and annual averages for South Pole air (blue triangles) (from figure 1 of Wolff 2013).

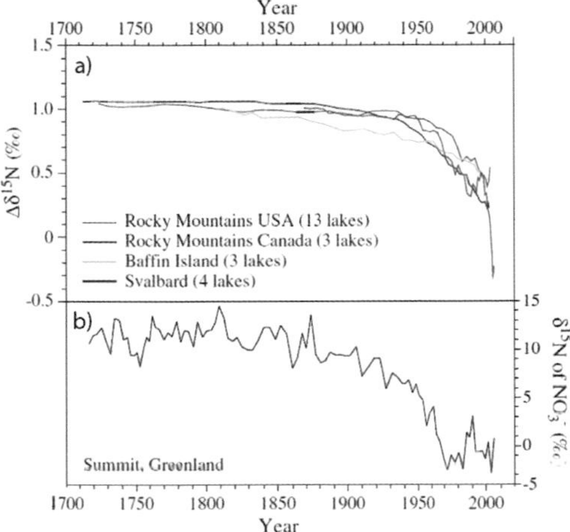

Figure 5.4.2 Changes in nitrogen-isotope ratios (a) in high-latitude northern lakes and (b) in nitrate deposition on the Greenland ice cap. Reprinted from figure 6 of Wolfe et al. (2013). ©2013, with permission from Elsevier.

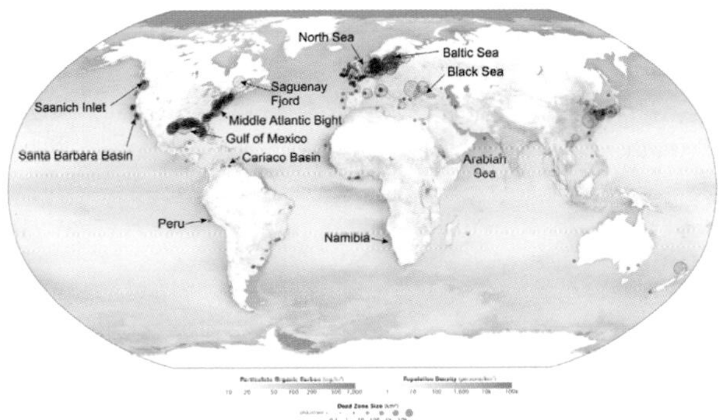

Figure 5.4.3 Distribution of marine dead zones globally (from NASA Earth Observatory https://earthobservatory.nasa.gov/IOTD/view.php?id=44677); Aquatic Dead Zones generated 17th July 2010).

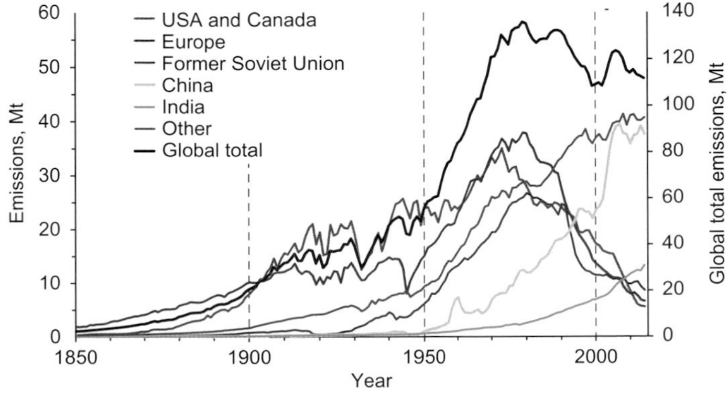

Figure 5.5.2 Emission inventories (in Mt SO_2) by region derived from bottom-up methodologies. Plotted from data in Hoesly et al. (2017).

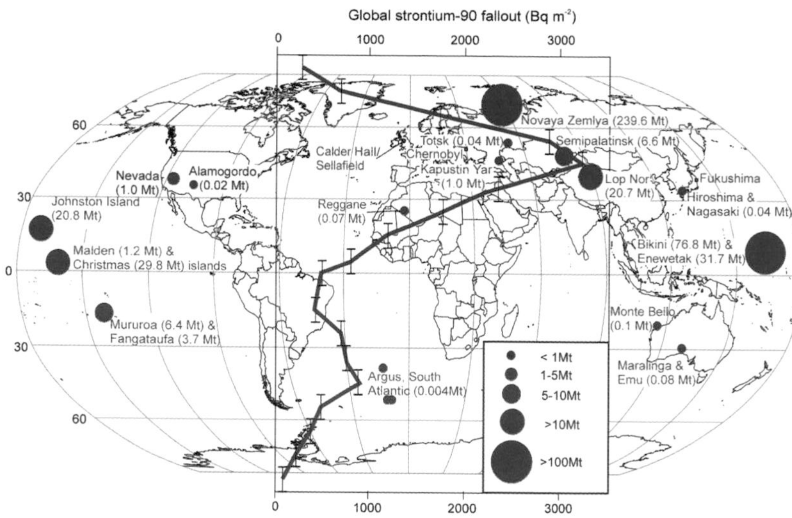

Figure 5.8.2 Distribution and total yields, in megatons, of atmospheric nuclear weapons tests (red); and location of significant nuclear accidents/discharges (blue); with superimposed latitudinal variation of global ^{90}Sr fallout, in becquerels per square meter, which is highest in midlatitudes in the Northern Hemisphere. From figure 2 of Waters et al. (2015), with data sourced from UNSCEAR 2000. Reprinted by permission of Taylor & Francis Ltd (http://www.tandfonline.com).

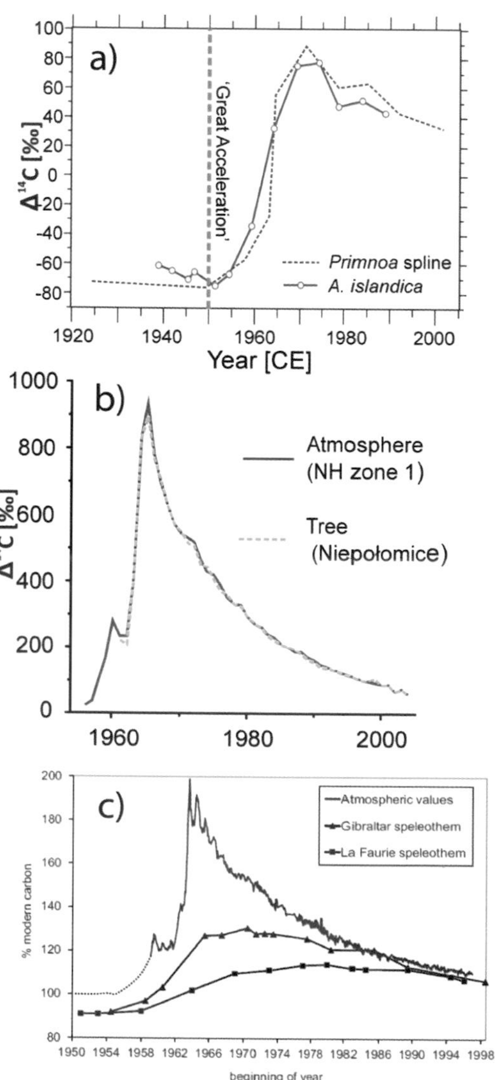

Figure 5.8.6 Examples of ^{14}C signals within different environments. (a) Plot of Δ^{14}C vs. age for seven different colonies of the deep-sea gorgonian coral *Primnoa resedaeformis* (Sherwood et al. 2005b) and for the bivalve mollusc *Arctica islandica* (Weidman & Jones 1993); (b) Δ^{14}C in tree rings (*Pinus sylvestris*) samples from 1960 to 2003 CE at Niepołomice (Poland) showing changes of radiocarbon concentration (from Rakowski et al. 2013) ©2013, with permission from Elsevier; (c) comparison of Northern Hemisphere atmospheric δ^{14}C values (from a German-Austrian composite; see Smith et al. 2009) expressed as percent modern carbon (defined as 1950 activities) within speleothems from La Faurie Cave, southwest France (Genty & Massault 1999) and Lower Saint Michael's Cave, Gibraltar (Mattey et al. 2008). The speleothem records show differing degrees of attenuation of the atmospheric signal, and the La Faurie record has a lagged peak. However, the initial rise is seen by 1958 in both records. and this can be regarded as a reliable marker for correlation (Hua et al. 2013). Figure sourced from Fairchild & Frisia (2014).

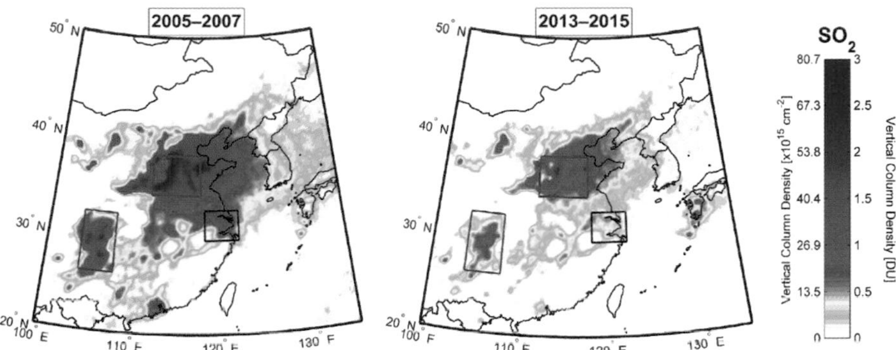

Figure 5.5.3 Satellite-derived OMI images of changing sulphur emissions over China. Emissions in hotspots are clearly falling between 2005–2007 and between 2013–2015, reversing the strong rise in total emissions seen in earlier years (Figure 5.5.2). From part of figure 5 of Krotkov et al. (2016).

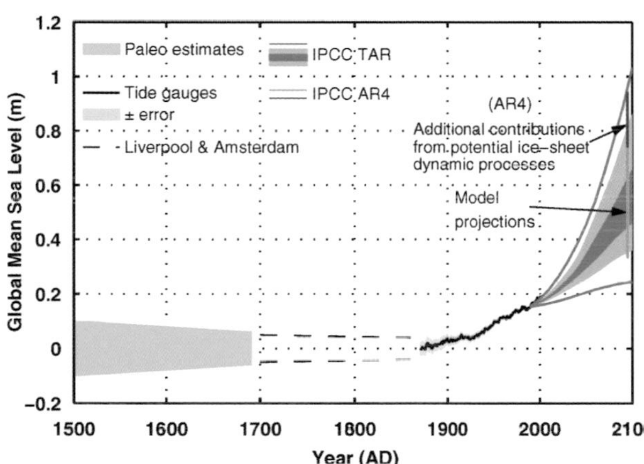

Figure 6.3.1 Sea level from 1500 to 2100 CE. The blue bar indicates the range of geological observations (mainly microfossil-based transfer functions), the dashed lines from 1700 to 1860 CE indicate the range of sea levels inferred from Europe's longest tide-gauge records, the dark line from 1870 to 2006 CE indicates the global average sea level based on multiple tide-gauge records, and the curves from 1990 to 2100 CE are based on satellite altimetric data and the projections from IPCC (taken from Church et al. 2008). Reprinted by permission from Integrated Research System for Sustainability Science and Springer, ©2008.

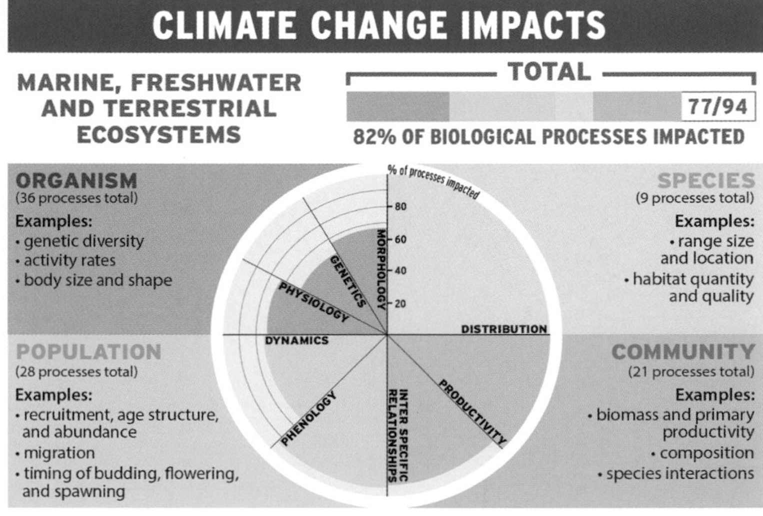

Figure 7.6.2 Climate-change impacts on ecological processes in marine, freshwater and terrestrial ecosystems (Scheffers et al. 2016).

Figure 7.6.3 Global distribution of studies documenting climate-induced widespread forest dieback events and consequences (filled circles). From figure 3 in Anderegg et al. (2012).

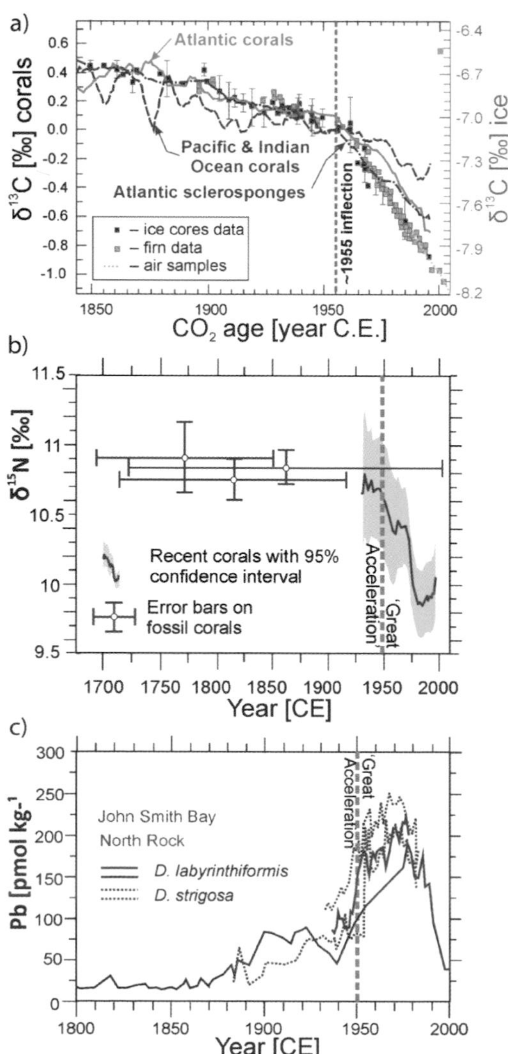

Figure 7.8.2 (a) Changes in $\delta^{13}C$ with respect to age for corals from the Atlantic and the Pacific/Indian Oceans compared to published data from sclerosponges, averaged after removing the mean $\delta^{13}C$ value of the coral skeleton from 1900 CE to the present day and shown as a five-year running mean (modified from figure 2 of Swart et al. 2010) ©2010, American Geophysical Union; (b) $\delta^{15}N$-AA depletion in deep-sea gorgonian corals in the northwest Atlantic (modified from figure 5 of Sherwood et al. 2011) ©2011 National Academy of Sciences; (c) Pb concentration in Bermuda corals (modified and amalgamated from figures 3 and 4 of Kelly et al. 2009). ©2009, with permission from Elsevier.

5.6.1 Sources and Environmental Fate of Metals

Metals are constituents of pollutants released to the environment from various anthropogenic sources (Figure 5.6.1). Coal combustion is one of the most important sources of chromium, manganese, mercury, tin and thallium (Pacyna & Pacyna 2001). Nickel and vanadium originate mostly from oil combustion, whereas the major source of lead is gasoline combustion. Other common pollutant metals, such as Cd, Cu and Zn, are emitted from non-ferrous metal production (Pacyna & Pacyna 2001).

Metals are present in the environment in different forms (in elemental form or as inorganic compounds, metal-organic compounds, ions, complexes and so on). They can undergo transformation after they are released from the source. Some of the processes that cause changes in the form of metals are mediated by microorganisms. An example of such processes is methylation of inorganic mercury, cadmium or lead in sediments (Pongratz & Heumann 1999). The chemical species in which metal occurs in the environment is critical to its mobility. Thus, it is important to understand processes that lie behind biogeochemical cycles of metals.

Long-range atmospheric transport is responsible for metal deposition in remote areas located far from pollution sources. Erosion and weathering of rocks and minerals influence mobilisation and fluvial transport of metals from the Earth's crust to water, soil and biota (Garret 2000). Atmospheric and fluvial transport are the main routes of metal transport in the environment, whereas microorganisms and plants contribute to metal mobilisation from the geosphere. According to many authors, the natural biogeochemical cycles of metals have been modified locally or regionally by human impact to such an extent that they should be considered as anthrobiogeochemical cycles (Rauch & Graedel 2007; Rauch 2012; Sen & Peucker-Ehrenbrink 2012; White & Hemond 2012). The role of anthropogenic activity in global biogeochemical metal cycles is both in mobilisation of metals and in their transport. Moreover, humans not only contribute to metal cycles in terms of material flow but can indirectly influence the mobility of metals. A good example of such an influence is increased mobilisation of metals from soil as a consequence of acid deposition (Wilson & Bell 1996).

Rauch (2012) has estimated that human activity controls approximately half of the mass movement of a given metal. Of nine metals (Ag, Al, Cr, Cu, Fe, In, Pb, Ni and Zn) in which anthrobiogeochemical cycles have been analysed so far, copper is of particular interest, because its anthropogenic flows by far dominate over natural ones. Biogeochemical cycles of metals have been changed on different scales. Mercury, lead and tin are metals whose biogeochemical cycles have been changed on a global scale (Pacyna & Pacyna 2016). Of these metals, lead seems to be the most important in terms of Anthropocene chemostratigraphy.

5.6.2 Metals in Different Environmental Compartments

Metals occur in every environmental compartment; however, their distribution is not uniform (Table 5.6.1). Rain, snow and glacial ice are depleted in metals, whereas particulates and sediments show ability to sequester metals due to their sorption onto clay minerals, metal oxides and organic matter. The fate of metals in the environment depends on many factors, of which the most important are pH, redox potential and interactions with other substances and with living organisms.

In the marine environment, metals are commonly concentrated in organic-rich sediments that form beneath highly productive surface waters from which organic remains have settled to the seabed within the ocean's oxygen-minimum layer, turning that layer anoxic through their decomposition (e.g., off Peru, Chile, California, Namibia). The metals are preserved by the anoxic conditions. These modern sediments are the equivalent of the metal-rich and organic-rich black shales common in the stratigraphic record of Paleozoic times. Metals such as Cu, Cr and Ni are also concentrated within the ferromanganese nodules that form under highly oxidising conditions at the

Table 5.6.1: **Typical concentrations of selected metals in different reservoirs**

	Cu	Cr	Ni	Pb	Zn
			mg/kg		
Bulk continental crust	27	140	59	11	72
Upper continental crust	28	92	47	17	67
Terrestrial sediments	40	74	40	17	65
Marine sediments	75	79	71	20	86
River sediments	100	100	90	35	250
Suspended matter in world rivers	75.9	130	74.5	61.1	208
Lake sediments	45	62	66	34	118
Pre-industrial lake sediments	34	48	40	22	97
Soil	39	130	25	27	48
Greenland ice	0.00006	0.000038	0.000058	0.000028	0.00022
Antarctic ice	0.000008	0.000007	0.00015	0.000005	0.00003

Note: From Callender (2007); Rauch and Pacyna (2009); Viers et al. (2009); Salomons and Förstner (2012).

seabed in the deep ocean, within sulphide ore deposits forming currently from volcanic exhalations (hydrothermal fluids) at the axes of the global mid-ocean ridge system, and at seafloor-spreading centres within back-arc basins around the Pacific margin.

Except for mercury, where the major fraction in the atmosphere is gaseous, other metals are present in the air as forms condensed onto airborne particles. In the case of mercury, the volatile form (elemental Hg^0) has a residence time in the atmosphere of about one year, while that of non-volatile mercury is from a few days to a few weeks (Rolison et al. 2013). The particulate phase carries a major fraction of the metals that are transported through the atmosphere. The transport distance of very fine particles can range over thousands of kilometres (Haygarth & Jones 1992), causing a global dispersion of metals that is most clearly seen for mercury, lead and tin (Pacyna & Pacyna 2016). During atmospheric transport, the form of metals in the aerosols may change. They can be dissolved or complexed with organic matter (Baker & Jikells 2015). The residence time of metals present in airborne aerosols strongly depends on the size of the particles and varies from days to weeks (Raes et al. 2000).

Deposition of atmospheric aerosols is an important source of metals in surface waters. Metals occur in water as free ions and complexes with inorganic and organic ligands (mobile forms). In aquatic systems they are also bound to suspended particulate matter and sediments (immobile forms). Processes of sorption/desorption and complexation play an important role in the distribution of metals between aqueous and solid phases, and dissolution of continental-shelf and slope sediments has been shown as a significant source of trace metals for seawater, larger than the atmospheric sources for elements such as iron or lanthanides, and for contaminants too (Tagliabue et al. 2014; Jeandel & Oelkers 2015).

Metals are readily adsorbed from the water column; they can also enter aquatic sediments by direct precipitation from water as carbonates and sulphides. In bottom sediments, metals participate in various biogeochemical processes, such as redox reactions, adsorption/desorption, dissolution of minerals, microbial transformations and so on. These processes are responsible for metal speciation and mobility in the aquatic environment (Van Cappellen & Wang 1995).

Metals are components of soils, in which they can undergo dissolution, precipitation, sorption/desorption, complexation, oxidation/reduction, migration and volatilisation (Kabata-Pendias 2010). Part of the total metal content in soil is always inherited from the parent rock, while other parts can be derived from anthropogenic and biogenic sources. This makes the study of the origin of metals in soil extremely complex. Moreover, metal concentrations in soil profiles are not uniform, with O- and B-horizons (see Figure 2.7.5 and Section 2.7.4) showing the highest metal contents due to metal binding with organic matter (O horizon) and adsorption by Fe and Mn oxides/oxyhydroxides (B horizon) (Sucharovà et al. 2012). The high metal-accumulation potential of peat and peat soils is used in reconstructing the history of atmospheric metal deposition (Shotyk et al. 2002, 2016). The half-lives of metals in soil, in comparison with other environmental compartments, are very long. Depending on the metal and soil properties, the residence time in soil can be from tens to thousands of years (Kabata-Pendias 2010).

Several metals are essential to living organisms (Co, Cr, Cu, Fe, Mn, Mo, Ni and Zn); however, they can also become toxic if their concentrations exceed optimal levels. Mercury, lead and cadmium are non-essential elements. Metals are taken up by plants from soil and translocated from roots to their above-ground parts. Eating of plants by herbivores causes metals to enter the food chain. There are three routes of exposure of animals and humans to metals, i.e., inhalation, ingestion and skin absorption. Once the metals are absorbed, they are not metabolised and can be either excreted from the body or bioaccumulated. Some of these metals have an extremely long biological half-life in a human body; for example, in the case of cadmium, it exceeds 30 years.

5.6.3 Metals as Signals of the Anthropocene

From a perspective of potential use of metals as markers of the Anthropocene, anthropogenic inorganic fly-ash particulates originating from combustion processes seem important (Oldfield 2015; Rose 2015; Swindles et al. 2015; Section 2.3.1 herein). Such particles are silicates and Fe oxides of <50 μm diameter. The study of particles showing magnetic properties derived from coal combustion and deposited in sediment cores enables us to reconstruct historical trends in pollution, which may be useful for distinguishing Anthropocene from Holocene strata (Hounslow 2017). In terms of Anthropocene chemostratigraphy, the metal-concentration patterns in dated sediment and peat cores, Arctic ice and snow cores, speleothems, corals and tree rings are of special interest (see Section 7.8.4). The wide use of metals that are generally depleted in the Earth's crust (e.g., platinum-group metals, rare-earth elements) in high-tech applications causes an increase of their concentrations in recent sediments. However, this enrichment would not be recorded on a global scale (Gałuszka & Migaszewski 2018). For practical reasons, a selection of metals most suitable for being markers of Anthropocene strata should be based on such criteria as a wide distribution (preferentially global) and a well-documented anthropogenic emission history with prevalence of anthropogenic over natural metal fluxes. According to Rauch and Pacyna (2009), increased anthropogenic fluxes of Ag, Cr, Cu, Ni, Pb and Zn are evident. Of these elements, Cu, Pb and Zn are commonly determined in sediments and other environmental archives. Their use as markers of the Anthropocene is documented by the fact that during the first half of the 20th century, their atmospheric emission rates increased dramatically, and then, around the 1970s, they started to decrease (Callender 2007). The scale of this increase can be seen by comparison of mine production and consumption trends for Cu, Pb and Zn. About 90% of the mine outputs for these metals was consumed during the 20th century. Moreover, Birch et al. (2015) recommended that Cu, Pb and Zn should be used as indicator elements of anthropogenic pollution of sediments.

5.6.3.1 Copper as a Marker of the Anthropocene

The abundance of copper in the upper continental crust is 27 mg/kg (Hu & Gao 2008). It is one of the metals used by humans since prehistoric times. Native copper was used as early as 10,000 years ago, and copper smelting started about 7,000 years ago (Yang et al. 2017). Bronze Age records are widespread but differ in age from region to region depending on the timing of the evolution of civilisations (e.g., Breitenlechner et al. 2014; Pompeani et al. 2015). Nowadays, major copper-pollution sources are mining, agriculture, municipal and industrial wastewater discharges, waste incineration and fossil-fuel burning, whereas in the past, Cu smelting was responsible for global pollution. Historic trends of atmospheric copper pollution have been reconstructed from dated ice cores (Hong et al. 1996), peat cores (Mighall et al. 2009), lake sediments (Michelutti et al. 2009) and even seabird excrement (Yan et al. 2010). The natural biogeochemical cycle of copper, which is a relatively mobile element under oxidising conditions and has a very low mobility under reducing conditions, has been substantially changed by anthropogenic activity. It has been estimated that the ratio between human and natural mobilisation of copper varies from 5.9:1 to 24:1 (UNEP 2013).

A threefold increase of Cu content in Greenland snow from the 1750s to the mid-1960s has been observed (Boutron et al. 1995). From the 1960s to 1995, Cu concentrations increased about 1.3 times. A study of metal-concentration trends in ice core from the Siberian Altai showed that Cu emissions increased between 1935 and the 1980s and then dropped in the 1990s to reach a level similar to that in the 1950s (Eichler et al. 2014).

Copper has a very high affinity for organic matter. In fresh water, a small fraction of Cu is present as organic and carbonate complexes. Copper is readily adsorbed onto suspended particles in water systems because it has a low solubility in water. It is widely assumed that a high concentration of Cu in sediment reflects pollution of a water body (Chen et al. 2012). The majority of Cu in lakes and oceans originates from fluvial transport of copper (85%–98%), whereas the atmospheric input is in the range of 2%–15% (Callender 2007). Maximum copper concentrations in natural archives such as sediments, ice cores and snow cores were usually recorded in the second half of the 20th century, as a consequence of an increase in Cu production (Boutron et al. 1995). The time trend of copper pollution in the last hundred years is similar to that of other metals emitted mostly from anthropogenic sources. However, Cu is not as often determined in sediment cores as is lead and zinc (Figure 5.6.2).

Copper is commonly abundant in estuarine settings where industrial plants on riverbanks have discharged waste directly into rivers and harbours (e.g., New Bedford, Massachusetts). Because modern estuaries primarily act as sediment traps, the Cu tends to stay close to its source rather than spreading substantially onto the adjacent continental shelf. However, the process of estuarine sedimentary fill from the seaward end commonly involves the landward movement of silt and clay on the flood tide and the seaward movement of clay on the ebb tide, which takes fine particulates, including those contaminated with metals, out to sea (Summerhayes et al. 1985).

Copper has two stable isotopes, ^{63}Cu and ^{65}Cu. The use of Cu isotope ratios in the study of environmental pollution has gained interest in the recent decade (Bigalke et al. 2010; Fekiacova et al. 2015; Gonzalez et al. 2016; Novak et al. 2016). Most reports are enthusiastic about the application of stable Cu isotopes to pollution provenance. Nonetheless, there are certain limitations in the use of these isotopes for study of the history of pollution because of natural fractionation processes occurring in the environment (Bigalke et al. 2011; Fuji et al. 2013).

5.6.3.2 Lead as a Marker of the Anthropocene

Lead is one of the most common anthropogenic pollutants, originating from fossil-fuel combustion,

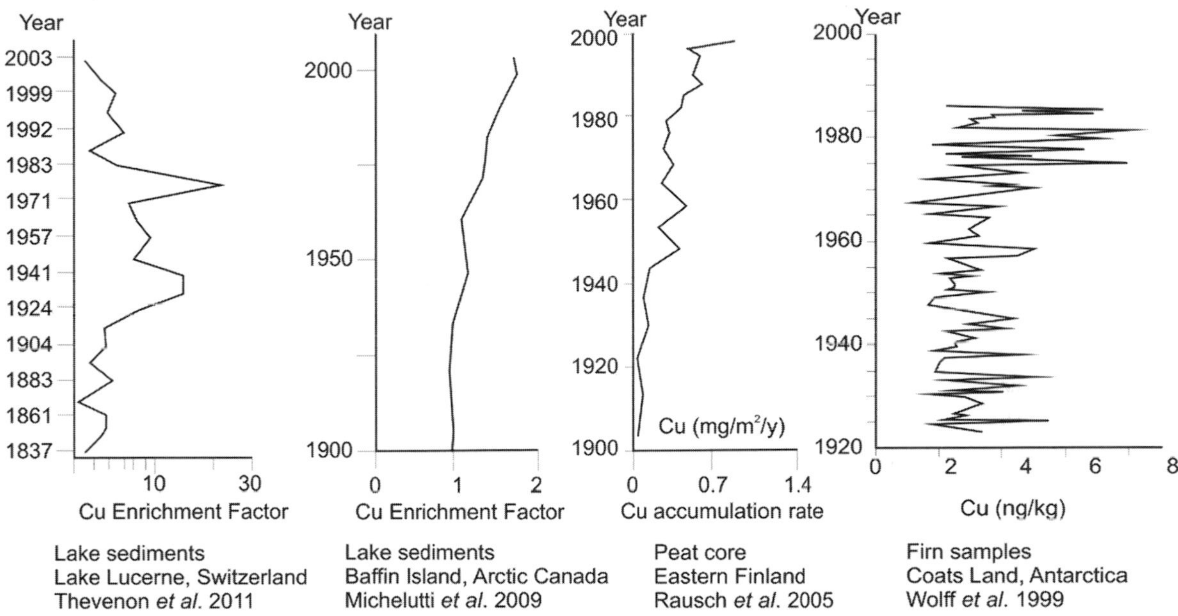

Figure 5.6.2 Examples of temporal changes in copper levels in different natural archives.

lead-ore mining and smelting, and improper disposal of waste containing lead (Gałuszka & Migaszewski 2018). According to Marx et al. (2016), in samples collected from natural archives (sediments, peat mires, ice cores) in which lead has been determined, including samples from Antarctica and Greenland, lead from anthropogenic sources dominates over natural Pb. In the Northern Hemisphere, enrichment factor values for Pb exceed 100 times, and in the Southern Hemisphere they are 4–15 times the Pb background concentrations (Marx et al. 2016). The broad-scale patterns of Pb pollution trends in the Northern Hemisphere show consistency and can be employed in Pb chemostratigraphy studies as an age proxy for sediments deposited up to some 3,000 years ago (Cook & Gale 2005; Zheng et al. 2007). Late Bronze Age and Iron Age lead peaks (Figure 5.6.3), resulting from Greek-Phoenician and Roman mining, can be distinguished from both near-field and distant archives, the latter including Arctic ice cores (Krachler et al. 2009; McFarlane et al. 2014; Wagreich & Draganits 2018). The signal of lead pollution is much weaker and appears later in the Southern Hemisphere. A distinct increase in Pb pollution from mining and smelting (especially in Australia) has been recorded in Antarctic ice from the period of 1890–1920, whereas the use of leaded gasoline and non-ferrous metal production contributed to the rise of Pb concentrations in ice from 1960 to 1980.

Organo-lead, in the form of tetraethyl lead, was used in large quantities as an anti-knock agent in leaded gasoline during 1940–1980. Exhaust, emitted from combustion of such fuel, contained lead in the form of oxides, chloride and bromide (Kakonyi & Ahmed 2013). The use of Pb in gasoline was banned in the United States in 1996, in Canada in 1992, in Japan in 1980, in other parts of Asia in the 1990s and in Europe in 2000 (Bollhöfer & Rosman 2001; Barbante et al. 2017). Several years before the ban, leaded gasoline was systematically phased out. For example, in the United States, three years before the final ban on leaded gasoline, the lead emissions from car exhausts were as low as 1% of those in 1970 (Bollhöfer & Rosman 2001).

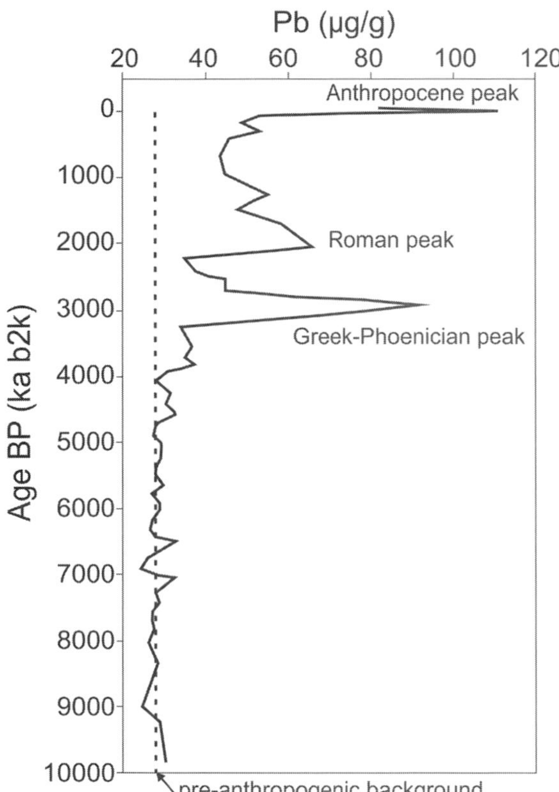

Figure 5.6.3 Temporal changes in lead levels from prehistoric to historic time (or during the Holocene) showing Greek-Phoenician, Roman and recent (Anthropocene) peaks as exemplified by lake sediments from Spain. (Garcia-Alix et al. 2013)

Lead stable isotope ratios are widely used for apportionment of pollution sources and for tracking pollution pathways in the environment because, unlike isotopes of copper and zinc, they do not undergo fractionation. Lead has four stable isotopes, namely ^{204}Pb, ^{206}Pb, ^{207}Pb and ^{208}Pb. The isotope ^{204}Pb is the least abundant (1.4%), and its determination by the most popular quadrupole ICP-MS technique is problematic. Thus, in many publications only ratios of $^{206}Pb/^{207}Pb$ and $^{206}Pb/^{208}Pb$ are reported. The values of the $^{206}Pb/^{207}Pb$ ratio in lead emitted from anthropogenic sources are in the range of 1.15–1.19, whereas in lead from natural sources, the $^{206}Pb/^{207}Pb$ ratio is approximately 1.5. Higher Pb concentrations in natural archives are accompanied by a decrease in $^{206}Pb/^{207}Pb$ ratio (Dean et al. 2014).

Lead accumulated in Greenland snow showed a 20-fold enrichment in the period from the 1750s to the mid-1960s (Boutron et al. 1995). Studies of ice cores from Greenland showed that over the second half of the 20th century the emissions of lead from coal burning decreased in Europe and North America (Barbante et al. 2017). Other natural archives, such as speleothems, provide high-resolution records of Pb pollution history. Studies of speleothems, though rarely conducted, show similar trends in Pb pollution to those recorded in sediment or ice cores (Allan et al. 2015). Peat cores are reliable archives of Pb pollution because the Pb record in peat is consistent and stable (Bindler 2006). Recently, Shotyk et al. (2016) in their study of peat bogs from Alberta, Canada, presented evidence for a decline of Pb contamination relative to the background level. A peak Pb level in natural archives (Figure 5.6.4) can be a solid marker of the Anthropocene.

5.6.3.3 Zinc as a Marker of the Anthropocene

Zinc is a metal of many uses and is a common pollutant. The major sources of zinc in the environment are the non-ferrous metal industry, mining, the rubber industry, waste incineration, coal burning, vehicle emissions, tyre wear and so on. Zinc is one of the metals that are most often determined in different environmental samples. This makes Zn a good candidate for an Anthropocene marker.

Anthropogenic Zn has been found in snow and ice cores from remote areas such as Antarctica (Planchon et al. 2002) and Greenland (Candelone et al. 1995). The sediment record of the 20th century shows that Zn levels increased from the 1950s until the 1990s, and then there was a slight decline in Zn concentrations (Figure 5.6.5 herein; Callender 2007).

Zinc has five stable isotopes, ^{64}Zn, ^{66}Zn, ^{67}Zn, ^{68}Zn and ^{70}Zn, of which two isotopes (^{64}Zn, ^{66}Zn) are the most important for tracing pollution sources.

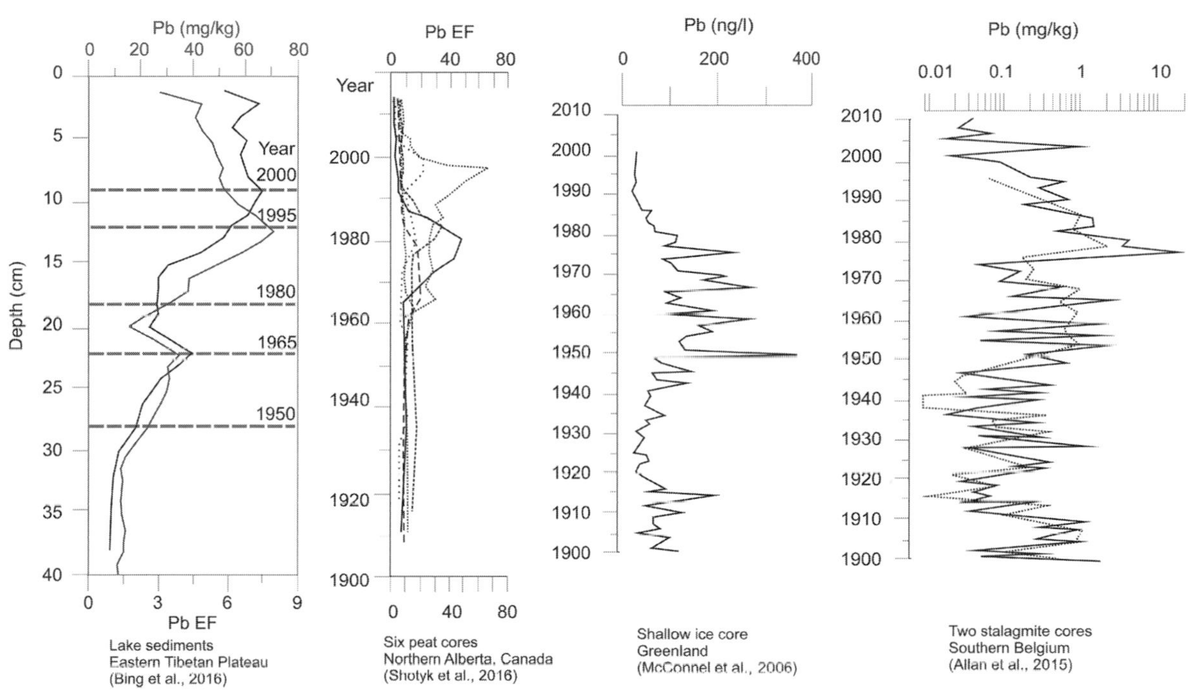

Figure 5.6.4 Examples of temporal changes in lead levels in different natural archives.

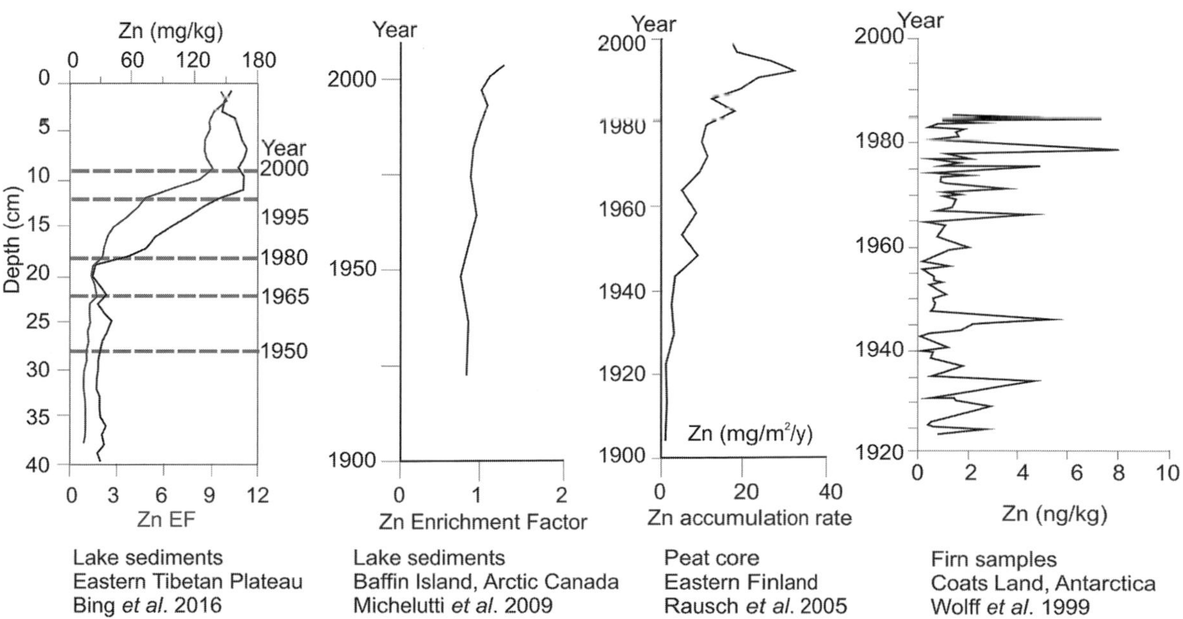

Figure 5.6.5 Examples of temporal changes in zinc levels in different natural archives.

Zn isotope ratios are expressed as $\delta^{66}Zn$ notation, being deviation of a $^{66}Zn/^{64}Zn$ ratio from a standard. Zinc isotopes have been determined in soils and sediments in order to trace several pollution sources, such as those related to traffic (Araújo et al. 2017). Combustion, electroplating and smelting are high-temperature processes leading to zinc isotope fractionation. During these processes, the light isotope is enriched in the gaseous phase, whereas the heavy isotope remains in waste or effluents. This can be used for identification of pollution sources (isotopic fingerprinting). The results of a recent study on estuarine sediment cores confirmed the usefulness of Zn isotopes in reconstruction of historic pollution (Araújo et al. 2017). As compared with the number of studies on lead isotopes in natural archive samples, zinc isotope data is very scarce, which limits its application as a marker of the Anthropocene.

5.7 Organic Compounds

Agnieszka Gałuszka and Neil Rose

Organic pollutants represent a large and diverse group of compounds released to the environment from natural and anthropogenic sources. These compounds show different structures, as well as physical and chemical properties, that influence their environmental fate and persistence. Many organic compounds are hydrophobic (they show a low solubility in water but are readily soluble in non-polar solvents), which causes their accumulation in living organisms and sediments. The presence of organic pollutants has been recorded in the most remote, uninhabited locations that are perceived as being pristine (Wania 2003; Jamieson et al. 2017). This evidence of global human impact on organic-pollutant levels in the environment has a potential application to the characterisation of Anthropocene strata.

One group of organic pollutants that has a very long residence time in the environment and shows toxic and bioaccumulative properties is the persistent organic pollutants (POPs) (Bacaloni et al. 2011). These compounds resist biodegradation and chemical and physical degradation (e.g., hydrolysis, oxidation, photolysis). There is no consensus on the criteria for identifying a compound as persistent, but most approaches use a threshold half-life of a substance in different environmental compartments. These define a persistent compound as one that has a minimum residence half-life of 180 days in soil/sediments, 60 days in surface water or 2 days in the atmosphere (O'Sullivan & Megson 2014).

A characteristic feature of POPs is their tendency to occur in different phases depending on physicochemical parameters. The occurrence of these compounds in the gas phase influences their long-range transport in the atmosphere, but they are also present and transported in the air, adsorbed to solid particles. Volatilisation, transport in the atmosphere and condensation are the main processes influencing the global distribution of POPs. These processes occur in cycles of evaporation-transport-condensation during global long-range transport from lower to higher latitudes and are often referred to as the 'grasshopper effect'. As this process is driven by temperature, a general rule is that the more volatile a compound is, the longer the distance from the source it is able to travel in the air (Jaspers et al. 2014). This results in a 'global distillation' whereby less volatile compounds are preferentially retained in surface waters and soils. This phenomenon can also be observed in mountainous regions, where volatile POPs preferentially move to higher altitudes with respect to less volatile compounds (Grimalt et al. 2001).

Atmospheric deposition of POPs, either in the vapour or particle phase, is the most important input route of these compounds to the marine environment. In aquatic systems, POPs bind to organic and inorganic particles, sink and are accumulated in bottom sediments. These compounds are very effectively bioaccumulated, especially in benthic fauna, and an efficient transport pathway for the deposition of POPs from the water column is through faecal pellets (Berrojalbiz et al. 2009). A recent study has shown that POP levels in marine amphipods from two of the deepest marine locations, the Kermadec and Mariana Trenches, were comparable to or even higher than those reported for fauna in nearby heavily industrialised areas (Jamieson et al. 2017).

POPs are also biotransported with migrating animals; for example, birds accumulate contaminants while feeding at sea and then transport these to terrestrial systems via faeces accumulated below nesting cliffs. The same occurs for anadromous fish, which deposit POPs in fresh water after a period at sea. After returning to spawn, these fish often die, depositing the contaminants they have accumulated into upland lakes and rivers (Blais et al. 2007; Gerig et al. 2015). POPs may also be adsorbed onto the surfaces of plastic debris and microplastic particles. In the marine environment they can be transported over long distances by ocean currents (Zarfl & Matthies 2010), during which time they may become incorporated within biota via ingestion.

Diagenetic processes cause magnification of POPs in sediments enriched in organic matter (de Bruyn & Gobas 2004). Diagenetic magnification depends on the structure of individual compounds, as well as on temperature and precipitation, favouring retention preservation of heavier compounds (Cappelletti et al. 2014). In soils, POPs are bound to soil organic matter and are relatively immobile. Atmospheric deposition, intentional use as pesticides and unintentional spills are typical sources in soils (Jaspers et al. 2014). POPs may evaporate from soil, depending on individual compound properties, soil characteristics and environmental conditions. While some POPs are solely synthetic chemicals, intentionally produced for certain uses, others occur naturally in the environment as a result of wildfires or biosynthesis. Wildfires may also re-emit POPs back into the atmosphere from storage in soils and vegetation. According to one classification, three groups of POPs can be distinguished – namely, pesticides, industrial chemicals and by-products (O'Sullivan & Megson 2014). More detailed categories of anthropogenic sources of POPs are the combustion of fossil fuels in public power, heat generation and residential, commercial and industrial locations; solvent and product use; road and non-road transport; waste incineration; and agriculture (van der Gon et al. 2007).

5.7.1 Polycyclic Aromatic Hydrocarbons

A single molecule of polycyclic aromatic hydrocarbon (PAH) contains at least two benzene rings, and those with lower molecular weight show highest volatility. PAHs originate from both natural (e.g., wildfires, low-temperature diagenesis of organic matter, biosynthesis, volcanic eruptions) and anthropogenic sources (e.g., coal and gasoline combustion, volatilisation from pavement materials such as asphalt). A characteristic feature of PAHs, similarly to several classes of POPs, is their occurrence as mixtures including both low-molecular-weight (e.g., two-ringed naphthalene) and heavier-molecular-weight PAHs (e.g., the six-ringed benzo [ghi]perylene). Different diagnostic ratios of PAH concentrations in these mixtures can be used for emission-source provenance (Tobiszewski & Namieśnik 2012). Because PAHs are emitted from numerous sources, they are currently ubiquitous in the environment. Soils and sediments are generally considered as long-term reservoirs of these pollutants, although natural sources of PAHs mean that background concentrations will also be present in pre-industrial periods. Elevated concentrations of PAHs in sediment cores have been recorded in sections representing periods of high industrialisation and traffic intensity (Bigus et al. 2014). Peak concentrations of PAHs in sediments are dated to the 1950s in Europe and North America, the 1960–1970s in South America, the 1950s–1980s in China and the 1940s in Japan (Heim & Schwarzbauer 2013; Bigus et al. 2014). Examples of sedimentary records of PAHs are shown in Figure 5.7.1.

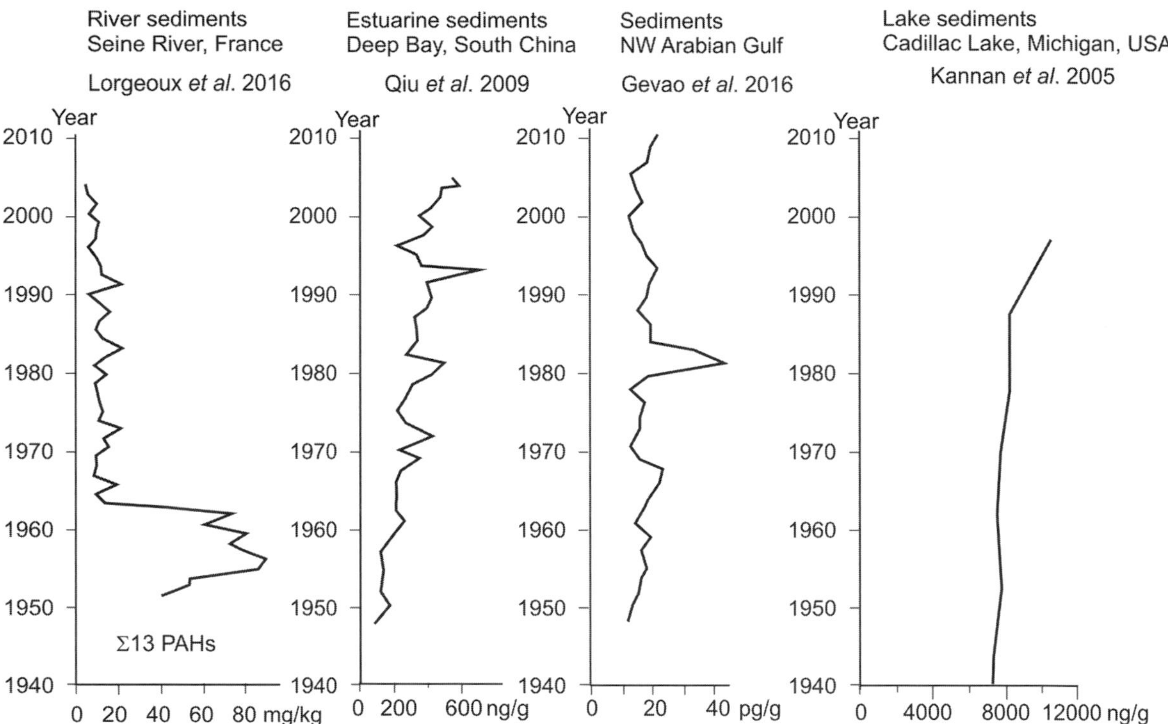

Figure 5.7.1 Time trends of total PAH concentrations in dated sediment cores.

5.7.2 Organochlorine Pesticides

Organochlorine pesticides were used as insecticides in agriculture and to control insect-borne diseases. Some of these compounds were first synthesised in the late 1800s (for example, dichlorodiphenyltrichloroethane [DDT] in 1874), but their commercial production started in the 1940s. In many countries, organochlorine pesticides have been successively banned since the 1970s, after which their concentrations in the environment have drastically decreased. In 2001, a United Nations treaty, the Stockholm Convention on Persistent Organic Pollutants, was signed, which aims to eliminate and restrict the production and use of selected POPs, including a number of organochlorine pesticides (DDT, chlorane, dieldrin), as well as PCBs and dioxins. However, due to their persistence, organochlorine pesticides and/or their metabolites are still detectable in natural environmental archives, such as peat or ice cores. Organochlorine pesticides stored in glacier ice are now being released to the environment from retreating glaciers (Bogdal et al. 2009; Schmid et al. 2010), resulting in concentrations in glacial-fed waters and sediments remaining elevated despite the ban in their use, giving rise to concern in the context of predicted climate changes (Kallenborn et al. 2012).

Organochlorine pesticides, which are most often determined in environmental samples, include DDT, dieldrin, endrin and hexachlorocyclohexane isomers (HCH) (Figure 5.7.2). Estimated half-lives of organochlorine pesticides depend on their environmental compartment and are up to 7 days in the air and up to 2,300 days in water, soils and sediments (O'Sullivan & Megson 2014). Results of sediment-core studies conducted in different parts of the world show that in Europe and North America organochlorine-pesticide concentrations peak around

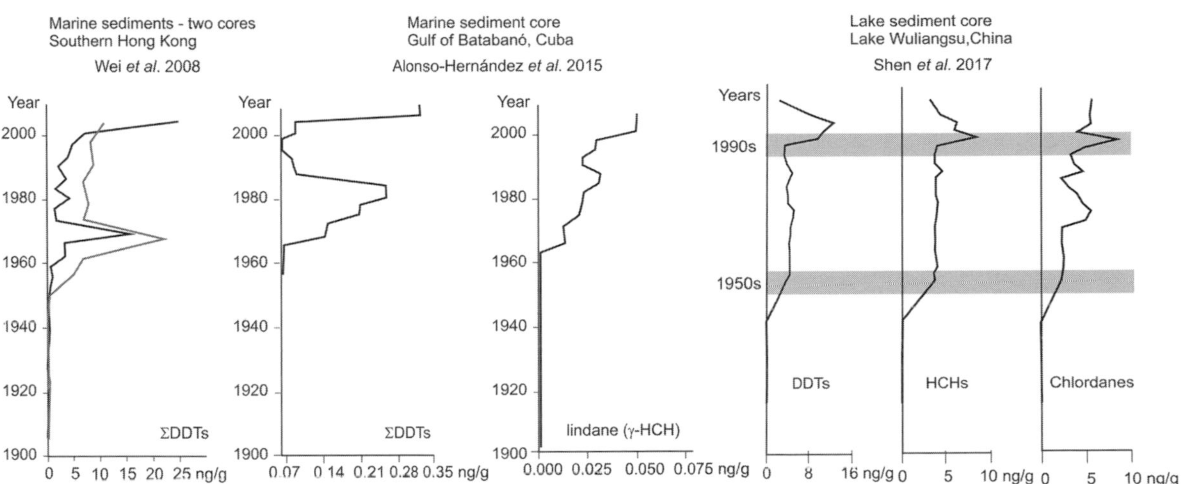

Figure 5.7.2 Time trends of selected organochlorine-pesticide concentrations in dated sediment cores.

the late 1960s–1970s and then decline. In Asia (China, South Korea, Hong Kong) a peak in organochlorine-pesticide concentrations occurs in the 1970s, and in some studies, a second peak is also reported for the 1990s, explained by either illegal use of these pesticides or remobilisation of legacy pollutants (Bigus et al. 2014). Furthermore, these compounds may be found at elevated concentrations in environments remote from their production and use, suggesting efficient and widespread transport and deposition over thousands of kilometres (e.g., toxaphene in remote Scottish lake sediments; Rose et al. 2001).

5.7.3 Polychlorinated Biphenyls (PCBs) and Dioxins (PCDDs)

A molecule of polychlorinated biphenyl (PCB) consists of two phenyl rings in which hydrogen atoms have been substituted by chlorine atoms in different positions. The number and position of chlorine atoms in the molecule is different in each of the 209 compounds, often referred to as congeners. PCBs are intentionally produced POPs with a semi-volatile character. They were used from 1929 as filling fluids in capacitors and transformers and as components of lubricants, solvents, paints and plasticisers. The total amount of PCBs produced in the 1930s–1970s is estimated at 1.3 million tons (Jamieson et al. 2017). Like polycyclic aromatic hydrocarbons, PCBs occur in mixtures of congeners, but for PCBs, tetra- (4 Cl atoms) and higher chlorinated compounds are the most persistent, while the less chlorinated congeners are more volatile (Shields et al. 2014). The use of PCBs was banned in 1976 in the USA and in the 1980s in the EU because of their toxicity and very high bioaccumulation potential.

A review of PCB concentrations in environmental samples from different countries shows that peak concentrations in the 1940s–1970s corresponded with production trends (Bigus et al. 2014). Examples of PCB-concentration trends in dated sediment cores are shown in Figure 5.7.3. However, PCBs are known to degrade, and depending on the number of chlorine atoms, the half-life of PCBs in sediments ranges from 3 years for 224'-trichlorobiphenyl to 38.5 years for 22'344'55'-heptachlorobiphenyl (Sinkkonen & Paasivirta 2000). Like metals such as Cu, PCBs are commonly found in sediments in estuaries on the banks of which PCB factories or factories processing

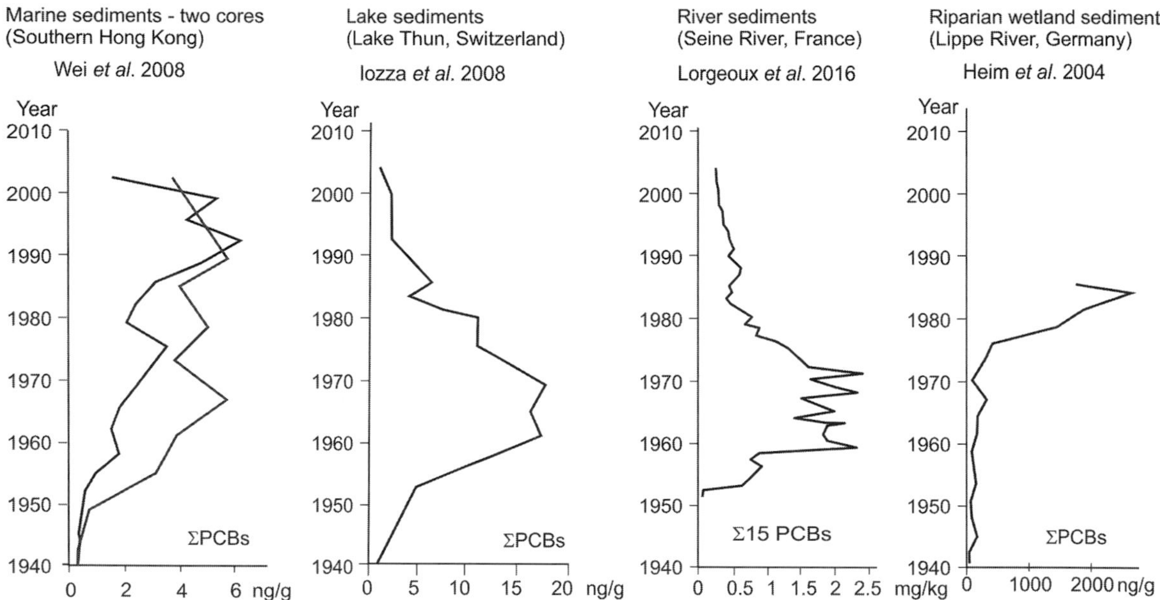

Figure 5.7.3 Time trends of PCB concentrations in dated sediment cores.

PCBs were located (e.g., New Bedford, Massachusetts; Martinez et al. 2017).

Unlike PCBs, dioxins (polychlorinated dibenzo-*p*-dioxins [PCDDs]) are not intentionally produced but are by-products of organic-material combustion in the presence of chlorine. Consequently, they also have natural sources, such as volcanic emissions or natural fires, as well as pre-industrial sources from, for example, the domestic combustion of coastal peats (Meharg & Killham 2003). In a dioxin molecule, two chlorine-substituted benzene rings are joined by a pair of oxygen atoms. The half-life of dioxins in sediments depends on the degree of chlorination and substitution position and is estimated at 64–150 years. An increasing trend in concentrations of dioxins in sediment cores from Europe and North America can be observed in sections older than the 1940s, with a maximum in the late 1960s and early 1970s (Bigus et al. 2014). Dioxins formed one of the original 'dirty dozen' compounds and groups of compounds listed under the Stockholm Convention.

5.7.4 Polybrominated Diphenyl Ethers and Fluorinated Compounds

Polybrominated diphenyl ethers (PBDEs) have been commercially produced since the 1970s as penta-, octa- and decabrominated diphenyl ethers (5-, 8- and 10-Br atoms respectively) and used mostly as flame retardants, antifungals and fumigants. Highly brominated compounds of this group were banned in Europe in the first decade of the 2000s, while a number of PBDE groups were added to a modification of the Stockholm Convention in May 2009. Their manufacture and use in North America has also been restricted (Lu et al. 2015). It has been estimated that brominated flame retardants were used as additives or constituents of more than 2.5 million tons of polymers produced annually (Morf et al. 2005), mostly used in electrical and electronic equipment, transportation, construction and manufacturing of textiles, carpets and furniture (Jinhui et al. 2017).

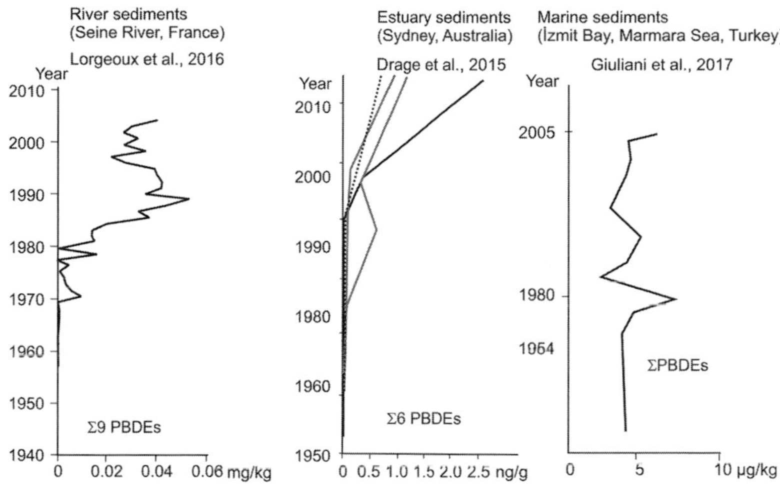

Figure 5.7.4 Time trends of PBDEs concentrations in dated sediment cores.

Like PCBs, PBDEs occur as 209 congeners. Photolytic debromination is the major abiotically mediated transformation process, resulting in the production of less brominated congeners, which both are more toxic and have a higher bioaccumulation potential than more highly brominated compounds. Because PBDEs are highly hydrophobic, they are readily accumulated in sediments. The production and use history of these compounds have been recorded in sediment cores. Restrictions on the use of PBDEs have been introduced only recently, and reductions in PBDE concentrations in sediment cores (Figure 5.7.4) are not as clear as for POPs that were banned in the 1970s. However, in some studies, peak concentrations are reported for the late 1990s/early 2000s (Drage et al. 2015; Yang et al. 2016; Lorgeoux et al. 2016).

Per- and polyfluorinated compounds (PFCs) have been produced since the late 1940s and are principally used as protective coatings on paper and fabrics and as fire-fighting foams (Zareitalabad et al. 2013). Production increased more rapidly from the 1970s, with global production of perfluorosulphonate acid (PFOS) increasing ninefold to 4,500 tons per year by the 1990s (Codling et al. 2014). The distribution of PFCs is now considered ubiquitous (Muir & de Wit 2010; Zareitalabad et al. 2013) in many environmental compartments, including water, sediments, biota and humans. Degradation products of PFCs are toxic to biota at environmentally relevant concentrations, and PFOS was added to the Stockholm Convention with PBDEs in 2009.

Temporal trends of PFCs in sediment records remain limited (Figure 5.7.5), but concentrations of Σ_{10}PFCs (the sum of 10 selected PFC compounds) in Lake Michigan show marked increases from the mid-20th century, especially since the 1970s, reaching maximum concentrations of 400–1,200 ng g^{-1} in most recent sediments. No recent decline in concentrations is observed (Codling et al. 2014). The same temporal pattern is recorded in sediments of Tokyo Bay, Japan, where concentrations for a range of PFCs increase from the 1970s onwards (Zushi et al. 2010).

5.7.5 Persistent Organic Pollutants as a Marker of the Anthropocene

A clear advantage of the use of POPs as markers of the Anthropocene is a well-developed database of historic trends in sedimentary records in industrial and pristine areas and consistency in production and sediment record histories. A correlation between changes in concentration levels in dated sediment

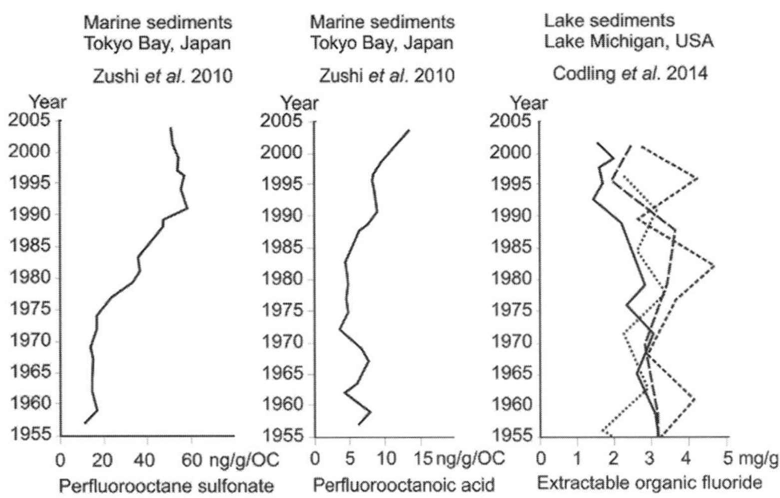

Figure 5.7.5 Time trends of PFC concentrations in dated sediment cores.

cores and the emission of POPs from anthropogenic sources confirms the usefulness of POPs for reconstruction of time trends, making them good candidates for signals of the Anthropocene (Dachs & Méjanelle 2010). Moreover, the long-range transport of POPs in the atmosphere, with their increased condensation and deposition in polar regions where there are few direct sources of these compounds, creates well-preserved archives of historic atmospheric pollution. Phasing out and banning the use of polychlorinated biphenyls and organochlorine pesticides since the 1970s and 1980s have left a recognisable stratigraphic signal in sediments. Depending on the structure of individual POPs and their physico-chemical properties as well as environmental factors, the half-lives of these pollutants can be over hundreds of years. However, the process of POP diagenesis in sediments and their preservation potential in future strata are poorly understood, and it seems that they may be a relatively short-time marker for the Anthropocene. On the other hand, the problem of persistence of emerging organic pollutants (e.g., fluorinated compounds, pharmaceuticals and their metabolites) in different depositional environments has not been studied in detail (Muir & Howard 2006), and perhaps it will be possible to find a well-preserved organic compound or its metabolite that could be a good marker for the Anthropocene and provide a far-future biomarker. Furthermore, because of their toxicity, the presence of POPs in the environment favours biotic species that have a higher tolerance to these compounds. For example, it has been documented that POPs stimulate cyanobacterial blooms (Harris & Smith 2016), and such changes in ecosystem structure will probably leave a distinct record in future fossil records, providing a potential secondary signal for a potential GSSP (see Section 7.8.3).

5.8 Artificial Radionuclide Fallout Signals

Colin N. Waters, Irka Hajdas, Catherine Jeandel and Jan Zalasiewicz

There are a number of naturally radioactive elements and isotopes, notably uranium and thorium. Where these are locally concentrated in minerals and ores, increased levels of natural radioactivity may be detected. In rare instances, fissionable elements can be concentrated sufficiently for chain reactions to be initiated, as in the natural underground 'reactors' that formed at Oklo, in Gabon, in Proterozoic times (Gauthier-Lafaye et al. 1996). Naturally radioactive substances have increased worldwide, in the air, in water, in ice and on the land surface, due to the mining of ore and the burning of coal and other fossil fuels. The widespread distribution of these radioactive contaminants began during the Industrial Revolution, with the burning of coal, and rapidly increased after 1945 during the Great Acceleration.

The discovery of radioactivity by Henri Becquerel in 1896 (Figure 5.7.1) was followed by its exploitation, firstly for radiometric dating (Holmes 1913) and subsequently for more or less controlled fission (and, later, fusion) reactions both in bombs and in nuclear reactors, together with local smaller-scale use, such as for tracers used for medical purposes. This development led, by various means, to the large-scale production and dissemination into the environment (air, water, soil, plants) of radioactive elements, many of which are rare in nature, such as ^{137}Cs (caesium-137), ^{241}Am (americium-241) and ^{239}Pu (plutonium-239), while the production of others that are moderately common in nature (such as ^{14}C, or carbon-14, continually naturally produced by the effect of cosmic rays on nitrogen in the atmosphere) also became measurably enhanced through human activities. Together, these mixtures of isotopes have produced a globally distributed, near-synchronous signal in sediments and other materials (such as ice, wood, corals and animal bone). This mid-20th-century signal constitutes an important anthropogenic marker within the Anthropocene stratal record (Hancock et al. 2014; Waters et al. 2015). Within this chemostratigraphic marker, a few specific signals can be distinguished that provide additional time control, such as the SNAP-9A satellite re-entry and burnup in 1964, which generated a clear ^{238}Pu signal, and the Chernobyl and Fukushima nuclear-power-station disasters of 1986 and 2011, respectively (Figure 5.8.1). However, these accidental discharges and the chemostratigraphic-marker signal in general are small compared with the fallout from atmospheric detonations of nuclear devices.

5.8.1 History of Nuclear Weapons Testing

The majority of anthropogenically sourced radionuclides present in the environment today were produced by atmospheric nuclear bomb tests. The tests began with the Trinity detonation at 5:30 a.m. on 16 July 1945 at Alamogordo, New Mexico, a time instant that has been suggested for a potential Anthropocene boundary, representing a Global Standard Stratigraphic Age (GSSA) (Zalasiewicz et al. 2015b). The Trinity test, and the subsequent Hiroshima and Nagasaki bombs, also in 1945, although devastating in terms of human lives lost, had a rather small impact on radionuclide abundances by comparison with subsequent thermonuclear test detonations (Figure 5.8.1).

In order to generate a widespread fallout signature, the detonations had to occur in the atmosphere. Between 1945 and 1951, the dissemination of artificial radionuclides as fallout was limited and local in extent. During this time, detonations were of fission weapons (commonly called atomic or A-bombs), which were relatively small; the Trinity detonation had an explosive yield equivalent to 21,000 tons of TNT (UNSCEAR 2000). Such detonations produced local to regional fallout along test-site latitudes via the lowest layer of the atmosphere (Aarkrog 2003).

Global fallout signals were produced by the more powerful atmospheric fusion ('thermonuclear') weapons tests (commonly called hydrogen or H-bombs), which began in 1952, typically with explosive yields in excess of 0.5 megatons (Mtons) TNT equivalent. The largest H-bomb, detonated by the USSR in 1961, had a yield of 50 Mtons (UNSCEAR 2000). These detonations produced high-altitude fallout that dispersed over the entire Earth surface, with a marked peak in fallout yields in 1961–1962, just after a testing moratorium in 1959–1960

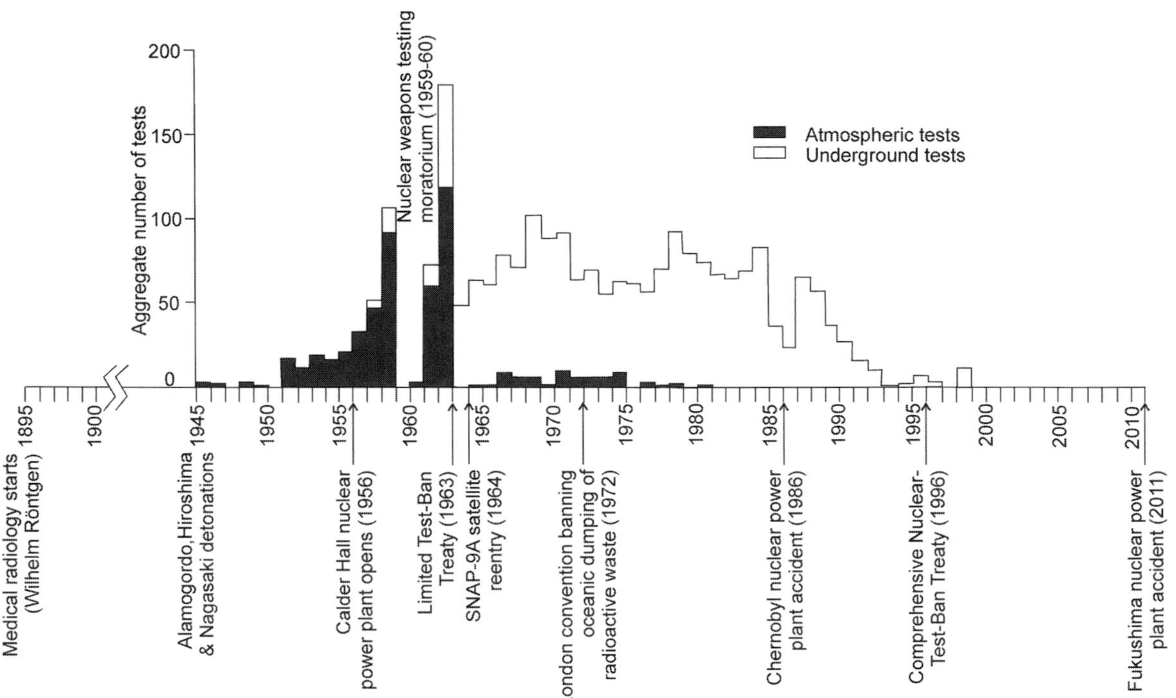

Figure 5.8.1 Timeline of important anthropogenic radiogenic events (from figure 1 of Waters et al. 2015). The frequency of atmospheric and underground nuclear weapons testing is sourced from UNSCEAR 2000. Reprinted by permission of Taylor & Francis Ltd (http://www.tandfonline.com)

(Figure 5.8.1). The delay reflects the time it takes for fallout from the stratosphere to reach the Earth's surface (much as we see in the effects of volcanic dust that reaches the stratosphere after the eruption of large volcanoes, like Mount Pinatubo in 1991, which cooled the global climate in 1992). The Limited Test-Ban Treaty of 1963, after which tests were mainly underground (Figure 5.8.1), led to a rapid decline in radionuclide fallout during the late 1960s, which effectively ceased in 1980.

Overall there were >2,000 nuclear weapons tests between 1945 and 1998, mainly in central Asia, the Pacific Ocean and the western USA (UNSCEAR 2000; Waters et al. 2015; Figure 5.8.2 herein). About three-quarters were underground and released little or no radiation to the atmosphere, although they caused substantial and long-lasting subterranean changes with the development of craters and deformed rock (Zalasiewicz et al. 2014c; Section 2.4.2.9 herein). The melting of quartz sand during detonations produced small amounts of radioactive glass, deposits of which in the USA are termed 'trinitite', named after the first test site (Eby et al. 2010), or 'kharitonchik' in Kazakhstan. Despite creating a long-lasting geological record, such deposits are limited to the extent of the detonation site.

5.8.2 Radioisotopes as Markers for the Anthropocene

Products of the nuclear detonations were spread worldwide but were affected by both wind and precipitation patterns (and, in the sea, by water depth and ocean currents – see the following paragraphs) as well as by test location. About three-quarters of all radionuclide fallout occurred in the Northern

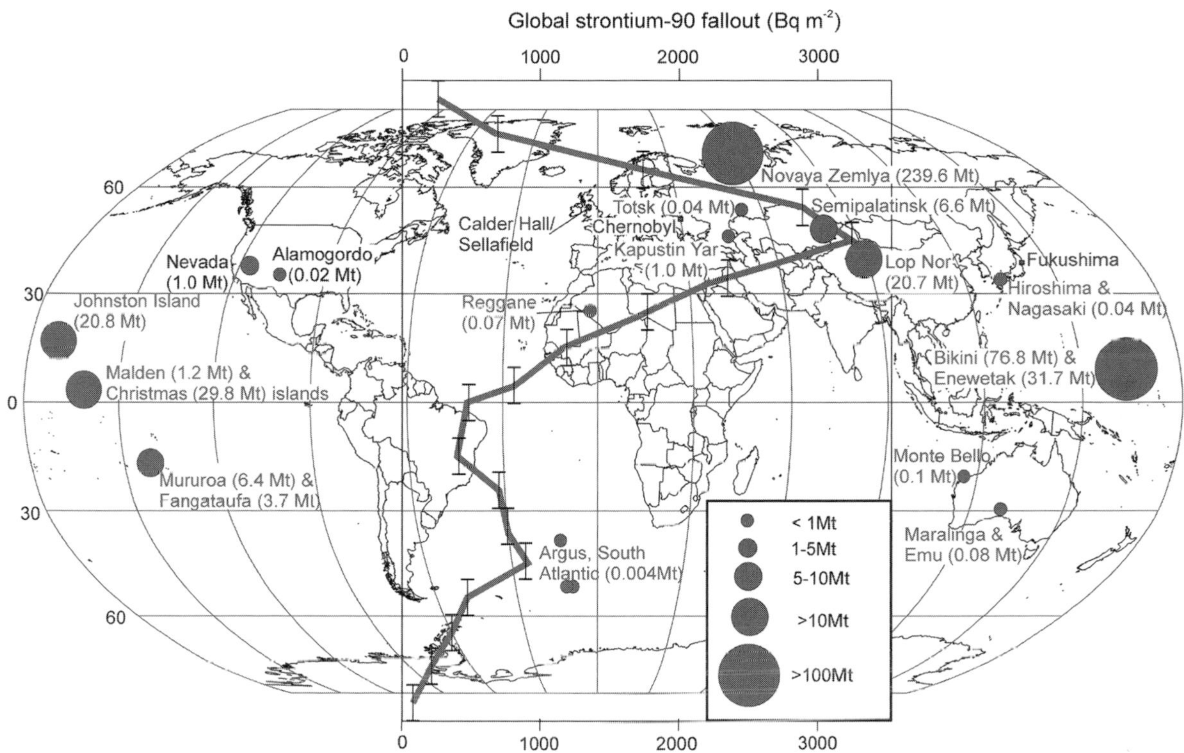

Figure 5.8.2 Distribution and total yields, in megatons, of atmospheric nuclear weapons tests (red); and location of significant nuclear accidents/discharges (blue); with superimposed latitudinal variation of global ^{90}Sr fallout, in becquerels per square metre, which is highest in mid-latitudes in the Northern Hemisphere. From figure 2 of Waters et al. (2015), with data sourced from UNSCEAR 2000. Reprinted by permission of Taylor & Francis Ltd (http://www.tandfonline.com). (A black-and-white version of this figure appears in some formats. For a colour version, please refer to the plate section.)

Hemisphere (Livingston & Povinec 2000), mainly in mid-latitudes, which were subject to the greatest number and magnitude of detonations (Waters et al. 2015). This is demonstrated in Figure 5.7.2 by the elevated ^{90}Sr (strontium-90) values, in Figure 5.7.3 by ^{137}Cs abundance, and in Figure 5.7.4 by atmospheric radiocarbon (^{14}C) excess. Some of the more abundant products include relatively short-lived isotopes (Table 5.8.1) such as tritium (^{3}H or hydrogen-3) (half-life of 12 years), ^{137}Cs (30 years), ^{90}Sr (29 years) and ^{241}Pu (14 years); these can still be used as chemostratigraphic markers (e.g., Hancock et al. 2011; Figure 5.7.5 herein), but after a century they will have decayed to undetectable levels. Longer-lived products include ^{241}Am, ^{14}C and ^{239}Pu, with half-lives respectively of ~432 years, 5,700 years and 24,110 years (Table 5.8.1); the last of these will remain detectable for ~100,000 years (Hancock et al. 2014) and beyond as the decay product ^{235}U (uranium-235). Note that because the peaks of ^{238}Pu and ^{239}Pu are not resolvable when alpha counting, they are commonly presented as $^{238+239}$Pu, which can be distinguished by using sensitive mass spectrometers.

The properties of the radioisotopes help to determine their stratigraphic use. Caesium is abundant and relatively easily detected, but it is also relatively soluble and so may migrate in groundwater. Thus, leaching in calcareous sediments may result in signal onset occurring below the level corresponding with the start of nuclear testing (Hancock et al. 2014). Similar mobility has also been recorded in marine

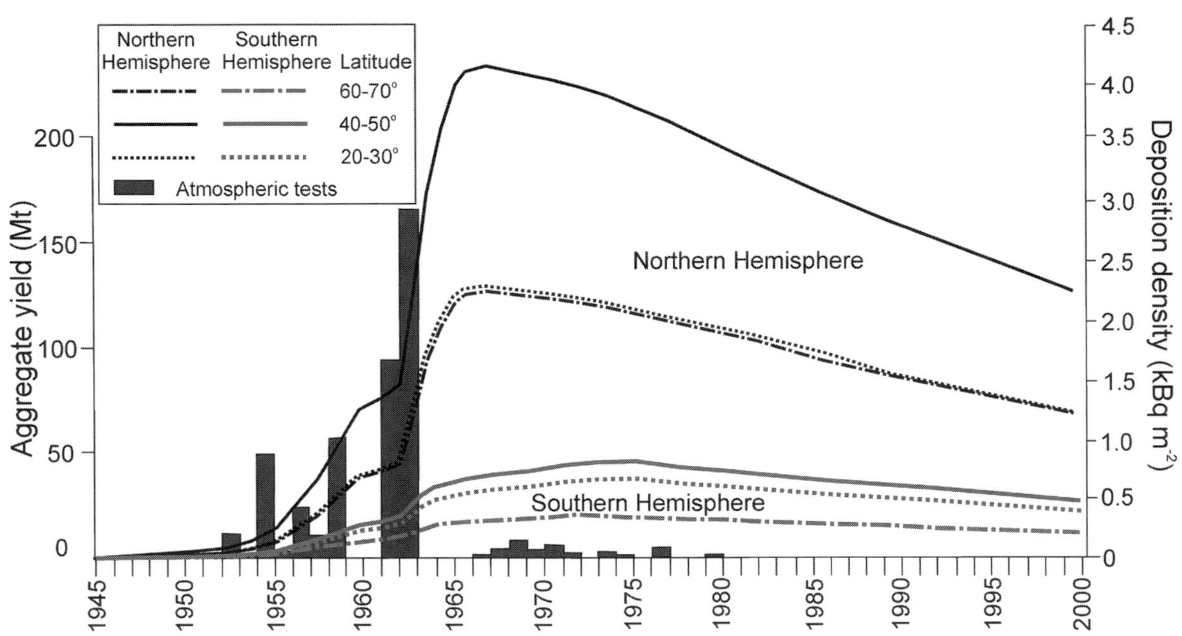

Figure 5.8.3 ^{137}Cs deposition density (kBq, kilobecquerel) at various latitudes plotted against the yield of atmospheric nuclear weapons testing (in dark vertical bars). Data sourced from UNSCEAR (2000).

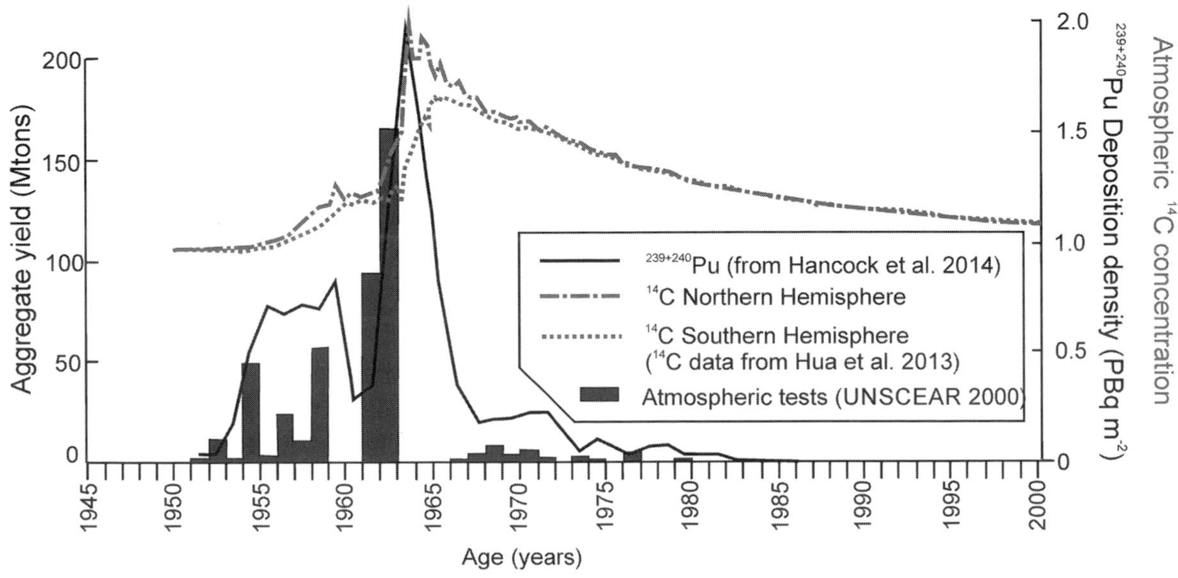

Figure 5.8.4 The atmospheric concentration of radiocarbon (Hua et al. 2013) and $^{239+240}$Pu (Hancock et al. 2014) radiogenic fallout from nuclear weapons testing (PBq, petabecquerel), plotted against annual aggregate atmospheric weapons test yields (as vertical bars) (UNSCEAR 2000). Sourced from figure 4 of Waters et al. (2016). Reprinted with permission from AAAS.

Table 5.8.1: **Fallout of radionuclides from atmospheric weapons testing**

Radionuclide	Half-life (years)	Global fallout (PBq)	Global fallout (g)
Hydrogen-3	12.3	186,000	–
Strontium-90	28.8	622	0.12×10^6
Caesium-137	30.1	948	0.30×10^6
Carbon-14	5,700	213	1.29×10^6
Plutonium-238	88	0.3	0.40×10^3
Plutonium-239	24,110	6.5	2.84×10^6
Plutonium-240	6,563	4.4	0.52×10^6
Plutonium-241	14.4	142	0.04×10^6
Americium-241	432.2	–	–

Note: Global fallout from ^{238}Pu does not include fallout from the SNAP-9A satellite re-entry, estimated at 0.6 PBq, and global ^{239}Pu and ^{240}Pu do not include proximal fallout from the Pacific Proving Grounds, estimated at 5.1 PBq. Modified from Hancock et al. (2014); original source UNSCEAR (2000).

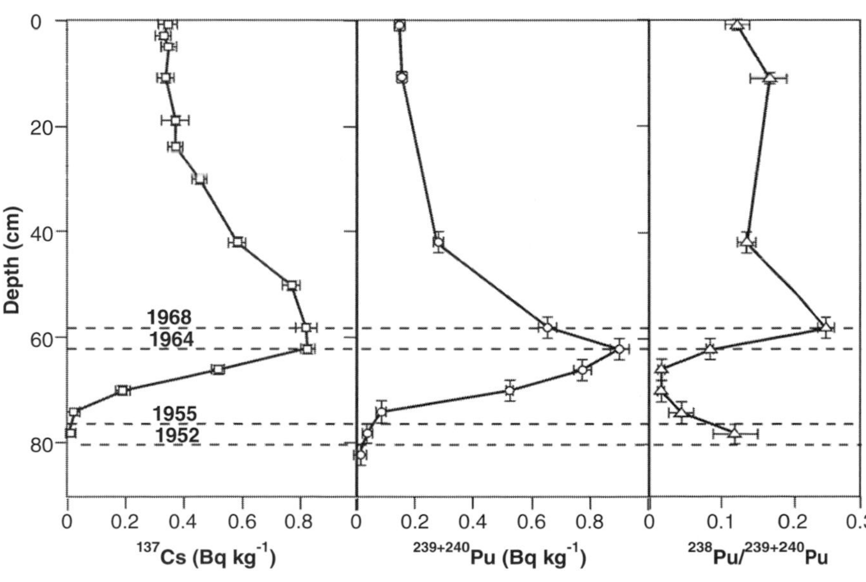

Figure 5.8.5 Radiogenic signature in a sediment core from Lake Victoria, Australia (reprinted from figure 5 of Hancock et al. 2011). Profiles of ^{137}Cs (closed squares) and $^{239+240}$Pu (open circles). ©2011, with permission from Elsevier.

(saline) sediments (Beasley et al. 1982; Sholkovitz & Mann 1984) and in anoxic conditions associated with organic-rich deposits such as peat and/or lake sediments (Jeandel 1981). Caesium can remain in solution within lake waters for longer than plutonium, which may result in the profile of ^{137}Cs being much broader than that of plutonium (Hancock et al. 2014; Figure 5.8.5 herein).

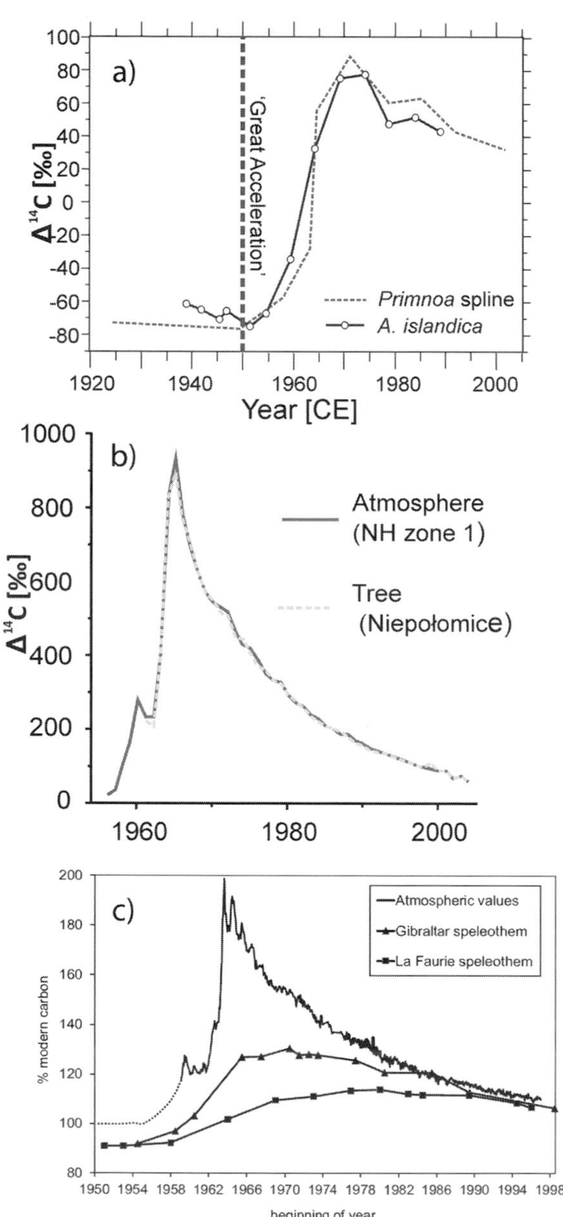

Figure 5.8.6 Examples of ^{14}C signals within different environments. (a) Plot of $\Delta^{14}C$ vs. age for seven different colonies of the deep-sea gorgonian coral *Primnoa resedaeformis* (Sherwood et al. 2005b) and for the bivalve mollusc *Arctica islandica* (Weidman & Jones 1993); (b) $\Delta^{14}C$ in tree rings (*Pinus sylvestris*) samples from 1960 to 2003 CE at Niepołomice (Poland) showing changes of radiocarbon concentration (from Rakowski et al. 2013); ©2013, with permission from Elsevier. (c) comparison of Northern Hemisphere atmospheric $\delta^{14}C$ values (from a German-Austrian composite; see Smith et al. 2009) expressed as percent modern carbon (defined as 1950 activities) within speleothems from La Faurie Cave, southwest France (Genty & Massault 1999) and Lower Saint Michael's Cave, Gibraltar (Mattey et al. 2008). The speleothem records show differing degrees of attenuation of the atmospheric signal, and the La Faurie record has a lagged peak. However, the initial rise is seen by 1958 in both records. and this can be regarded as a reliable marker for correlation (Hua et al. 2013). Figure sourced from Fairchild & Frisia (2014). (A black-and-white version of this figure appears in some formats. For the colour version, please refer to the plate section.)

Radiocarbon (^{14}C) has a long half-life and is one of the more abundant radionuclides dispersed in fallout. It is absorbed within growing organic matter (e.g., wood or bone), and the excess radiocarbon of the 'bomb spike' may be easily detected in corals (e.g., Sherwood et al. 2005b) and bivalves (e.g., Weidman & Jones 1993; Figure 5.8.6a herein), with tree rings (e.g., Rakowski et al. 2013; Figure 5.8.6b herein) producing a signal sufficiently sharp to have been suggested for a potential GSSP (Lewis & Maslin 2015). The ^{14}C signal is also clearly seen in some speleothems (Fairchild & Frisia 2014; Figure 5.8.6c herein). Nevertheless, ^{14}C can be relatively easily mobilised in many sediment types, where the signal may become blurred. The signal may also be diachronous, because its onset in the Northern Hemisphere is typically in 1955, with a peak in 1963–1964 (Hua et al. 2013; Figure 5.8.4 herein), but is delayed and attenuated in the Southern Hemisphere. For example, the onset of the signal in corals in the Abrolhos Archipelago of Brazil is delayed by ~1 year during the late 1950s compared with corals in Florida (Druffel 1996), which are located closer to the main latitude of detonations (Figure 5.8.2). It takes about a decade for the ^{14}C present within the atmosphere to equilibrate with the ocean surface, resulting in surface ocean concentrations lagging the atmosphere significantly (Key et al. 2004). In shallow-water corals, the lag is expressed in the peak radiocarbon signal, while the inception of the signal dates to the mid-1950s. At greater water depths, in excess of 1 km, the first

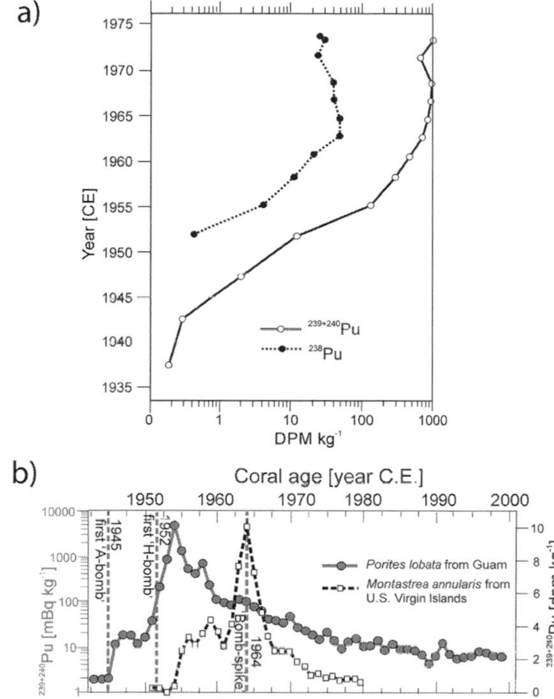

Figure 5.8.7 Examples of plutonium signals within different environments. (a) In the Santa Barbara Basin, USA (Koide et al. 1975); (b) $^{239+240}$Pu concentrations in annual growth bands in corals from Guam in the Pacific (Lindahl et al. 2011) and the US Virgin Islands in the Caribbean (Benninger & Dodge 1986); dpm kg^{-1} = decays per minute per kilogram; mBq kg^{-1} = megabecquerel per kilogram. ©2017, with permission from Elsevier.

record of bomb radiocarbon in gorgonian corals is at ~1980, the attenuated signal showing a delay to both the inception and peak of some 25 years compared with the atmospheric signal (Lee et al. 2017). Evidently, ^{14}C should not be used as a proxy in deep-marine sediments or corals but is suitable in shallow-water sediments and corals.

Plutonium binds to organic and clay particles and iron oxides (Chawla et al. 2010) and has low solubility, so it tends to be relatively immobile in many sediments (including ice). This generally preserves the stratigraphic signal, though it can migrate in anoxic conditions associated with organic-rich deposits such as peat (Jeandel 1981; Quinto et al. 2013). In the open ocean, most of the plutonium of the bomb spike is still held within the ocean waters (Lee et al. 2005) and is only slowly descending to the seafloor, with peak fallout signals descending at an average rate of 12.5 metres per year (Livingston et al. 2001). This will produce a 'smeared' deposit in marine sediments, as demonstrated in the Santa Barbara Basin (Koide et al. 1975; Figure 5.8.7a herein). In contrast, americium is carried down to the seafloor rather more quickly (Lee et al. 2005), although its short half-life (Table 5.8.1) makes this radioisotope unsuitable in the long term. In shallow seas, for instance in corals (e.g., Benninger & Dodge 1986; Lindahl et al. 2011; Figure 5.8.7b herein) and shallow lakes (e.g., Hancock et al. 2011; Figure 5.8.5 herein), this delay is not seen. The accumulation rate of ^{239}Pu is typically greater in ice, with greater accumulation in the Arctic, and shows less post-depositional alteration or mixing than in corals and lake sediments (Arienzo et al. 2016). A high-resolution profile from a glacier on Monte Rosa on the Italian-Swiss border shows an initial rise from 1954 to 1955 and sharp double peaks in 1958 and 1963, reaching a minimum in 1967 CE (Gabrieli et al. 2011).

In summary, high-resolution archives for these artificial radionuclides include the following for caesium and plutonium: polar ice cores, lacustrine, estuarine and coastal sediment and corals; and for ^{14}C: trees, terrestrial sediments, speleothems and corals. These archives provide clear and generally unambiguous indicators of post-~1952 deposition; comparable signals are not clearly resolved in abyssal sediments. Overall, ^{239}Pu has been suggested to provide the most reliable long-term signal (Waters et al. 2015, 2016), though natural decay will see this transform to a ^{235}U signal (with a half-life of ~700 million years) over longer geological timescales.

6 Climate Change and the Anthropocene

CONTENTS

We describe here the rapidly increasing rises in the greenhouse gases carbon dioxide (CO_2), methane (CH_4), nitrous oxide (N_2O) and ozone (O_3) that have taken place since the beginning of the 19th century, rises that are linked with a currently small but growing rise in temperature, as well as a yet smaller but also growing rise in sea level accompanied by increasing loss of ice from land. Natural changes cannot explain the warming since the late 1800s, which has reached levels higher than in the Holocene and is approaching those of past peak interglacial times in the Quaternary. Sea-level rise lags the rise in warming, reflecting the slow absorption of heat needed before land ice melts, along with the slow penetration of heat into the ocean interior. Due to this lag, sea levels are still up to 4–9 m lower than they were at past peak interglacial times in the Quaternary and may require several hundred years to equilibrate with the rise in temperature – or less time, if ice melt proves to be rapid. Continued anthropogenic emissions will likely also lead in the short term to prolonged rises in temperature above those typical of Quaternary interglacial phases, and above those regarded as desirable limits by the UN Framework Convention on Climate Change, bringing a variety of impacts including greater extremes of heat, drought and flooding.

6.1 Climate

Colin P. Summerhayes

6.1.1 Pre-Holocene Climate Developments

Since the days of James Hutton, all geologists have known that the present is the key to understanding the past. Rather fewer recall that Hutton also pointed out that the past is a guide to what we may expect in the future, given much the same circumstances (Hutton 1795). The geological record clearly shows that the greenhouse gas CO_2 declined throughout much of the Phanerozoic (the past 420 million years; Figure 6.1.1), primarily in response to changes in plate tectonics – principally the balance between volcanoes, as a source of CO_2, and chemical weathering, sedimentation and subduction, as a sink for CO_2 (Berner 2004; Royer 2006; Beerling & Royer 2011; Bender 2013; Ruddiman 2014).[1] That decline in CO_2 drove a general fall in global temperature (Figure 6.1.2), operating against the trend of solar radiation, which is calculated to have increased by about 5% during that interval (e.g., Foster et al. 2017). The trend of negation of the effects of solar forcing by the long-term decline in atmospheric CO_2 has also been enhanced since the Silurian by the additional negative feedback imposed by the rise of CO_2-consuming land plants over the past 420 million years (Beerling 2007; Foster et al. 2017). As a result, as Bertler and Barrett (2010) point out, the world has evolved over the past 100 million years from a warm and largely ice-free greenhouse Earth, with atmospheric CO_2 concentrations three to six times pre-industrial levels and sea level over 60 m higher, into a chilly icehouse Earth with lower CO_2 concentrations and present-day sea levels (Figure 6.1.2). Following Hutton (1795), then, this geological evidence suggests that if we continue to add greenhouse gases to the atmosphere, much as nature did in the past, we may expect to see the climate warm and sea level rise.

While the Sun provides the primary control on our climate, its effects are modified at the surface of the Earth not only by geologically driven changes in the abundance of greenhouse gases like CO_2 (Figures 6.1.1 and 6.1.2) and CH_4 but also by variations in the Earth's orbital characteristics. Fluctuations in the eccentricity of the orbit, in the tilt of the Earth's axis and in the slow precession of the Earth around its orbit, cause the amount of heat received by the Earth (referred to as orbital insolation) to change with time, causing climate to change with periods of about 100,000, 40,000 and 20,000 years, respectively. These changes in orbital insolation lie behind the relatively rapid oscillations between cold glacial and warm interglacial periods within the Pleistocene Ice Age of the past 2.6 million years (Figure 6.1.3 herein; e.g., Lüthi et al. 2008; Ruddiman 2014). The increasing tilt of the Earth's axis heats both polar regions simultaneously, increasing the probability of global ice melt. That helps to explain why much of the Pleistocene ice-core record shows glacial-interglacial cycles of 41,000 years. The recent work of Tzedakis et al. (2017) suggests that as ice sheets grew larger over the past million years, the attainment of full interglacial state required progressively more energy, which prevented some of the 41,000-year warm periods from becoming full interglacials, leaving us instead with a cycle close to 100,000 years that appears to be driven by the eccentricity cycle.

[1] Our understanding of the control of temperature by CO_2 is attributed to the experimental work of John Tyndall in the 1860s (Tyndall 1868). In the slow (geological) part of the carbon cycle (Berner 2004), volcanoes add new CO_2 to the atmosphere. The chemical weathering (dissolution) of freshly exposed silicate and carbonate minerals, especially in mountains, extracts CO_2 from the air and converts it to bicarbonate ions. Rivers transport these to the sea, where plankton use the dissolved carbon to build their soft tissues and skeletons of $CaCO_3$. Most dead calcareous plankton are recycled in the water column, but some form deep-sea carbonate ooze. When subducted by plate tectonic processes, these sediments melt beneath island arcs and resupply CO_2 to the air via volcanoes, completing the cycle. Chemical-weathering reaction rates double with every 10°C rise, consuming more CO_2 under hot conditions, thus providing Earth with a natural thermostat.

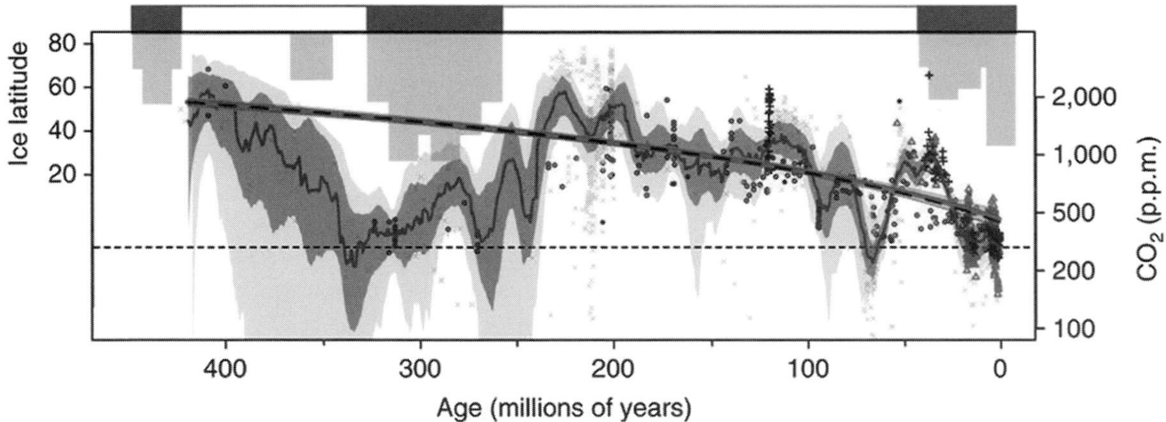

Figure 6.1.1 Distribution of CO_2 through time (dark line), with 68% and 95% confidence limits shown by dark and light shading. Curved heavy dashed line is the modelled best least-squares fit through the data. Horizontal dashed line is pre-industrial CO_2 (278 ppm). White and black bands at the top of the diagram represent greenhouse and icehouse climatic intervals respectively. Pale black shading indicates the latitudinal extent of continental ice deposits. Symbols represent various sources of CO_2 data. From Foster et al. (2017).

During interglacials, ocean warming driven by orbital change increases CO_2 in the atmosphere (Figure 6.1.3) because warm ocean water holds less gas. Changes in ocean circulation lead to additional release of CO_2 as upwelling brings more CO_2-rich deep water to the surface in the Southern Ocean when warming decreases the area of polar sea ice (e.g., Skinner et al. 2010; Hansen et al. 2016). The CO_2 then accentuates global warming through positive feedback (e.g., Petit et al. 1999; Wolff 2011; Parrenin et al. 2013; Marcott et al. 2014).

Evidently, then, changing CO_2 through plate tectonic processes and chemical weathering can cause temperature to change (e.g., the Eocene to Pleistocene experience) (Figures 6.1.1 and 6.1.2), and changing temperature through orbital change (the Pleistocene Ice Age experience) can cause CO_2 to change (Figure 6.1.3). In both cases the changes are accentuated through positive feedback. Although CO_2 and temperature seem to behave like inseparable twins, the actual relationship between them at any one time is dictated in addition by the behaviour of other variables (e.g., see Ruddiman 2014; Summerhayes 2015; Hansen et al. 2016), such as the Earth's albedo – e.g., via the extent of snow and ice on land and of ice at sea; the extent of more absorptive vs. more reflective vegetation (forests vs. grassy savannahs); the extent of deserts (which, like savannahs, have high albedo); the area of the absorptive ocean, which is greater when sea level is high and reflective sea ice is minimal; the eruption of volcanoes large enough to eject dust and sulphate aerosols into the stratosphere, where they will reflect sunlight (as in the case of the eruption of Mount Pinatubo in 1991, which cooled the world by 0.2°C in 1992); the operation of natural internal oscillations in the climate system (e.g., the El Niño–Southern Oscillation, or the Pacific Decadal Oscillation); and the natural variability of the Sun (e.g., the 11-year sunspot cycle and the 208-year Suess cycle). In recent times we must add the artificial effects of the emission of industrially produced aerosols that may reflect sunlight and of black carbon (soot) (see Section 2.4) and greenhouse gases (see Section 5.2) that absorb heat. The interaction of these various factors means we should not expect to see a simplistic 1:1 relationship between CO_2 and temperature (Figure 6.1.4).

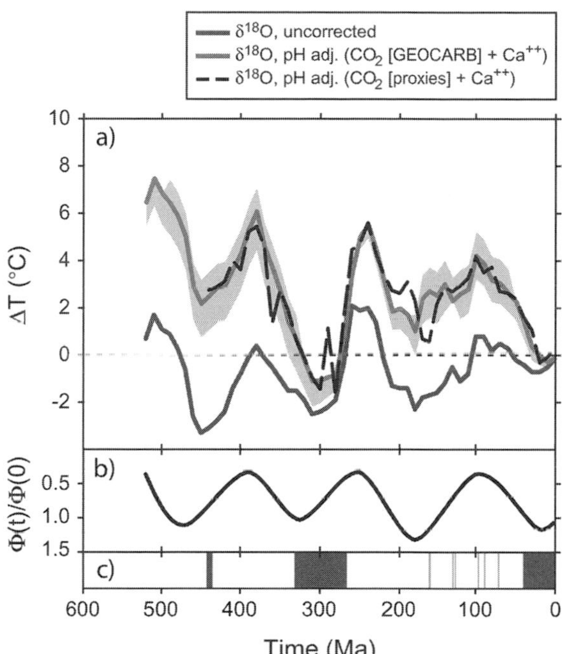

Figure 6.1.2 (a) In the lower part of the panel the dark line oscillation about the dashed line shows the temperature deviation from today calculated by Shaviv and Veizer (2003) from oxygen isotope data. In the upper part of the panel the oxygen isotope data have been corrected for pH effects attributable to changes in seawater Ca^{2+} and CO_2 concentration (by two slightly different methods) to produce two new curves of temperature deviation within an envelope representing one standard deviation. (b) Cosmic-ray fluxes relative to the present, as calculated by Shaviv (2002), which help Royer et al. (2004) to make the point that these fluctuations are not related to those of climate. (c) Intervals of glacial (in black) and cool climate (grey lines). From Royer et al. (2004) ©GSA.

Examining palaeotemperatures for the Cenozoic (the past 66 million years, or Ma), we find that overall global surface temperatures were warmest in the Early Eocene some 50 Ma ago. At the base of the Eocene there is evidence for a brief Paleocene-Eocene Thermal Maximum (the PETM) (Figure 6.1.5), the stratigraphic expression of which is discussed in Section 1.3.1.4. This event involved a major release of more than 2,000 gigatons (Gt) of carbon, making atmospheric CO_2 rise by about 700 ppm, which caused global surface temperature to rise by 4°C–6°C and sea level to rise by ~15 m, as well as making the ocean more acidic (Zachos et al. 2007; Zeebe et al. 2009, 2016). It took 100,000 years for the planet to recover from that event (Zeebe et al. 2009). Temperatures then slowly declined, and they fell more abruptly when the first great ice sheet formed on Antarctica 34 Ma ago, at the beginning of the Oligocene Epoch (Figure 6.1.2). For the next 20 Ma, temperatures changed little, reaching a temporary high during the Mid-Miocene climatic optimum about 17–14 Ma ago, when CO_2 rose to up to 300–500 ppm (Foster et al. 2012; Greenop et al. 2014; Levy et al. 2016; Shevenell 2016) and sea level rose to about 20 m above the present level (Kominz et al. 2008). Temperatures then began a moderately steep decline (from ~14 Ma ago) towards the coolness of Pleistocene times, a fall that was interrupted briefly during the Pliocene by a warm period around 3 Ma ago, when temperatures globally reached 2°C–3°C above those typical of the mid-20th century (Salzmann et al. 2011), CO_2 reached about 450 ppm (Lunt et al. 2012) and sea levels were up to about 10–20 m above today's level (Miller et al. 2012). Data from the Arctic indicate that Pliocene temperatures there were some 8°C above those of today, and their persistence through time helped to delay the full onset of Northern Hemisphere glaciation in the high Arctic until after 2.2 million years ago (Brigham-Grette et al. 2013). In the Southern Hemisphere, the East Antarctic Ice Sheet retreated substantially during the Pliocene, allowing deposition of diatoms in the Aurora Subglacial Basin, which were later blown into the Sirius Formation in the Transantarctic Mountains (Scherer et al. 2016).

During most of the Pleistocene, temperatures remained lower than our pre-industrial levels except during the last four Pleistocene interglacials starting about 400,000 years ago (Figure 6.1.3). In the most recent of these, the Eemian, about 125,000 years ago, surface temperatures were above those typical of the mid-20th century by 0.5°C ± 0.3°C globally (Hoffman et al. 2017) and by 2°C–3°C in Antarctica (Kopp et al. 2009; Figure 6.1.3 herein). CO_2 rose from 180 ppm to about 280 ppm in Eemian times, typical of Late Holocene pre-industrial levels (Lüthi et al. 2008;

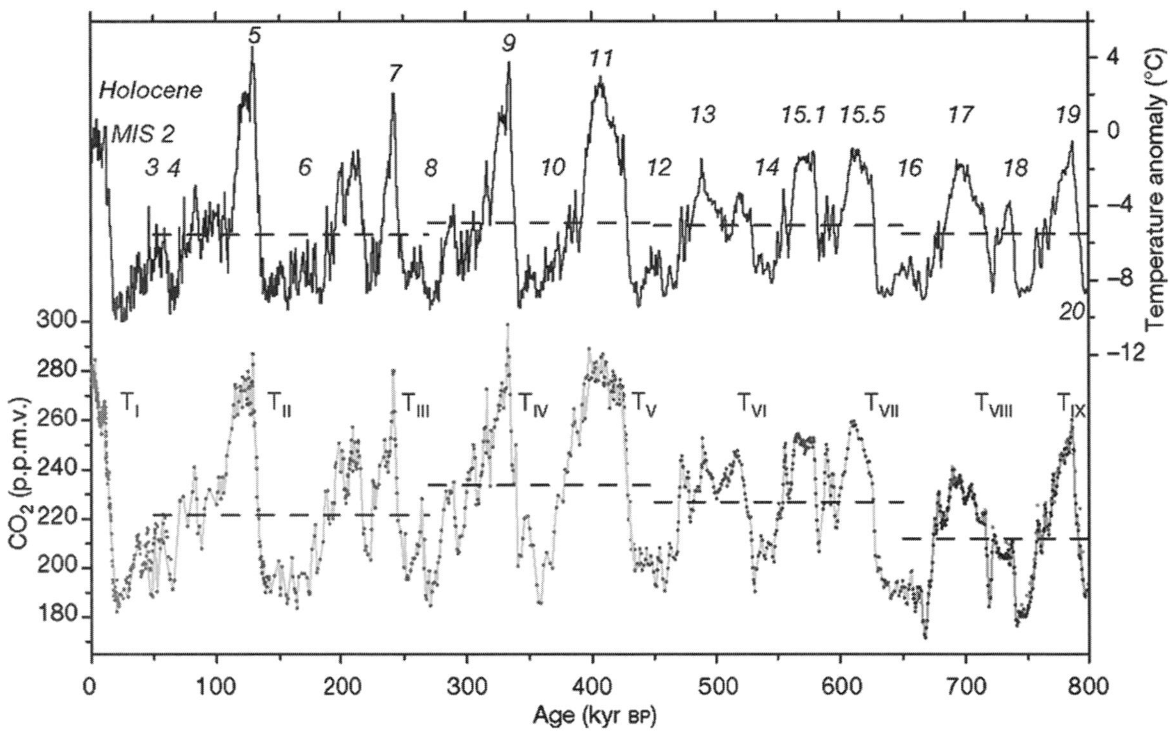

Figure 6.1.3 CO_2 records from Antarctic ice cores and palaeotemperatures from the EPICA Dome-C ice core in Antarctica. Horizontal lines represent mean values of temperature and CO_2 for specific time periods. Glacial terminations (T) are in Roman numerals (I through IX); Marine isotope stages (MIS) are given in Arabic numerals. From figure 2 in Lüthi et al. (2008).

Figure 6.1.3 herein). In response to the warming, sea level rose by 4–9 m above levels typical of the mid-20th century (Kopp et al. 2009).

During the last (Eemian) interglacial (130,000–120,000 years ago) Arctic temperatures were ~5°C warmer than at present, and almost all Arctic glaciers melted completely except for the Greenland ice sheet, which was substantially reduced in size (Miller et al. 2010a). Capron et al. (2014) found that during the Eemian the Southern Ocean and Antarctica began warming before the northern polar regions and stayed warmer for longer. This pattern of warming starting in the south also typifies the latest deglaciation (see below) as well as smaller-scale events within the last glaciation. Wolff et al. (2009) considered this pattern to be evidence for deglaciations being the result of a runaway seesaw effect led by Southern Ocean warming.

It is unlikely in the foreseeable future that plate tectonic processes will cause the positions of landmasses or ocean gateways to change sufficiently to have the kinds of effects on climate that they had in the past or that mountain altitudes will change sufficiently to affect rates of chemical weathering. Similarly, as we shall see in more detail below, it is unlikely in the next 1,000 years that the Earth's orbital parameters will change much from where they are now. However, the recent rise in CO_2 is encouraging plant growth, which, in turn, is helping to keep atmospheric CO_2 values (hence also air temperatures) slightly lower than they would be otherwise (Campbell et al. 2017).

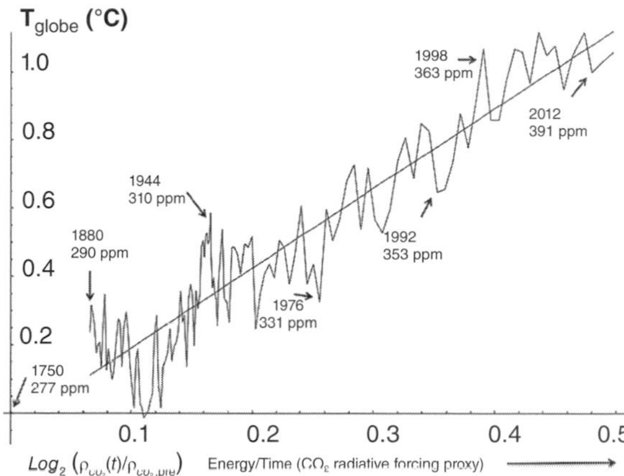

Figure 6.1.4 Global temperature (from NASA, for 1880–2013) plotted as a logarithmic function of CO_2 representing radiative forcing. Fluctuations represent natural variability. The slope of 2.33°C per CO_2 doubling is the 'effective climate sensitivity' (based on figure 1 in Lovejoy 2015, which is based on Lovejoy 2014a). The post-war cooling was likely due not only to a negative Pacific Decadal Oscillation but also to the accumulation of industrial aerosols prior to the development of Clean Air Acts; the 'pause' from 1998 to 2012 is thought to be the result of a further negative phase of the Pacific Decadal Oscillation (e.g., Lovejoy 2014b; Summerhayes 2015). Since this graph ended in 2013, warming has risen to progressively higher values in 2014, 2015 and 2016.

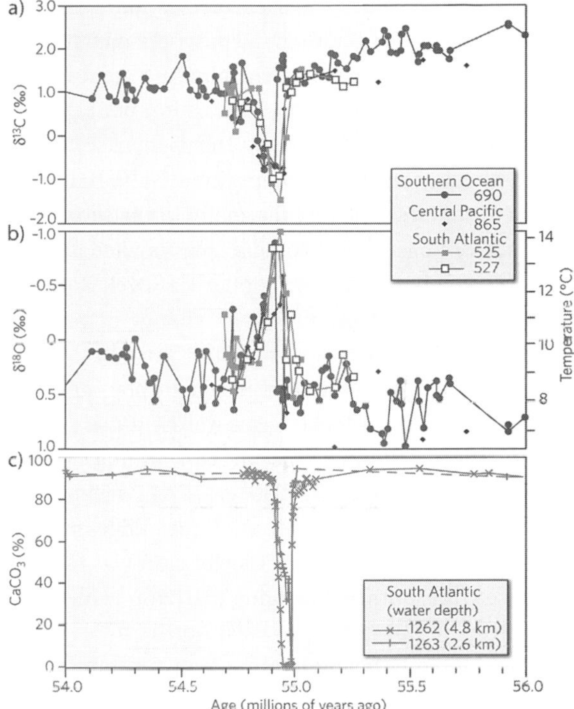

Figure 6.1.5 The PETM as recorded in deep-sea cores, showing (a) carbon isotope and (b) oxygen isotope records based on benthic foraminifera, plus inferred bottom-water temperature; (c) $CaCO_3$ values in deep-sea sediments. The decrease in $CaCO_3$ reflects dissolution caused by a substantial decline in seawater pH (i.e., an increase in ocean acidification) induced by the increase in atmospheric CO_2 and its exchange with the ocean. From figure 3 in Zachos et al. (2008).

6.1.2 The Deglaciation and the Holocene

6.1.2.1 The Last Deglaciation

The last glacial maximum about 20,000 years ago ended with a pronounced rise in orbital insolation in the Northern Hemisphere. Temperature, CO_2 and CH_4 rose in concert (Parrenin et al. 2013), despite some minor deviations, including the short cold Younger Dryas interval of the Northern Hemisphere centred on 12,000 years ago. The most likely source of the CO_2 was the warming ocean (Skinner et al. 2010), while that for CH_4 was increasing organic decomposition driven by growing warmth in largely tropical wetlands. The immediate result was the melting of the great Northern Hemisphere ice sheets covering Canada, the northern USA, Scandinavia, Britain and the Siberian coastal region and continental shelf, as well as the melting of the Antarctic ice back from the edge of the Antarctic continental shelf, accompanied by a rise in sea level of 130 m (for more details on ice, see Section 6.2).

While Northern Hemisphere summer insolation rose from a low 22,000 years ago to a peak around 11,700 years ago and then declined, Southern Hemisphere summer insolation almost did the opposite, falling to a low point around 13,000 years ago and then rising to the present, though on a much flatter trajectory than in the north (Shakun et al. 2012). While one might expect that this would make the climate of the two hemispheres out of phase, Huybers and Denton (2008) realised that Southern Hemisphere climate was responding not primarily to the absolute values of Southern Hemisphere summer insolation but rather to the longer duration of Southern Hemisphere summers, which meant, somewhat counterintuitively, that the two hemispheres operated thermally in phase rather than out of phase. The two hemispheres were also phase locked by changes in global sea level. Both hemispheres experienced the same change in axial tilt and hence the same exposure to the Sun's heat (cf. Tzedakis et al. 2017). The net result, without going into detail, was that local ocean warming and CO_2 emission in the Southern Ocean led warming in the Northern Hemisphere (see also, e.g., Toggweiler 2008; Wolff et al. 2009; Shakun et al. 2012; Fudge et al. 2013; Marcott et al. 2014). CO_2 emission was largely in phase with southern warming (Parrenin et al. 2013; Marcott et al. 2014) but preceded Northern Hemisphere and global warming by a substantial amount (Figure 6.1.6), not least because of the delay caused by the use of solar energy to slowly melt the great Northern Hemisphere ice sheets (Shakun et al. 2012; Marcott et al. 2014), something that was completed by about 6,000 years ago, with the exception of Greenland (Clark et al. 2016).

Independent confirmation of the deglacial history of West Antarctica comes from recent analyses of temperatures in deep boreholes (Cuffey et al. 2016). The region warmed by 11.3°C ± 1.8°C (about two to three times the global average) after the last glacial maximum, and warming was complete earlier than in the Northern Hemisphere. In summary, global warming at the deglaciation was driven by a combination of increasing orbital insolation, increasing CO_2 and CH_4, and northward heat transport via winds and through the ocean currents of the global thermohaline conveyor system.

The ocean plays a key role in controlling the climate by transferring heat and salt around the world through the global thermohaline conveyor system. Much of the heat exchange between the hemispheres takes place through the Atlantic meridional overturning circulation (AMOC), in which warm, salty surface water moving north in the Benguela and Florida Currents and the Gulf Stream eventually cools and sinks in the Norwegian-Greenland Sea, to return south at depth as North Atlantic Deep Water, which eventually wells up around Antarctica to continue the cycle. The process is augmented by the sinking of cold and salty Antarctic coastal water to form Antarctic bottom water that moves north beneath

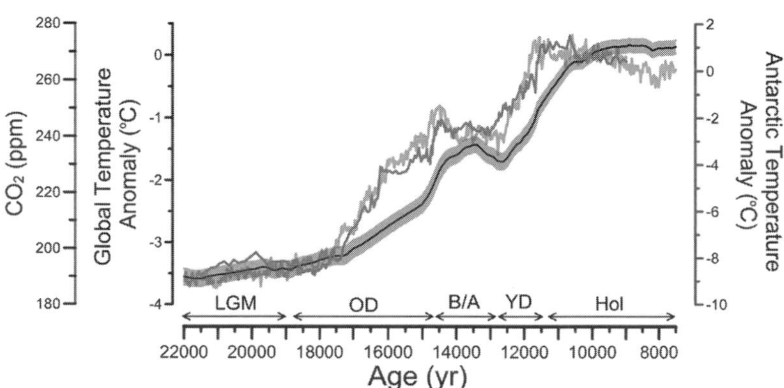

Figure 6.1.6 CO_2 abundance in the West Antarctic Divide ice core (WDC) (thin dark line) compared with Antarctic temperature stack based on East Antarctic ice cores (pale zigzag line) and with a global-temperature reconstruction (dark line with grey band representing 1σ uncertainty envelope), from extended data in figure 7 in Marcott et al. (2014).

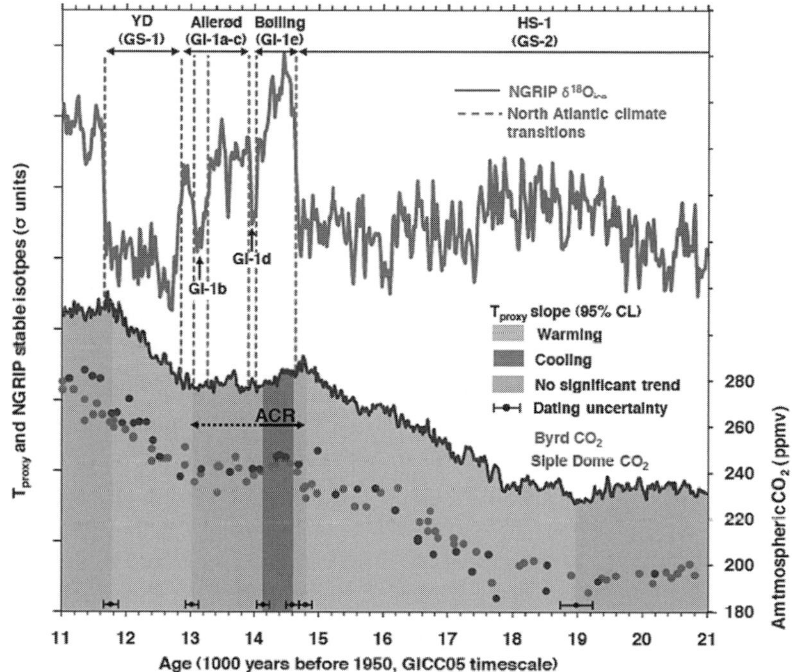

Figure 6.1.7 Synchronous proxy temperature (T_{proxy}) and atmospheric CO_2 signals in the last deglaciation at Byrd and Siple coring sites, displaying the bipolar seesaw. Significant warming and cooling trends in T_{proxy} are represented by shaded vertical bands. Climate in the North Atlantic region is represented by the NorthGRIP ice core $\partial^{18}O$ record at top. Changes in the slope of Antarctic T_{proxy} are synchronous with climate transitions in the North Atlantic (vertical dashed lines), within relative dating uncertainties (horizontal error bars). The deglacial increase in CO_2 occurs in two steps, starting at 19 ka and 13 ka and corresponding to significant warming trends in T_{proxy}. A pause in the CO_2 rise is aligned with a break in the Antarctic warming trend during the Antarctic Cold Reversal (ACR). Within the core of the Antarctic Cold Reversal, significant cooling in T_{proxy} (dark shaded band) coincides with an apparent decrease in CO_2. Note that the North Atlantic cools while Antarctica warms from 19 to 14.8 ka, and then the North Atlantic warms into the Bølling-Allerød warm stage as the Antarctic Cold Reversal begins. Antarctica then warms as Younger Dryas cooling takes place. The Younger Dryas ends as Antarctic warming stops, at 11.8 ka. Fast-acting inter-hemispheric coupling mechanisms linking Antarctica, Greenland and the Southern Ocean are required to satisfy these timing constraints. From figure 3 in Pedro et al. (2012).

south-flowing North Atlantic Deep Water to reach the North Atlantic before mixing with the overlying North Atlantic Deep Water and returning south. These processes are augmented by the sinking of northward-moving Southern Ocean surface water at the Polar Front to form Antarctic Intermediate Water that circulates through much of the ocean (as far north as 20°N) at depths of around 700–1,200 m. These processes ensure that there are much closer links between the northern and southern polar regions than one might expect from considering orbital insolation alone, although the timing of complete ocean circulation (~1,000 years) ensures that there are lags in the arrival of interhemispheric signals.

It seems highly likely that variations in the AMOC caused a seesawing of heat between the hemispheres during the last deglaciation (Shakun et al. 2012) and earlier (Buizert & Schmittner 2016; Figure 6.1.7 herein). The AMOC was weak and heat transfer between the hemispheres was reduced during the Younger Dryas cold interval immediately prior to the Holocene, but the AMOC was stronger and transported heat from south to north during the Bølling-Allerød warm interval prior to the Younger Dryas and later during the Holocene. This seesawing of heat between the hemispheres helps to explain the contrast between the coincidence of the rise of temperature and CO_2 during the deglaciation in the south and the lag of Northern Hemisphere temperature behind CO_2 in the north (cf. Shakun et al. 2012; Marcott et al. 2014). Emphasising the interhemispheric difference, we now know that the Bølling-Allerød warm interval in the north correlates with the Antarctic Cold Reversal in the south and that New Zealand glaciers melted back during the north's Younger Dryas cold interval (Kaplan et al. 2010). More importantly for considering the controls on modern and future temperature change, it appears that the variations in the strength of the AMOC on timescales between 35 and 120 years may affect global climate (Crucifix 2012). The AMOC is thought to have strengthened during periods of low solar activity, like the Maunder Minimum of the 1600s, warming the North Atlantic, and to have weakened during times of high solar activity, cooling the North Atlantic (Kobashi et al. 2013).

The seesawing of heat between the hemispheres was also apparent during the heart of the last glacial phase, and it led to the development in Greenland of so-called Dansgaard-Oeschger events and associated ice rafting; more on that in Section 6.2.2.

6.1.2.2 The Holocene

The beginning of the Holocene interglacial epoch has been placed (Walker et al. 2009; see Section 1.3.1.6 herein) at 11,700 years b2k, when both the Northern and Southern Hemisphere climates stabilised. Summer insolation then began a gradual fall towards the present in the Northern Hemisphere, while in the south it began a slow rise. Winter insolation began a slow rise in both hemispheres.

Global temperatures rose slowly to an optimum around 7,000–8,000 years ago and then declined before rising again after about 1850 (Marcott et al. 2013; Figure 6.1.8 herein). Northern Hemisphere

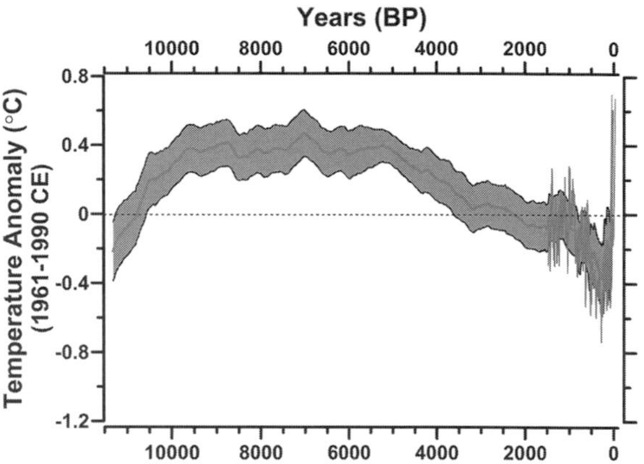

Figure 6.1.8 Global temperature stack for 73 cores largely from marine archives. '0 BP' refers to 1950 CE, and temperature anomalies are calculated with respect to the instrumental mean for 1961–1990. From figure 1B in Marcott et al. (2013). Reprinted with permission from AAAS.

temperatures stayed more or less flat while the great ice sheets melted, and then they rose slightly to a Holocene thermal optimum between 9,000 and 6,000 years ago. In the Southern Hemisphere, temperatures had reached a Holocene thermal maximum between about 11,700 and 9,500 years before present (Shakun et al. 2012; Marcott et al. 2014; Figure 6.1.6 herein). Continued slow decline of Northern Hemisphere insolation following the thermal optimum then led to the advance of Northern Hemisphere glaciers in what became known as the Holocene neoglacial. In contrast, the glaciers of the Southern Hemisphere retreated during the Holocene, in line with the slow rise in Southern Hemisphere insolation there (Mayewski et al. 2017). That slow rise in orbital insolation changed to a decline that increased in intensity from the equator to the pole during Southern Hemisphere summers over the past 2,500 years (Wanner et al. 2008).

During the warm Early Holocene, the Intertropical Convergence Zone lay far enough north that rains fell on the Sahara to feed extensive rivers and lakes (Ruddiman 2014). The neoglacial cooling in the Northern Hemisphere pushed the Intertropical Convergence Zone south, drying the Sahara 5,000 years ago and bringing to an end the North African Humid Period (Ruddiman 2014).

As Northern Hemisphere summer insolation declined during the second half of the Holocene, Arctic glaciers readvanced and sea ice cover grew, sea ice reaching its maximum extent during the Little Ice Age (~1250–1850 CE) (Masson-Delmotte et al. 2012). Late Holocene Arctic cooling reached its nadir during the Little Ice Age (Miller et al. 2010a). Miller et al. (2010b) found that summer cold and ice growth began abruptly in the Little Ice Age between 1275 and 1300 CE. Explosive volcanic activity around 1275 CE accentuated the cooling caused by a decline in solar intensity following the Medieval Warm Period (Miller et al. 2012).

In the Southern Hemisphere, the trend towards glacial retreat induced by increasing summer insolation continued in opposition to the Northern Hemisphere's neoglacial trend. But summer insolation stopped increasing about 3,000 years ago and was in slight decline by about 2,500 years ago. Glaciers readvanced there during the Little Ice Age, and long-term cooling is evident in most parts of Antarctica for the past 2,000 years (Stenni et al. 2017), where the decline in summer orbital insolation was most intense (Wanner et al. 2008).

In association with the Holocene decline in Northern Hemisphere summer insolation, the abundances of CO_2 and CH_4 in the atmosphere initially began to fall (see Section 5.2; Figures 5.2.2a and 5.2.3b). In most previous interglacials, the association between falling insolation, temperature and these two greenhouse gases continued, but not in the Holocene, where Ruddiman et al. (2015b) noted that CO_2 began to rise slightly around 6,000 years ago and CH_4 around 3,000 years ago. Those authors thought that the rise in CO_2 may have been due to the growth in land clearances as agriculture began to expand (e.g., the burning of forests releasing CO_2 and their replacement by grassland), a suggestion that is controversial, while the rise in CH_4 reflected the growth of rice cultivation. In both cases the rises were small and ran counter to the continuing decline in insolation. In effect the rising greenhouse gases and falling insolation seem to have cancelled one another out, as pointed out by Marcott et al. (2013). It is uncertain whether these small rises were sufficient to prevent an ice age from recurring: Berger and Loutre (2002) suggest not, though Ganopolski et al. (2016) suggest this as a possibility.

In the past, the 208-year Suess cycle and 88-year Gleissberg cycle in the Sun's output led to slight increases above or below the thermal baseline driven by orbital change (Steinhilber et al. 2012). For example, an increase above baseline occurred during the Medieval Warm Period between about 1000 and 1250 CE, which was experienced especially in the Northern but not in the Southern Hemisphere, and a decrease occurred during the Little Ice Age, between about 1250 and 1850 CE, which was experienced in both hemispheres (Neukom et al. 2014). These solar variations led to warm periods even within the Little Ice Age, centred on 1300–1350, 1560–1640, 1740 and 1810 (Steinhilber et al. 2012; Clette et al. 2014). From the abundance of the ^{10}Be and ^{14}C isotopes in ice cores and tree rings, which represent sunspot activity, it seems likely that the Medieval Warm Period was no warmer than the mid-20th century (Bard et al. 2000). The preceding Roman Warm Period, representing an earlier increase in solar output, was warmer than the Medieval Warm Period because it occurred when the underlying orbital insolation was higher, which explains why there were Roman vineyards near York.

6.1.2.3 The Current Trajectory

Post-1500 palaeoclimate records and model simulations confirm that sustained industrial-era warming of the tropical oceans first developed during the 19th century, nearly synchronously with continental warming in the Northern Hemisphere. Warming appears to have been delayed in Southern Hemisphere palaeoclimate reconstructions but not in model simulations. These various findings show that in some regions there has been about 180 years of industrial-era warming. Clearly, the instrumental records, which cover no more than about 150 years, are too short to assess anthropogenic climate change comprehensively (Abram et al. 2016). Greater clarity emerges from the data when results from coastal oceanic-upwelling regions are separated from non-upwelling regions: enhanced oceanic upwelling (of cold water from intermediate depths) plausibly explains recent cooling trends at some upwelling sites, consistent with theories that climate warming could, in some locations, strengthen the surface winds that drive coastal upwelling. In that context, the strengthening of westerly winds over the Southern Ocean caused both enhanced upwelling and the northward advection of any surface warming signal and so is likely to have delayed development of sustained industrial-era warming over Antarctica (Abram et al. 2016). Similarly, sustained ocean warming in the North Atlantic is now delayed or characterised by cooling due to the very recent slowdown of the AMOC. The finding that warming began in Antarctica during the mid-19th century when forcing by rising greenhouse gases was small 'suggests that Earth's surface temperature may respond to even small increases in greenhouse gas forcing more rapidly than previously thought' (Abram et al. 2016). Significant Antarctic warming since 1900 CE is apparent on the Antarctic Peninsula, the West Antarctic Ice Sheet and the Dronning Maud Land coast (Stenni et al. 2017).

Since 1900 the lower atmosphere has warmed by 1.2°C (latest independent analyses by NOAA, NASA and the UK Met Office),[2] while the stratosphere has cooled, the ocean has warmed to progressively greater depths, sea level has risen, sea ice is melting, most mountain glaciers are in retreat, both the Antarctic and Greenland ice sheets are losing mass, albedo is falling and the ocean is becoming more acidic (see Section 6.2 for ice-related details and 6.3 for sea level). The only factor that consistently explains these variations is the rise in greenhouse-gas emissions (CO_2, CH_4, N_2O and O_3) caused by human activities, enhanced by the evaporation of progressively more H_2O from the warming ocean as time has gone by. Much the same changes were experienced at the PETM, 55 Ma ago (Figure 6.1.5), accompanied by the dissolution of deep-ocean carbonate sediments (i.e., ocean acidification), mass extinction of benthic foraminifera, proliferation of exotic planktonic foraminiferal taxa and 'the dispersal and subsequent radiation of Northern Hemisphere land plants and animals' (Zachos et al. 2001). In keeping with that development, we already see plants, insects and animals moving poleward of their usual ranges, especially in the Northern Hemisphere (Juniper 2013).

As mentioned earlier, during the Holocene the neoglacial warming in the Southern Hemisphere caused glaciers there to retreat, while cooling in the Northern Hemisphere caused glaciers there to advance. In contrast, glaciers in both hemispheres retreated synchronously post-1850 to 1900 with the anthropogenic global warming that began at the end of the Little Ice Age, e.g., in New Zealand (Mackintosh et al. 2017), in Patagonia (Masiokas et al. 2009; Glasser et al. 2011) and globally (Kaser et al. 2006; Mernild et al. 2013).

In West Antarctica, Turner et al. (2016) noted that an ice core from James Ross Island east of the Antarctic Peninsula showed several periods of rapid warming and cooling over the past 1,000 years. In that context the warming trend of the past 50 years along the peninsula was highly unusual, though not unprecedented. Moreover, an ice core from Ferragno, at the southern end of the west coast of the peninsula, showed a 50-year period in the 18th century when surface air temperatures increased at a faster rate than at the Vernadsky Station on the Peninsula during the late part of the 20th century. 'These studies', they noted, 'suggest that the rapid warming on the AP since the 1950s and subsequent cooling since the late 1990s are both within the bounds of the large natural

[2] For the purpose of comparing the modern climate with pre-industrial conditions (those prior to the Industrial Revolution or prior to the time when humans began 'to demonstrably change the climate through combustion of fossil fuels'), Hawkins et al. (2017) define the pre-industrial period as 1720–1800. It was likely 0.55°C–0.80°C cooler than 1986–2005 and more than 1°C cooler than 2015.

decadal-scale climate variability of the region'. Clearly the evidence for anthropogenically driven warming in Antarctica is not as clear cut as it is in other regions. Nevertheless, the latest data show not only that the post-1900 CE warming trends for the Antarctic Peninsula, West Antarctica and Dronning Maud Land are robust but also that this century-scale trend is unusual in the context of natural variability over the past 2,000 years on the Peninsula (Stenni et al. 2017). The warming trend goes along with an increase in snowfall since 1900 CE (Thomas et al. 2017).

Nevertheless, confirmation that East Antarctica has warmed in recent times comes from ice-core data analysed by Ekaykin et al. (2016), who found a 1°C warming over the past three centuries, interrupted by a particularly cold period from the mid-18th to mid-19th century coincident with the latter part of the Little Ice Age. Peak cooling occurred in 1840. Ekaykin et al. (2016) also found that inter-annual variability in East Antarctica was primarily governed by the Antarctic Oscillation and the (tropical) Interdecadal Pacific Oscillation modes of atmospheric variability, with lower frequency variability (>27 years) being related to variation in the Indian Ocean Dipole.

In contrast, the Arctic warmed by two to three times the global average surface-temperature rise during the 20th century. Its glaciers receded, its terrestrial ecosystems advanced north and its perennial sea ice diminished (see Section 6.2). Miller et al. (2013) used radiocarbon dates on rooted tundra plants exposed from beneath melting Canadian ice caps to show that 5,000 years of Holocene summer cooling (by ~2.7°C) has now been reversed and that average summer temperatures of the last 100 years are now higher than during any century in more than 44,000 years, including the peak warmth of the Early Holocene.

Data from ice cores show that Greenland temperatures are now at about the same level they were in 1000 CE (Vinther et al. 2009; Masson-Delmotte et al. 2012). A surface-temperature reconstruction from the recently drilled NEEM (North Greenland Eemian Ice Drilling) ice core shows warming by 2.7°C ± 0.33°C in the period 1982–2011 compared with the long-term average for 1900–1970 (−28.55°C ± 0.29°C) (Orsi et al. 2017). The warming is due to increased downward long-wave heat flux, which is underestimated by 17% in atmospheric reanalyses. Greenland's meteorological data reveal a sharp rise in surface air temperature starting in 1993, with 2001–2010 being the warmest decade since the onset of meteorological measurements in the 1780s, surpassing the generally warm 1920s–1930s by 0.2°C (Masson-Delmotte et al. 2012). The year 2010 CE was exceptionally warm, with temperatures at coastal stations three standard deviations above the 1960–1990 climatological average. This warming was particularly pronounced in West Greenland and associated with a record melt over the Greenland ice sheet; it was associated with a very negative North Atlantic Oscillation (NAO) during 2010 and 2011. During the last two decades, sea surface temperatures in the southwest sector of Greenland have risen by approximately 0.5°C in winter and approximately 1°C in summer as the influx of heat to the Irminger Sea between southwest Greenland and Iceland has increased (Masson-Delmotte et al. 2012). These changes are likely to result from development of the positive phase of the AMOC since the mid-1990s. Current coastal surface air temperatures are comparable to the mean surface air temperatures of the Mid-Holocene 4,000–6,000 years ago, following which temperatures declined at about 0.4°C per 1,000 years (Vinther et al. 2009; Masson-Delmotte et al. 2012). Nevertheless, as pointed out earlier, the AMOC recently entered a negative phase that is now leading to minor local cooling.

The global rise in CO_2 is now about 20 ppm/decade (see Section 5.2.1), which is 100 times faster than most rises in CO_2 during the past 800,000 years (Wolff 2011). For example, the overall rise in CO_2 since the Last Glacial Maximum was about 10 ppm/1,000 years (or 0.1 ppm/decade) (Figure 6.1.6), but it was punctuated by three short bursts of CO_2 reaching 10–15 ppm. They represented injections of no more than 0.5 billion tons of carbon (GtC) per year (Marcott et al. 2014), significantly less than the present emissions of 10 GtC per year.

Although the PETM has some parallels with the present anthropogenic increases of atmospheric CO_2, carbon release from anthropogenic sources reached ~10 GtC/yr in 2014, which is an order of magnitude faster than that estimated for the PETM by Zeebe et al. (2016). However, opinions differ, since Bowen et al. (2015) suggest that the rates of carbon release during the PETM may have been within an order of magnitude of (and may even have approached) the rate associated with modern anthropogenic carbon emissions. A more recent study by Gutjahr et al. (2017) suggests that the total amount of carbon released may have been 10,000 Gt, at least double that estimated by Zeebe et al. (2016). Gutjahr et al. (2017) also suggest that much of the CO_2 came from volcanic activity associated with the North Atlantic Igneous Province, rather than from the destabilisation of seabed methane hydrates. Despite the modern rise in CO_2 emissions, recent years have seen a slight decline in the growth rate of CO_2 in the atmosphere, possibly due to enhanced uptake by terrestrial plant growth (Keenan et al. 2016), contributing to the 'greening of the planet' (Mao et al. 2016; Campbell et al. 2017).

The changing carbon isotopic composition of atmospheric CO_2 indicates that the source for the rise is rich in ^{12}C, a clear indication that the main source is emissions from the burning of fossil fuel (see Figure 5.2.2 in Section 5.2.1). The present emission rate is unprecedented in the past 66 Ma and puts us in a 'no-analogue' state that 'represents a fundamental challenge in constraining future climate projections' (Zeebe et al. 2016). CO_2 reached 400 ppm in 2016, and temperatures rose to their highest level since records began, 1.2°C above those of 1860–1900. This level of CO_2 is 120 ppm higher than the maximum global atmospheric-CO_2 levels of past interglacials, an increase of 43%. The 1.2°C average temperature anomaly for 2016 (relative to 1860–1900) was 0.4°C higher than the average global temperature estimated for the Holocene thermal optimum of 9,000–6,000 years ago by Marcott et al. (2013; Figure 6.1.8 herein). Marcott's data are supported by a recent reanalysis of the changing temperatures of Greenland (Vinther et al. 2009). Earth's climate is now close to the warmest temperatures of the last interglacial, which were above those typical of the mid-20th century by 0.5°C ± 0.3°C globally (Hoffman et al. 2017). Mann et al. (2017) use statistical means to establish that the recent temperature highs of 2014 through 2016 had a negligible likelihood of occurrence in the absence of anthropogenic global warming. This temperature rise may be linked to other climate-related phenomena: for instance, in Europe winter floods now seem to occur later around the North Sea, while spring floods occur earlier (Blöschl et al. 2017).

In addition to the observed rise in CO_2 since the Industrial Revolution, there has also been a significant (162%) rise in the more powerful greenhouse gas methane (CH_4), from 700 ppb in past interglacials (Loulerge et al. 2008) to 1,834 ppb in recent years (Blasing 2016; Nisbet et al. 2016; see Figures 5.2.1 and 5.2.3 and Section 5.2.2 herein). Along with the rise in CH_4 abundance since 2007, the $\partial^{13}C$ of its carbon became significantly more negative (see Figure 5.2.3a in Section 5.2.2), suggesting a substantial increase in the supply of biogenic methane from tropical wetlands, which have been experiencing more rainfall in association with warming (Nisbet et al. 2016).

Methane is now seen bubbling up from the Arctic continental shelf (Westbrook et al. 2009; Shakova et al. 2010, 2013), most likely due to melting of permafrost there by warming Arctic waters (Wadhams 2016). However, there is no conclusive proof that methane derived from the dissociation of seabed methane hydrate is reaching the atmosphere at present (Ruppel & Kessler 2017).

A recent re-evaluation of sunspot abundance (Clette et al. 2014) shows that solar output in the late 20th century peaked in 1980–1990 (Figure 6.1.9). That sunspot peak was at about the same level as those of 1780–1790 (during the Little Ice Age) and 1860–1870 (immediately postdating the end of the Little Ice Age), which means there was nothing particularly unusual about solar conditions at the

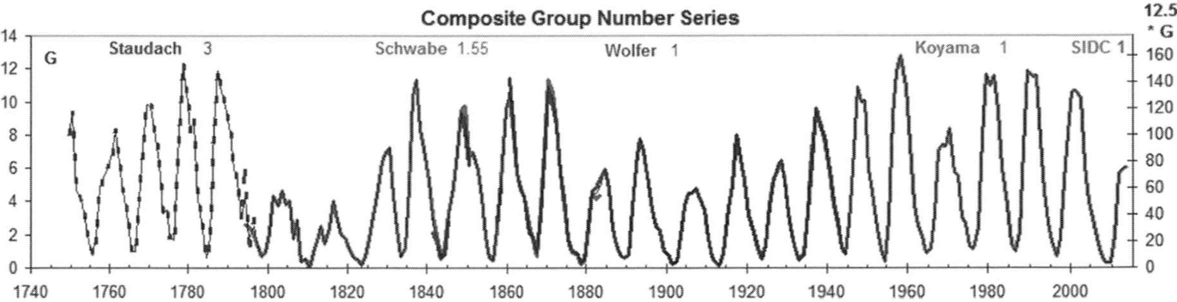

Figure 6.1.9 Sunspot activity 1749 to 2014. From figure 28 in Clette et al. (2014). Left-hand scale shows group counts, multiplied on the right-hand side by 12.5 to get most probable sunspot numbers. For subsequent updates see the NOAA Space Weather Prediction Web Site at www.swpc.noaa.gov/products/solar-cycle-progression.

end of the 20th century. Total solar irradiance (TSI) increases with increasing numbers of sunspots up to a sunspot number of about 150 (Hempelmann & Weber 2012). Hence, while it is possible that some of the initial warming from 1900 to about 1960 may have been caused by a slight increase in solar output over that period, solar activity then flattened while the temperature continued to rise. Indeed, the current trend in sunspots has been downward following the peak in 1990 (Clette et al. 2014; Figure 6.1.9 herein), which is opposite from what one would expect from the global rise in temperature.

These various data show that Earth's climate has now been pushed off its natural trajectory of the last 11,700 years into what is clearly a different climatic regime (cf. Bender 2013; Hay 2013; Ruddiman 2014; Summerhayes 2015; Waters et al. 2016; Steffen et al. 2018). We have moved out of the Holocene climatic envelope. Temperatures continue to rise consistent with a global warming trend, on which is superimposed random, stationary, short-term variability or 'noise' (Lovejoy 2014a and b; Ramstorf et al. 2017; Figure 6.1.4 herein). The causes of that natural noise are elaborated on below.

As pointed out earlier, the change in temperature is not as uniform as one would expect if the relationship between temperature and CO_2 were 1:1 (Figure 6.1.4). That is because temperature is affected not just by the rise in CO_2 but also by changes in the abundance of other gases, especially methane and water vapour, by changing albedo, by the emission of aerosols, by dust and sulphur dioxide (SO_2) gas from large volcanic eruptions and by natural oscillations in the climate system (e.g., see Ruddiman 2014; Summerhayes 2015; Hansen et al. 2016). For instance, El Niño events (like that in 1997–1998) temporarily enhance global warming (see Figure 6.1.4) by releasing heat stored in the upper ocean in the western Pacific Ocean, while intervening La Niña events temporarily lead to cooling below the global warming trend. These are part of the 3–7-year-long El Niño–Southern Oscillation, or ENSO. Natural change also takes place within the ocean-atmosphere system at decadal or multi-decadal scales in particular regions, for example through the North Atlantic Oscillation, the Arctic Oscillation, the Pacific Decadal Oscillation (PDO) and the Indian Ocean Dipole. The roughly 20-year-long PDO appears to have a global effect, its negative (cold) phase (with a cool equatorial Pacific and warm Gulf of Alaska) being associated with a slowdown in global warming from about 1945–1970 (e.g., Chavez et al. 2003; Yan et al. 2016) and again in the period from 2000 to 2013 (Yan et al. 2016; see Figure 6.1.4 herein). The precise cause of these internally driven oscillations remains the topic of research.

Much of the lack of warming in the 1950s and 1960s also appears to have been the result of massive emissions of reflective aerosols from enhanced industrial output in the years before aerosols were decreased in the atmosphere by the application of Clean Air Acts; e.g., see Budyko (1977) and Smith et al. (2011). Nevertheless, at least some of the global cooling or warming stasis then and in the 1998–2013 period may be attributable to 'the increased uptake of heat energy by the global ocean during those years' (Yan et al. 2016) – i.e., during the negative (cool) phase of the PDO.

Aside from the effects of change in the AMOC, mentioned earlier, there is also an Atlantic Multi-decadal Oscillation (AMO), during which the North Atlantic warms by about 0.2°C during the positive (strong) phase and cools by the same amount during the negative (weak) phase, at intervals of about 50–70 years (Knudsen et al. 2011). The positive phase of the AMO may have contributed to the American 'Dust Bowl' conditions of the 1930s. It is not to be confused with the North Atlantic Oscillation (NAO), which is mainly an atmospheric phenomenon driven by changes in air pressure between the Iceland low- and Azores high-pressure cells. These and other quasi-periodic internal oscillations like the PDO and ENSO supply much of the background high-frequency noise in the climate spectrum (Figure 6.1.4), as do occasional large volcanic eruptions, like Mount Pinatubo (1991), mentioned in Section 6.1.1.

6.1.3 Projections for the Future

Given that orbital conditions are likely to remain more or less stable for the next 1,000–5,000 years (Berger & Loutre 2002), the main drivers for climate change into the future are likely to be the large and rapid ongoing anthropogenic greenhouse-gas emissions that have already shifted Earth into an Anthropocene climate state, the ultimate scale of which will depend on the amount of future emissions (Clark et al. 2016; Hansen et al. 2016; Steffen et al. 2018), as shown vividly in Figure 6.1.10. This trajectory will be modified by short-term variations in solar output within the 208-year Suess cycle and 88-year Gleissberg cycle and by occasional volcanic activity (cf. Summerhayes 2015). We can expect these solar cycles and natural oscillations to continue to create short-term warmings and coolings above and below the underlying upward trend of warming driven primarily by emissions of greenhouse gases, much as shown in Figure 6.1.4. On the longer timescales of orbital change, Ganopolski et al. (2016) calculated that greenhouse-gas emissions are now abundant enough to keep the Earth from descending into another glaciation for 50,000 years and that a moderate further addition of CO_2 is likely to keep the Earth from moving into another glacial period for 100,000 years.

Inevitably, there are lags within the climate system as it moves towards a new equilibrium state. One of these involves CO_2, because although it rose in phase with temperature in the Southern Hemisphere (rising insolation warmed the ocean, releasing CO_2 almost immediately to enhance temperature rise; Parrenin et al. 2013; Marcott et al. 2014), it was lagged by temperature in the Northern Hemisphere, where solar energy was used to melt the great Northern Hemisphere ice sheets (Shakun et al. 2012). As we shall see in more detail in the section on sea level (Section 6.3), sea-level rise also lags temperature rise, not least because it takes a great deal of time to melt the great ice sheets on land and for heat to penetrate the ocean to full ocean depth.

Land plants introduce a further important lag in the climate system. The northern extent of coniferous forests is not moving north across Siberia as rapidly as one might expect from global warming. It appears that the growth of shallow-rooted larch trees, on top of soil hardened by permafrost, preserves ice below ground from melting, making it impossible for deeper-rooted conifers to move north. This kind of 'vegetation-climate lag' was once thought to last no more than a few centuries, but now it appears to be persistent for millennia (Herzschuh et al. 2016). Given continued or persistent warming on the multi-millennial scale, we can expect a slow northward migration of larch and then conifer and spruce into what is now Arctic tundra, darkening the surface,

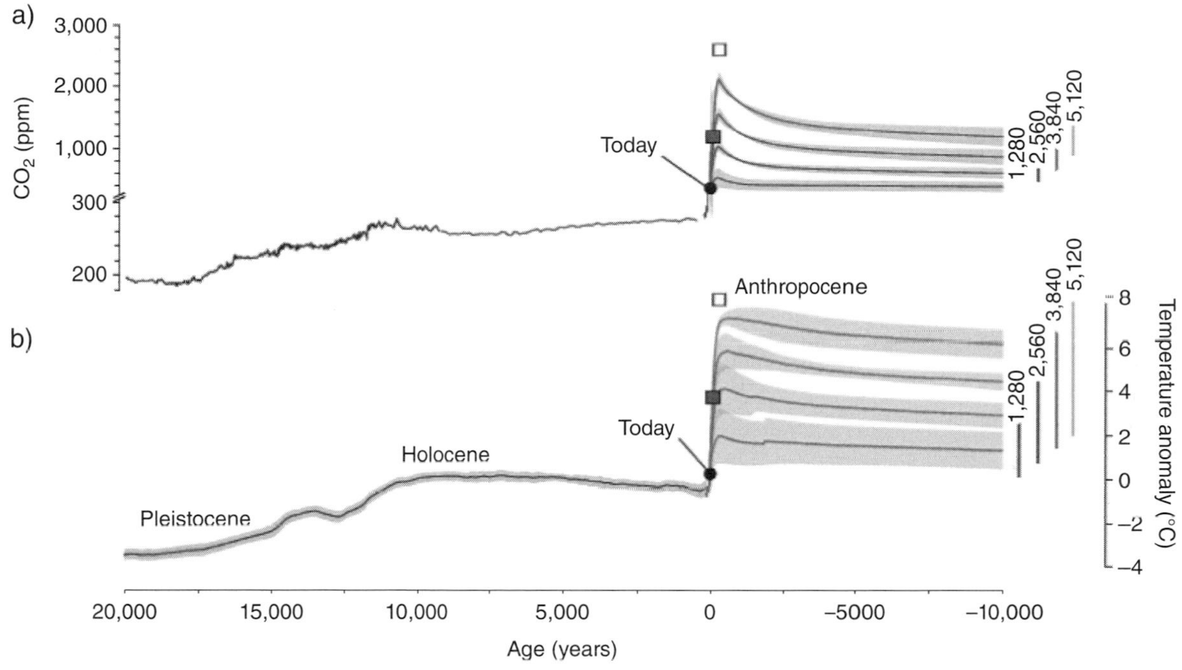

Figure 6.1.10 Past and future changes in concentration of (a) atmospheric CO_2 from ice cores and for four future emission scenarios (1,280, 2,560, 3,840 and 5,120 PgC), with changes in CO_2 for each emission scenario (mean and one standard deviation). CO_2 levels for RCP8.5 for 2100 (filled square) and its extension for 2300 (open square) are shown for comparison (values are CO_2 equivalent). Vertical grey bars show range of CO_2 increase for the four emission scenarios based on a range in equilibrium climate sensitivity (1.5°C–4.5°C). Note change in y-axis scale at 300 ppm. (b) Global temperature (mean and one standard deviation) reconstructed from palaeoclimate archives for the past 20,000 years and from four simulations for each of the four emission scenarios for the next 10,000 years, based on an equilibrium climate sensitivity of 3.5°C. Temperature anomalies are relative to the 1980–2004 mean. Vertical grey bars show range of temperature increase for the four emission scenarios (1,280, 2,560, 3,840 and 5,120 PgC) based on a range in equilibrium climate sensitivity (1.5°C–4.5°C). Temperature projections (mean values) for RCP8.5 for 2081–2100 (filled square) and its extension for 2281–2300 (open square) are shown for comparison. From figure 1 in Clark et al. (2016).

which will have a further positive feedback effect on warming.

As a further example of what may be expected with continued warming on the millennial scale, it seems possible that the southward movement of the northern edge of the Intertropical Convergence Zone and the associated monsoons, which began with the cooling forced by the decline in Northern Hemisphere insolation, may eventually reverse, bringing moisture back to the Sahara once temperature and sea level have reached a new equilibrium some hundreds of years from now. This implies an expansion in the northern branch of the Hadley Cell, which will cause the aridity typical of the descending limb of the Hadley Cell to move north into the Mediterranean, as projected, for example, in recent IPCC reports (IPCC 2013).

We can also expect further natural changes in ocean circulation. ENSO is likely to continue at intervals of three to seven years, the Pacific Decadal Oscillation at intervals of one to two decades. Fluctuations will also continue at somewhat longer intervals within the AMOC. Nevertheless, while natural oscillations will undoubtedly continue as the Anthropocene progresses, they seem unlikely to contribute global temperature changes larger than about 0.2°C. The same applies to the occasional large volcanic eruption, like that of Mount Pinatubo in 1991.

Considering the present behaviour of the AMOC, the North Atlantic immediately south of Greenland and on the track of the northern branch of the Gulf Stream (the North Atlantic Current) is showing signs of cooling in contrast with the warming observed where other major western boundary currents (e.g., the Kuroshio Current, Brazil Current, Agulhas Current, East Australia Current) approach the polar regions. This is because the warm and salty Gulf Stream water arriving in the warming Norwegian-Greenland Sea is no longer cooling enough to achieve the increase in density required to sink and feed the northern descending limb of the global thermohaline conveyor system (the AMOC). That decrease in sinking rate diminishes the northward pull on the Gulf Stream, which thus transports less warm water into the northern North Atlantic, causing it to cool relative to its warming surroundings (Yang et al. 2016). This pattern will persist, keeping the North Atlantic region (and adjacent northwest Europe) relatively cool compared with surrounding warming areas (Hansen et al. 2016). In contrast, warming will be accentuated where the other major western boundary currents are taking more heat poleward (Yang et al. 2016).

Is the AMOC likely to collapse if warming continues? Liu et al. (2017) consider the possibility that present-day climate models contain biases favouring a stable AMOC, while 'observationally based freshwater budget analyses suggest that the AMOC is in an unstable regime susceptible for large changes in response to perturbations'. Correcting the model biases, they show that the AMOC may collapse in 300 years if CO_2 is doubled from its 1990 values. This would cool the northern North Atlantic and adjacent seas, increase sea ice in the Norwegian-Greenland Sea and south of Greenland and cause the rain belt to migrate south over the tropical Atlantic.

Stratification of the ocean plays a key part in the story (Hansen et al. 2016). As land ice melts, it supplies a cool freshwater lid to the polar ocean. That slows the formation of Antarctic Bottom Water and the AMOC and increases the area of sea ice. In turn, that temporarily slows the growth of global warming. The lid prevents loss of heat to the atmosphere from upwelling deep water, warming the deep ocean more and causing sea level to rise as well as melting more ice shelves from beneath, further accelerating land-ice discharge. Most of this happens in the Southern Ocean, which holds 80% of the world's deep water. Gradual warming of the global ocean's deep water will warm the Southern Ocean south of the Polar Front on centennial scales (Armour et al. 2016).

The Southern Ocean dominates climate change by controlling ventilation of the enormous deep-ocean CO_2 reservoir. This ventilation may slow when sea ice expands, but as Hansen et al. (2016) note, that expansion will be a temporary pause in what they refer to as the Hyper-Anthropocene phase of climate change. If these developments continue, they could lead to shutdown of the North Atlantic overturning circulation, bringing substantial cooling to the North Atlantic coastal regions. Here we see a potential negative-feedback aspect of the global thermostat. Hansen et al. (2016) suggest that these changes will be accompanied by increasingly powerful storms, like those inferred to have affected the Bahamas and Bermuda in the Eemian interglacial.

How fast might these changes be? Hansen et al. (2016) remind us that during the Eemian interglacial, sea level rose to ~4–9 m above current levels, while Earth was less than 1°C warmer than today. Hence the possibility of rapid sea-level rise ('several metres over a timescale of 50–150 years') rather than a gradual change in ice-sheet response to continued rapid warming cannot be ignored. In sum, 'expectation of non-linear behaviour is based in part on recognition of how multiple amplifying feedbacks feed upon each other and thus can result in large rapid change' (Hansen et al. 2016).

Is rapid change likely? Analysing the output of models used in the Fifth Assessment by the IPCC, Drijfhout et al. (2015) found evidence for 37 forced regional abrupt changes arising in response to global temperature increase, but with no single threshold. They occurred at temperatures above and below a warming level of 2°C but were more common where warming was most extreme.

If CO_2 emissions reach 720 ppm by 2100, surface air temperatures in Greenland are expected to reach something like 3.3°C ± 1.3°C above the levels of 1900 (Masson-Delmotte et al. 2012). If Greenland's surface air temperatures reach approximately 5°C above the levels typical of 1900, they will be comparable to temperatures reached during past interglacials like the Eemian, 130,000–120,000 years ago. That rate of increase is not unusual for the last glacial period; it is comparable to the mean rate of increase (5°C/century) for the Dansgaard-Oeschger (D-O) warm events during the last ice age (Masson-Delmotte et al. 2012; for more on D-O events, see Section 6.2.2 herein). It is, however, unusual for the Holocene.

Given that temperature is rising, land and sea ice and snow are melting, albedo is falling, sea level is rising, and the ocean is becoming more acidic, we may expect future changes to take us towards those experienced at the PETM, 55 Ma ago (Figure 6.1.5). Adding CO_2 to the ocean will in due course (another one of those lags) make it sufficiently acidic to dissolve deep-ocean carbonate sediments, raise the carbonate-compensation depth and cause the extinction of benthic foraminifera. If the PETM is a reasonable guide, we can also expect the proliferation of exotic planktonic foraminifera and other microplankton in the ocean, as well as substantial effects of a changing climate upon the dispersal and radiation of land plants and animals that have already begun (cf. Zachos et al. 2001). Corals may die off with extreme tropical warming, as happened in previous biological extinctions associated with warming events in the distant past (e.g., at the end of Permian time).

We are already seeing the 'greening of the planet' as added CO_2 encourages plant growth (Mao et al. 2016; Zhu et al. 2016; Campbell et al. 2017). However, plants are also susceptible to significant changes in temperature, humidity and soil moisture that may reduce plant growth beyond certain limits. Those limits may change regionally if, as is forecast, currently dry areas expand in some places (e.g., the Mediterranean) and rainfall is enhanced in others (e.g., central Europe) as warming continues. While the Arctic tundra is greening on the North Slope of Alaska, in southern Canada and in central and eastern Siberia, it is 'browning' in western Alaska, the Canadian Arctic archipelago and western Siberia as trees die or are affected by infestations of pine-bark beetle (Richter-Menge et al. 2016).

As it has in the past, polar amplification will likely force Arctic temperatures to exceed the Northern Hemisphere average by a factor of three or four, with concomitant reductions in ice mass and associated rises in sea level (Miller et al. 2010b). Winners may therefore include Arctic countries like Canada and Russia, which will be able more easily to exploit the resources of the north, although the extent to which agriculture may move north in places like Canada will also depend on the extent to which the movements of the great northern ice sheets in the Pleistocene scraped the soil off the underlying bedrock. The development of agriculture in the far north will also depend on the extent to which permafrost disappears.

By 2100, the warming of Greenland is projected to increase by 5°C above the 1961–1990 average, which was itself 0.5°C above the average for the last millennium (Masson-Delmotte et al. 2012). Arctic precipitation is likely to change, with less snow and more rain. Currently, rain accounts for ~35% of Arctic precipitation, a fraction likely to increase in winter to 50%–60% by 2100 (Bintanja & Andry 2017). These projected changes are likely to affect river discharge, the melting of permafrost, sea ice and ecology (with knock-on effects on food availability).

Further lagged shocks may be in store when the 1,400 Gt of carbon currently frozen as organic matter in permafrost decays as the permafrost melts and as methane hydrates trapped in Arctic sediments are released (Archer et al. 2009). Some have suggested that this may supply as much as 50 Gt of CH_4, which could raise global surface air temperatures a further

0.6°C (Whiteman et al. 2013; Wadhams 2016). That could be the case if decomposition takes place in wetlands in the absence of oxygen, but if oxygen is readily available, CO_2 would be produced instead, and it is a weaker greenhouse gas than methane.

For how long will global warming persist? Evidence from the PETM and other hyperthermal events in the geological record suggests that after the injection of CO_2 into the atmosphere stops, it takes around 100,000 years for conditions to return to normal. Much the same estimate emerges from models of the residence of anthropogenic carbon in the atmosphere (e.g., Archer et al. 2005; Solomon et al. 2009). And as already mentioned, Ganopolski et al. (2016) (see also Clark et al. 2016) calculate that the present level of emissions is sufficient to postpone the next glaciation for 50,000 years. Beyond that, what happens will depend on the extent to which humans continue to be able to find and burn fossil fuels and upon the impact on humanity of continued orbitally induced climate change, with the potential for a return to full glacial conditions if CO_2 levels eventually fall below those typical of pre-industrial levels (Berger & Loutre 2002). Evidently the future trajectory of the climate will depend on humanity's ability to control emissions by decarbonising (e.g., through carbon capture and storage, or CCS), but if that does not prove adequate (as implied by Heard et al. 2017), some form of geoengineering may be adopted either to take CO_2 out of the air (e.g., via fertilising the ocean) or to shade the planet (e.g., through a veil of aerosols in the stratosphere or through the artificial formation of clouds over the ocean; Crutzen 2006; Royal Society 2009; Morton 2015). This is taking a leaf out of nature's book, where volcanic emissions of SO_2 lead to the natural formation of H_2SO_4 droplets in the stratosphere that shade the planet temporarily. Most people don't realise that so-called sulphate aerosols are derived by the injection of SO_2 gas into the stratosphere to combine with water vapour to form reflective sulphuric acid (H_2SO_4) droplets. Does society want a shade made up of H_2SO_4? After all, in recent years several governments legislated against the emission of SO_2 from coal-fired power-station chimneys so as to stop acid rain.

Where will climate go in the future? The latest analysis based on geological data from the last 420 million years of Earth's history (Foster et al. 2017) suggests that 'humanity's fossil-fuel use, if unabated, risks taking us, by the middle of the twenty-first century, to values of CO_2 not seen since the early Eocene (50 million years ago)'. Furthermore, 'unabated fossil-fuel use therefore has the potential to push the climate system into a state that has not been seen on Earth in at least the last 420 Myrs'.

Evidently, future humans will need to adapt to the warmer greenhouse world of a more or less permanent (on human timescales) 'super-interglacial' or 'greenhouse' state, although they may seek to reduce its effects through geoengineering (cf. Steffen et al. 2018).

6.2 Ice

Colin P. Summerhayes

Although during Cenozoic time the first great ice sheet formed on Antarctica 34 Ma ago, the great Northern Hemisphere ice sheets did not form until about 2.6 Ma ago, at the start of Pleistocene time. The Antarctic ice sheet shrank in the warm periods of the Mid-Miocene climate optimum and Late Pliocene (Mid-Piacenzian warm period), when CO_2 was as abundant as it is today or more so. During cold periods in the last glaciation, Greenland experienced sudden warming periods that lasted 500–1,000 years (Dansgaard-Oeschger events) that were linked to iceberg outbreaks (Heinrich events). These warm events were preceded by warming in the Antarctic and occurred as Antarctica cooled (the seesaw effect). Whether or not these changes were driven by solar activity remains to be seen. Ice sheets in both hemispheres are currently being melted from below by warm ocean deep water. Greenland also experiences melting from warm air, and the last remnants of the

North American ice sheet, on Baffin Island, are rapidly melting as the Arctic air warms. Surface melting is less common in Antarctica, because strong winds around the continent prevent warm air from the tropics from reaching it. Arctic sea ice has been shrinking since the 1950s, but the area of Antarctic sea ice is largely static, probably because melting ice shelves supply fresh surface water that freezes quickly. The world's glaciers are now declining rapidly, and Greenland and West Antarctica have been losing ice at an increasing rate. Continued melting of Antarctic ice shelves will lead to the loss of ice from both West and East Antarctica, as in Pliocene times. Arctic summer sea ice may be gone by 2050–2080.

6.2.1 The Cryosphere

The temperatures and pressures at the Earth's surface are such that water can exist in all three of its forms – vapour (in the atmosphere), liquid (oceans, lakes and rivers) and solid (mostly ice). The parts of the planet made of solid water form the cryosphere, which includes the great ice sheets of Greenland and Antarctica and their surrounding floating ice shelves, together with ice caps like that of Baffin Island, mountain glaciers, lake ice, river ice, sea ice, snow and frozen ground including permafrost, which may occur beneath both the land surface and the seabed. Currently the largest area of solid ice occurs in Antarctica (14 million km^2, or about 10% of the planet's land surface), with somewhat less in Greenland (1.8 million km^2) and yet lesser amounts in the great mountain ranges – the Himalayas and the Tibetan Plateau, the Alps, the Southern Alps of New Zealand, the Andes and the Rockies. Hidden from view, permafrost lies beneath 54 million km^2 of exposed land in the Northern Hemisphere, including the Tibetan Plateau (1,300 km^2). Winter sea ice around Antarctica covers an area of 17–20 million km^2, significantly extending the area of the continent's ice. But because there is more land in the Northern Hemisphere, there is more snow, ice and permafrost there. The permafrost is up to 600 m thick along the coast of the Arctic Ocean. Although the cryosphere covers a larger area of the Northern Hemisphere, Antarctica contains some 30 million km^3 of ice, representing 70% of all freshwater on the planet. Ice is up to 4,776 m thick beneath the Polar Plateau of East Antarctica, providing sufficient weight to depress the underlying land surface by 2 km. In contrast, the volume of Greenland ice, which is locally up to 3 km thick, is a mere 2.8 million km^3.

A key issue in the quest for establishing a lower boundary for the Anthropocene is whether a sufficiently meaningful signal (or signals) exists within the ice record. Such a record (in the oxygen and hydrogen isotopes) was found within a Greenland ice core to mark the lower boundary of the Holocene (Walker et al. 2009). Submitting an article on this topic in 2013, Wolff (2014) offered the suggestion that we were currently in the transition to the Anthropocene and that it was perhaps too early to identify a boundary for the end of the Holocene. But one might also take the view that the transition to the Anthropocene began with the evolution of humans on the planet and with their subsequent growing influence. In this book, we focus on the 'Great Acceleration' in humanity's activities in the mid-20th century as the principal marker of the Anthropocene.

6.2.2 The Geological History of Ice on Earth from Cretaceous to Holocene

Considering the geological history of the Earth since the end of the Cretaceous Period (66 Ma ago), it seems somewhat counterintuitive that there was no great ice sheet on Antarctica until the start of the Oligocene 34 Ma ago, even though Antarctica had drifted to its south polar position 90–100 Ma ago (Bertler & Barrett 2010). There are signs, in the form of dropstones in Southern Ocean deposits, that mountain glaciers must have existed on the continent as far back as the Early Eocene (50 Ma ago). But evidently Earth's greenhouse climate in Paleocene and Eocene times was warm enough to prevent the formation of the first great ice sheet until CO_2 levels had fallen to the point (possibly between 420 and 560 ppm) where conditions were cold enough for mountain glaciers to coalesce into ice sheets (Pollard & DeConto 2005).

Oxygen isotopic data from marine sediment cores suggest that the Antarctic ice sheet remained more or less intact through Oligocene and Miocene times but occasionally experienced some melting – for example, during the Mid-Miocene climatic optimum (Zachos et al. 2001). Evidence gained from ANDRILL drilling-programme cores in the Ross Sea region of Antarctica shows that during the Early to Mid-Miocene (20–14 Ma ago) there were repeated fluctuations in climate leading to ice-sheet expansion and contraction and attendant changes in sea level, some of them at about 17 Ma ago related to the Mid-Miocene climatic optimum recognised from oxygen isotope data (Fielding et al. 2011). These cores contain abundant evidence of orbitally driven Milankovitch climate cycles (see Section 6.1).

The Antarctic ice sheet also experienced some significant melting during the Pliocene, when, for instance, the Ross Ice Shelf disappeared, allowing the Ross Sea to accumulate diatomaceous oozes (Naish et al. 2009; Pollard & DeConto 2009). Sea levels rose by up to 22 m at the time (Miller et al. 2012). That amount of rise implies loss of ice from East Antarctica as well as from West Antarctica and Greenland. In confirmation, Cook et al. (2013) found evidence off the coast of Adélie Land, East Antarctica, for both increased oceanic productivity – associated with elevated Southern Ocean temperatures – and active erosion of continental bedrock from within the now ice-covered Wilkes Subglacial Basin, indicating retreat of the ice sheet by several hundred kilometres inland. That melting is likely to have added between 3 and 10 m to global sea level (Cook et al. 2013). The Totten Glacier behind the Totten Ice Shelf drains East Antarctica's Aurora Subglacial Basin with a volume equivalent to at least 3.5 m of global sea-level rise (Greenbaum et al. 2015). Geological analyses of the interior behind the Totten Ice Shelf show that it has been subject to repeated erosion in the Sabrina Subglacial Basin, caused by advance and retreat of the Totten Glacier (Aitken et al. 2016). Clearly, then, the great East Antarctic Ice Sheet is capable of a dynamic response to modest amounts of warming.

The great Northern Hemisphere ice sheets did not form until the beginning of the Pleistocene, 2.6 Ma ago (Bertler & Barrett 2010). Even so, it seems likely from the distribution of dropstones offshore that there may have been an ice sheet (or at least mountain glaciers) on Greenland for as much as 18 million years before present (Thiede et al. 2010). Indeed, there is evidence for ice caps on the mountains of East Greenland and southern Greenland during the Pliocene, with evergreen woodland in parts of northern Greenland (Dolan et al. 2011). Analysis of marine strata from beneath the Arctic Ocean indicates intermittent presence of perennial sea ice as early as 44 Ma ago (Stickley 2014).

From 2.6 Ma ago, much of northern North America, including the Canadian Arctic islands, was covered during glacial periods by the great Laurentide ice sheet, which mostly extended as far south as latitude 48°N but, in the mid-continent, reached as far south as 38°N. Amongst other things, it gouged out depressions now filled by the multitude of lakes of the Canadian Shield, including the depths of the five Great Lakes. It left substantial end moraines at its edges, including those forming Cape Cod, Massachusetts, and Long Island, New York. Canada's Baffin Island ice cap is its last remaining remnant. The ice sheet was up to 3.2 km thick, and its weight depressed much of Canada, especially in the area of Hudson's Bay, which is shrinking in area as the land continues to rise there through glacial isostatic adjustment.

While the ice sheet was shrinking following the last glacial maximum, large glacial lakes formed along its southern edge, occasionally breaking out either into the Arctic Ocean via the Mackenzie River or into the Atlantic via the Saint Lawrence River. These 'eruptions' flooded the northern seas with a freshwater lid that prevented northward penetration of the Gulf Stream, leading to extreme cooling like that of the Younger Dryas cool period 12,000 years ago and a smaller, briefer cooling 8,200 years ago.

Pleistocene ice sheets also covered much of northwestern Eurasia (Svendsen et al. 2004). The western segment covered the British Isles north of London, as well as much of northern Europe north of Berlin and Moscow, reaching as far south as about 52°N in the west and 50°N in the east. In the eastern segment, extending from longitudes 40°E–115°E, the ice sheet extended less far south, mostly being located north of 60°N. It covered not only northern Russia, including northwestern Siberia, but also the continental shelves of the Barents and Kara Seas, including the offshore islands of Svalbard, Frantz Josef Land and Severnaya Zemlya (Patton et al. 2017). These areas too are now rising. As in Canada, glacial processes carved out depressions now represented by lakes, which are widespread in Finland, for instance. A whole host of landforms attributable to the action of ice or run-off from the ice sheet characterises all of these formerly glaciated regions.

Throughout the Pleistocene, the great ice sheets fluctuated in size under the influence of cycles of eccentricity, precession and axial tilt (Section 6.1.1). The fluctuations in ice volume also resulted in fluctuations in sea level (Section 6.3). During the cold periods between the interglacials of the past 400,000 years, the Northern Hemisphere ice sheets around the Atlantic experienced periods of binge (growth) and purge (decay), with the decays represented as sudden collapses of the ice sheet that sent armadas of icebergs out across the North Atlantic from Canada to Iberia (MacAyeal 1993). Their remains are identified in swathes of commonly dolomitic rock fragments scattered in a broad band along the iceberg track (Ruddiman 1977). The major iceberg-outbreak events were named Heinrich events after their discoverer, Hartmut Heinrich (Heinrich 1988; Bond et al. 1992). Paradoxically, they coincide with periods of significant cooling recorded in Greenland ice – the cold phases of millennia-scale climate oscillations known as Dansgaard-Oeschger (or D-O) events (Bassis et al. 2017). D-O events, in turn, are linked to similar warming and cooling events in Antarctica (Figure 6.2.1). The ice-core records from Greenland and Antarctica can be matched precisely in time for the past 130,000 years through the use of the methane signal in bubbles of fossil air trapped in the ice (see Section 5.2.2), because methane (CH_4) is rapidly mixed through the atmosphere. Those records show that warming in Antarctica preceded the warm phases of the D-O events in Greenland (Figure 6.2.1). During D-O events, Greenland warmed by as much as 6°C–10°C within less than 50 years, and perhaps in as little as 2.5 years, after which warm temperatures remained for about 500 years before cooling to glacial conditions began again in Greenland at the same time as warming began in the Antarctic (Steffensen et al. 2008). This out-of-phase pattern of alternating warming in the south and cooling in the north and vice versa was described as the bipolar seesaw (Broecker 1998; Stocker 1998). The seesaw idea is that iceberg outbreaks put a lid of freshwater on the North Atlantic that prevents the formation and sinking of North Atlantic Deep Water, which stops the northward transfer of heat by the northern branch of the Gulf Stream. As a result, heat builds up in the South Atlantic, warming Antarctica. Once the northern iceberg armadas disappear, the AMOC reasserts itself, taking heat north. Antarctica then cools while Greenland warms.

In the marine-sediment record, Heinrich events appear to be the extreme end members of a series of ice-rafting episodes indicative of shrinking phases in the Northern Hemisphere ice sheets on either side of the northern North Atlantic. Those episodes have become known as Bond events after their discoverer, Gerard Bond. They seemed to recur at intervals of roughly 1,500 years, an interval that initially, at least, did not appear to correspond with any known variation in solar output, and there was some evidence that they persisted, with much lower amplitude, into the Holocene (Bond et al. 1997). Despite the lack of any known variation in

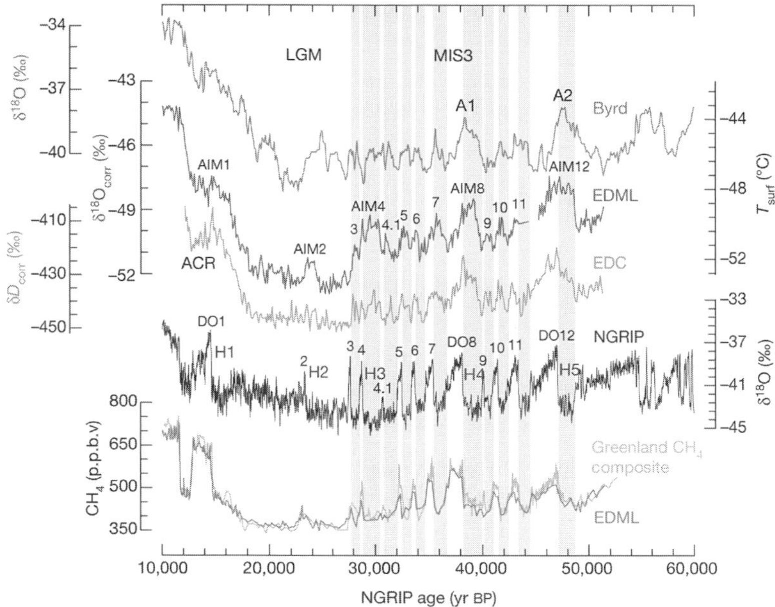

Figure 6.2.1 Ice cores from EPICA Dronning Maud Land (EDML), Antarctica and North Greenland Ice Coring Project (NGRIP), synchronised using their methane, reveal a one-to-one link between Antarctic warm periods and Greenland cold periods (stadials) (grey bars) during the last glaciation, supported by data from the EPICA Dome C (EDC) and Byrd cores from Antarctica. NGRIP numbers represent Dansgaard-Oeschger events (DO1 through DO12). Heinrich events, which coincide with the cold phase of D-O events, are numbered H1 through H5. ACR is the Antarctic Cold Reversal (for more detail, see Figure 6.1.7). AIM1 through AIM12 are Antarctic (oxygen) Isotope Maxima, representing warm conditions. The temperature scale applies to the EDML oxygen isotope data. From figure 2 in EPICA Community Members (2006) (see also Ahn & Brook 2007, figure 1).

solar output with a frequency of 1,500 years, Bond later found that there was a close association between ice-rafting events and the abundance of the isotopes of Carbon-14 (^{14}C) and Beryllium-10 (^{10}Be), which form when the solar wind is weak, allowing more cosmic rays to penetrate the outer atmosphere (Bond et al. 2001; Figure 6.2.2 herein). Although careful examination of the data by Richard Alley and colleagues suggested that Bond's 1,500-year cycle was due to 'stochastic resonance' rather than any external forcing (Alley et al. 2001), subsequent research suggests that there could indeed be a link between ice-rafting events and solar output (Braun et al. 2005).

The sudden D-O warm phases of Greenland tend to occur in groups of three, each one slightly less warm than its predecessor, with its associated cold phase being progressively colder and with the Heinrich events occurring during the coldest phase (Figure 6.2.1). Many of Bond's ice-rafting events coincide with the cool phases of D-O events in Greenland, with cool periods in the North Atlantic and with warm periods in the South Atlantic and Antarctica. They represent the operation of a natural internal oscillator within the climate system, but one that has only weak representation in the Holocene.

Bassis et al. (2017) used an ice-sheet model to show that the magnitude and timing of Heinrich events can be explained by the same processes that account for the reduction in Antarctic ice shelves – namely, subsurface ocean warming causing melting of the ice shelves from below, triggering rapid retreat and increased iceberg discharge (Figure 6.2.3). Subsequently, the removal of the weight of lost ice allows the land to rise isostatically, isolating the ice base from warm water and allowing the ice to expand seaward again. It appears that the first response of northern ice shelves to the reassertion of the AMOC (hence resupply of heat from the south) is to melt from beneath, causing ice to discharge in Heinrich events while Greenland is still cold, following which the ongoing warming causes Greenland temperatures to rise abruptly as the warm phase in the classic D-O cycle (Figure 6.2.1). As Bertler and Barrett (2010) point out, 'the observations indicate that slow, gradual changes in climate forcing can lead to abrupt climate shifts both within and beyond major regions of the Earth, which shows why we need to better understand processes and interactions between all parts of Earth's climate system'.

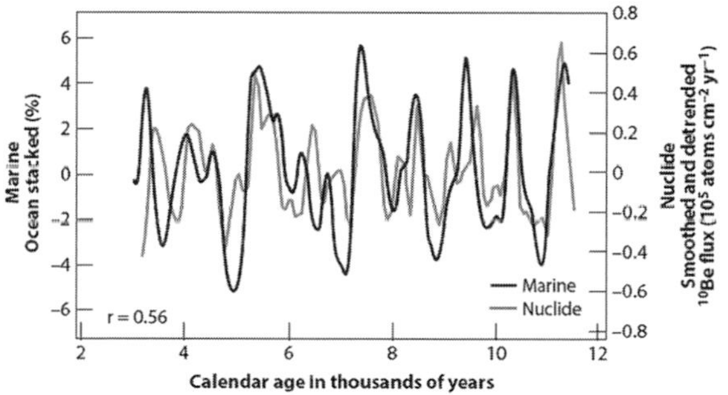

Figure 6.2.2 The record of ^{10}Be extracted from Greenland ice cores and the stacked record of ice-rafted minerals extracted from a number of ocean-sediment cores in the North Atlantic. (Bond et al. 2001, figure 3F), also reproduced in Haigh and Cargill (2015, figure 6.2). Reprinted with permission from AAAS

During the Eemian interglacial Arctic, surface temperatures were ~5°C warmer than at present, and almost all Arctic glaciers melted completely except for the Greenland ice sheet, which was substantially reduced in size (Miller et al. 2010a). Indeed, recent evidence in the form of cosmogenic isotopes in

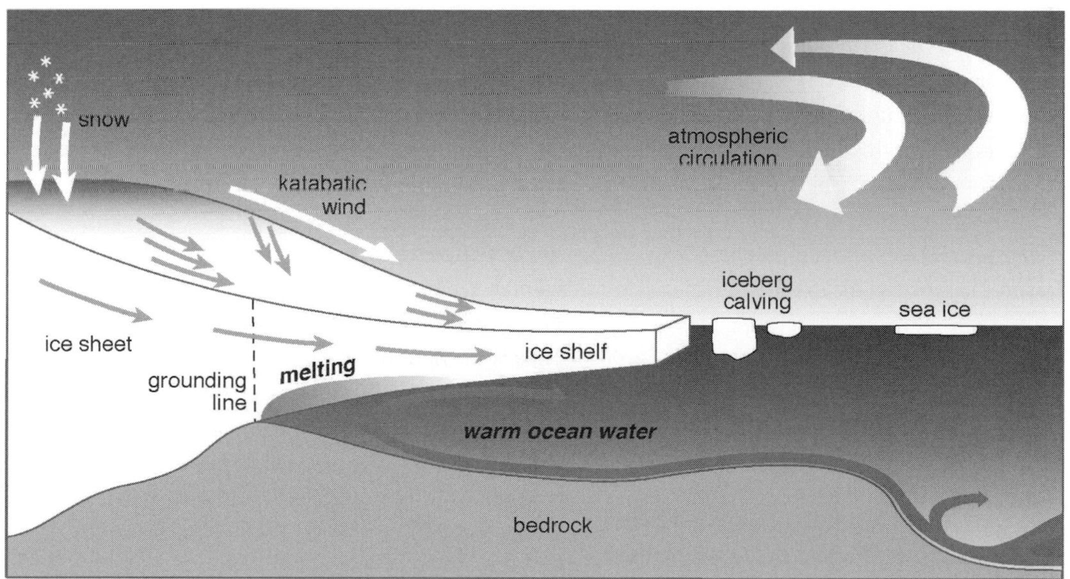

Figure 6.2.3 A schematic cross section through Antarctica and the adjacent Southern Ocean showing the displacement and thinning of inland ice towards the coast leading to the formation of ice shelves, which are then eroded from beneath by warm ocean water welling up from below, which weakens their buttressing effect, allowing more ice to move from land to ocean. Surface waters, which freeze to form sea ice in winter, would typically reach temperatures of −1.8°C, while bottom waters would be 1°C or slightly more. (from NASEM 2015). Reprinted with permission from the National Academies Press, Washington, D.C

bedrock suggests that the Greenland ice sheet was reduced to less than 10% of its present size at times during the warm periods of the Pleistocene (Schaefer et al. 2016).

Evidence for the extent of the ice sheets during the Last Glacial Maximum comes from marine geological observations off Greenland and Antarctica, which show that ice sheets advanced to the edge of the continental shelf. For instance, scour channels like the Palmer Deep crossing the Antarctic continental shelf just south of Anvers Island represent the paths of ice streams within the ice sheet over the continental shelf (Domack 2006). A comprehensive analysis of such features confirms that the ice sheet reached the shelf edge at the Last Glacial Maximum and that discharge from ice streams at the shelf edge led to the development of submarine fans on the continental slope (Cofaigh et al. 2014).

In the north, ice recession was well underway by 16,000 years ago, and most of the Northern Hemisphere ice sheets had gone by 6,000 years ago. Solar energy reached a summer maximum 9% higher than at present ~11,000 years ago, and the extra energy elevated Arctic surface temperatures by 1°C–3°C above the 20th-century average – enough to melt most Northern Hemisphere glaciers, but leaving the Greenland ice sheet largely intact (Miller et al. 2010a, b). Warming air was accompanied by warm water from the North Atlantic, helping to keep sea-ice cover minimal (Miller et al. 2010a, b; Masson-Delmotte et al. 2012).

6.2.3 Ice in Holocene and Recent Times

6.2.3.1 Ice Sheets, Ice Shelves and Glaciers

The great ice sheets of the Northern Hemisphere melted away more or less completely during the deglaciation following the Last Glacial Maximum 20,000 years ago and had gone from everywhere there except Greenland by about 7,500 years ago (Clark et al. 2016). Over that same time interval, the Antarctic ice sheet melted back to its present position from the edge of the Antarctic continental shelf. Removal of the weight of the ice sheets was accompanied by uplift of the formerly depressed land surface by glacial isostatic adjustment, which was associated with a temporary increase in volcanic activity that would have contributed to the rise in CO_2 that accompanied the postglacial warming (Huybers & Langmuir 2009). While there is adequate evidence for this association for the latest deglaciation, it seems reasonable to assume that it is also a feature of earlier deglaciations.

Last remnants of the Laurentide Ice Sheet of North America include the Barnes and Penny Ice Caps on Baffin Island. Both are in retreat, and it is estimated that the Barnes Ice Cap will melt away over the next 300 years (Gilbert et al. 2017). Only in the three warmest interglaciations of the past 2.6 Ma has the Barnes Ice Cap been as small as it is presently. After 2,000 years of little change in its dimensions, recent observations show that the ice cap is now losing mass at all elevations, despite the continued decrease in summer insolation, and this mass loss is most likely in response to the exceptionally warm 21st-century Arctic climate.

As pointed out in Section 6.1.2, orbitally controlled summer insolation began declining in the Northern Hemisphere from about 10,000 years ago, taking that hemisphere into a 'neoglacial' period from 5,000 to 6,000 years ago, during which mountain glaciers readvanced. The readvance fluctuated with time as solar output waxed and waned, with some melt-back during the Medieval Warm Period and growing advance during the cold phases of the Little Ice Age.

Glacial retreat from the Little Ice Age in the Northern Hemisphere began around 1850 (e.g., Vinther et al. 2009) and sped up after 1950 (e.g., Thompson et al. 2006; Thompson 2010). Using data compiled by the World Glacier Monitoring Service, Zemp et al. (2015) demonstrated that glaciers around the world are now showing unprecedented global decline. Intermittent readvance at regional and decadal scale, exhibited by a small subset of glaciers, has not come close to achieving the positions of maximum advance of the

Holocene in general or the Little Ice Age in particular. The rates of early 21st-century mass loss are without precedent on a global scale for recorded history (Zemp et al. 2015).

In the Southern Hemisphere, very slow glacier retreat began within the Little Ice Age (e.g., around 1600 CE) in Patagonia. That rate of retreat increased significantly there after 1900 and markedly after 1950 (Newman et al. 2009; Glasser et al. 2011). Most of the 244 glaciers of the Antarctic Peninsula also retreated markedly after about 1950 (Cook et al. 2005). Those in the southern part of the region show more retreat than those in the north because they terminate in warm upwelled Circumpolar Deep Water that is melting the glacial termini from below (Cook et al. 2016).

Examination of high-resolution records of ice rafting shows that small changes in subsurface ocean temperature during the Holocene resulted in the increased discharge of icebergs around Antarctica, which amplified multi-centennial variability in the climate system both regionally and globally (Bakker et al. 2017). This is further evidence for the dynamic nature of the Antarctic ice sheet.

In more recent years, the 200–800 m thick ice shelves that represent land ice floating on the ocean around the Antarctic coast have started to disappear, with marked losses from the 1950s onwards, apparently associated with the increased warming of the environment. East of the Antarctic Peninsula, the ice shelf blocking the Price Gustav Channel between the Antarctic Peninsula and James Ross Island on the western coast of the Weddell Sea began to decay in the 1950s and broke out completely in 1995. The adjacent Larsen A ice shelf lost much of its area in 1995, and the Larsen B ice shelf farther south broke out in 2002 (losing an area the size of the state of Rhode Island). The next one south is the Larsen C ice shelf, covering an area of 50,000 km^2 and averaging around 350 m thick. It has been receding in area since 1980 (Tollefson 2016), and in September 2017 an iceberg some 175 km long by 50 km wide, an area twice the size of Luxembourg, and weighing ~1,300 Gt, nearly as much as all of the new icebergs formed annually in Antarctica (Rackow et al. 2016), broke away from the ice shelf and started drifting north (Marchant 2017). It is likely to drift through the Southern Ocean for about eight years and will mainly melt from the bottom up. Ice shelves west of the Peninsula have also lost substantial areas, including the Müller, Jones, Wordie, Wilkins and George VI Ice Shelves (Cook & Vaughan 2010). Most recently a large crack, propagating at 1.7 km/yr, appeared in the Brunt Ice Shelf on the east coast of the Weddell Sea, forcing the British Antarctic Survey (BAS) to relocate its new Halley-VI research station, established in 2013.

The disappearance of ice shelves, which buttress glaciers descending from the mountainous interior, allows those glaciers to speed their delivery of land ice to the ocean (Rignot et al. 2004; Figure 6.2.3 herein). The loss of the Larsen B ice shelf was accentuated by the development of ponds of water on its surface during summer months, with the water eventually melting its way to the base of the ice shelf and causing the development of fissures that allowed the shelf to break up into numerous small icebergs. Meltwater ponds have been seen on the surface of the Larsen C ice shelf (Tollefson 2016), much like those that appeared on the surface of the Larsen B shelf before its collapse in 2002. However, in some places such surface water may simply run off ice shelves through drainage networks, as in the case of the Nansen Ice Shelf in the Ross Sea or the ice shelf in front of the Petermann Glacier in Greenland (Bell et al. 2017).

All of the ice shelves east and west of the Peninsula show signs of thinning due to the incursion of warm water from below (Pritchard et al. 2012; Figure 6.2.3 herein). Such incursions also account for the thinning of the ice shelves along the coast of the Amundsen Sea (Pritchard et al. 2012). The rate of loss of ice has been accelerating along this coast since the 1970s. As a consequence, the ice streams formerly buttressed by the coastal ice shelves have sped up: for example, in the Pine Island, Thwaites and Smith Glaciers, draining the hinterland of the West

Antarctic Ice Sheet. These glaciers cross Marie Byrd Land, which is volcanic. High heat flow at certain points beneath the ice sheet indicates that some of the Marie Byrd Land volcanoes are active (Fisher et al. 2015). But none of the high-heat-flow regions are close enough to these glaciers to have caused them to speed their flow to the sea. Konrad et al. (2017) showed that the glaciers of West Antarctica thinned extensively during the period 1995–2015. They found that thinning was not uniform but spread upstream along the cores of the Pine Island and Thwaites Glaciers at 13–15 km/yr, twice as fast as elsewhere in the region. While this latest episode of thinning began in about 1980 in the Pope-Smith-Kohler basin, about 1990 in the Pine Island Glacier, and about 2004 on the Thwaites Glacier, Smith et al. (2016) found evidence to suggest that retreat of the Pine Island glacier/ice-shelf complex began around 1945. New examination of the process of rifting and retreat of the Pine Island Glacier shows that rifts giving rise to icebergs originate from basal transverse crevasses in the centre of the ice shelf, most probably initiated by ocean warming (Jeong et al. 2016). The crevasses are associated with valleys in the ice shelf where the ice has thinned. As an example of what we may expect to see more of in the future as warming continues, a large iceberg measuring 45 km by 15 km broke off from the end of the Pine Island Glacier in September 2017.

Although at present most of East Antarctica and its adjacent ice shelves seem to be stable, the rise of deep-ocean water onto the continental shelf there is causing some of its ice shelves to melt from beneath, just as happens in West Antarctica. This is already apparent for the Totten Ice Shelf (Rintoul et al. 2016), which buttresses the Totten Glacier, where warm water is funnelled beneath the shelf through a newly discovered deep channel. Loss of ice from East Antarctic ice shelves tends to vary in response to local trends in air temperature and sea ice that are driven by the dominant mode of atmospheric variability (the Southern Annular Mode); although the losses are mainly local at present, they suggest that East Antarctica may be more vulnerable to external forcing than previously recognised (Miles et al. 2013).

In response to the continued warming of the modern era, the Greenland ice sheet continued to melt through the summer of 2016 (Richter-Menge et al. 2016). The 2016 melt season lasted 30–40 days longer than usual in the northeast and 15–20 days longer along the west coast compared with the 1981–2010 average. GRACE (Gravity Recovery and Climate Experiment) satellite data obtained using the methods of Velicogna et al. (2014) show that from April 2015 to April 2016, Greenland lost 191 gigatons (Gt) of ice, about the same as during April 2014–2015 but slightly less than the average April-to-April mass loss (232 Gt) for the GRACE period of record (2002–2016) (Richter-Menge et al. 2016).

Greenland loses ice not only by sublimation and surface melting but also (as in Antarctica) by the melting back of glaciers that terminate at the coast (Hanna et al. 2013). Based on LANDSAT and ASTER satellite data, in 2016 coastal terminating glaciers lost 60.6 km^2, the largest loss since 2012; 22 glaciers retreated, losing 100.8 km^2, and 11 advanced, gaining 40.9 km^2 (Richter-Menge et al. 2016). Regional effects are important. GRACE data show that 70% of the ice loss in Greenland for the period 2003–2013 (280 $\pm$ 58 Gt/yr) came from the southeast (40%) and northwest (30%), driven by ice dynamics; elsewhere losses were driven more by changes in surface mass balance (sublimation and melting) (Velicogna et al. 2014). Khan et al. (2016) considered that these values may underestimate actual losses, suggesting that by failing to take into account the isostatic rise of land due to ice loss, the model outputs based on the GRACE satellite data may be low by some 17 billion tons per year. Adjusting for these factors, they concluded that Greenland had provided 4.6 m of sea-level rise since the Last Glacial Maximum, or 44% more than previously estimated. Whatever the precise figure, there is no doubt that Greenland has been losing ice at an increasing rate as the Arctic has warmed (Rignot et al. 2011; Shepherd et al. 2012; Hanna et al. 2013).

In Antarctica most of the total loss (180 ± 10 Gt/yr) during this same period (2003–2013) came from the Amundsen Sea sector (74%) and the Antarctic Peninsula (17%), mainly driven by ice dynamics. Queen Maud Land was the only sector in East Antarctica with a significant gain in mass (63 ± 5 Gt/yr) due to changes in surface mass balance (Velicogna et al. 2014).

Calculations of decadal-scale changes in ice-shelf thickness around Antarctica, based on recent satellite radar altimeter observations, show that the loss of ice-shelf volume accelerated from negligible (25 ± 64 km^3 per year) for 1993–2003 to rapid (310 ± 74 km^3 per year) for 2003–2012 (Paolo et al. 2015), confirming the trend noted by Rignot et al. (2011) and Shepherd et al. (2012). West Antarctic losses increased by 70% in that last decade, and earlier gain by East Antarctic ice shelves ceased. 'In the Amundsen and Bellingshausen regions, some ice shelves have lost up to 18% of their thickness in less than two decades' (Paolo et al. 2015).

While these changes are impressive, there remains a slight possibility that they may be part of a longer-term pattern. For instance, Steig et al. (2013) found that the $\partial^{18}O$ characteristics of the past 50 years of West Antarctic precipitation also occurred about 1% of the time over the past 2,000 years. Modern values of $\partial^{18}O$, although rare in the 2,000-year record, thus seemed to be not unprecedented, though they are near the upper limit of natural variability. Steig et al. (2013) found that the $\partial^{18}O$ variability in ice cores showed a strong link between West Antarctic ice and the behaviour of the tropical Pacific, which suggested to them that 'recent trends in $\partial^{18}O$ and climate in West Antarctica cannot be distinguished from decadal variability that originates in the tropics' (something also suggested in the case of the behaviour of Antarctic sea ice; see Section 6.2.3.2).

Nevertheless, the current trend of increasing ice loss from West Antarctica (Paolo et al. 2015) tends to suggest, in contrast with the precipitation data, that continued warming will lead to ice-shelf collapse, which implies that we have now moved out of the envelope of natural variability typical of the past 2,000 years. Much the same picture comes from a study of ice melt over the past 1,000 years in an ice core from James Ross Island, just east of the northern end of the Antarctic Peninsula, where intensification of ice melt has largely occurred since the mid-20th century, and 'summer melting is now at a level that is unprecedented over the past 1,000 years' (Abram et al. 2013). Abram et al. (2013) noted that this pattern was part of a longer-term trend beginning about 1460 CE but concluded that 'ice on the Antarctic Peninsula is now particularly susceptible to rapid increases in melting and loss in response to relatively small increases in mean temperature'. A similar result emerged from 300-year records of snow accumulation in ice cores drilled in Ellesworth Land, West Antarctica, near the base of the Peninsula. There, Thomas et al. (2015) found dramatic increases in snow accumulation during the 20th century that were linked to deepening of the Amundsen Sea Low, to tropical sea surface temperatures and to large-scale atmospheric circulation. The increase during the late 20th century was unprecedented in the context of the past 300 years and part of a longer-term trend.
A similar result comes from a study of the advection of warm Circumpolar Deep Water onto the Amundsen Sea continental shelf, which has intensified with the poleward shift in the Southern Hemisphere's westerly winds since about 1950; it is this CDW that is melting ice shelves from beneath, causing their collapse (Hillenbrand et al. 2017). Much the same process decimated the continent's ice shelves during the Holocene Thermal Optimum in the Southern Hemisphere between 10,400 and 7,500 years ago (Hillenbrand et al. 2017).

6.2.3.2 Sea Ice

Antarctica: Sea-ice extent around Antarctica bounced along with little change since satellite records began back in 1980. Amounts grew and declined slightly about a trend line that suggested a 1% increase over time, which Turner et al. (2009b) and Walsh (2009) thought could have been caused by an increase in surface wind strength driven by the development of the ozone hole (westerly winds drive surface waters and sea ice north around the continent). Although NASEM (2017) found from a study in January 2016 that 'it is not clear what mechanisms best explain the observed variability and the slight increase in overall Antarctic sea ice extent', that same year sea-ice extent decreased significantly to values 2% lower than the 1980–2010 mean (http://nsidc.org/data/seaice_index), and on 1 March 2017, Antarctic late-summer sea-ice extent dropped to just over 2 million km^2, the smallest amount observed in a record starting in late 1978. This change shows how variable the Antarctic climate system can be on annual and decadal timescales, making it difficult to reliably quantify the contribution that increasing greenhouse-gas concentrations may have made to changing concentrations of sea ice (Walsh 2009; Turner et al. 2017).

While pre-satellite-era data suggest a larger extent of sea ice in those days, confidence in prior measurements is low. Using data from the expeditions of early explorers between 1897 and 1917, Edinburgh and Fay (2016) concluded that the southern summer ice edge lay between 1° and 1.7° farther north than in 1989–2014 in the Weddell Sea area but was not much different from present elsewhere. Estimates from whaling ships suggest that the sea-ice edge was also significantly farther north than at present in 1931–1961 but shrank back from 1973 to 1978 (NASEM 2017). In contrast, Hobbs et al. (2016) found that sea-ice coverage was significantly less in the decades before the late 1950s to early 1960s. And there was very little sea ice off the coast of Queen Maud Land in the southern summer of 1939 (Lüdecke & Summerhayes 2012).

The lack of a significant decrease in sea-ice area from the late 1970s to the present, in contrast to what is observed in the Arctic, is explained by (1) melting of coastal ice shelves freshening the sea surface, stratifying it and making it more likely to freeze (Bintanja et al. 2013); (2) the increase in reflective ice area keeping surface air and water cool close to the Antarctic coast, in contrast to the warming and decrease in sea ice experienced in the Arctic (Bintanja et al. 2013); (3) strengthening offshore winds blowing surface water (and sea ice) further to the north, increasing the sea-ice area (Holland & Kwok 2012) (in part the increase in westerly wind strength was caused by the development of the ozone hole [Turner et al. 2009b]); control by the winds helps to explain why Antarctic sea ice shows different patterns depending on the region, growing in all sectors except the Bellingshausen and Amundsen Seas west of the Antarctic Peninsula and West Antarctica; (4) much of the slight growth in sea ice coincides with cooling in the equatorial Pacific caused by the development of the negative phase of the Pacific Decadal Oscillation, demonstrating the influence of tropical conditions on Antarctic sea ice (Meehl et al. 2016); (5) finally, upwelling of cool deep water driven by the strong westerly winds keeps surface waters cool, a process lacking in the Arctic.

Ice cores can be used to study the change in sea ice through time via the abundance of methanesulphonic acid (MSA), derived by the oxidation of the dimethylsulphide emitted by the marine plankton living in the sea-ice zone. MSA records from the Antarctic Peninsula show that the decline in sea ice in the Bellingshausen Sea in recent decades is part of a long-term regional trend that occurred throughout the 20th century, is consistent with evidence for 20th-century warming on the Antarctic Peninsula and may represent progressive deepening of the Amundsen Sea low-pressure region caused by the combination of greenhouse-gas-induced warming and an increase in the Antarctic ozone hole (Abram et al. 2010). Countering that trend, since 1990 a change to more easterly in the direction of the winds east of the Peninsula led to a slight increase in sea ice around the tip of the Peninsula (Turner et al. 2016). This local change reflects 'the extreme natural internal variability of the regional atmospheric circulation' rather than any change in global warming (Turner et al. 2016). It has not affected the climate of East Antarctica, which remains cool due to the effects of the ozone hole, which are superimposed on the warming trend of the past 300 years there (Ekaykin et al. 2016). Closure of the ozone hole over the next 50 years should lead to renewed warming in East Antarctica.

Arctic: Studies of Arctic Ocean sediment cores from the Chukchi and East Siberian Seas show that sea ice was minimal during the Early to Mid-Holocene (10,000–7,500 years ago), when Northern Hemisphere insolation was strongest, and increased significantly during the past 4,500 years as insolation waned and neoglacial conditions developed, much as found in other Arctic marginal seas (Stein et al. 2016). In the Mid-Holocene, sea-ice coverage was highly variable, probably driven by variations in solar output. Another major factor controlling sea ice in these two seas was the rate of entry of warm water from the Pacific (Stein et al. 2017).

In contrast with Antarctica, Arctic sea-ice area declined significantly since reaching a peak of ~11.5 million km^2 in summer in 1950 (Figure 6.2.4). Between 1900 and 1950 it fluctuated about a mean of ~11 million km^2, with lows of around 10.5 million km^2 between 1900 and 1910 and again between 1920 and 1945, and highs of ~11.5 million km^2 between 1910 and 1920 and again between 1945 and 1951. The slightly low summer sea-ice area between 1920 and 1945 corresponds to a prior period of Arctic warming (Overland et al. 2011) but was nowhere near as low as it has been seen since 1951 (Figure 6.2.4). That early-20th-century period of warming is thought to have been an 'essentially random climate excursion imposed on top of the steadily rising global mean temperature associated with anthropogenic forcing' (Wood & Overland 2010). It is associated with a slight rise in sunspot activity (cf. Clette et al. 2014).

In 2008, Christophe Kinnard published time series of maximum and minimum Arctic sea-ice extent dating from 1870 to 2003, showing that the area of winter sea ice was essentially constant from 1870 to 1950, after which it steadily declined, much as shown in Figure 6.2.4, confirming that the climate forcing of the early 20th century differed from that of the late 20th century. By 2011 he had used proxy data to extend the

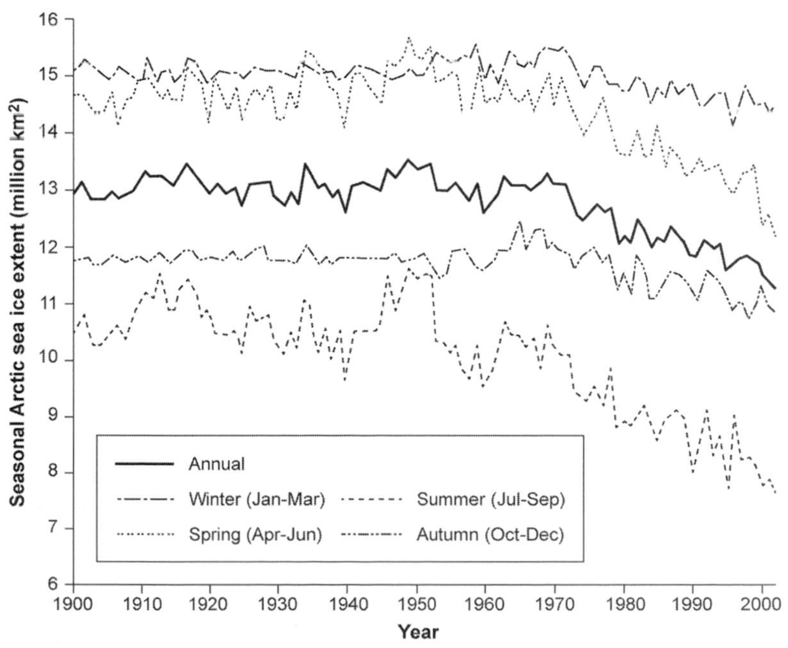

Figure 6.2.4 Arctic sea-ice area from 1900 onwards. From figure 6.1 in Wadhams (2016), modified from figure 16.14 in ACIA 2005, based on Walsh & Chapman 2001, figure 4.

record of the extent of sea ice back by 1,450 years, from which it was clear that 'both the duration and magnitude of the current decline in sea ice seem to be unprecedented for the past 1,450 years' (Kinnard et al. 2011). To their surprise, Kinnard et al. (2011) found that there had been less summer sea ice during the Little Ice Age between about 1500 and 1850 (about 9.5 million km^2) than there was between 1850 and 1950, or between 1200 and 1500 (10.5 million km^2) and that sea ice was also moderately less during the Medieval Warm Period (10 million km^2) (800–1200 CE), though not as low as it is today. As far as they could tell, the driver for periods of declining sea ice seemed to be the advection of warm Atlantic water to the Arctic. The warming in medieval times and at the end of the 20th century was related to an increase in the positive index of the North Atlantic Oscillation, which pushed westerly winds north between Scotland and Iceland, funnelling warm air and surface water towards the Arctic. His team concluded that anthropogenic warming explained the record warming and sea-ice loss of recent decades. The current sea-ice decline shows no sign of decreasing despite year-to-year fluctuations in sea-ice area.

The decline in sea-ice extent since the early 1950s is particularly evident from the satellite records that began in 1979, which show a steep decline from 8 million km^2 to 4 million km^2 at present (NSIDC 2017; Ding et al. 2017; Swart 2017). At the same time the ice thinned from an average of 3.29 m thick to an average of 1.25 m (Pearce 2017). The area of thick multi-year ice (more than one year old) shrank to 22% of the total, compared with 45% in 1985 (Richter-Menge et al. 2016), with the thickest ice tending to be packed up against the north coasts of Greenland and the Canadian islands. The rest is mostly annual and about 1 m thick and tends to be relatively fragile, being easily broken up by wind and currents.

As a result of both the shrinking area and the thinning, Peter Wadhams and colleagues – using data obtained from upward-looking sonars on submarines and other sources – calculated that the volume of Arctic ice declined by about 10 km^3 between 1979 and 2016 (Wadhams 2016). Minimum Arctic sea-ice extent at the end of summer 2016 tied with 2007 for the second lowest in the satellite record that began in 1979 (33% lower than the 1981–2010 average), a pattern reflected by the spring snow-cover extent in the North American Arctic in 2016, which was the lowest in a satellite record that began in 1967 (Richter-Menge et al. 2016). These various declines were due to the average Arctic surface air temperatures for the year ending September 2016 being at their highest since 1900, by 3.5°C, with a heat wave in November 2016 reaching 15°C above the 1981–2010 norm. In March 2017, the sea ice reached its maximum winter extent at 14.42 million km^2, the smallest winter area in 38 years of satellite records (Pearce 2017).

It had long been thought that most of the decline in sea ice was entirely due to global warming, but as Walsh (2009) pointed out, 'model simulations indicate that the Arctic changes have been shaped largely by low-frequency variations of the atmospheric circulation, superimposed on a greenhouse warming'. Ding et al. (2017) confirmed that interpretation, finding that naturally induced changes in atmospheric circulation caused about half of the ice loss. Year-to-year change is associated with the atmospheric vagaries of the North Atlantic Oscillation, which are well understood and allow sea-ice cover to be predicted with some accuracy. Ding et al. (2017) found that near-surface warming and sea-ice loss correlate strongly with changes in upper-level atmospheric circulation over Greenland, which, it turns out, is a driver of sea-ice loss and not a response to it and could be responsible for up to 40% of the ice loss since 1979. The other 60% of the loss was attributable to global warming. Ding's finding would help to explain why models that consider only the effect of global warming suggest that the decline in Arctic sea ice should be occurring much more slowly than is the case.

A further link between sea ice and climate is evident from the correlation between the decline in sea ice in the Barents and Kara Seas and the extremely cold winters of 2005–2006 in Europe and northern Asia (Petoukhov & Semenov 2010). This association arises because shrinkage of the sea-ice area warms the lower troposphere in the eastern Arctic, resulting in strong anticyclonic activity over the polar ocean and anomalous easterly advection of cold air over northern continents, lowering average temperatures by up to −1.5°C. Counter-intuitively, then, global warming can result in severe winters in areas adjacent to the Arctic.

6.2.4 Projections for the Future

The Greenland ice sheet is likely to continue thinning (by up to 100 m) along the ice-sheet margin; that in turn should cause additional melting as ice from higher elevations inland slides to lower and warmer coastal elevations (Hanna et al. 2013). Dynamical changes in glacier discharge due to enhanced lubrication, calving and ocean warming remain difficult to predict, not least because as the ice margin retreats, calving seems to decrease in relative importance (Hanna et al. 2013). Nevertheless, several of Greenland's large ice streams (e.g., the Northeast Greenland Ice Stream and Jacobshavn Isbrae) are located in deep troughs with bases well below sea level that extend far into the ice sheet; destabilisation of these ice streams by melting from beneath could lead to massive and rapid mass losses (US CLIVAR Project Office 2012).

If the thinning of the cores of the Pine Island and Thwaites Glaciers continues upstream towards the interior, it provides a mechanism for draining the ice of the West Antarctic Ice Sheet. Reassessing the potential for the collapse of the West Antarctic Ice Sheet, which partially collapsed in previous interglacials, Bamber et al. (2009) concluded that it would contribute a sea-level rise of about 3.3 m rather than the oft-quoted potential for a rise of 5–6 m. 'If the present climate forcing is sustained, we expect a drastic reduction in volume of the rapidly thinning ice shelves at decadal-to-century time scales, resulting in grounding-line retreat and potential ice-shelf collapse. Both these processes further accelerate the loss of buttressing, with consequent increase of grounded-ice discharge and sea-level rise' (Paolo et al. 2015).

The various findings reported above confirm that some parts of the East Antarctic Ice Sheet are dynamic. Careful consideration of the geometry of the bedrock topography underlying the East and West Antarctic Ice Sheets suggests that certain ice drainage basins may respond more sensitively than others to environmental forcing. Combined with a range of climate models, those observations suggest that most future ice loss from East Antarctica will likely come from the Recovery subglacial basin in the eastern Weddell Sea (Golledge et al. 2017).

However, we also have to bear in mind the observations of Steig et al. (2013) on the recurrence of modern $\partial^{18}O$ values in West Antarctic ice, which make predictions of future change in West Antarctica somewhat uncertain and rather dependent on our ability to predict the behaviour of the climate of the tropical Pacific. As Steig et al. (2013) point out, 'internal variability associated with the tropics is a significant source of uncertainty in climate change projections at high latitudes, especially for the atmospheric circulation'. Therefore, 'projections of the contribution of the WAIS to future sea-level rise that are based on present rates of ice-sheet mass loss should be treated with caution'. Despite that reservation, Turner et al. (2016) agreed that 'climate model projections forced with medium emission scenarios indicate the emergence of a large anthropogenic regional warming signal, comparable in magnitude to the late twentieth century Peninsula warming, during the latter part of the current century'.

While the caution urged by Steig et al. (2013) is understandable, the data from the likes of Abram et al. (2013) and Paolo et al. (2015), including the dramatic loss of ice shelves and melting back of glaciers, suggest that the Antarctic Peninsula is now showing signs of an intensified response to warming that is larger than what has been seen before and that is in agreement with what would be expected from global warming – though local fluctuations imposed by regional variability and the ozone hole may tend (at least temporarily) to obscure the response to a global signal.

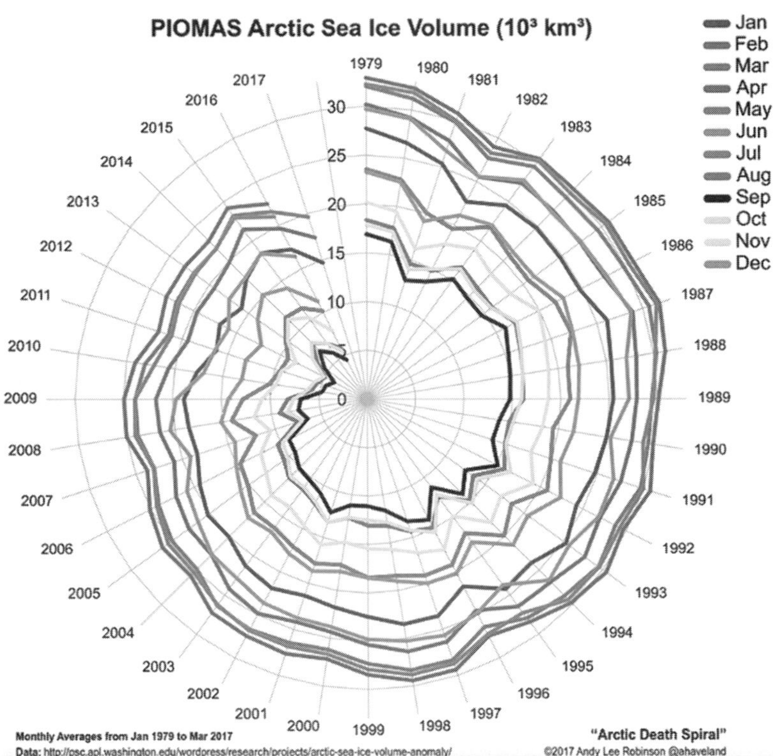

Figure 6.2.5 The so-called Arctic death spiral, showing sea-ice volume calculated for each month of each year since 1979, on a polar plot, so that declining ice volume is seen as a spiral moving towards the centre of the graph. From plate 16 in Wadhams (2016), published with permission of Andy Lee Robinson, and derived from Haveland-Robinson Associates' (www.haveland.com/).

In any case, the geological perspective tells us that given longer-term warming the melting of the West Antarctic Ice Sheet will affect sea level, as happened in the distant past. For instance, based on what is known of the behaviour of Antarctica during the Pliocene and during the last interglacial (130,000–115,000 years ago), De Conto and Pollard (2016) calculated that 'Antarctica has the potential to contribute more than a metre of sea-level rise by 2100 and more than 15 metres by 2500, if emissions continue unabated. In this case atmospheric warming will soon become the dominant driver of ice loss, but prolonged ocean warming will delay its recovery for thousands of years'. It is equally likely, based on data from the Holocene, that future ocean-ice-sheet interactions around Antarctica will 'amplify small subsurface ocean temperature variations to produce substantial near- and far-field centennial-to millennial climate responses' (Bakker et al. 2017). The magnitude of simulated periodic Holocene discharge from the Antarctic ice sheet is similar to that estimated in recent years as being due to global warming, leading Bakker et al. (2017) to conclude that the future behaviour of the ice sheet will represent a combination of the effects of long-term anthropogenic warming and multi-centennial natural variability.

Away from the two poles, further loss of ice from mountain glaciers is inevitable in response to warming and will occur even if the climate remains stable at its present warm level (Zemp et al. 2015).

Numerical models of the effect of global warming on sea ice suggest that Antarctica may lose 33% of its sea ice by 2100 (Bracegirdle et al. 2008).

But there is a significant dispute about the adequacy of those numerical models (Turner et al. 2009a).

As part of the Pan-Arctic Ice Ocean Modeling and Assimilation System (PIOMAS) project at the University of Washington, it has been possible to visualise the changing volume of Arctic ice month by month since 1979. Figure 6.2.5 shows the sea-ice volume for each month slowly decreasing. According to Wadhams (2016), the trend lines in Figure 6.2.5 can be used to predict two ice-free summer months in 2016, three in 2017 and five in 2018. Ice-free in this context means the disappearance in midsummer of the bulk of the thin sea ice that covers most of the Arctic Ocean. Pockets of thick ice will remain plastered to the north coasts of the Canadian islands and Greenland and filling some of the channels amongst the islands. That picture differs markedly from the one supplied by the Intergovernmental Panel on Climate Change (IPCC 2013), which suggested that the summer sea ice would last until 2050–2080. In contrast, Wadhams (2016) argues that we should expect to see the disappearance of summer sea ice by 2020 at the latest.

Bearing in mind the conclusion by Ding et al. (2017) that some 30%–40% of the decline in sea-ice area was likely due to atmospheric effects rather than global warming, Swart (2017) points out that 'predictions of the first year in which the Arctic summer will be ice-free are irreducibly uncertain to within ±20 years, due to the influence of random internal variability'. Hence Wadhams's prognosis of complete decline by 2020 might easily turn into 2040, in agreement with the prediction by Wang and Overland (2009) that we can expect a nearly sea-ice-free Arctic by 2037. Clearly, the prospect of an ice-free central Arctic Ocean in summer remains highly probable for the second half of this century and into the future. Continued reduction in Arctic sea ice is expected to lead to a reduction in the number of Arctic winter storms and a poleward shift of the midlatitude storm tracks (strengthening the storm track north of the British Isles) (Bader et al. 2011).

6.3 Sea Level

Alejandro Cearreta

6.3.1 Sea-Level Change

Sea-level change is one of the most important consequences of Quaternary climate variability. The association of human population centres with lowland coastal regions dates back to early civilisations, when people congregated at river mouths, deltas and estuaries because of abundant and accessible food sources (Stanley & Warne 1997). The emergence and rapid expansion of settled societies as sea-level rise decelerated at approximately 7.0 ka allowed coastal communities to become permanently established (Day et al. 2007).

The coastal zone changed profoundly during the 20th century, as human populations grew and coastal environments were reclaimed for agriculture, urbanisation, industry and so on. Much of the world's population lives in the coastal areas, at densities about three times higher than the global average, while most megacities are also located on the coast, many of the largest located on deltas (Table 2.8.2; Section 2.8.4). However, contemporary sea-level rise is anomalous in a Late Holocene context, and many low-lying locations are now threatened (Mee 2012).

Sea-level science is a highly interdisciplinary field, in which Quaternary scientists, geophysical modellers and instrumental specialists work closely together (Gehrels 2010b). Its importance is increasing. In the First IPCC Assessment Report from 1990, reference to sea-level changes previous to the last century was absent. By the Third Assessment Report published in 2001, sea-level change over longer timescales was incorporated, but only in 2007 with the Fourth Assessment Report was palaeoclimate information compiled into a single chapter (Caseldine et al. 2010), as also in the last assessment report (Masson-Delmotte et al. 2013).

Sea level is not a constant, planar surface but shows changes in time and space at many scales. Earth scientists define sea level as a measure of the position of the sea surface relative to the land – both of which may be moving relative to the centre of the Earth – and it reflects both climatic and geophysical factors (Masson-Delmotte et al. 2013). Climate influences on sea level are ocean temperature and salinity, causing sea water to expand or contract (steric changes in volume); the size of glaciers and ice sheets, leading to glacio-eustatic changes in water mass; changes in the storage of water on and in land; and the pattern of ocean currents. Geophysical factors affecting sea level include tectonic uplift and downwarp of land masses, glacial isostatic adjustments via vertical land motions in response to changes in the amount of loading by ice, and subsidence of coastal plains and deltas (see Section 2.8.4). Local and regional changes in these climate and geophysical factors produce significant deviations from a global average rate of sea-level change (Masson-Delmotte et al. 2013).

Rising ocean heat content, causing ocean thermal expansion, is currently an important part of global sea-level rise. The remaining contributions principally involve the transfer of liquid water from the land to the oceans as land ice melts – most importantly the major ice sheets of Antarctica and Greenland. There are some additional contributions from changes to the storage of water in lakes, wetlands, artificial reservoirs and groundwater (Church & White 2011; Reager et al. 2016). Other processes, such as ocean dynamics, tectonics or glacial isostatic adjustment, are spatially variable and cause sea-level rise to vary in rate and magnitude between different regions (Kopp et al. 2015).

Sea level can be measured with respect either to a surface of the solid Earth (called relative sea level) or to a geocentric reference ellipsoid (called absolute sea level). Relative sea level is the more relevant as regards the coastal impacts of sea-level change; it has been measured using tide gauges during the past three centuries and estimated for longer time spans from geological records (Figure 6.3.1). Absolute sea level has been measured since the 1990s by means of satellite altimetry (Church et al. 2013).

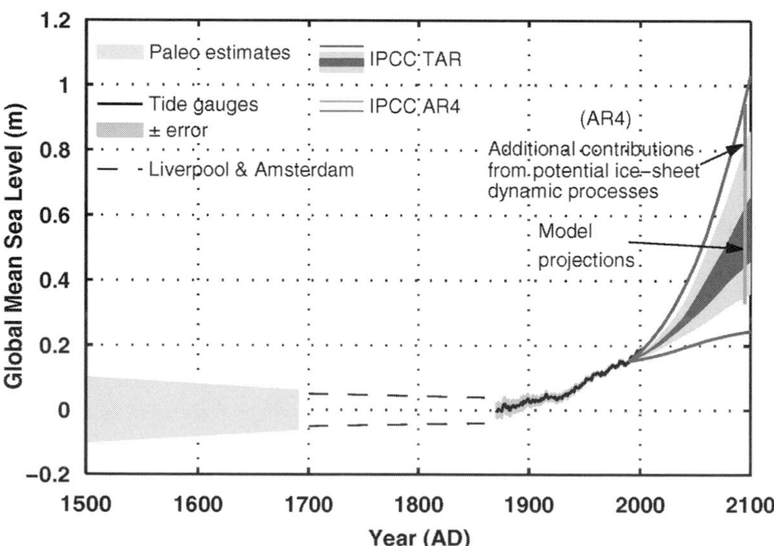

Figure 6.3.1 Sea level from 1500 to 2100 CE. The pale grey bar indicates the range of pre-1700 CE geological observations (mainly microfossil-based transfer functions), the dashed lines from 1700 to 1860 CE indicate the range of sea levels inferred from Europe's longest tide-gauge records, the dark line from 1870 to 2006 CE indicates the global average sea level based on multiple tide-gauge records, and the curves from 1990 to 2100 CE are based on satellite altimetric data and the projections from IPCC. (taken from Church et al. 2008). Reprinted by permission from Integrated Research System for Sustainability Science and Springer, ©2008 (A black-and-white version of this figure appears in some formats. For a colour version, please refer to the plate section.)

The idea that relative sea-level changes were a simple combination of isostatic and eustatic changes motivated, in the 1960s, the search for a single eustatic sea-level curve that could be applied worldwide, the most influential of these being that of Fairbridge (1961). But by the end of the 1970s, the search for a postglacial global eustatic curve had been abandoned: it was realised that regions within the same ocean basin could be represented by postglacial sea-level curves with different shapes, affected by such factors as far-field glacial isostatic adjustment. This led to more sophisticated conceptual models of sea-level change (Gehrels & Shennan 2015). Far-field adjustments came about, for example, when the Laurentide Ice Sheet depressed the North American landmass during the Last Glacial Maximum, causing a lateral displacement of material in the hot asthenosphere below the Earth's crust, which led to a 'balancing' peripheral bulge further afield, roughly beneath the southern United States and the Caribbean islands. That process is now in reverse as Hudson's Bay rises in the north, freed from its former weight of ice. Similarly, removal of the Scandinavian and British Ice Sheets mean that Scandinavia and Scotland are now rising, while the south of England is sinking. Islands like Barbados, sited on the peripheral bulge, are now sinking as the Earth's crust adjusts isostatically to the rise of deglaciated northern lands (Gehrels 2010a).

Systematic sea-level observations began in Amsterdam in 1682 CE, and the five longest tide-gauge records in the world are all in northwest European ports (Amsterdam, Brest, Liverpool, Stockholm and Świnoujście). These used visual observations of the heights and times of high and low waters, before the invention of automatic tide gauges in the 1830s that enabled the measurement of the full tidal curve and a computation of mean sea level (Woodworth et al. 2011). Southern Hemisphere measurements started only in the late 19th century, when similar instruments were installed worldwide at most major ports. The Permanent Service for Mean Sea Level (PSMSL) in Liverpool is the global databank for long-term sea-level information from tide gauges, though this record remains biased towards Northern Hemisphere records and is most complete for the second half of the 20th century. All these tide-gauge data are expressed relative to the height of a benchmark on the nearby land. In order to convert any changes they show into absolute terms, they need to be corrected for vertical land movements (Woodworth et al. 2011). Vertical land movements at tide-gauge stations can now be measured by GPS (Pugh & Woodworth 2014).

Satellite altimetry provides a measure of absolute height of the ocean surface relative to the centre of the Earth. Precision altimetry began with the launch of TOPEX/Poseidon in late 1992 CE and was later followed by Jason-1 (2001 CE) and Jason-2 (2008 CE). Consequently, a precisely calibrated continuous record of global sea-level change between 66° latitude of both hemispheres from 1992 CE to the present is now available (Woodworth et al. 2011; Pugh & Woodworth 2014). Two other observing systems complement these altimetric data. The Argo network, with its series of autonomous floats, has monitored temperature and salinity within the top 1–2 km of the ocean from 2000 CE. And the Gravity Recovery and Climate Experiment (GRACE) satellite mission, launched in 2002 CE, until recently provided monthly measurements of water-mass distribution with a spatial resolution of ~500 km anywhere on the planet and monitored ice-mass loss from the Greenland and Antarctic ice sheets and mountain glaciers (see Section 6.2.3). Thus, it is now possible to determine how much of the sea-level rise observed from satellite altimetry and tide gauges is attributable to melting ice and how much is due to thermal expansion (Milne et al. 2009; Rietbroek et al. 2016).

Satellite altimeter data have shown that rates of sea-level change can vary regionally, on decadal timescales, due to regional variations in the density of the ocean. Nevertheless, attempts have been made to produce a global average sea-level time series for the instrumental period (Figure 6.3.1), though the coverage of tide-gauge and satellite measurements is too short to capture earlier trends of sea level. This is where geological data become necessary to place instrumental estimates into a longer-term context. Reconstruction of Quaternary sea-level trends based on detailed litho- and biostratigraphic analyses of coastal deposits was developed in a succession of IGCP projects started in 1979 CE (Woodroffe & Horton 2005), and since the late 1990s, methods to create sea-level index points (i.e., dated sediments which can be related to a former sea-level position) have become increasingly quantitative (Gehrels 2010b).

Many sea-level reconstructions from coastal sediments rely on indicators that are found in contemporary salt-marsh and mangrove environments, which are limited to narrow sea-level ranges. The surface of a salt marsh, for instance, is close to the high-tide level, and when sea level rises over decades, sediments accumulate and the salt-marsh surface builds up vertically as it keeps pace with the rate of sea-level change. In this way, salt-marsh sediments can provide excellent archives of past sea-level changes.

Quantitative sea-level reconstructions make use of different groups of microfossils such as foraminifera (Scott & Medioli 1980), diatoms (Zong & Horton 1999) and thecamoebians (Charman et al. 1998), different species of which reflect different parts of a salt marsh, defining specific heights within the tidal range. Using the known tolerances of modern forms as a comparator, fossil examples in the sedimentary record can be used for accurate sea-level reconstruction of former sea level over decadal to millennial timescales (Gehrels et al. 2001), allowing for various correction factors such as palaeo-tidal changes, sediment compaction and local sedimentation regime (Barnett et al. 2017).

The specific marginal marine conditions of a salt marsh, allied to narrow elevational ranges, exist on a worldwide scale, and the same few species of foraminifera are ubiquitous worldwide (Scott et al. 2001). Their fossil assemblages in the stratigraphic record can be dated by radiometric analyses (^{14}C, ^{210}Pb) and stratigraphic-marker techniques (e.g., ^{137}Cs, pollen, PAHs, Pb isotopes, metal concentrations, spheroidal carbonaceous particles). By these means, salt-marsh sediment successions can produce proxy records with a precision of $\pm$10 years in the 19th century and $\pm$5 years in the 20th century (Gehrels & Shennan 2015). This is less precise than instrumental records but extends the tide-gauge data both geographically and backward in time. Their vertical resolution is impressive, though, constraining Late Holocene relative sea-level changes to within $\pm$5 cm (Gehrels & Woodworth 2013; Table 6.3.1 herein). Salt-marsh proxy records for recent times are validated using tide-gauge records and provide independent evidence of significant sea-level rise during the 20th century compared to preceding centuries (García-Artola et al. 2015).

Average global sea levels are currently higher than at any point within the past ~115,000 years, since the termination of the last interglacial of the Pleistocene Epoch (Church et al. 2013). The start of the Holocene Epoch reflects the transition from the last glacial phase into an interval of warming accompanied by ~120 m of sea-level rise, at an average rate of about 10 mm/yr until about 7,000 years ago (Lambeck et al. 2002). However, very high rates of sea-level change (>40 mm/yr) occurred at about 14.6–14.3 ka during the Meltwater Pulse 1A (MWP-1A; corresponding to the Bølling warming in the Northern Hemisphere – see Section 6.1.2), associated with a phase of rapid ice-sheet disintegration during this transition (Deschamps et al. 2012). Evidence for other meltwater pulses is still being debated, but it seems increasingly likely that the rapid cooling that occurred around 8.2 ka in the North Atlantic region was caused by an outburst of meltwater from the Laurentide Ice Sheet. This provoked a rise rate of about 11 mm/yr over 500 years (Gehrels & Shennan 2015). These high rates of rise caused widespread inundation of coastal areas, and sedimentary facies back-stepped (retrograded) landward (Waters et al. 2016).

The past 7,000 years of the Holocene Epoch, when ice volumes stabilised near present-day values

Table 6.3.1: **Methods of sea-level reconstruction over various timescales and the time and height precision associated with each**

Time period (kyr BP)	Sea-level indicator	Chronology	Maximum resolution (yr)	Estimated vertical precision (±m)	Maximum rate (m per century)
0–470	Oxygen isotopes	AMS^{14}C, palaeomagnetism, tuning	200	12	2.5
0–30	Corals	U/Th	400	5	4
0–20	Sediment facies, microfossils	AMS^{14}C	200	3	4
0–16	Isolation basin stratigraphy	AMS^{14}C	200	0.2–1.0*	n/a
0–10	Basal peat	AMS^{14}C	200	0.2–0.5*	0.2
0–7	Microatolls	^{14}C	200	0.1–0.2*	0.2
0–7	Biological indicators on rocky coasts	^{14}C	200	0.05–0.5*	0.1
0–2	Archaeology	Historical documentation	100	0.1–0.5*	0.1
0–0.5	Salt-marsh microfossils	AMS^{14}C, ^{210}Pb, ^{137}Cs, Pb isotopes, pollen, chemostratigraphy	20	0.05–0.3*	0.2

Note: The maximum rate of sea-level rise is given for each time period and reconstruction method (rates have been corrected for land motion). Asterisk indicates uncertainties that are tidal-range dependent. Modified from Milne et al. (2009). Reprinted by permission from Nature Geoscience, ©2009.

(e.g., Clark et al. 2016; Section 6.2.3 herein), provide the baseline for discussion of anthropogenic contributions. Relative sea-level records indicate that from 7.0 to 3.0 ka, global mean sea level likely rose 2–3 m to nearly present-day levels at a rate of 0.5–0.75 mm/yr (Church et al. 2013). Sea-level records spanning the past 2,000 years suggest that fluctuations in global mean sea level during this interval have not exceeded 0.25 m on timescales of a few hundred years (rate of ±0.125 mm/yr). Other estimates of global sea-level change over the Common Era (CE), based upon statistical synthesis of a global database of regional sea-level reconstructions, indicate that pre-20th-century variability was very likely between ±0.07 and ±0.11 m in amplitude. Sea level rose at 0.1 ± 0.1 mm/yr over 0–700 CE, was nearly stable from 700 to 1000 CE and fell 0.2 ± 0.2 mm/yr over 1000–1400 CE associated with a 0.2°C global mean cooling (Kopp et al. 2016). Other variations, such as a global sea-level rise from 1400 to 1600 CE at 0.3 ± 0.4 mm/yr and a fall over 1600–1800 CE at 0.3 ± 0.3 mm/yr, are reconstructed from the western North Atlantic but not found worldwide (Kopp et al. 2016). As a consequence of these Late Holocene very low rates of rise, coastlines have been more or less fixed, and

Table 6.3.2: **Estimates of the total rise (in m) and average rate (in mm per year) of global mean sea-level change for four selected time intervals during the Holocene**

Time interval	Total rise (m)	Average rate of rise (mm per year)
7–3 ka	2–3	0.5–0.75
2 ka–present	0.25	0.125
1901–2010 CE	0.19 ± 0.2	1.7 ± 0.3
1993–2010 CE	0.05 ± 0.01	3.2 ± 0.4

Note: Adapted from data in Church et al. (2013).

sediment has built outward from beaches, estuaries and deltas (Waters et al. 2016).

Instrumental and geological records show that mean global sea level is now rising above Late Holocene rates (Table 6.3.2). Depending on the trajectory of future anthropogenic forcing, this trend may reach or exceed the envelope of Quaternary interglacial conditions (Waters et al. 2016), which show a maximum global mean sea level of 6 m higher than present during the last interglacial (MIS5e, 129–116 ka; Masson-Delmotte et al. 2013). The most robust signal, captured in salt-marsh records from both the Northern and Southern hemispheres, indicates a transition from relatively low rates of change during the Late Holocene (<1 mm/yr) as sea level began to rise above the Late Holocene background rate in the late 19th to early 20th century (Gehrels & Woodworth 2013; see Figure 6.3.2 herein) to reach current rates of 3.2 ± 0.4 mm/yr from 1993 to 2010 CE (Church et al. 2013; Table 6.3.2 herein).

During the Holocene, the lag between maximal/maximum insolation (i.e., warming) about 11.7 ka and maximum sea-level rise was of 4,500 years. The lag between the rise of temperature in the atmosphere and the rise in sea level occurs because the latter depends on heat penetrating deep within the ocean, which inevitably takes considerable time (Clark et al. 2016). Hence it is expected that the sea-level rise

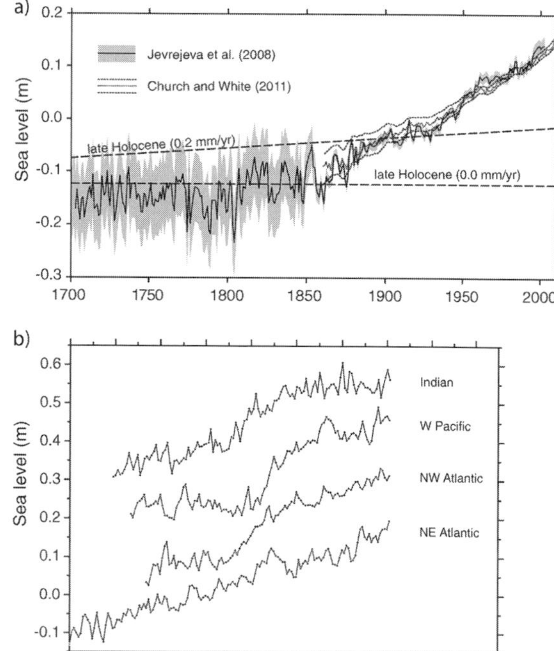

Figure 6.3.2 (a) Global tide-gauge compilations from Jevrejeva et al. (2008) and Church and White (2011). Increased error bands before 1850–1900 CE reflect the low number of tide-gauge records that cover the 18th and 19th centuries. The range of global Late-Holocene sea-level trends (0–0.2 mm/yr), as proposed by Jansen et al. (2007), is also shown. (b) Compilations of tide-gauge records for four relevant oceanic regions from Milne et al. (2009), based on data from Jevrejeva et al. (2006). (reprinted from Gehrels & Woodworth 2013, ©2013 with permission from Elsevier)

consequent upon current global warming will show the same kind of delay.

The acceleration of sea-level rise in the late 19th to 20th centuries has been regionally non-synchronous (Long et al. 2014), reflecting factors such as the influence of land-ice changes, ocean-temperature changes and long-term ocean dynamics (Milne et al. 2009), and has also been non-linear (Kemp et al. 2015; see also Figure 6.3.2 herein).

The current rate of rise since 1992 CE has been faster (by almost a factor of two) than the average rate of rise during the 20th century, which, in turn, was an order of magnitude faster than the rate of rise over the two millennia prior to the 18th century (Church et al. 2008). This rate increase is likely to largely reflect anthropogenic climate change (Kopp et al. 2016). Ocean thermal expansion and glacier melting have been the dominant contributors to 20th-century global mean sea-level rise, while the contribution of the Greenland and Antarctic ice sheets has increased since the early 1990s (Church et al. 2013). The timing of the inflections in this rising sea-level curve matches, with a delay of about a decade, the stepped changes in CO_2 concentrations. The sea-level signal of the Anthropocene is not yet as strongly expressed as other stratigraphic changes described in this book, as it lags climate warming, which in turn lags atmospheric greenhouse-gas rise on human timescales. Nevertheless, its trajectory has clearly been changed. A major question is how that trajectory will continue.

Over the past century, sea level has risen by some 30 cm. The magnitude of eustatic sea-level rise predicted for the coming century is a subject of considerable debate. The IPCC AR5 projections, based primarily on physical-ocean and ice-sheet models, predict a global mean sea-level rise for the late 21st century (relative to the reference period of 1986–2005 CE) of 28–61 cm for a scenario of drastic greenhouse-emissions reduction and of 52–98 cm if there is unmitigated growth of these emissions (Church et al. 2013). This is a substantial upward revision compared to the previous IPCC report published in 2007 (Figure 6.3.3). Semi-empirical estimates, based on the past statistical relationship between the rate of sea-level rise and temperature, projected a 70–140 cm rise between 2000 and 2100 CE under a business-as-usual scenario (Miller et al. 2013). Other recent estimates suggest that global mean sea-level rise could even exceed 200 cm by 2100 CE (Oppenheimer & Alley 2016). This range reflects uncertainties about how the Greenland and Antarctic ice sheets will behave dynamically, with the potential for a non-linear sea-level response to climate change. If the Greenland ice sheet were to melt, it would raise sea level by about 7 m; if the West Antarctic ice sheet melted, it would raise sea level by 3–5 m (Hoegh-Guldberg & Bruno 2010), a scenario more consistent with a transition to a period-scale change (Berger & Loutre 2002). Such rises are comparable with the 4–9 m typical of the Eemian interglacial or the 10–20 m rise typical of Pliocene times (see for example figure 2a in Clark et al. 2016). According to Hansen et al. (2016), such rises may occur sooner rather than later. If the East

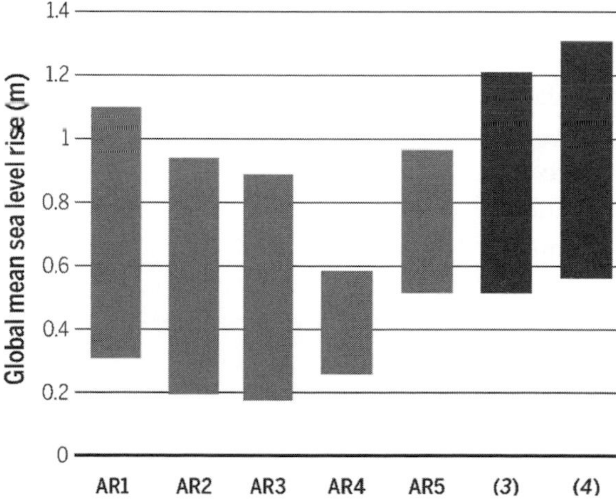

Figure 6.3.3 Sea-level projections from models for year 2100 CE have changed markedly since the IPCC published its First Assessment Report in 1990. Recent projections (3, 4) are based on ice models and other approaches that may capture the ice-sheets' contribution better than in the past, but large uncertainties remain. (taken from Oppenheimer & Alley 2016). Reprinted with permission from AAAS

Antarctic ice sheet melted as well, sea level would rise something like 65 m; although this is highly unlikely in the short to medium term, this would ultimately result in sea level more typical of the greenhouse climate of Cretaceous times some 100 Ma ago, when there were no great ice sheets.

6.3.2 Impacts of Near-Future Sea-Level Rise on Coastal Areas

Sea-level rise has a wide range of effects on coastal areas, including submergence and increased inundation of coastal land; coastal wetland loss and change; saltwater incursion into coastal areas, aquifers and many urban centres; erosion of beaches and soft cliffs; degradation or removal of salt marshes and sand dunes; and general coastal retreat and the displacement of coastal ecosystems landward.

The extent of these impacts will be determined by the relative sea-level change, reflecting not only the global mean trend in sea level but also regional and local variations in sea level and in geological uplift and subsidence. Many of the present concerns associated with rising sea level represent the cumulative effects of natural and anthropogenic processes that have been ongoing for decades, such as reduced or exhausted sediment supplies and human activities like groundwater extraction or land reclamation (FitzGerald et al. 2008). In addition to inundating low-lying coastal areas, rising sea level increases the vulnerability of coastal regions to flooding caused by storm surges, tsunamis and extreme astronomic tides. As sea level rises, ocean surges produced by storms of a given magnitude reach higher elevations and flood more extensive coastal areas (FitzGerald et al. 2008).

Six hundred million people live within 10 m of modern sea level (Rohling et al. 2013), with between 100 and 150 million people living within 1 m (Ericson et al. 2005 and references therein; Church et al. 2008). This concentration of humanity is in large part a Holocene legacy: the flat, fertile delta tops and coastal plains that have built out over the last several thousand years, graded to a level just at or above sea level (Section 2.8.4). Large parts of these are now at or just below sea level, because of local to regional ground subsidence of up to several metres caused by drainage and groundwater or hydrocarbons extraction, causing irreversible compaction of the formerly soft, fluid-soaked sediments below (Syvitski et al. 2009; Higgins 2016), and locally by removal of metres-thick surface peat layers by draining, deflation and oxidation (Smith et al. 2010). Places like the Netherlands polders (Figure 2.5.7; Section 2.5.2.8), the English Fenland and the city of New Orleans are now below mean sea level, the sea (usually) being held back by walls or dykes.

Increasing vulnerability of social infrastructures and resulting socio-economic disruption provide major challenges to policymakers, as global populations concentrated in coastal cities will need to adapt to the effects of these rising sea levels. Measures specifically designed to be mindful of evolving predictions of sea-level rise include building defensive structures resilient to periodic flooding (e.g., the Thames Barrier in the United Kingdom and the MOSE Project in Italy) and retreat from exposed areas, combined with enhancement of natural defences such as wetlands (Oppenheimer & Alley 2016).

Given the low gradient of the coastal strip, just a small rise in sea level will flood a very large area. It is the first 1–2 m rise in sea level that will submerge the largest area of land, with subsequent increments flooding smaller though still significant areas. So as regards societal adjustment to large-scale changes in land-sea geography, it is the next 1–2 centuries that will arguably provide the largest challenge.

In detail, patterns of inundation – and its delay – will be controlled by factors including sediment supply and the redistribution of sediment to the coast to build out new land (currently, much potential sediment is held behind the large dams that now traverse most major rivers on the Earth; see Section 2.8.3); patterns of coastal vegetation, notably salt-marsh grass and mangroves, that trap sediment; and control of the factors affecting local subsidence (e.g., ceasing the extraction of coastal groundwater and hydrocarbons).

Rising sea levels will also contribute to the recession of the world's sandy beaches, 70% of which have been retreating over the 20th century, with less than 10% prograding (Bird 1993). On sandy coastlines, beach erosion commonly occurs at tens to hundreds of times the rate of sea-level rise and will degrade or remove protective coastal features such as sand dunes and vegetation, further increasing the risk of coastal flooding (Church et al. 2008).

The reduced sediment flux to major deltas, combined with increasing extraction of groundwater, hydrocarbons and sediments (for aggregates), has caused many large deltas to subside at rates substantially faster than current sea-level rise (Waters et al. 2016; see Section 2.8.2 herein). Deltaic regions suffer from a combination of existing anthropogenic problems (e.g., sediment entrapment by upstream dams leading to a lack of fluvial accretion) and sea-level rise (Church et al. 2008). The response of river-dominated systems like deltas and estuaries is largely dependent upon how much sediment is supplied downriver from the catchment and deposited within these environments (McKee et al. 2012).

Salt marsh stability depends critically on the balance of competing sediment supply and erosional processes. These wetlands are amongst the world's most valuable providers of 'ecosystem services', such as storm protection, water filtering and seafood production. They also capture as much as 450 billion metric tons of carbon globally, acting as a carbon sink that prevents CO_2 and CH_4 from leaking into the atmosphere (Kintisch 2013). Over long time periods, coastal wetlands seem to have kept pace with sea-level changes. As sea level rises, coastal wetlands have the ability to maintain their position in the intertidal zone through the accumulation of both organic and inorganic materials. This ability is largely governed by the complex interaction of vegetation, flooding and sediment supply. For salt marshes to survive future sea-level increases, they must be able to grow vertically at a rate great enough to offset the rate of sea-level rise (Baustian et al. 2012). Their ability to do so presupposes an appropriate rate of supply of sediment to the coastal zone.

Coral-reef systems grow within the photic zone, their accretion normally keeping pace with sea-level rise. Estimates of the vertical accretion rate of coral reefs during the period of rapid postglacial sea-level rise in the Early Holocene suggests an upper rate of about 14 mm per year for corals living very near sea level (Buddemeier & Smith 1988 and references therein). A healthy and functional reef system should be able to grow upward to compensate current rates of rise. However, reef systems are now being adversely impacted by factors such as ocean warming, acidification, pollution and overfishing (see Section 3.4 on reef systems), all of which are already severely affecting the ability of corals to grow and hence add to reef mass. It is unclear to what extent other elements of the reef system (e.g., coralline algae, foraminifera) can continue to thrive and secrete calcium carbonate to maintain and build the reef structure, even as corals decline. In spite of these concerns, it has to be recognised that many coral reefs clearly did survive the rapid rates of sea-level rise between the Last Glacial Maximum and the Holocene. The principal dangers to coral reefs, then, are more likely to be rising temperatures combined with ocean acidification and exacerbated by sea-level rise.

7 The Stratigraphic Boundary of the Anthropocene

CONTENTS

Here we outline the basis on which a formal proposal should be made for potential inclusion of the Anthropocene in the Geological Time Scale, examining the scale and rate of human change to the Earth System to help recognise the point at which anthropogenic impacts became of sufficient scale to allow discrimination of the Anthropocene as a geological unit. This examination covers such factors as impacts from early hominin species, the first human artefacts, early ecosystem modification through agriculture, deforestation, the domestication of animals, urbanisation, metal mining and smelting and early globalisation. The Industrial Revolution, starting in the UK in the 18th century, and the global Great Acceleration of the mid-20th century, are investigated, as both provide popular narratives that explain the Earth System changes indicative of the Anthropocene, with the latter producing the near-synchronous stratigraphic signals most consistent with an effective geological time boundary. We assess which hierarchical level – age, epoch, period, era or eon – seems most suitable for the Anthropocene, and suggest that epoch (= series) level is conservative and appropriate. The Anthropocene might be defined via a Global Standard Stratigraphic Age or a Global boundary Stratotype Section and Point, with the latter being most appropriate. Finally, we assess the kinds of geological environments, including anoxic marine basins, annually banded coral and bivalve skeletons, estuaries and deltas, lake floors, peat mires, anthropogenic deposits, polar ice, speleothems and tree rings, in which such a physical reference level might be placed.

7.1 Geological Validity of the Anthropocene

Jan Zalasiewicz, Colin N. Waters, Mark Williams and Colin P. Summerhayes

'The Anthropocene' was originally improvised by Paul Crutzen as a geological time term, with the suggestion that one may regard as an implicit hypothesis that conditions of the Holocene Epoch have terminated. Working within the Earth System science community, his case (Crutzen & Stoermer 2000; Crutzen 2002; see Chapter 1 herein) was based on process changes in the Earth System: population growth, energy output, pollution of atmosphere and ocean, biosphere changes and so on.

A formal geological time unit is based, though, on a sufficiency of evidence within the stratal record, as for almost all of Earth time this represents our only evidence, in proxy form, for Earth history and process. Therefore, for the Anthropocene to be considered on the same terms as the units of the Geological Time Scale, it must comprise a recognisable and clearly distinct body of strata. Testing the Anthropocene as a material, stratigraphic unit has therefore been the central task of the Anthropocene Working Group.

The data assembled and summarised in this book clearly shows that there is substantial evidence from stratigraphic sources, as well as from Earth System science, for an Anthropocene unit. The stratigraphic evidence ranges from novel minerals, rock types, particulates and modified sedimentary successions (Chapter 2) to a biosphere transformed largely through species invasions and agriculturally driven assemblage changes rather than (yet) by mass extinction (Chapter 3) to an extraordinary diversity of highly fossilisable artefacts or 'technofossils', the products of a burgeoning technosphere (Chapter 4) to a variety of chemical signatures (Chapter 5), linked to accelerating changes to climate, sea level and the cryosphere (Chapter 6). Various combinations of the resultant stratigraphic signatures allow an Anthropocene chronostratigraphic unit to be recognised across virtually all of the sedimentary environments of the globe.

The Anthropocene, therefore, has geological as well as Earth System reality, justifying formal proposal for its inclusion in the Geological Time Scale (Zalasiewicz et al. 2017c). To help clarify the basis on which such a proposal should be made, we examine the history of human impact on the Earth System to help recognise the point at which anthropogenic impacts became of sufficient scale to allow discrimination of the unit; provide an assessment of which hierarchical level seems most suitable (i.e., age, epoch, period, era or eon); discuss whether the Anthropocene should be defined via a Global Standard Stratigraphic Age (GSSA) or Global Boundary Stratotype Section and Point (GSSP); and, for a GSSP, discuss the relative merits of the different environments in which such a physical reference level might be placed. We begin by examining the scale and rate of human change to the Earth System through time.

7.2 The Early Stratigraphic Record of Humans

Mark Williams and Eric Odada

Early species of *Homo* (the genus encompassing our species *Homo sapiens*) and of their hominin relatives, the australopithecines, have left a stratigraphic signal in fossil skeletons, traces of activity in rock deposits, and stone-tool artefacts, stretching back over 3 million years (Figure 7.2.1). Body fossils and stone tools of the earliest *Homo* species (*Homo habilis, Homo rudolfensis*) are exceptionally rare, with all uncontested fossil sites prior to 2 Ma being in Africa. The earliest biostratigraphic record of *Homo* is a mandible with teeth from rocks between 2.8 and 2.75 million years old at Lee Adoyta in the Ledi-Geraru area, Afar, Ethiopia (Villmoare et al. 2015). These postdate the earliest evidence of hominin stone tools (likely made by *Australopithecus*) by some 0.5 million years (Harmand et al. 2015). The stratigraphic level of

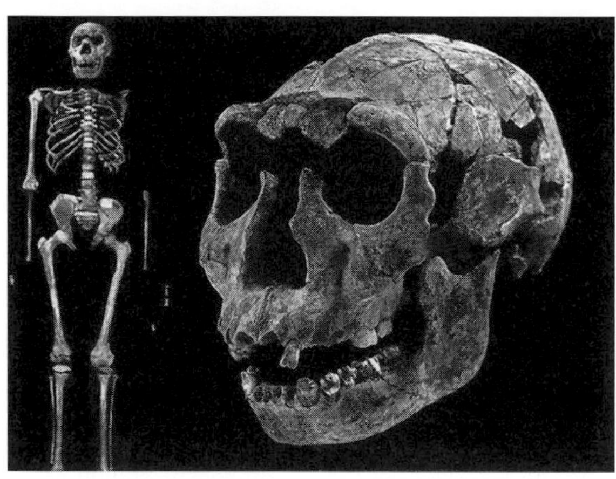

Figure 7.2.1 This skeleton of a *Homo erectus* about 160 cm in height is colloquially known as Turkana Boy and was sourced from Pleistocene deposits (~1.5 Ma) at Nariokotome, near Lake Turkana, Kenya. Left, the skeleton as preserved, and right, the cranium. The image is courtesy of the National Museums of Kenya, Nairobi.

the mandible (sample LD350–1) is in a sedimentary succession constrained by an absolute age of 2.82 Ma in a tuff bed ~10 m below the fossil horizon. This places the earliest signal of *Homo* within the Piacenzian Stage of the Upper Pliocene Subseries.

The migration of *Homo* out of Africa into Eurasia, providing a more geographically widespread stratigraphic record of the genus, is widely regarded to have occurred after 2 Ma and is underpinned by detailed palaeontological and geochronological studies at Dmanisi, Georgia, on fossils of *Homo* within a sedimentary succession above basalt (dated at 1.85 Ma) and within volcanic ash layers that are dated at 1.81 Ma (e.g., Garcia et al. 2010). This level is close to the boundary between the Gelasian and Calabrian stages of the Lower Pleistocene Subseries, in the upper part of the Olduvai palaeomagnetic Subchron of the Matuyama Chron (Figure 1.2.7). It also coincides with a marked palynological change that signals more arid environmental conditions, which might have influenced the migration of *Homo* (Messager et al. 2010). *Homo* (especially *H. erectus*) became widespread from east to west Eurasia between 1.8 and 0.75 Ma, with early fossil sites in, for example, China (e.g., Zhu et al. 2004), although with a patchy geographical and stratigraphic overall occurrence. *Homo erectus* likely persisted in parts of East Asia until a few tens of thousands of years ago. Several taxa of *Homo* overlap in time with *H. erectus* but with a more regional distribution, and these include *Homo heidelbergensis* and *Homo neanderthalensis*. Hence, multiple *Homo* species that were relatively long lived (over tens to hundreds of thousands of years) overlapped in time, but in such small numbers that they had little impact on the Earth's environment.

The origin of our own species, *H. sapiens*, has been contentious, with both 'African' and 'multi-regional' models of evolution proposed, although a general consensus has formed around an African origin (see Weaver & Roseman 2008). The earliest records of *H. sapiens* as body fossils associated with stone tools date to ~300,000 years ago, from Jebel Irhoud, Morocco (Richter et al. 2017; Hublin et al. 2017). Other early records are East African and comprise body fossils of *H. sapiens* from the Omo Kibish site in Ethiopia, which has been dated to about 195,000 years ago (McDougall et al. 2005). There is also early evidence of *H. sapiens* in the sedimentary succession of the Upper Herto Member of the Bouri Formation of Ethiopia, in rocks dated to between 160,000 and 154,000 years old (Clark et al. 2003), together with stone implements of Acheulean and other types.

As with the fossil evidence of earlier *Homo* taxa, that of *H. sapiens* is fragmentary during most of the Mid to Late Pleistocene, though the species went on to colonise all continents except Antarctica by 14,000 years ago, reaching Australasia by ~65,000 years ago, spreading into Europe 45,000 years ago (Clarkson et al. 2017) and arriving in the Americas by ~15,000–14,000 years ago. *Homo sapiens* subsequently colonised most of the world's inhabitable islands and finally reached the Antarctic region in the late 19th century.

A body-fossil record of the developing cognitive abilities of early *Homo* is revealed by increases to brain capacity preserved in the brain cases of skulls and by the artefacts that are associated with hominin sites. The earliest Lomekwian stone tools, from Turkana, Kenya, predate species of *Homo* and were likely made by australopithecines. They show a hominin perception of stone fracture properties (Harmand et al. 2015). Lomekwian technologies were followed by more sophisticated stone technologies, including those of Oldowan, Acheulean, Mousterian, Aurignacean and microlithic types, reflecting the increasing skill of toolmakers over 2 million years of time. It seems highly likely that these technological developments went along with anatomical changes, especially the development of the third metacarpal styloid process (TMSP), a structure at the base of the middle finger that helps the hand to lock into the wrist. This structure confers strength to the hand and allows it to manipulate stone tools – or a power drill – with great dexterity. The TMSP is absent from the hands of all non-*Homo* primates and from the earliest *Homo* too. Evidence for the TMSP is found in fossils of *Homo* from ~1.4 million years ago in Kenya, at a fossil site called Kaito, in West Turkana (Ward et al. 2014). The earliest stone tools of Acheulean type are also found in this region, although from somewhat older rocks, suggesting that anatomical evolution in *Homo* was proceeding in tandem with the evolution of tools.

Stone tools such as those of the Oldowan culture are geographically widespread across the Old World. Oldowan stone tools persisted from their origins (in Africa) ~2.6 million years ago through more than 2 million years to leave a time-transgressive fossil signature of *Homo* technology in Africa and Eurasia. Oldowan tool sets persisted into the Mid-Pleistocene (e.g., Clark et al. 1994) and overlap in time with the development of more sophisticated tool sets, such as the Acheulean type. The latter originated later but also show a time-transgressive distribution within a mosaic of stone-tool types that reflects local variation in culture, geography, ecology and raw materials. As a result, there is no simple or gradual transition from one type of stone-tool culture to another (Clark et al. 2003).

Anatomically modern *H. sapiens* developed Upper Palaeolithic stone-blade technologies from ~50,000 years ago (Bar-Yosef 2002), extending their use and impact as they migrated across the Old and New World, invading Australasia from about 65,000 years ago (Clarkson et al. 2017) and the Americas from about 15,000–14,000 years ago (Koch & Barnosky 2006; Barnosky et al. 2014). Dating of human artefacts of the Middle Stone Age in level BBC M1 in the Blombos Cave in South Africa provides evidence for the origins of modern behavioural traits, signalled by engraved abstract representations on pieces of ochre (Henshilwood et al. 2002). Thermoluminescence dates of five burnt lithic samples from archaeological phase BBC M1 have a mean age of 77,000 $\pm$ 6 ka (Henshilwood et al. 2002), and specimens from phase BBC M3 date to ~100 ka (Henshilwood et al. 2009). The development of modern behavioural traits, such as large-scale social organisation and forward planning, allowed humans to expand their impact on their surroundings, which is a key characteristic of the Anthropocene biosphere (Ellis 2015). The Blombos Cave is therefore significant from an Anthropocene perspective in signalling a marked change in the social behaviour of humans during the Late Pleistocene Subepoch. As with earlier stratigraphic signals of significant changes in human evolution, the fossil record of behaviourally modern humans is piecemeal and time transgressive in its development throughout Africa, Eurasia and elsewhere.

These landmarks in the evolution of *Homo* represent key stages in the evolution of our capacity to influence the Earth System. They help to demonstrate that the Anthropocene had a long and gradual development over some 2 million years from a world-state that lacked human influence to one that is strongly influenced by humans. By the Late Pleistocene, *H. sapiens* had assembled the physical and psychological toolkits needed to develop a complex material civilisation and were well placed to respond to the development of the warm interglacial climate of the Holocene. This human toolkit included the ability to kill and butcher large animals, a deep-rooted skill present in pre-*H. sapiens* hominins and evident in Ethiopia with the earliest fossil *H. sapiens*, where hippopotamus was a major prey animal (Clarke et al. 2003).

7.3 Pre–Industrial Revolution Start Dates for the Anthropocene

Michael Wagreich, Mark Williams, Erich Draganits, Jan Zalasiewicz, Colin N. Waters and Matt Edgeworth

7.3.1 Early Ecosystem Modification by Humans

A long interval of gradually increasing impact of humans on the environment was acknowledged by Foley et al. (2013) in coining the informal term 'Palaeoanthropocene' that includes the Holocene and much of the Pleistocene, from about the time of the first appearance of the genus *Homo* until the Industrial Revolution. The start of (significant) ecosystem modification by humans is difficult to pin down, but its progress is evident from the spread of developments including the mastery of fire, agriculture, domestication of animals, urbanisation, the invention of writing, and industry, combined with the effects of population growth. All of these impacts are markedly time transgressive. Regionally, significant continental ecosystem transformation, including the effects of large-scale hunting, biomass burning and deforestation, may already have begun during the Late Pleistocene, perhaps as early as 60,000 years ago (Balter 2013). By the beginning of the Holocene Epoch, humans were well equipped to capitalise on environmental stability and on the greater range of habitats that opened for colonisation as land ice retreated. Early evidence for open-water seafaring has been found in the Aegean (Strasser et al. 2010), and seafaring technology must have facilitated the occupation of Australia 65,000 years ago. Experimentation with crop cultivation, perhaps even during the Late Pleistocene (Snir et al. 2015), led to centres of agriculture and urban civilisation developing during the earlier Holocene as far apart as China, the Americas and the Near East (Ruddiman 2013 and references therein) at about the same time as animals were being domesticated (see Section 3.3.2.2). The stratigraphic record of the Late Pleistocene and Early Holocene shows significant changes in human technology, such as the widespread use of ceramics for storing and processing food (see Sections 2.2.3, 2.5.2 and 4.2). The spread of agriculture was causally linked to the evolution of new kinds of technofossils such as digging sticks, hoes and ploughs, though the early stratigraphic record of these and traces of ploughing are scant. The substantial impacts of humans on the extinction of megafauna during the Late Pleistocene have been covered in Section 3.2.1.

While the Palaeoanthropocene was proposed as an informal, non-chronostratigraphic unit to contrast with the sharper and more globally synchronous changes ascribed by Foley et al. (2013) to the Anthropocene, it creates a potentially confusing overlap between the two terms. Changes in agriculture and land use or the beginnings of globalisation of trade in the early modern period have also been used to suggest either synonymy of the Anthropocene with the Holocene or an 'early' start date for the Anthropocene.

7.3.2 Anthropogenic Charcoal in the Palaeoanthropocene

Glikson (2013) speculated that the ability of a primate species to ignite fire led to a turning point in human evolution in the Pleistocene at ~1.8 Ma, and he used this event as the start for his 'Early Anthropocene'. Such early signs of the mastery of fire are restricted to a few sites with discolouration of mammalian bones suggestive of burning. Anthropogenic charcoal, as robust evidence for a widespread use of fire, dates from about 125 ka onwards (Glikson 2013). Even so, the charcoal record and its interpretation regarding widespread use of and human adaption to fire remains controversial (e.g., Roebroeks & Villa 2011; Gowlett 2016).

7.3.3 'Human Niche Construction' as the Start Point of the Anthropocene

A proposal that equates the base of the Anthropocene to that of the Holocene – and thereby negates the need for a separate stratigraphic definition for the Anthropocene – was proposed by Smith and Zeder (2013). This notion has been supported by some in the archaeological community (e.g., Erlandson 2013), who noted the record of 'significant human ecosystem engineering' through domestication of plants and animals and equated that to the interval between 11,000 and 9,000 years ago. While there is a significant stratigraphic signal associated with this Neolithic agricultural revolution, the transition from a hunter-gatherer lifestyle to one of agriculture in the terrestrial record, including from the fossil record of cultivars, is markedly time transgressive, with multiple centres developing at different times from about 11,000 years ago (Ruddiman 2013). Indeed, the globalisation of agriculture did not develop until much later, for example with the Polynesian colonisation of Pacific Islands including New Zealand and Easter Island in the 13th century CE (Wilmshurst et al. 2011). Animal domestication, too, likely began in the latest Pleistocene in the case of dogs and occurred over time in multiple Eurasian centres (Larson et al. 2012; Section 3.3.2.2 herein). The development of agriculture and livestock first appeared near the start of the Holocene. It included the development of different crop types (e.g., maize in the Americas, rice in East Asia, wheat in the Fertile Crescent) and animals (e.g., turkeys and llamas in the Americas, pigs and chickens in East Asia, sheep and goats from the Fertile Crescent). Those developments do not provide a globally correlatable terrestrial signal until the early modern period. Instead they provide a blurred and growing signal that extends from the megafauna extinction (beginning about 50,000 years ago; see Section 3.2.1) to the Pleistocene/Holocene boundary (11,700 years before the year 2000 CE) and then grades into the 'Early Anthropocene' (~8,000 and 5,000 years ago) of Ruddiman (2003). These signals extend diachronously across the terrestrial record, leaving little or no trace in the marine realm.

7.3.4 The 'Early Anthropocene'

The 'Early Anthropocene' hypothesis of Ruddiman et al. (2003, 2013) is founded on two observations. Firstly, Holocene atmospheric CO_2 levels preserved in Antarctic ice cores show a slight increase through the past 7,000 years, following a slight decline of the preceding few thousand years of the Early Holocene. Secondly, atmospheric methane increased through the past ~5,000 years, following (like CO_2) a decline through the Early Holocene (Section 5.2). These two patterns contrast with the trend in previous interglacials, where a steady reduction of these greenhouse gases ushered in the next glacial phase (see Sections 5.2 and 6.1 and Figures 5.2.2 and 5.2.3). Ruddiman (2003, 2013) has argued that the trends in CO_2 and CH_4 are related to the spread of agriculture, the rise in atmospheric CO_2 resulting from protracted deforestation, while the rise in CH_4 was linked to cultivation of rice in East Asia and the spread of livestock farming through Africa and Asia. In Ruddiman's (2003) model, the anomalous rise in atmospheric CO_2 during the Holocene reached ~80% of the natural emissions by 2,000 years ago, suggesting a slow but persistent accumulation, accompanied by widespread deforestation by the Greek/Roman classical period. His model relates several minor dips in atmospheric CO_2 during the past 2,000 years to epidemics that reduced human population, with consequent regrowth of natural vegetation absorbing atmospheric CO_2. While the growth of methane does appear related to the growth in rice cultivation, alternative causes may apply to the growth in the CO_2 signal. These include, for example, on the one hand a reduction in terrestrial biomass linked to regional climate changes in the African Sahel (Ruddiman 2014) and on the other hand development of a higher (more alkaline) ocean pH resulting from initial forest growth in high latitudes following ice retreat. This forest growth formed a major sink for carbon, reaching its peak capacity about 8,000 years ago, after which ocean pH was reduced (less alkaline) as more carbon accumulated in the ocean/atmosphere than in forest biomass (Broecker 2006).

A key strength of Ruddiman's 'Early Anthropocene' hypothesis is that its potential start date is recorded in the slow rise in atmospheric CO_2 detected in ice cores (Section 5.2.1). It would be acceptable to use ice-core data for the basal boundary of the Anthropocene, since that is what was used for defining the base of the Holocene (Walker et al. 2009; Section 1.3.1.6 herein) and the Middle Holocene (Walker et al. 2012). However, the inflexion in the CO_2 curve is so gradual that it would be difficult to pick a precise point for a GSSP. Opinion is divided as to whether the human population was large enough to produce the deforestation required by the Ruddiman hypothesis. In any case, a Holocene/Anthropocene boundary within the Middle Holocene would not rest comfortably with the tripartite subdivision of the Holocene (Walker et al. 2012), with boundaries placed at sharply defined but small climatic events at ~8.2 and ~4.2 ka.

7.3.5 Early Metal Mining and Smelting Signals

Mining and smelting for metals started in Eurasia more than 7,000 years ago (Radivojević et al. 2010), although these left few and only local geochemical anomalies in geological archives. Copper and lead are elements that can be used to discern early anthropogenic contamination of the environment (see Section 5.6.3). Copper was one of the first metals used and thus has a longer and older record than other metals in many of the archives (Hong et al. 1996), locally dating back into the Middle Holocene (Pompeani et al. 2015). However, Cu is usually preserved only as regional or local records, its depositional pathway being complex and its signal being modified, especially in peat bogs (e.g., Bobrov et al. 2011).

The first recognisable anthropogenic pollution in data from Arctic ice cores takes the form of elevated heavy-metal concentrations. Lead anomalies and later lead isotope fluctuations were amongst the first metal signals to be connected to early mining activities (Hong et al. 1994). Mining and smelting were and are one of the main anthropogenic processes enriching heavy metals in geological archives (Gałuszka et al. 2014), with transport into remote areas taking place in dust and aerosol particles, at least in the Northern Hemisphere. Care must be taken to distinguish anthropogenic peaks in the concentrations of lead and other metals from natural fluctuations in metal concentrations (Gałuszka et al. 2014). Smelting of sulphide ores by ancient civilisations created atmospheric contamination throughout the Northern Hemisphere and has been described as defining the 'dawn of the Anthropocene' (Krachler et al. 2009). Early anthropogenic trace-metal pollution has been found in several geological archives (Marx et al. 2016), including lakes (Renberg et al. 1994), ice cores (e.g., Hong et al. 1994; Zheng et al. 2007), ombrotrophic peat bogs (Shotyk et al. 1998), fluvial-estuarine sediments (Négrel et al. 2004) and coastal sediments (García-Alix et al. 2013). Two early pre-industrial peaks have been distinguished in Arctic ice cores from Devon Island, Canada (Hong et al. 1994; Zheng et al. 2007), and are closely comparable with signals seen in Spanish lakes (Garcia-Alix et al. 2013): (1) a ~3,000–2,600 years peak related to Greek-Phoenician mining (Krachler et al. 2009); (2) a more distinct ~2,000 years peak representing intensive Roman metal production, especially from Spanish mines (see Section 5.6.3.2 and Figure 5.6.3).

Using a lead peak to define an 'Early Anthropocene' could fulfil the requirements of stratigraphic markers (Remane et al. 1996; Wagreich & Draganits 2018) for a GSSP (see Section 7.8.1), where lead pollution was distributed supra-regionally, at least in the Northern Hemisphere. A GSSP picked on that basis may be practical, and it would mean that an Anthropocene so defined would include a substantial material record. However, a means would have to be found of integrating such a boundary into the subdivision of the Holocene (Walker et al. 2012), as noted above, and it is unlikely that this level would prove to be as clearly, widely and precisely correlatable as is the mid-20th-century boundary described in this book. Nor would it represent profound Earth System changes, as do later phenomena such as the major perturbations of the carbon, nitrogen and phosphorus cycles.

7.3.6 A Soil-Based Anthropocene?

Various types of soils and archaeological deposits, such as shell middens marked by the accumulation of mollusc shells, may be used to provide anthropogenic stratigraphic signatures (Erlandson 2013; see Section 2.5.2.3 herein). Certini and Scalenghe (2011) have proposed a start date for the Anthropocene as defining the last ~2,000 years, based on the stratigraphic record of anthropogenically modified soils of Roman times. Stratigraphic signals of human activity in soils are both physical and chemical, including ploughed soil, plough marks, human artefacts such as pottery, the addition of organic material to make *terra preta de indio*, and phosphorus from fertilisers (see Section 2.7.2). Certini and Scalenghe (2011) argued that the development of civilisation by the 'Christian era' resulted in a global signal of anthropogenic soils.

There are four fundamental problems with the soil record as a marker for the start of the Anthropocene. Firstly, the stratigraphic record of anthropogenic soils unfolds over several millennia, beginning with the inception of cultivation some 8,000 years earlier than the suggested start date of ~2,000 years ago. Thus, the stratigraphic signal of human-modified soils, like that of agriculture itself, is time transgressive. It cannot be used to define a GSSP. Secondly, many of the stratigraphic characteristics of soils, such as their artefact content, chemistry and bioturbation, are specific to particular regions. Thus, *terra preta de indio* is typical for the Amazon Basin of South America and is associated with an artefact record from that region. Although there are similarities of process with the development of *terra preta australis* in Australia, the artefact records of these sites are different, and they cannot be correlated to provide the isochronous surface demanded by a GSSP. Thirdly, soil units develop by progressive alteration of the substrate, and the boundary between the soil, subsoil and substrate is typically complex and gradational, with the traces of many generations of roots and burrows extending down into the substrate and with pockets of unaltered substrate within soil and subsoil (see Section 2.7.4). This sedimentary architecture is inherently incapable of providing the kind of high-precision stratigraphic resolution (to annual level) that will be needed for a potential Anthropocene GSSP. Last but not least, ploughed soil is for obvious reasons internally rather chaotic and shows an erosive base – neither property recommends its use to define a GSSP.

The growth of human-modified ground is one of the major stratigraphic developments to have taken place in the Holocene and forms the background to this discussion. Cultivation soils, dark earths and shell middens are just some of the types of anthropogenic deposits that contribute to the overall build-up: also relevant here are urban occupation debris, reclaimed land, quarried materials transported and redeposited elsewhere and so on (see Section 2.5). The processes of coalescence of such deposits to form a global-scale stratigraphic entity, still forming and growing at accelerating rates today, are described in Edgeworth et al. (2015). Together, such deposits effectively form a new layer covering large parts of the terrestrial ice-free surfaces of the Earth.

Like all the evidence reviewed in this section, most accumulations of human-modified ground are pervasively time transgressive, containing no isochronous surfaces of global extent. Nevertheless, these substantial deposits do contain a wealth of artefactual, ecofactual, chemical and other evidence (e.g., first appearances of specific types of artefacts or novel materials, or chemical signals such as radioactive fallout in recently formed strata), which allow dating of and temporal correlation within this complex physical stratigraphic record and direct or indirect correlation with other successions, such as those in lake basins and on ice sheets. This stratigraphic index of human-environment interaction, time transgressive though it may be, is likely to prove indispensable in the identification of a GSSP marking the start of the Anthropocene.

7.3.7 Early Globalisation: The Colonisation of the Americas by Europeans during Early Modern Times

The colonisation of the Americas and the resulting early globalisation (Columbian Exchange) provides a major phase or step in the human impact on and anthropogenic changes to the Earth System. Several suggestions for start dates of the Anthropocene are connected to this pre-industrial phase, related either to geological archives (Lewis & Maslin 2015) or to non-geological evidence derived from the social sciences, using socio-economic models (e.g., at 1500 CE; Fischer-Kowalski et al. 2014).

Lewis and Maslin (2015) used a ~10 ppm dip in atmospheric CO_2 recognised in ice-core records to date the inception of the Anthropocene at 1610 CE (see Figure 5.2.2b and Section 5.2.1). They linked this dip to the colonisation of the Americas by Europeans and the ensuing epidemic spread of disease, which caused the decline of indigenous populations and thus led to forest regrowth, which lowered atmospheric CO_2 levels. Despite the intrinsic attraction of this idea, this CO_2 signal is not outside the range of Holocene variability (for example, see Ruddiman 2013) and may not be linked to New World colonisation (Rubino et al. 2016). It is also not recognisable in other archives. In addition, associated signals (e.g., the spread of transported species such as maize discussed in Section 3.3.2.1) are also time transgressive over centuries (Zalasiewicz et al. 2015a). This suggestion, like those for earlier start dates noted above, is thus not optimal as regards chronostratigraphic definition.

7.4 The Industrial Revolution and the Anthropocene

John McNeill

7.4.1 Introduction to the Industrial Revolution

The Industrial Revolution was one of the two most revolutionary transitions in world history. It began in the 18th century in Britain and in some sense is still underway in places such as Vietnam and Mali, although scholars thinking in British terms often give it an end date of 1870 or 1914.

The term 'Industrial Revolution' was first used by French and German observers in the 1830s to describe what they saw underway in Britain: the rise of a new form of manufacturing based not on hand tools but upon machinery driven by water or steam power. The term entered the English language in the 1880s to refer to the British experience. Here, however, we use it in a broad and global sense to refer to the industrialisation of many parts of the world, beginning with Britain but continuing until about 1900. For what began in Britain in the late 18th century soon had echoes on the European continent, especially in Belgium and Germany and in eastern North America. Before the end of the 19th century, industrialisations had occurred in Japan and Russia as well. We consider all these as covered by the term Industrial Revolution.

Taking 1900 as an end point for the Industrial Revolution is arbitrary. It corresponds roughly with the advent of petroleum as a key fuel, with the rise of electricity use, and with a handful of other technical advances that signalled a gradual evolution away from coal and iron to newer bases of industrial society. But processes of industrialisation are still in train here and there, if different in character from that which began in Britain. So the choice is arbitrary, not much better or worse than any other possible end date.

In broad terms it is fair to say that the Industrial Revolution was the most momentous change in human history since the Neolithic agricultural revolution a hundred centuries before. The Neolithic agricultural revolution, which consisted of domestication of plants and animals and the advent of farming, opened the way to dense settled populations, cities, hierarchical societies and indeed civilisation (see Section 7.3.3). The Industrial Revolution opened the way to high-energy societies in which economic growth generally has outpaced population growth, enriching at least a segment of almost every society around the world – and the entirety of some fortunate societies that have banished hunger and extreme poverty. The Industrial Revolution also opened the way to new degrees of inequality, as some people could now amass wealth on a scale unavailable even to pharaohs and emperors in previous times. Many of us live today more comfortably than medieval kings.

The Industrial Revolution gave rise to more efficient transport and communications, which tightened existing societal linkages wherever railroads, steamships, telegraph lines and later technologies took root. It changed the nature of many of those older linkages, especially in terms of economic specialisations. More than ever before, some regions specialised in raw-material production – cotton, iron ore, leather – while others concentrated on making these raw materials into shirts, ships and shoes. Thus the Industrial Revolution paved the way for a more globalised world economy, with higher degrees of specialisation, exchange and consumption. That brought greater economic disturbance to the biosphere and the lithosphere, in the form of demand for fibres, ores, lubricants and fuels – and all the other raw materials that went into (and still go into) industrial production.

What made the Industrial Revolution truly revolutionary was steam power, vastly expanding the amount of work, production and consumption a society could achieve. What made steam power work was the energy locked up in coal.

7.4.2 Energy and the Industrial Revolution

At its heart the Industrial Revolution was an energy revolution (Wrigley 2010). Like the Neolithic agricultural revolution that began in Syria some 11,000 years ago, the Industrial Revolution ratcheted up the quantities of energy available for human use. Fossil fuels made it possible for the first time ever for more than a tiny fraction of people to break out of poverty.

At first, the energy in question came from flowing water. In Britain, the textile industry until 1830 scaled up to factory production with waterwheels. Soon, however, the best riverside locations for siting mills (mostly in the north of England) were taken. So by the 1830s, the next generation of textile mills typically used steam power.

People had used coal and fossil fuels before, of course, at least in a few places. In North China during the 11th and 12th centuries, for example, coal had fuelled a metallurgical industry. In the Dutch lands from the 14th century onwards, people used peat (semi-fossilised remains of bog mosses) to fuel brewing, salt making, sugar refining and other energy-intensive pursuits. From the 16th century, Londoners used coal for home heating and for a handful of energy-guzzling industries such as beer brewing or glass making. So the use of fossil fuels for heat energy was not genuinely novel. Queen Elizabeth I complained about the 'smoake' from burning sea coal in London's fireplaces and industries.

What was new with the Industrial Revolution was converting the chemical energy of fossil fuels into mechanical energy. This involved steam engines, which used coal's energy to heat water into steam. The expansion of hot steam and its contraction as it cooled served as the motor force behind pistons, shafts and belts that carried power to looms, pumps or wheels. The advantage of steam over water power was that production was no longer tied to fast-moving rivers like those of Britain's Pennines; it could in theory take place anywhere, although initially much new production was located in or near coal-mining regions.

Coal worked its transformations only because of the steam engine. People in France (and by some accounts China as well) experimented with rudimentary steam engines. But in Britain, because it had a coal industry already, steam engines proved more useful. The first steam engines built in France and Britain were so fuel hungry the only place they could be used was at the mines themselves, where scraps of coal lay around unused. Only Britain had enough coal mines to make it worthwhile to improve steam engines. Between 1712, when Thomas Newcomen revealed an early model, and 1812, steam engines mainly worked to pump water out of mine shafts.

In the 18th century, dozens of engineers made small improvements to steam engines so that they became more than 10 times more fuel efficient (1712–1812). The most famous was the Scot James Watt (1736–1819), born into an educated family. He tinkered with steam engines while employed to maintain astronomical/mathematical instruments at the University of Glasgow, and his first steam engines went into commercial operation in 1776. Watt had no head for business – he said he'd rather face a loaded cannon than close an account – but he found partners who did. Together they designed, built and marketed much more efficient steam engines, starting in the 1780s. They also sued everyone else who built good steam engines until their key patent ran out in 1800.

The first factory to make use of a steam engine was a cotton mill, in 1795. Water power still remained more efficient for most factories until the 1830s, when steam engines' fuel requirements shrank and transport links grew to the point where it was practical to use steam engines far from coal pits. By 1870, water power was effectively extinct in British manufacturing, and steam engines reigned supreme. Coal use in Britain skyrocketed in the 19th century, as Table 7.4.1 indicates.

Coal-fired industrialisation soon appeared elsewhere in Western Europe and in North America, Russia and Japan. By 1870, the human population used more coal energy annually than all the green plants on the face of the Earth could capture from the Sun through photosynthesis. By 1914 the global appetite for coal had tripled consumption from its 1870 level (see Figure 7.4.1).

Table 7.4.1: **Coal production in Britain, 1700–1913 (in million tons)**

Year (CE)	Coal production
1700	3
1750	5
1830	30
1870	125
1913	287

Note: From Griffin (2010, p. 109).

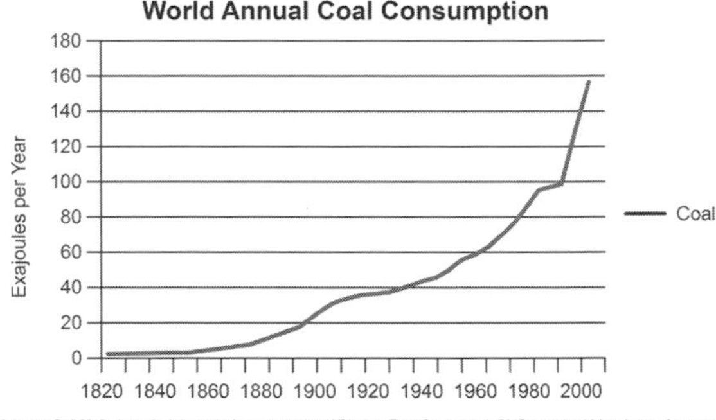

Figure 7.4.1 The growth of global coal use, based on Smil (2010). An exajoule is a unit of energy equal to 10^{18} joules. The US economy in 2015 used about 94 exajoules of energy.

The environmental implications of such reliance on coal were mainly of three kinds. First and most conspicuous was local pollution. Coal combustion and related industrial activity led to enormous local contamination of air, water and soil. In 1837, the German chemist Justus von Liebig visited England. He described the country between Leeds and Manchester as 'one big smoking chimney' (quoted in Weightman 2007, p. 388). In 1900 the pH of rain in Manchester averaged 3.5 (somewhere between the acidity of tomato juice and wine). Water quality was no better in industrial zones. Rivers and canals everywhere served as industrial sewers. Mischievous boys amused themselves setting polluted canals on fire. The water of the River Calder made 'tolerably good ink' in 1864 according to a parliamentary commission (McNeill 2001, p. 131).

Second, and much less conspicuous, the Industrial Revolution ratcheted up the quantities of fibres, ores, lubricants and fuels used in manufacturing, provoking environmental changes. For example, broad areas of forest in the American South were converted to cotton lands (Rothman 2007). Some 10 million hectares of South American forest disappeared (1870–1910) to provide dyestuffs from the heartwood of the *quebracho colorado* tree (Gori 1965).

Third, and least conspicuous of all, industrialisation and specifically coal combustion released ever-larger quantities of CO_2 and other greenhouse gases into the atmosphere. Gradually, the mounting concentrations of CO_2 enhanced the heat-trapping capacity of the atmosphere (see Section 6.1.2) and began, at first imperceptibly, to warm the Earth, as Arrhenius anticipated in 1896. It is this connection between the Industrial Revolution and the carbon cycle that inspired some observers to consider the dawn of industrialisation as the dawn of the Anthropocene.

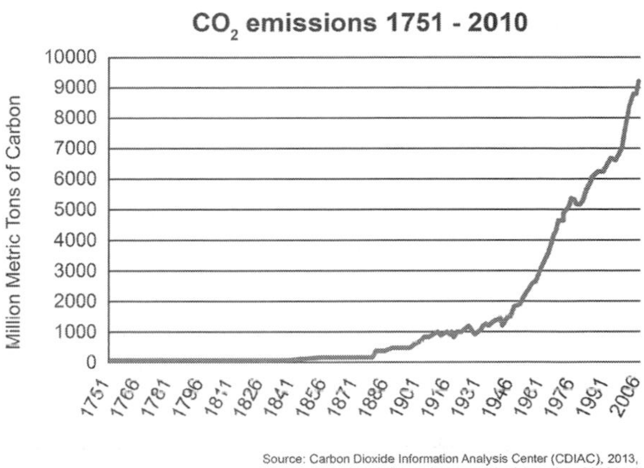

Figure 7.4.2 Growth in global carbon dioxide emissions.

7.4.3 The Anthropocene and the Industrial Revolution

When Paul Crutzen and Eugene Stoermer began to shape the definition of the Anthropocene in 2000, they considered the Industrial Revolution as the Anthropocene's starting point (Crutzen & Stoermer 2000). This was also the conclusion of initial stratigraphic study (Zalasiewicz et al. 2008). On occasion, Crutzen even offered the specific date of 1784, the date of one of Watt's most important patents (Crutzen & Stoermer 2000). That argument has its merits, but it requires a focus on specific causes rather than on stratigraphic effects in defining the Anthropocene. The reason for that is that industrialisation, even as it spread from Britain to continental Europe and eastern North America, had minimal impacts on global atmospheric chemistry or any biogeochemical cycles until after 1870. CO_2 emissions remained too modest for too long, as Figure 7.4.2 shows.

It took several decades of industrialisation (and simultaneous massive forest clearance in North America and Russia) to boost the CO_2 concentration in the atmosphere appreciably above its long-term Holocene average of ~280 ppm. Nearly a century after Watt's key patent, the level reached 290 ppm in 1879. It took another three decades to reach 300 ppm in 1910. Until the 1870s, the net increase in CO_2 never exceeded 1 ppm per decade. Not until the 1970s would it increase by more than 10 ppm per decade (NASA 2007). Thus, the Industrial Revolution's impact on the global carbon cycle remained negligible for several decades after the deployment of steam engines.

The debates, both within the Anthropocene Working Group and at large, about the best start date for the Anthropocene turn in part on a philosophical question: Should one invest greater significance in the beginning of a process or the beginning of its discernible impact? The process that led to elevated CO_2 (and other greenhouse gases) began with the Industrial Revolution in the late 18th century. But the discernible impact on atmospheric chemistry and biogeochemical cycles began much later, in the 1870s at the earliest and in a pronounced and clearly globally correlatable way only in the mid-20th century.

Whether or not one prefers to see the Industrial Revolution as the origin of the Anthropocene, the formal requirements of stratigraphy pose no insurmountable challenge. Potential levels exist for a GSSP that corresponds to the late 18th or early 19th century. One could choose, for example, the first appearances of spheroidal carbonaceous particles (SCPs), which are microscopic signatures of high-temperature coal (or oil) combustion (see Section 2.3.1). Rose (2015) and Swindles et al. (2015) independently found sharp upturns in the presence of SCPs starting about 1950 and judged that this finding supports an interpretation of the Anthropocene that begins about then. One might, though, consider the debut (or 'first appearance datum', to use the stratigraphic term) of SCPs rather than the moment when they proliferated, which would (to judge by Rose's data) give a result of about 1830. There would be some comparison here with the way that stratigraphers consider the temporal distribution of fossils, where first appearance carries more weight than any moment of proliferation, because first appearance is usually inferred to be a better proxy for a synchronous level than a proliferation event (see Section 3.1); here, though, the empirical evidence suggests that the upturn in proliferation is more nearly globally synchronous than is the first appearance of SCPs.

If the Anthropocene endures as a geological unit, most efforts at distinction between the Industrial Revolution and the Great Acceleration as suitable start points may eventually lose their meaning. Few people know or care whether the Cenozoic began 66 million years ago or 65.99 ± 0.12 million years ago (Vandenberghe et al. 2012), a difference roughly equivalent to that between 1784 and 1950 (though see stratigraphic discussion of this boundary in Section 1.3 herein, where inferred global diachroneity of the only hours or days of the spread of iridium-rich bolide debris was taken into account in the definition of this stratigraphic boundary). In discussions where the conventions of stratigraphy are unnecessary, one can simply consider the Anthropocene as having stages in its development rather than any precise moment of inception (Steffen et al. 2007). So while a legitimate argument may be made in favour of an Anthropocene that begins with the Industrial Revolution at the end of the 18th century, there are others in favour of seeing that moment as, so to speak, a precursor to the Anthropocene. In that view, the Anthropocene begins in the mid-20th century. Section 7.5 makes this position clear.

7.5 Mid-20th-Century 'Great Acceleration'

Will Steffen

The Industrial Revolution ushered in a new interval of human development. However, this development pathway was not smooth, beginning in the late 1700s and continuing smoothly up to the present, but rather experienced a slow build-up until the mid-20th century, when a number of factors led to a sharp increase in the rate of development in virtually all spheres of an increasingly globalised human society.

The quantification of this abrupt, rapid increase in human activity – now often referred to as the 'Great Acceleration' – and its impacts on the structure and functioning of the Earth System was first carried out during 2000–2003 as part of the synthesis of the first decade of research under the auspices of the International Geosphere-Biosphere Programme, IGBP (Steffen et al. 2004). The term 'Great Acceleration' was coined during a Dahlem workshop in 2005 (Hibbard et al. 2006) and first appeared in a journal article in 2007 (Steffen et al. 2007). It has since become widely used as a shorthand term for the sharp increase in the magnitude and rate of change of human activities and their impacts around the mid-20th century (McNeill 2001; McNeill & Engelke 2016).

The two Great Acceleration figures, one for human societies and the other for the Earth System, were first published in Steffen et al. (2004) and were updated in 2015 (Figures 7.5.1 and 7.5.2 herein; Steffen et al. 2015). Each of the figures consists of 12 individual graphs using indicators designed to capture the fundamental features of complex human societies and how they develop, as well as the major features and processes of the Earth System. The time interval for all of the graphs was chosen to begin at 1750 to capture the start of the Industrial Revolution. This choice arose directly from the original proposal by Paul Crutzen in 2000 for the Anthropocene as a new geological epoch in Earth history with a start date at the beginning of the Industrial Revolution (Crutzen & Stoermer 2000; Crutzen 2002).

The Great Acceleration stands out clearly in Figure 7.5.1. The socio-economic indicators capture many important features of contemporary society – population, economic activity, energy consumption, resource use, transport and communication. All 12 of the indicators show a sharp increase in their rate of change around 1950; the two indicators most closely associated with globalisation – foreign direct investment and telecommunications – increase their rate of change later in the 20th century.

Only 1 of the 12 indicators, the construction of large dams, shows any sign of levelling off, largely because nearly all of the world's large rivers have now been dammed and do not flow unimpeded to the ocean. Water use, however, has not levelled off, reflecting the increasing use of groundwater resources.

The socio-economic indicators are of considerable interest in their own right, and together they show the explosion of human activity and connectivity since the mid-20th century. From the perspective of the Anthropocene, however, the critical question is whether or not the Great Acceleration of human population and activity has significantly altered the structure and functioning of the Earth System. Addressing this question was central to the work of the IGBP, and so the 12 graphs of Figure 7.5.1 were accompanied by a companion set of 12 graphs in Figure 7.5.2, showing changes in the structure and functioning of the Earth System over the 1750–2010 period. The 12 Earth System graphs are presented as two groups of six, the top six representing the geosphere and the bottom six the biosphere. At its most fundamental level, the Earth System can be conceptualised as the co-evolution of these two components over about four billion years (Lenton 2016).

As for the socio-economic graphs, nearly all of the Earth System graphs show an increase in the rate of change of the indicator around the mid-20th century. The only exception is domesticated land – that is, land that has been significantly modified for human purposes and no longer resembles the original ecosystem(s) in structure or functioning. The rate of increase of domestication of land slowed somewhat after the mid-20th century, most likely reflecting the switch from extensification of agriculture (clearing more land) to intensification (increasing the productivity of already cleared land) as the Green Revolution spread to many agricultural areas. This intensification is reflected by the very rapid increase in the flow of nitrogen through the coastal zone after 1950, the result of a sharp increase in the use of fertilisers as a key component of the Green Revolution (see Section 5.4).

A similar phenomenon can also be seen in the marine biosphere. Marine fish capture exploded from the mid-20th century onwards with the mechanisation of the fishing industry (see Section 2.8.5.2), until catches plateaued around 1990. This plateau does not represent a shift to a sustainable wild fishery industry but rather the exhaustion of most of the commercial fish stocks around the world. Seafood consumption continues to increase, however, with the phenomenal post-1990 rise in aquaculture, represented here by shrimp aquaculture, which itself brings ecological problems through the destruction of coastal mangrove forests to make way for shrimp farms.

Socio-economic trends

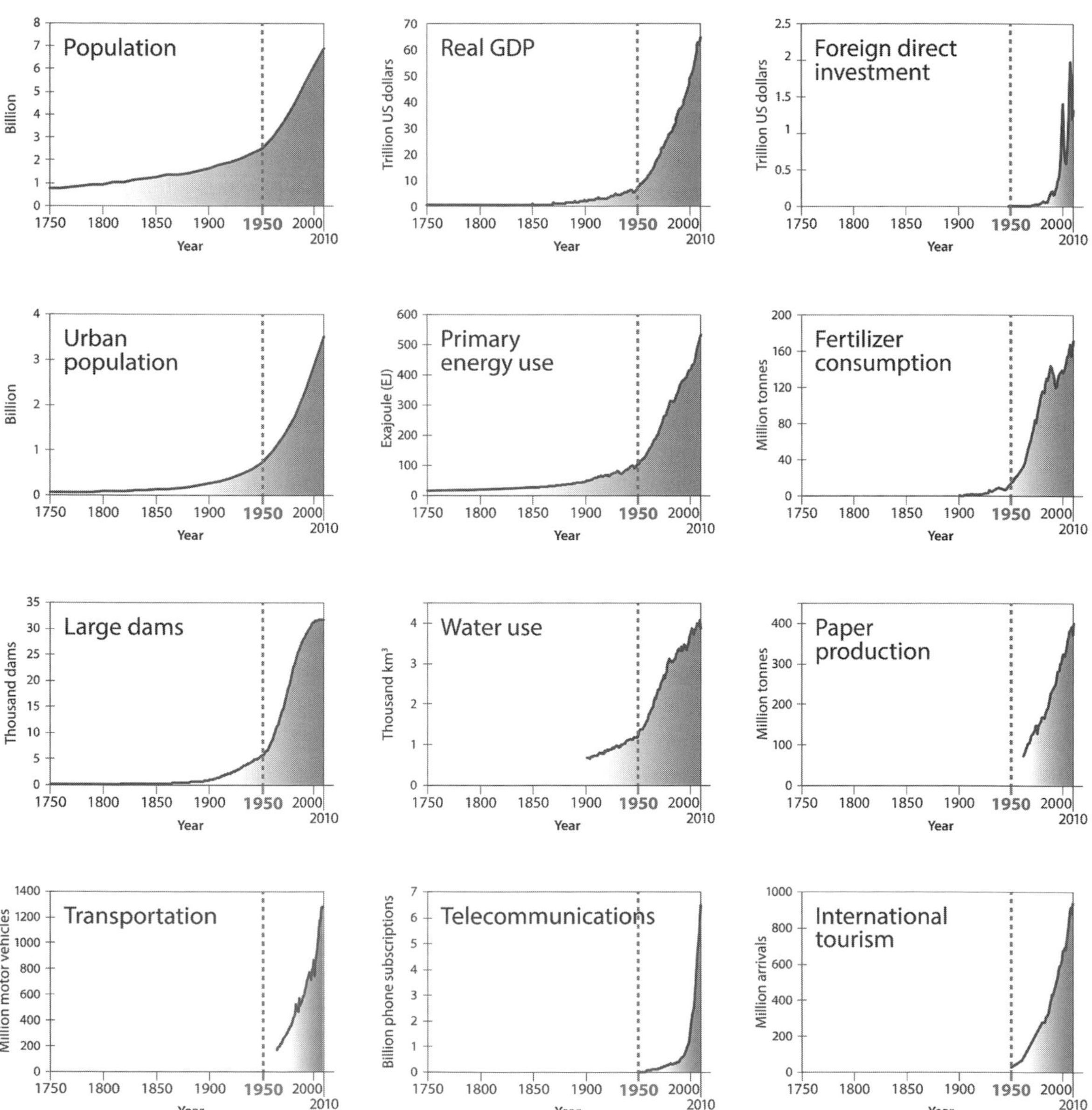

Figure 7.5.1 Trends from 1750 to 2010 in globally aggregated indicators for socio-economic development. From Steffen et al. (2015), which includes references for the individual data sources.

Earth System trends

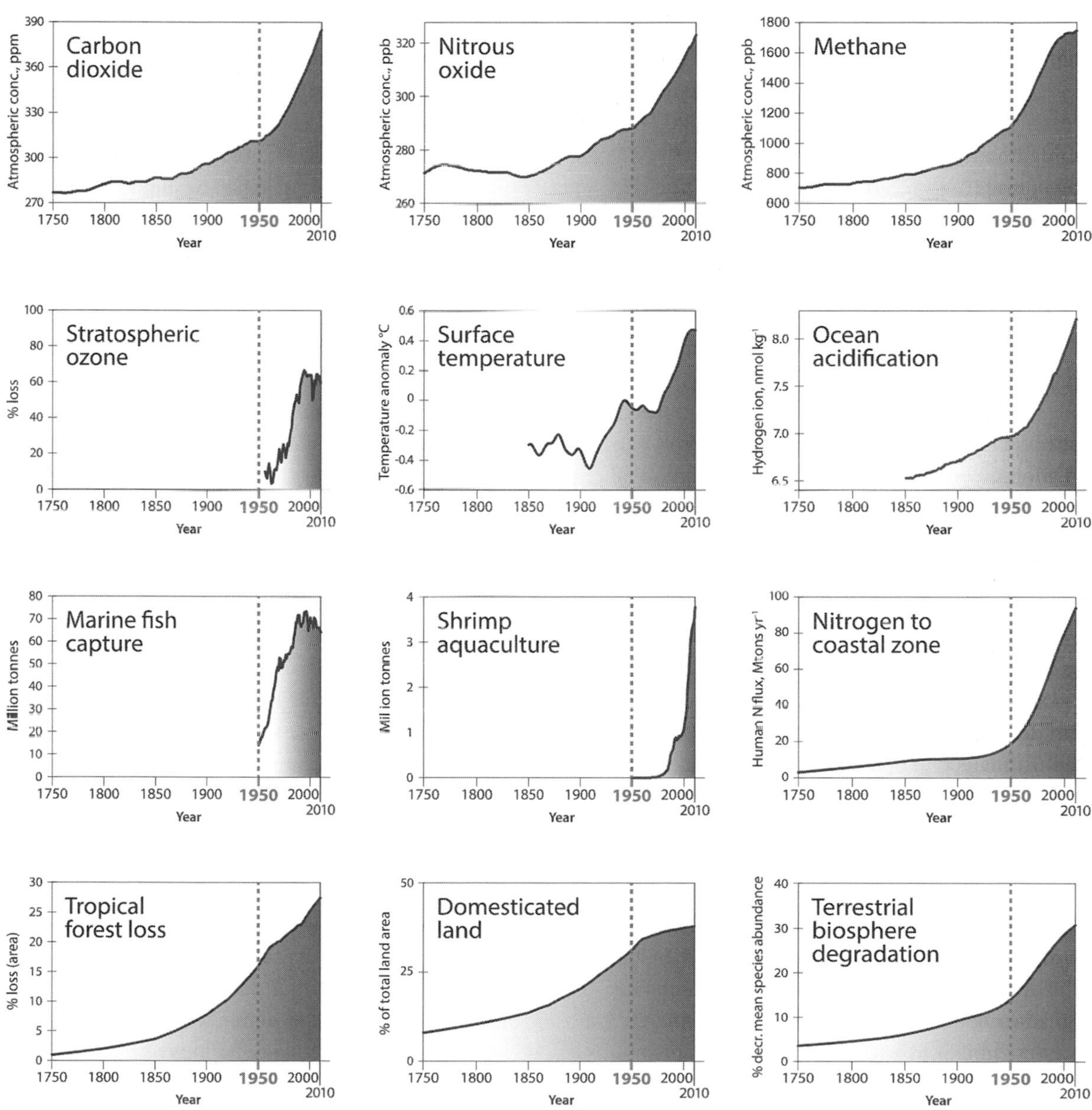

Figure 7.5.2 Trends from 1750 to 2010 in indicators for the structure and functioning of the Earth System. From Steffen et al. (2015), which includes more detail on each of the indicators and on the treatment and source of the individual data used to generate the figures.

The levelling off of stratospheric-ozone depletion from about 1990 does, however, represent a success story. This has resulted from the widespread adoption of the Montreal Protocol on ozone-depleting substances, which came into force in 1989. Apart from stratospheric ozone, the other indicators for the geosphere – atmospheric concentrations of greenhouse gases, the physical climate system and ocean acidity – all show increased rates of change from the mid-20th century onwards. There are other success stories – for instance, the clearing up of acid rain by the requirement of the Convention on Long-Range Transboundary Air Pollution (1979) that sulphur scrubbers be placed on the chimneys of coal-fired power stations to remove the SO_2 gas that led in the atmosphere to the formation of sulphuric acid that was then transported across national boundaries by upper-level winds.

Two features of the Earth System trends stand out: Firstly, that in all cases, the indicators are already well outside Holocene norms. Even global average surface temperature, which in 2016 reached 1.2°C above the level of 1900 (NOAA 2017), seems recently to have surpassed the Mid-Holocene global maximum, estimated to be 0.5–1.0°C above the pre-industrial baseline (Marcott et al. 2013), and is virtually certain to rise further (Collins et al. 2013; see Section 6.1.2.3 herein). Secondly, that there is a wealth of scientific evidence that the primary drivers of these observed changes in the Earth System are of human origin and are not due to natural variability (e.g., Steffen et al. 2004; Williams et al. 2016; IPCC 2013).

All of the individual graphs in Figure 7.5.1 are global aggregates because, from an Earth System perspective, it made sense to treat all of humanity as a single entity in terms of its activities and their consequences for the Earth System. However, such a treatment masks very large inequalities amongst various countries, societies and groups within countries. This is an important point, as noted by Malm and Hornborg (2014), amongst others, who argued that 'humankind as a whole' was not responsible for the Anthropocene.

In response to these criticisms, Steffen et al. (2015) introduced more detail into the socio-economic trends of the updated Great Acceleration graphs. These are shown in Figure 7.5.3 as splits in the individual graphs into three groups of countries – (1) the wealthy OECD countries; (2) large countries with rapidly developing economies – the so-called BRICS countries (Brazil, Russia, India, China and South Africa) and (3) the rest of the world, that is, the poorest countries. The data allowed these splits to be implemented for 10 of the 12 individual socio-economic trends; the exceptions were primary energy use and international tourism.

In terms of the equity issue, the two most relevant trends are population and economic growth, as measured by the trend in global GDP. The latter is an excellent proxy for consumption, which is the proximate driver of many of the human impacts on the Earth System. The differences amongst the three groups of countries for these two trends are striking. Nearly all of the population growth from 1950 to 2010 occurred in the BRICS and poor countries. On the other hand, even with the rapid rise of the Chinese economy in the first decade of the 21st century, most of the world's economic activity and hence consumption still resided in the OECD countries. In 2010, the 18% of the world's population that lives in OECD countries accounted for 74% of global economic activity. Thus, the Malm/Hornborg hypothesis that industrial capitalists of the wealthy countries, not 'mankind as a whole', are largely responsible for the Anthropocene, as seen in the Great Acceleration patterns, is borne out by the data. In this book we are not concerned with 'responsibility' for these Earth System trends but with the net result of those trends in stratigraphic terms.

The relationship amongst the three groups of countries is somewhat different for other socio-economic trends, especially those directly related to resource use. For example, the sharp increase in resource use by the BRICS countries as they developed is clearly shown in the figures. For fertiliser consumption, water use and paper production, the

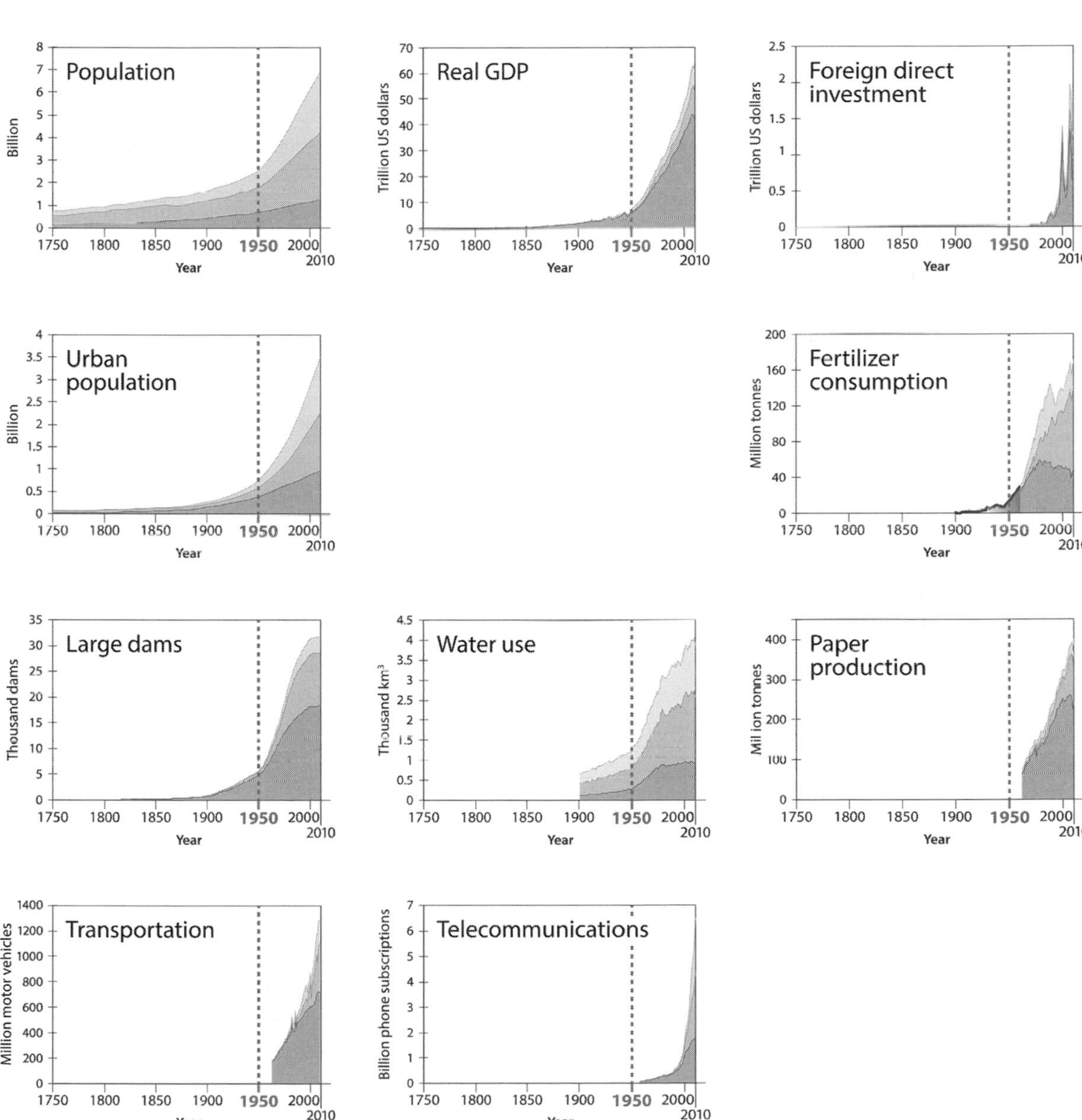

Figure 7.5.3 Trends from 1750 to 2010 for 10 of the socio-economic graphs of Figure 7.2.5 (excluding primary energy use and international tourism) with three splits for the OECD countries, the so-called BRICS (Brazil, Russia, India, China [including Macau, Hong Kong and Taiwan where applicable] and South Africa) countries and the rest of the world. From Steffen et al. (2015).

trends for the OECD countries levelled off over the last two or three decades or even declined somewhat, but the trends for the BRICS countries showed rapid increases. The shift from near-complete domination of fertiliser consumption by the OECD countries around 1960 to the domination by the BRICS countries a half century later is probably the result of both increased efficiency of use in the OECD world and the enormous increase in food production in the BRICS countries as the Green Revolution took off.

The Great Acceleration trends for telecommunications are an excellent example of technological 'leapfrogging'. The telecommunications curve was virtually non-existent before 1950, started to increase slowly thereafter and has been undergoing a striking acceleration since 1990. Even more striking, though, is the post-2000 split. While there has been some increase in mobile-phone subscriptions in the OECD countries, the bulk of the sharp increase over the last decade has occurred in the BRICS and poor countries. In nearly all of these countries, mobile technologies are the first telecommunication technologies in the country, completely bypassing the landline phase that typified the evolution of telecommunications in the OECD countries. Such smartphone technologies may even provide a technostratigraphic interval within the Anthropocene (Section 4.2). Such leapfrogging in other technologies – renewable energy, for example – holds great promise for avoiding some of the impacts on the Earth System driven by the development pathways of the OECD countries.

In summary, the Great Acceleration graphs capture the explosion of socio-economic activity since the mid-20th century as well as the many significant changes in the structure and functioning of the Earth System that this explosion has driven. Many of the indicators they show can be directly linked with stratigraphic proxy signals (Zalasiewicz et al. 2017a). Taken together, the socio-economic and Earth System Great Acceleration graphs provide a coherent, convincing body of evidence for the existence of the Earth System Anthropocene and for its beginning around the middle of the 20th century.

7.6 Current and Projected Trends

Will Steffen

Both complex human societies and the Earth System are in a period of very rapid change, with this trajectory, strikingly different from that of the Holocene, being one of the key characteristics of the Anthropocene. It is difficult, though, to project the trends of their commonly used indicators into the future. This is a particular challenge for human societies, where the pace of technological change, especially in information and communication technologies, is extraordinarily fast (Friedman 2016).

7.6.1 Socio-economic Trends

Trends in human population can be projected with some confidence for a few decades, given the age structure of the current population. From a current population of ~7.5 billion, the global population is projected to rise to 9.7 billion by the middle of the century (UN DESA 2015, median projection). Projections beyond that become less secure, as assumptions need to be made about future fertility rates, especially in Africa. However, it is likely that the strong trend towards higher urban populations will continue over the new few decades at least, with about 60% or more of the human population projected to live in urban centres by mid-century (UN DESA 2014). In effect, population growth will add to the current human population by 2050 as many people as existed on the planet in about 1950. Most of these people will live in cities, whereas before 1950 most people still lived on the land (www.theguardian.com/news/datablog/2009/aug/18/percentage-population-living-cities).

Energy use will continue to grow with human population, but it is difficult to predict the rate or the type of generation (fossil fuel, nuclear, renewables, etc.). Per capita energy use is very likely to increase strongly in the least developed countries as they become wealthier, but in the developed world and the emerging economies, there are signs of a relative decoupling between energy use and further economic growth. This trend is particularly important in China, where the energy intensity of the economy (the amount of energy used per unit economic growth) is decreasing, primarily due to the rapid phasing out of coal-fired electricity plants and the development of wind and solar as sources of electricity (IEA 2017).

Rather than trying to predict future trends in socio-economic indicators, many researchers and planners are using scenarios of various types. These provide plausible pathways for the future based on assumptions concerning technological developments, policy settings and value judgements by societies as the century progresses. Various sets of assumptions generate contrasting scenarios for socio-economic development into the future. Four well-known sets of scenarios of relevance to the future trajectory of the Earth System are briefly described below.

The International Energy Agency (IEA), an agency established by 29 member states (primarily OECD countries), provides authoritative research and analysis, including outlooks, to promote energy security amongst its member states. Its most recent outlook, published in 2017 (IEA 2017), provides a scenario of energy production and use and hence an upstream scenario for greenhouse-gas emissions (see Section 7.6.2) out to 2040. The scenario predicts a 30% increase in energy demand over that period, with an increase in consumption of all modern fuels. A massive $44 trillion in investment is required for the increase in energy supply needed to meet the demands of the growing population, with 60% of that investment going into the extraction and use of oil, gas and coal. Given that these fossil fuels comprised 70% of energy investment in the 2000–2015 period, the IEA scenario represents a shift away from fossil fuels, but not nearly enough to limit the rise in global average temperature to less than 2°C, which is driven by the emission of CO_2 from the burning of fossil fuels (Section 6.1.3).

Companies in the private sector also build scenarios for socio-economic development into the future. Perhaps the most well-known of these are the Shell scenarios, the most recent of which were published in 2013 (Shell International 2013). These scenarios go far beyond the energy challenge to explore possible societal pathways based on qualities such as leadership, connectivity and efficiency. Their analysis leads to two fundamentally different futures: (1) 'Room to Manoeuvre', which is a sustainability pathway based on visionary leadership, knowledge-based societies and integrated solutions to energy, transport, land use, etc. and (2) 'Trapped Transition', which is based on the primacy of market forces, failure to tackle 'wicked problems' (like global warming) and a mass of individual, ad hoc solutions. 'Trapped Transition' leads to worsening financial, social and political conditions and ultimately leads to either a forced reset or a collapse.

The Millennium Ecosystem Assessment (MA), a multi-year global project that involved 1,360 experts from 95 countries around the world, assessed the impacts of human activities on the biosphere based on an ecosystem-services framework (MA 2005). As part of the MA, four comprehensive scenarios were developed to explore alternative futures, particularly with respect to the future relationship between human societies and the rest of the biosphere. Both quantitative models and qualitative analysis were used in developing the scenarios, each of which is formulated around a coherent storyline for the future. The four scenarios – 'Global Orchestration', 'Order from Strength', 'TechnoGarden' and 'Adapting Mosaic' – represent future trajectories of the Earth System based on contrasting approaches to changes in ecosystem services (either reactive or proactive) and future trends in the organisation of human societies (more globalised or more regionalised societies).

The IPCC (Intergovernmental Panel on Climate Change) also develops a suite of socio-economic scenarios as input to its model-based projections of future climate change (Collins et al. 2013). Although these scenarios are designed to provide inputs into the physical climate models that simulate future changes in the climate system, they must make assumptions about the fundamental development of human societies and economies on a multi-decadal timescale. The most recent IPCC assessment uses 'Representative Concentration Pathways' (RCPs) that are projected trajectories of greenhouse-gas concentrations in the atmosphere. The key difference between these and the earlier emission scenarios is that the RCPs are 'internally consistent sets of time-dependent forcing projections that could potentially be realised with more than one underlying socioeconomic scenario' (Collins et al. 2013). Integrated Assessment Models are often used to link RCPs to plausible socio-economic scenarios that are based on internally consistent assumptions regarding population growth and economic and social development.

7.6.2 Greenhouse Gases

Emissions of the most important of the long-lived greenhouse gases, carbon dioxide (CO_2), have flatlined over the past three years, while the global economy has continued to grow (Le Quéré et al. 2016) – giving us perhaps the first signs of decoupling of emissions from economic growth. However, more time is required to determine whether this is a long-term trend. If so, it would be an important development, as emissions need to peak within the next few years and then fall if we are to have any chance of meeting the 1.5°C Paris target of December 2015 (Figueres et al. 2017). The initial data for 2017, however, indicate a renewed growth of about 2% in emissions of CO_2 from the combustions of fossil fuels (Le Quéré et al. 2018).

Regardless, atmospheric CO_2 concentration has continued to grow and will likely continue to do so. It crossed the 400 ppm (parts per million) level within the last two years and is increasing at 2–3 ppm per year (Le Quéré et al. 2016; Sections 5.2.1 and 6.1.2.3 herein). The rate of increase of atmospheric CO_2 concentration is now more than 100 times greater than the rate during the last deglaciation, when it rose from about 180 to 280 ppm over 7,000 years (Wolff 2011).

The atmospheric concentration of methane, a shorter-lived but more potent greenhouse gas than CO_2, has begun to rise more rapidly again after two decades of very slow rises. Since 2014, these growth rates are approaching those for the most greenhouse-gas-intensive scenarios. The cause for the recent rate increase is not clear, but it appears to be primarily from biological sources, most likely from agriculture, with smaller contributions from fossil-fuel use and wetlands (Saunois et al. 2016).

7.6.3 Climate System

As discussed in Section 6.1, global average surface temperature, commonly used as an indicator for the state of the climate system, continues to rise strongly and is now 1.1°C–1.2°C above a pre-industrial baseline (NOAA 2017; Figure 7.6.1 herein). The rate of the rise in temperature since 1970 is about 170 times the baseline rate of the last 7,000 years of the Holocene, and in the opposite direction (Marcott et al. 2013; NOAA 2017). Given inertia both in the climate system and in the human systems that are driving emissions of greenhouse gases, global average surface temperature will continue to rise for several decades at least (IPCC 2013). It is almost certain that the Paris 1.5°C target will be breached, and it will take rapid and very deep reductions in greenhouse-gas emissions over the next three decades to cap temperature rise at the 2.0°C Paris target (Rockström et al. 2017; Figueres et al. 2017).

As pointed out in Section 6.3.1, global sea level has risen about 20 cm since the beginning of the 20th century. The rate of sea-level rise has increased from 1.7 mm per year for the 1901–2010 period to 3.2 mm per year between 1993 and 2010. The rise in sea level is due to a combination of factors, including the thermal expansion of a warming ocean and additional water from melting continental glaciers and polar ice sheets. Global mean sea level near the end of the 21st century relative to 1986–2005 will likely be in the range 0.26–0.98 m, depending on the level of greenhouse-gas emissions through this

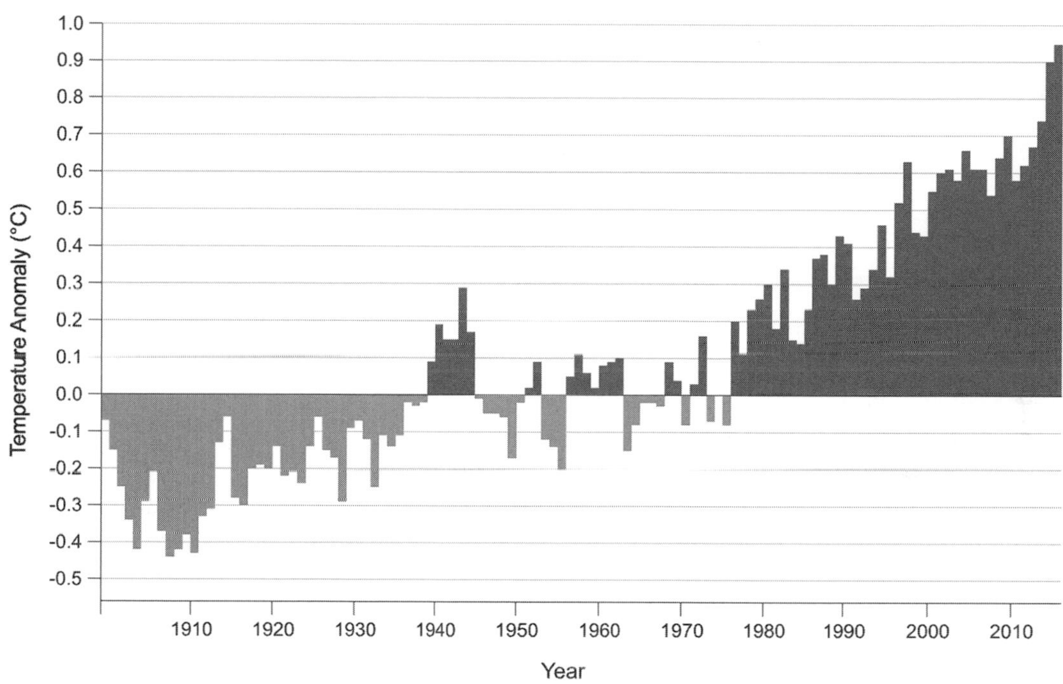

Figure 7.6.1 Trend in annual global temperature anomalies through 2016, relative to the global average temperature for the 20th century. From NOAA (2017).

century (IPCC 2013). Contributions from collapse of marine-based sectors of the Antarctic ice sheet, allowing faster seaward flow of land-based ice, could drive sea level towards the upper end of or somewhat beyond this range (e.g., DeConto & Pollard 2016; Rintoul et al. 2016; for more detail see Section 6.3 herein).

The oceans are absorbing about 25%–30% of the additional CO_2 emitted to the atmosphere from human activities, changing ocean chemistry and making the ocean slightly less alkaline (i.e., more acidic) (see Section 5.3.1). The pH has decreased by nearly 0.1 unit (about a 30% increase in acidity) since the mid-20th century (IPCC 2013). The current rate of increase in ocean acidity is estimated to be faster than at any other time over the past 300 million years (Hönisch et al. 2012).

Other climate-related indicators are also changing. There is evidence, with varying degrees of confidence, that since 1950 some extreme weather events are increasing in frequency and/or intensity under the influence of climate change (IPCC 2013). For example, it is very likely that warm spells and heatwaves have increased in frequency and/or duration over most land areas, and significant rises (depending on emissions scenarios) have been projected in the proportion of the Earth's population that will be intermittently subject to a lethal combination of heat and relative humidity over the next century (Mora et al. 2017). Increases in the intensity and/or duration of drought are likely in some regions. It is likely that, since 1970, there has been an increased incidence and/or magnitude of extreme high-sea-level events. Evidence is also emerging to suggest that there has been an increase in the incidence of floods in Europe, for example (Blöschl et al. 2017).

7.6.4 Changes in the Biosphere

Biosphere changes are somewhat more difficult to predict than changes in the climate system. Williams

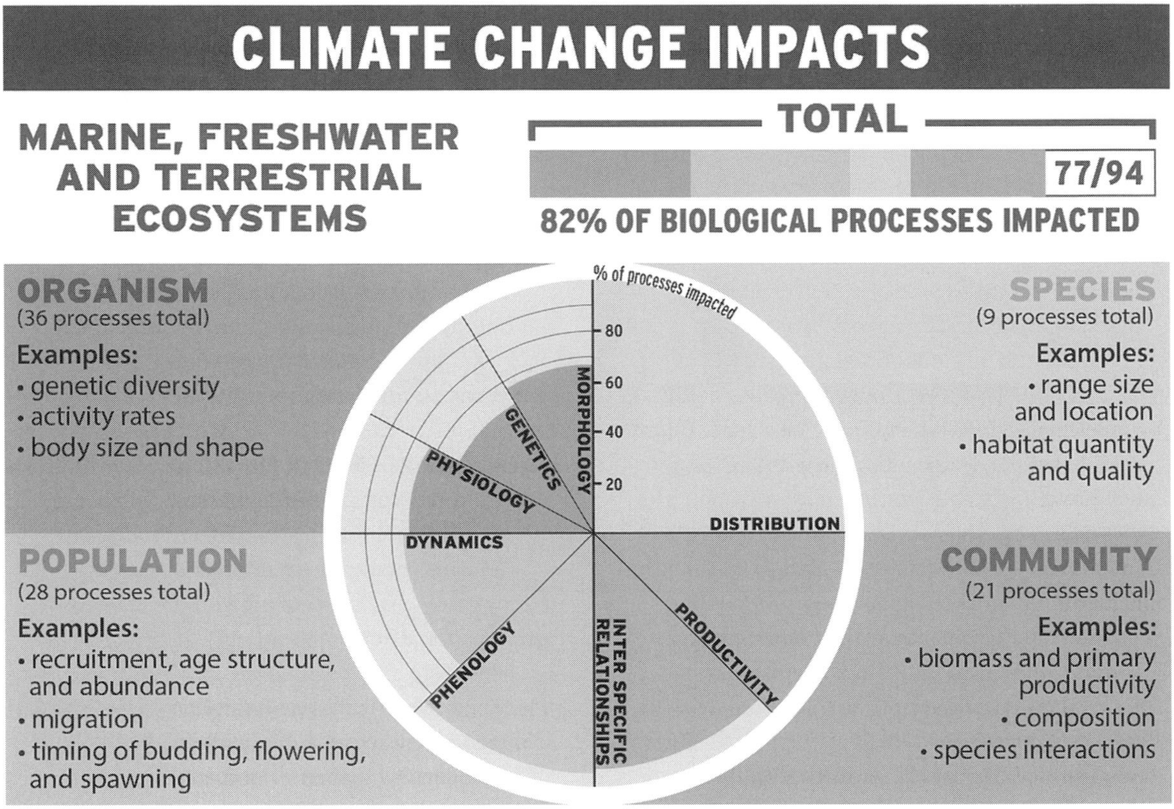

Figure 7.6.2 Climate-change impacts on ecological processes in marine, freshwater and terrestrial ecosystems (Scheffers et al. 2016) (A black-and-white version of this figure appears in some formats. For a colour version, please refer to the plate section.)

et al. (2015b) reviewed trends in the contemporary biosphere and suggested that current changes are significant enough to be considered as a third fundamental stage in the evolution of the biosphere. Contemporary trends include global homogenisation of flora and fauna, the appropriation of 25%–50% net primary biological production by *H. sapiens*, human-directed evolution of other species, and a growing interaction between the biosphere and the complex technological system created by humans.

While direct human impacts have long been the dominant factor in biospheric trends, climate change has recently begun to exert a strong influence on the global biosphere, with a very wide range of impacts, from shifts in species ranges and changes in phenology and population dynamics to more disruptive impacts on a wide range of scales from genes to biomes (Scheffers et al. 2016). The Scheffers review focused on core ecosystem processes and found that the 82% of the 94 processes considered showed evidence of impacts from climate change (Figure 7.6.2)

Strong trends are also evident in several of the Earth's large biomes, some driven primarily by climate change and others by the interaction of climate change with direct human pressures. A well-known example of the latter is the continuing deforestation of the Amazon rainforest by humans, but the potential for a tipping point related to both climate change and deforestation is well established, resulting in the conversion of the forest to a savannah or grassland. In particular, deforestation of the Amazon reduces rainfall from the evapotranspiration-precipitation cycle, leading to a decline in resilience of the remaining forest (Zemp et al. 2017). Model-based simulations suggest that tipping points exist for deforestation alone when deforestation exceeds 40% of the area or when the rise in global average temperature is beyond 3°C–4°C (Nobre & de Simone Borma 2009). Interaction between these two drivers could result in a tipping point at a lower temperature change coupled with a lower level of deforestation. The Amazon rainforest has not yet crossed such a tipping point. However, severe droughts coupled with extensive fires in 2005 and again in 2010, resulting in an emission of carbon to the atmosphere equivalent to a decade of carbon uptake, suggest that we may be nearing such a tipping point in the Amazon (Feldpausch et al. 2016).

Boreal forests can also be converted to savannah or grassland biomes through a combination of heat, fire and drought, although there does not appear to be a potential tipping point as for the Amazon forest but rather a more gradual process. Conversion of boreal forest to steppe grassland at its southern boundary is possible, with model simulations showing that a large area of boreal forest in southern-central Siberia and a smaller area southwest of Hudson Bay in Canada could be converted to savannah/woodland or grassland by 2100 in a moderate climate-change scenario (Joos et al. 2001). The likelihood of such model-simulated dieback in the future is supported by observations in Canada, where an increase in disturbance regimes associated with a warming climate (insect-induced tree-stand mortality and wildfires) from 1970 to 1989 drove a significant decrease in carbon storage (Kurz & Apps 1999).

Amazon and boreal forest dieback are part of a larger, global trend of climate-change-driven forest dieback from drought and extreme heat. All major types of forest biomes – tropical, temperate and boreal – as well as woodlands and savannahs have been affected (Anderegg et al. 2012; Figure 7.6.3 herein). The consequences of such dieback include not

Figure 7.6.3 Global distribution of studies documenting climate-induced widespread forest dieback events and consequences (filled circles). From figure 3 in Anderegg et al. (2012). (A black-and-white version of this figure appears in some formats. For a colour version, please refer to the plate section.)

Table 7.6.1: **Increase in global coral-bleaching events: 1960–2010**

Time period	Number of bleaching events
Pre-1980	12
1980s	236
1990s	1,874
Post-2000	5,094

Note: From Donner et al. (2017).

only emission of carbon to the atmosphere but impacts on a wide range of ecosystem functions and services. We note that this warming-driven trend in the plant world is in the opposite direction of the trend towards more plant growth induced by the enrichment of the atmosphere in CO_2 (the so-called greening of the planet).

Marine biomes are also being affected by climate change, none more so than coral reefs, which display a striking global trend of rapidly increasing numbers of mass-bleaching events over the last half century (Donner et al. 2017; Table 7.6.1 and see also Section 3.4 herein). Australia's Great Barrier Reef (GBR), the world's largest reef system, has been particularly hard hit, with massive bleaching in 2016 and 2017 leading to a 66% mortality rate on the GBR's pristine 1,000 km long northern section. The distinctive geographical footprints of mass bleaching on the GBR were determined by the patterns of high sea temperatures, leaving no doubt as to the primary cause of the bleaching (Hughes et al. 2017a, b). Such a mass bleaching at an increase in global average temperature of about 1.2°C above pre-industrial suggests that the tipping point for global mass coral bleaching, estimated at 1.5°C –2.0°C, may actually lie at the bottom end of that range (Schellnhuber et al. 2016).

An overarching trend in the biosphere as a whole is the increasing loss of biodiversity, estimated to be at least tens of times above background levels (Barnosky et al. 2011). Although actual species loss over the past 500 years has been only a few percent at most, much higher fractions (25% of mammals, 41% of amphibian species, etc.) are recognised as being threatened with extinction (IUCN 2014). The pattern of growing pressure on biodiversity is similar to that of the five past mass extinction events in Earth history (Barnosky et al. 2011), so although we are not yet in the sixth mass extinction event, continued direct pressure from human activities and indirect pressure from accelerating climate change could, over coming centuries, drive such a mass extinction event.

7.7 Hierarchy of the Anthropocene

Colin N. Waters, Jan Zalasiewicz and Martin J. Head

In Section 1.3 the two parallel means of classifying Earth history were described: geochronological classification representing time intervals within which certain events and processes took place; and chronostratigraphic classification of the material record (i.e., of strata) that preserves the evidence of that history. This duality of classification has a long history, having been introduced during the Second International Geological Congress in Bologna, 1881. Both schemes are hierarchical, and smaller-scale units are grouped together into large ones. Thus, the geochronological units are nested, with the largest – eons – divided in turn into eras, periods, subperiods, epochs, subepochs, ages and potentially subages. The precise material chronostratigraphic counterpart comprises eonothems, erathems, systems, subsystems, series, subseries, stages and potentially substages, with the stage being regarded as the fundamental chronostratigraphic unit for which the GSSP is formulated (see Section 7.8). The implication is that the higher the rank attributed to the Anthropocene, the greater the change that has occurred between it and the preceding stratigraphic unit (Gibbard & Walker 2014). When Paul Crutzen first used the term Anthropocene (see Section 1.2), it was at least in part a spur-of-the-moment improvisation, and hence the term was not constructed with the technicalities of the stratigraphic hierarchy in mind. However, the use of the '-cene' suffix has been used consistently during the Cenozoic to denote epoch/series rank, and Crutzen's case was based on his assessment that the Holocene was no longer an adequate descriptor of the state of the Earth System. Can this hierarchical level be justified?

The lengths of previously defined geological epochs/series are highly variable; they do not have a fixed duration. In the Cenozoic Era, there are epochs that range in duration from more than 20 million years for the Eocene to less than 12,000 years for the Holocene, and beginning with the Pliocene, epochs become successively shorter (see Section 1.3). However, there is criticism voiced by Finney and Edwards (2016) that 'with 1945 as the beginning, it would be a geologic time unit that presently has a duration of one average human life span', with the intimation that epochs/series should be longer than such a span. However, the Anthropocene lies along a trend that reflects the increasing resolution of the geological record as one approaches the present. Furthermore, Zalasiewicz et al. (2017d) suggested that the key issue is not how long epochs are but whether the geological record that allows characterisation and correlation of the Anthropocene is already sufficiently distinct, whether that record will persist for at least many millennia, and whether changes already in train will inevitably affect the future course of Earth history and therefore the pattern of the future stratigraphic record. All of those conditions appear to be fulfilled with the Anthropocene.

The observed signals of palaeoenvironmental changes associated with a possible base of the Anthropocene in the mid-20th century are described in detail by Waters et al. (2016). These include large-scale production of novel minerals and materials (including fuel ash, plastic and concrete); marked acceleration of rates of erosion and sedimentation, in part relating to human terraforming of the landscape (described fully in Chapter 2); accelerated biotic changes including increased rates of extinctions and extirpations and unprecedented levels of species invasions across the Earth (see Chapter 3); large-scale chemical perturbations to the cycles of carbon, nitrogen, phosphorus and other elements, including those associated with radiogenic fallout (see Chapter 5); and the inception of significant change to global climate and sea level (see Chapter 6). Many of these palaeoenvironmental signals are, in the geological context, essentially synchronous and globally distributed. Furthermore, many of the changes associated with these markers, such as atmospheric CO_2, CH_4 and N_2O levels, changes in carbon stable isotope ratios, and the presence of reactive nitrogen and phosphorus in soils, already exceed both Holocene and Quaternary natural variability; increased nitrate levels are at levels higher than for the Holocene and Late Pleistocene. Other changes, such as the temperature and sea-level rises, exceed Holocene but not Quaternary climate variability. Not only are these markers significantly greater than the changes proposed to subdivide the Holocene into subseries (Walker et al. 2012), they are also comparable to or greater in scale than equivalent signals associated with the base of the Holocene (Walker et al. 2009). Overall, it therefore seems both reasonable and conservative to consider the Anthropocene at a rank of epoch/series (Zalasiewicz et al. 2017d).

Head and Gibbard (2015, p. 24) noted that the Anthropocene defined at the rank of series/epoch would truncate the Holocene, which 'has included modern deposits since its original inception in the late nineteenth century (Gervais 1867–1869)'. This would cause the Holocene to lose its traditional and historical context as the geological time of the present and would require stratigraphers to discriminate Anthropocene from Holocene strata, which at times may be impractical. The Anthropocene defined at the rank of stage or even substage, these authors suggested, would overcome such difficulties.

However, when Gervais introduced the term Holocene in 1869, he was living at a time soon after the end of the Little Ice Age with environmental conditions occupying Holocene norms and could not have foreseen the unprecedented changes occurring from the mid-20th century. Units of the Geological Time Scale, too, are defined only by their base, the top of the Holocene having until now remained essentially undefined. Defining the Anthropocene at the rank of series/epoch would therefore not cause a change to the existing formal definition of the Holocene. And the question of systematic recognition of chronostratigraphic units affects all of the geological column: where this is problematic, combined terms such as Permo-Triassic and Plio-Pleistocene have been used widely (Zalasiewicz et al. 2017d).

One might consider, conversely, that the Anthropocene could be of period/system (or even era/erathem) rank, which would require that the Quaternary Period/System has terminated. If current trends of habitat loss and rates of species extinctions are maintained, the planet may face the sixth mass extinction event (with ~75% of species extinct) in the next few centuries (Barnosky et al. 2011; Ceballos et al. 2015). The previous mass extinction events form the basis for, even though they do not necessarily coincide with the GSSP definition of, the Silurian, Carboniferous, Triassic, Jurassic and Paleogene periods/systems; two of these also coincide with the start of the Mesozoic and Cenozoic eras/erathems (see Section 1.3). If such a comparable mass extinction event does take place over the next few centuries, then – *at that future time* – consideration of a hierarchical level above that of epoch may well become appropriate.

Similarly, there exists the possibility that the end of the succession of Northern Hemisphere glaciations may lie ahead as a result of human actions, and if so, this could potentially signal the end of the Quaternary Period/System (Wolff 2014).

As the final scale of the changes related to the transition from the Holocene to Anthropocene is unknown, some have argued to wait until the full effects are stabilised (Wolff 2014). But given that considerable, irreversible change has already taken place to the Earth System and to the stratigraphic record, and as the trajectory of the Earth System and stratigraphic change has been sharply modified, there is currently a strong case to make a proposal to establish the Anthropocene formally now as a series/epoch (Figure 7.7.1). No formalisation decision is irrevocable, as it has only to stay in place for a minimum of 10 years (Remane et al. 1996). For the current and next generations of Earth scientists, such an arrangement should be appropriate and effective for working use. It will be up to future generations to decide on any appropriate revisions to the rank of the Anthropocene, depending on the circumstances and conditions that will arise.

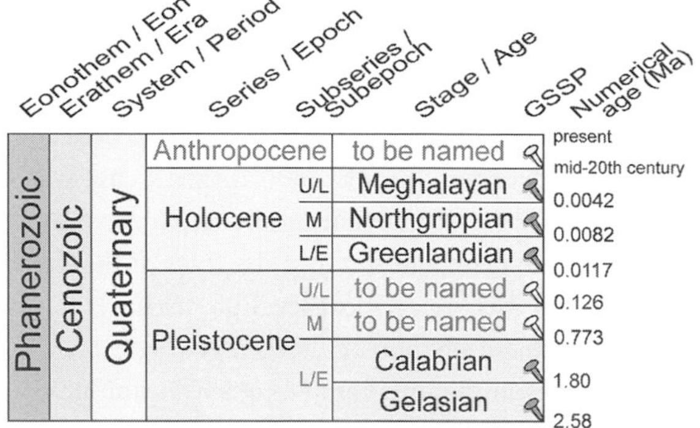

Figure 7.7.1 The Quaternary timescale as currently preferred by the Anthropocene Working Group, with the Anthropocene shown at the rank of series/epoch. Black type indicates names officially approved and ratified by the International Commission on Stratigraphy (ICS)/Executive Committee of the International Union of Geological Sciences (IUGS EC). Names in grey type have yet to be officially sanctioned by ICS/IUGS EC. The rank of subseries (Head et al. 2017) has recently been recognised in the International Chronostratigraphic Chart for the Quaternary. Note that if the Anthropocene is proposed as depicted here, then one or more stages/ages corresponding to the Anthropocene should also be proposed for approval. Dark grey spikes = approved/ratified GSSPs; pale grey spikes = GSSPs awaiting submission or approval. Updated from Zalasiewicz et al. (2017c). ©2017, with permission from Elsevier.

7.8 Potential GSSP/GSSA Levels

Colin N. Waters

By convention (Remane et al. 1996), each chronostratigraphic/geochronological unit that makes up the Geological Time Scale is defined only at its base, giving rise to the so called 'topless stage'. The base of each unit then effectively defines the top of the subjacent unit. In the case of the Anthropocene, and until now at least the Holocene, the top is not defined. Two kinds of stable reference points can be used to provide formal definition of a unit of the Geological Time Scale: a physical boundary known as a Global Boundary Stratotype Section and Point (GSSP or 'golden spike'); or a numerical age known as a Global Standard Stratigraphic Age (GSSA). Here we consider the process by which geological time units are formalised, what may represent a suitable Anthropocene GSSA event, the process by which GSSPs are defined and what may be suitable environments within which an Anthropocene GSSP may be defined. Finally, we weigh the arguments for and against both methods.

7.8.1 Process of Defining a Geological Time Unit

The Anthropocene Working Group (AWG) was established as a task group in 2009 by the Subcommission on Quaternary Stratigraphy (SQS), a constituent subcommission of the International Commission on Stratigraphy (ICS). Its remit was to assess the evidence that there are environmental signatures preserved in geological strata that can be attributed uniquely to an Anthropocene chronostratigraphic unit and, if so, to propose to the SQS the definition of such a unit via a GSSP at a type locality or via a GSSA. The initial stage of analysis was for the AWG to determine the following: whether the Anthropocene possessed geological 'reality' (summarised in Section 7.1); the most suitable timing of its beginning (outlined in Sections 7.2–7.6); what hierarchical level (rank) the unit should have (see Section 7.7); and which of the GSSP or GSSA options should be used if a formal proposal were to be compiled.

In 2016 the AWG reported their majority agreement that there is validity in the concept of the Anthropocene as a chronostratigraphic unit, with substantial evidence accumulated and published (e.g., Williams et al. 2011; Waters et al. 2014), to the effect that the optimum level lies around the mid-20th century, that epoch status is appropriate (Waters et al. 2016) and that the group should work towards such a proposal based on a GSSP (Zalasiewicz et al. 2017c).

For a unit of the International Chronostratigraphic Chart/Geological Time Scale to be established and ratified, it has to follow established protocols, outlined by Remane et al. (1996), Remane (1997, 2003), Finney (2014) and Head and Gibbard (2015). The process begins with the preparation of a formal proposal to define the boundary. The AWG can formulate its own proposal, or it can consider a proposal formally submitted to it by an external research group. Whether one or more proposals are submitted, the AWG must then vote on the proposal(s). For the preferred proposal to be approved, the AWG must obtain a supermajority ($\geq$60%) vote, as required for all ICS voting. The AWG then submits the approved proposal to its parent subcommission, the SQS. Finney (2014) emphasised the importance that the proposal be aimed to inform SQS and ICS members who have not been involved in the detailed research, and so any proposal must not only explain and justify its recommendation but also address all questions that could likely be raised against it (see Zalasiewicz et al. 2017d). Decisions on the submitted proposal are made by supermajority votes first by voting members of SQS, from which a formal recommendation arises, and then by the voting members of the ICS (three executive officers and chairs of the 16 subcommissions). Final ratification by the IUGS Executive Committee of a recommendation approved by ICS is not guaranteed and will require assurance that ICS policies and procedures were followed (Finney 2014).

7.8.2 Proposed GSSA Definition

For the Archean and most of the Proterozoic stratigraphic records, fossils are scarce or absent, and strata are often strongly deformed, making unambiguous correlation by means of relative dating difficult (see Section 1.3). Therefore, boundaries in rocks predating ~630 Ma are mostly defined in terms of GSSAs, based upon nominal ages that produce eons, eras and periods of broadly equal duration. For example, the Mesoproterozoic Era is subdivided into three periods, each of 200 Ma duration, essentially recognised by radiometric dating. At the other end of the Geological Time Scale, until 2009 the beginning of the Holocene was defined chronometrically at 10,000 ^{14}C years BP (although as no formal proposal for this was submitted for ratification, it was technically not a GSSA). However, a GSSA is a potential means to define an Anthropocene beginning.

Zalasiewicz et al. (2015b) suggested that for the Anthropocene, a boundary defined by a GSSA might offer some advantages. They proposed a GSSA based on the precise time of the detonation of the world's first nuclear bomb explosion of the Trinity A-bomb, at 05:29 (local time) on 16 July 1945 at Alamogordo, New Mexico. This was a precisely defined moment of considerable historical importance, coinciding with (and produced by) the first dissemination of bomb-produced radionuclides (a strong candidate for primary signal of the Anthropocene) and occurring at around the beginning of the Great Acceleration of human population growth, economic activity and energy production (see Section 7.5).

Alternatively, a 'neutral' date for a GSSA might be selected, such as 1950. To some, this has an advantage in not linking the start of the Anthropocene to a negative and highly emotive event. It also neatly bisects the century and has geological significance in that radiocarbon dates are quoted as ages relative to BP, where BP (before present) is taken as 1950.

Zalasiewicz et al. (2015b) considered this proposal could be superseded in the future when the greater duration of the Anthropocene signal could allow reassessment of the relationship. However, the Alamogordo detonation was associated with only localised fallout, and recognition of a global radioactive signal only becomes practicable by 1952 with the first detonation of the thermonuclear H-bombs, which were capable for the first time of transmitting the fallout signal into the stratosphere, permitting its rapid global dispersion (Waters et al. 2015 and Section 5.8 herein).

Other candidates have not been proposed directly as potential GSSAs, though some suggestions for the start of the Anthropocene are described in terms of numerical ages. For instance, a 2,000-years-BP Anthropocene beginning was proposed by Certini and Scalenghe (2011), based upon an extensive anthropogenic soil horizon of about this age, though soils tend to be time transgressive and difficult to date precisely, the chosen date therefore being partly arbitrary (see discussion in Section 7.3.6).

It should be noted that the original intent of GSSAs was to subdivide the Proterozoic, where they were first introduced in 1989 (Plumb 1991; Remane et al. 1996), and not to replace the function of GSSPs in more recent deposits. Even for the Proterozoic and much of the Archean, efforts are underway or envisioned to replace the GSSA concept with GSSPs (Van Kranendonk et al. 2012). For the Anthropocene, a GSSP-based boundary would be consistent with this philosophy.

7.8.3 Proposed GSSP Definition

Established stratigraphic procedures for deciding on a GSSP have been described by Salvador (1994), subsequently outlined by Remane et al. (1996) and Remane (1997, 2003) and summarised by Smith et al. (2014). Deciding on a GSSP is a complex process that requires (1) an initial selection of usually one primary stratigraphic marker (e.g., the lowest occurrence of a fossil species, a geochemical change or a palaeomagnetic boundary), but with one or more secondary markers; (2) selection of a stratotype section from a number of potential sections in which the key markers must be represented, ideally together with some auxiliary stratotypes in which the same level is represented by similar or other proxy signals in different parts of the world and (3) definition of the precise point within stratified rock or sediment (or glacial ice, in the case of the Holocene) that fixes the chronostratigraphic boundary to represent a precise moment of time. Once ratified, formalisation of a GSSP provides a stable reference, since a GSSP cannot be subsequently modified for at least 10 years. Importantly, the exact level chosen remains the reference point, even if the key marker is later found to have appeared elsewhere in the stratal succession at the same location (as happened, for instance, at the Cambrian GSSP – see Section 1.3.1.1).

Criteria for formulating an International Commission on Stratigraphy (ICS) proposal includes the requirement for stratigraphic completeness across the GSSP level, with adequate thickness of strata both above and below the boundary in order to demonstrate the transition. Therefore, the presence of discontinuity, such as an erosion horizon or marked palaeosol at or near to the proposed boundary, would render it unsuitable. The section should also be accessible for subsequent investigations, ideally with provision for conservation and protection of the site.

7.8.4 Suitable Environments for a GSSP

Most ratified Phanerozoic GSSPs use species of marine fossil as primary markers (see Section 1.3). However, some recent GSSPs are defined using physico-chemical markers, while the Holocene GSSP, based on isotopic signals, also broke with precedent by being sited within glacial ice (Walker et al. 2009) rather than in more typical rock strata.

A proposed GSSP succession does not need to include numerically dated levels, but it is considered an advantage to have them. For some GSSPs this takes the form of radiometric dating of volcanic ash layers that lie close to the boundary, whereas others may be directly dated using astrochronology. Such dates include analytical and other uncertainties, which is why correlation is done using an extensive and approximately isochronous environmental marker rather than absolute time. However, as any Anthropocene boundary is geologically very recent, high-precision dating of some kind will be required so that the correlative signals used are indeed as synchronous as possible and to provide confidence that there are no missing strata.

In modern deposits, high-precision dating, potentially to annual or subannual resolution, can be achieved in seasonally layered sediments, ideally with layer counting supported by radiogenic dating techniques (e.g., ^{210}Pb dating) or unambiguously dated events, such as known large volcanic eruptions. Such annually or seasonally resolved laminations can be recognised in many different environments. Anoxic marine basins, some associated with eutrophic dead zones with little or no benthic disturbance, or mud patches developed seaward of a sinking marine delta, perhaps provide the environments most typically considered suitable for a GSSP, because the absence of oxygen prevents bioturbation and so can preserve annual layering. Such environments represented at rock exposures have been used for GSSPs, such as those for the Ordovician-Silurian and the Paleocene-Eocene boundaries (see Section 1.3.1.4), although to date no GSSP has been located in any marine core taken from the present ocean floor (Smith et al. 2014). Varved sediments are often found in hypoxic, glacially influenced or hypersaline lakes and so provide an alternative to annually laminated marine successions, with the precedent that four of five auxiliary stratotypes for the Pleistocene-Holocene GSSP were located in lacustrine deposits (Walker et al. 2009). Annually layered glacial snow and ice layers (see Section 1.3.1.6) also provide high-resolution geochemical signals suitable for potentially defining an Anthropocene boundary. Annually laminated speleothems provide another potential environment for the location of a GSSP, and a stalagmite from Mawmluh Cave, India, has been selected as GSSP for the Middle/Upper Holocene boundary (Walker et al. 2012). No biotic organism, whether living or fossil, has yet been suggested as a location for a GSSP, but for the Anthropocene, annual growth layers in corals and bivalves or in tree rings might provide highly resolvable geochemical signals. Other environments that lack clear annual laminae, such as anthropogenic deposits and ombrotrophic peat bogs, can nevertheless contain useful signals marking the transition from Holocene to Anthropocene and also need to be considered.

Key potential archives are briefly discussed below with regard to factors that may or may not favour their use as candidate GSSPs, illustrated with examples, sourced from Waters et al. (2018). Amongst potential primary markers, the plutonium signal of the 'bomb spike' has been noted (Waters et al. 2015, 2018; Zalasiewicz et al. 2017c).

7.8.4.1 Marine Anoxic Basins

Suitable marine successions need to include seasonal rhythmic varves that are undisturbed by storm or current activity, animal burrowing or human disturbance – like bottom trawling (which now precludes most continental shelves). They should be characterised, too, by sufficiently high rates of sedimentation to accumulate an adequate thickness of Anthropocene strata (typically precluding abyssal depths). Continental slopes, where turbidite fans and contourite drifts build up thick sedimentary accumulations, are prone to intermittent erosion by submarine currents. The combination of these prerequisites is rare. Deoxygenated dead zones currently cover ~0.7% of the oceans, although many of these, where hypoxia has been induced by eutrophication (see Section 5.4.1), only date back from the 1960s (Diaz & Rosenberg 2008). Furthermore, dead zones usually occur on continental shelves, where storm waves can stir the sediment and prevent the development of undisturbed annual layering even in the absence of bioturbation. Hence the most promising marine GSSP candidates occur in anoxic basins well below wave base, in which annual layering is preserved as a consequence of long-term oxygen deficiency caused by thermal or chemical density stratification of the water and/or high rates of accumulation of organic matter (Schimmelmann et al. 2016).

A promising candidate area is the Santa Barbara Basin off California, in which annual laminae are related to winter influx of terrigenous materials alternating with summer planktonic blooms that go back at least two millennia (Schimmelmann et al. 2013). These laminated sequences have been studied in detail to extract histories of climate and biota, notably from the remains of planktonic foraminifera and diatoms (Field et al. 2006; Figure 7.8.1c herein). Modified sediment supplies derive from climate-driven changes to fluvial systems in the California hinterlands, which may enhance or reduce terrigenous sediment supply, which in turn changes the frequency of turbidity currents. The abundant organic and clay content typical of such deposits scavenges contaminants from the water column, including metals (Schmidt & Reimers 1991; Figure 7.8.1b herein), radionuclides (Koide et al. 1975) and persistent organic pollutants (POPs) (Hom et al. 1974). In the Santa Barbara Basin, however, the water depth of ~600 m can cause decadal-scale signal time lags such that although the onset of the Pu fallout signal is clearly expressed in 1950–1954 CE varves, there is a marked smearing of the peak signal expected from the 1963–1964 CE bomb spike (Koide et al. 1975; Figure 5.8.7a herein). Nonetheless, comparable marine settings occur around the world in other depressions on the continental shelf (e.g., the Cariaco Basin off Venezuela), in fjords (e.g., Saanich Inlet and Saguenay Fjord of Canada) and in enclosed seas (e.g., the Black and Baltic Seas).

7.8.4.2 Coral and Bivalve Skeletons

Long-lived biota that lay down annual or seasonal laminae in their skeletons provide a high-resolution, if unorthodox, option for locating a GSSP for the Anthropocene. Suitable biota include shallow-water corals, limited to tropical waters (<100 m depth), and cold-water corals and benthic bivalves that occupy a range of water depths and latitudes. Such skeletal

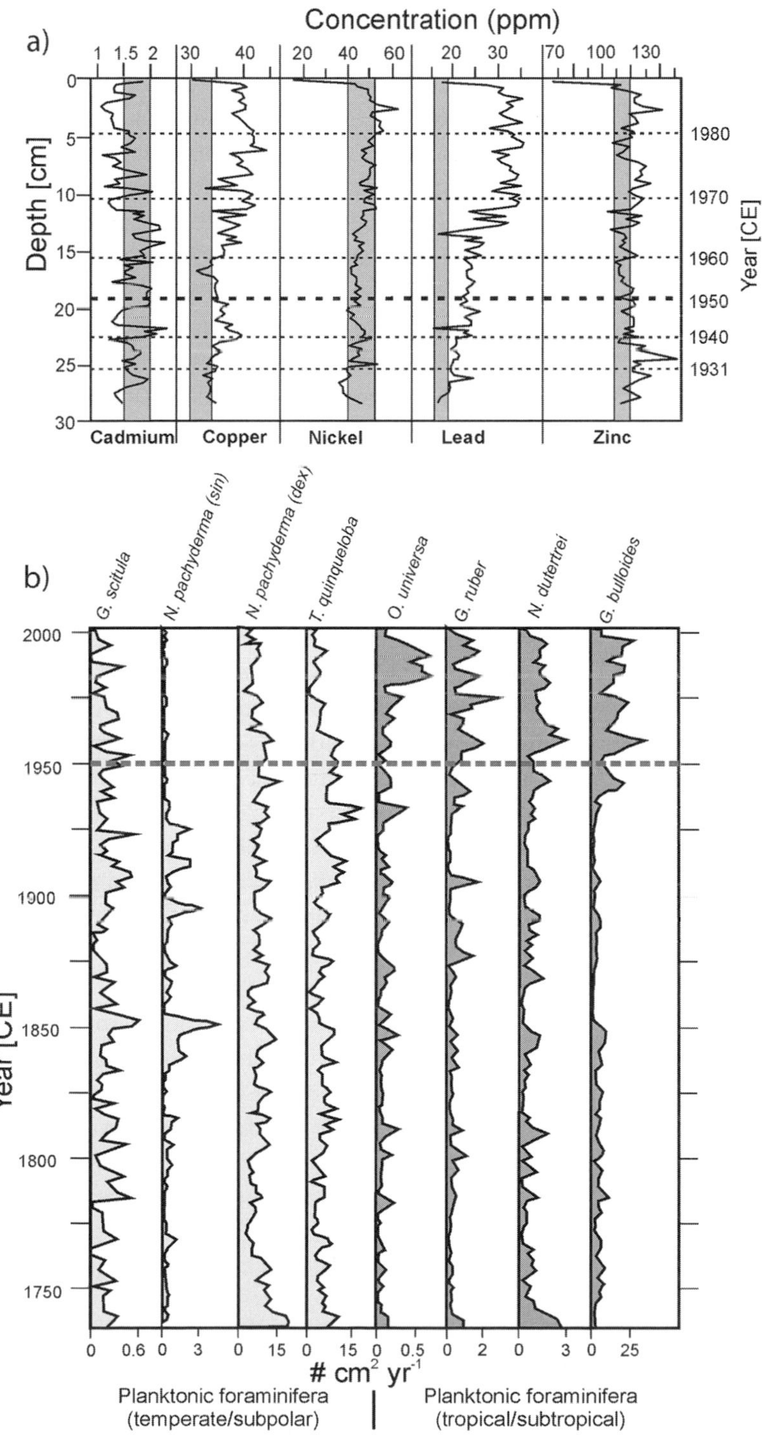

Figure 7.8.1 Key signals in marine cores from the Santa Barbara Basin, with (a) selected heavy metals (Schmidt & Reimers 1991) and (b) planktonic foraminifera (Field et al. 2006). Reprinted from figure 4 of Waters et al. (2018). ©2018, with permission from Elsevier

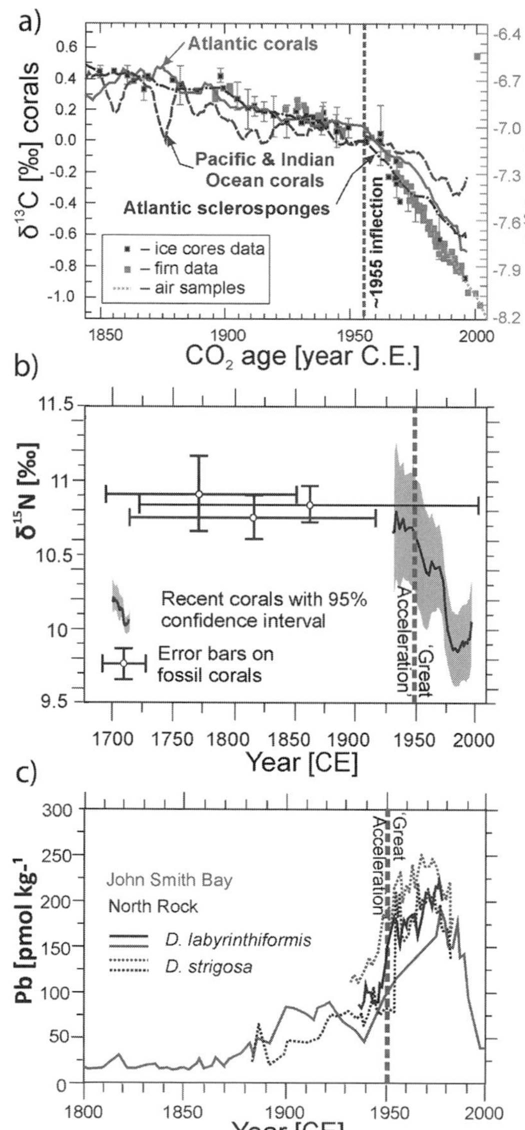

Figure 7.8.2 (a) Changes in δ^{13}C with respect to age for corals from the Atlantic and the Pacific/Indian Oceans compared to published data from sclerosponges, averaged after removing the mean δ^{13}C value of the coral skeleton from 1900 CE to the present day and shown as a five-year running mean (modified from figure 2 of Swart et al. 2010) ©2010, American Geophysical Union; (b) δ^{15}N-AA depletion in deep-sea gorgonian corals in the northwest Atlantic (modified from figure 5 of Sherwood et al. 2011) ©2011 National Academy of Sciences; (c) Pb concentration in Bermuda

laminae imprint the Suess effect of declining stable isotope ratios of carbon (Swart et al. 2010; Figure 7.8.2a herein) and nitrogen (Sherwood et al. 2011; Figure 7.8.2b herein), increased Pu (Benninger & Dodge 1986; Lindahl et al. 2011; see Section 5.8.2 and Figure 5.8.7b herein) and radiocarbon (Weidman & Jones 1993; Section 5.8.2 and Figure 5.8.6a herein) fallout signals and heavy-metal signals (Kelly et al. 2009; Figure 7.8.2c herein). Shallow-water corals, in particular, provide proxies for increasing sea surface temperatures since the mid-20th century through increased thicknesses of annual bands that equate to growth rate (Lough & Barnes 2000) and decreases in δ^{18}O (Cole 1996; Hetzinger et al. 2010; see Section 3.4 herein).

Shallow-water corals have rapid growth rates and are less likely to show a lag in Pu, Δ^{14}C and heavy-metal records than deeper-water fauna. Furthermore, the anthropogenic-CO_2 ocean inventory shows the greatest values at shallow water depths, most notably within the Atlantic Ocean (Swart et al. 2010). This may support selecting a potential GSSP site within a shallow-water coral in the Caribbean. The Caribbean is also a long way from nuclear testing facilities, so corals there will record a simple global fallout signal (Benninger & Dodge 1986), while near to the Pacific Proving Grounds the fallout records are dominated by local detonations (Lindahl et al. 2011).

7.8.4.3 Estuaries and Deltas

Major river estuaries and deltas are present on all continents except Antarctica, forming the transition between the fluvial and marine realms. This represents the zone through which most terrestrial sediments, including those associated with human signals, are transported to the oceans. The trapping of river sediments behind major dams, which have been constructed on nearly all major rivers in recent

Figure 7.8.2 (*cont.*) corals (modified and amalgamated from figures 3 and 4 of Kelly et al. 2009). ©2009, with permission from Elsevier (A black-and-white version of this figure appears in some formats. For a colour version, please refer to the plate section.)

decades (Syvitski & Kettner 2011; see Section 2.8.3 herein), has greatly reduced the flux of sediments reaching the coast, while projected absolute sea-level rise may be expected to drown existing estuarine and deltaic deposits (see Section 6.3.2). This will leave a clear response in the shape of transgressive, typically upward-fining deposits, likely with excellent preservation potential.

Estuarine and deltaic deposits are prone to erosion caused by storms or floods, and so the most suitable potential location for a GSSP might therefore be in mud depocentres that occur on the continental shelves in front of important estuarine and deltaic areas. They show a moderate sedimentation rate of 1–5 mm/yr (Hanebuth et al. 2015), though the deposits are prone to reworking through bioturbation, storms and anthropogenic disturbance, including trawler fishing and seabed dredging (Mahiques et al. 2016).

Estuaries and deltas are sites of widespread translocation of invasive aquatic species, especially diatoms, dinoflagellates, foraminifera and ostracods, which can be associated with marked declines in the abundance of indigenous species (Wilkinson et al. 2014; Section 3.3.1 herein). This biotic replacement has accelerated during the 19th and 20th centuries through the establishment of global shipping routes and resultant transfer of species in ballast water and from hull fouling, especially close to major ports (McGann et al. 2000). Loss of natural ecosystems, most notably of mangroves, and introduction of aquaculture, which also leads to displacement of native species and eutrophication of the water column, also contribute to local species-assemblage changes (Martinez-Porchas & Martinez-Cordova 2012).

The Clyde Estuary in Scotland (Figure 7.8.3) provides an example of how estuarine and deltaic sediments are typically highly reflective of industrial activities. The sediments there contain elevated anthropogenic organic chemicals (Figure 7.8.3b), including total petroleum hydrocarbons (TPHs),

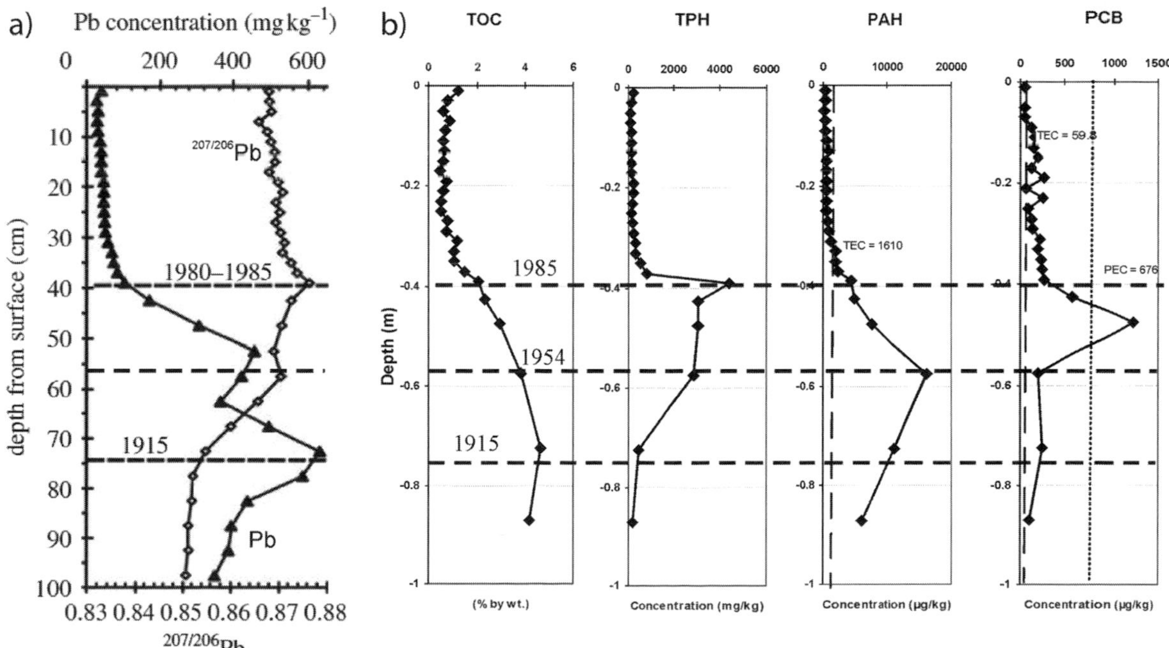

Figure 7.8.3 Example of data from a single core from the Clyde Estuary showing (a) Pb concentrations and $^{207}/^{206}$Pb isotope ratios and (b) PAH, TPH and PCB organic chemical signatures (Vane et al. 2011).

polycyclic aromatic hydrocarbons (PAHs), polychlorinated biphenyls (PCBs) and brominated flame retardants (polybrominated diphenyl ethers, PBDEs), while lead concentrations and Pb isotope ratios related to heavy industry and Pb additives in petrol provide a very fine-resolution chronometer of pollution (Vane et al. 2011; Figure 7.8.3a herein). Pb concentrations, TPHs and PAHs provide a local signal, whereas the Pb isotopes (peak in 1980s) and PCBs (1965–1977 peak) provide a more widespread and consistent, although not precisely coincident, global signal that shows a marked decline over recent decades (Vane et al. 2011). Such successions are rich records of Anthropocene processes, but because the successions lack annual sedimentary rhythms and likely contain cryptic erosional levels, they are unlikely to provide convincing GSSP candidates.

7.8.4.4 Lakes

Lakes, like estuaries and deltas, are also sinks for fluvial sediment and so can contain pollution and land-use histories. They are also traps for airborne particles, and through this they can provide a record of both local and global events. The most suitable lake environment for a potential GSSP is one where annual varves are present, where a flat lake floor limits sediment flow and disturbance and, ideally, where there is little or no bioturbation, water movement or gas emission from buried organic material to disturb the laminae (Zolitschka et al. 2015). Lakes are abundant, particularly in the boreal and Arctic latitudes of the Northern Hemisphere. There, varves can occur within glacial lakes or lakes with stratified water columns prone to hypoxia, a state encouraged by enhanced nutrient input (see Section 5.4.1). Varves can also develop in saline lakes in response to seasonal evaporation and be preserved because the saline waters are too hostile for benthic life, but such lakes may also dry up completely and so be prone to missing laminae. Widespread airborne contaminants, e.g., radiogenic fallout (Hancock et al. 2014), nitrates (Holtgrieve et al. 2011; Wolfe et al. 2013), fly ash (Rose 2015), black carbon (Han et al. 2016), POPs including chlorinated pesticides and polychlorinated biphenyls (PCBs) (Muir & Rose 2007), Hg (Yang et al. 2010) and Pb (Renberg et al. 2000), can show consistent sedimentary patterns in lakes across continents. High-density microplastics are a common feature of lakes that have effluent sources from nearby conurbations (Zalasiewicz et al. 2016a). Lakes are also prone to rapid assemblage changes through invasive species such as zebra mussels, while their diatom assemblages are sensitive indictors of climate change, hypoxia and changing acidity (Wilkinson et al. 2014). Pollen and spore records can change markedly in response to the clearance of native forests, the introduction of domestic plants and changing agricultural practices (e.g., Crawford Lake, Canada; Ekdahl et al. 2004).

Many lake sedimentary successions have undergone multi-proxy analysis, the type required to characterise and ultimately define an Anthropocene boundary. A good example is Lilla Öresjön (southwest Sweden), located in a high-sulphate deposition area resulting in acute lake acidification (Renberg & Battarbee 1990). Such areas tend to be associated with atmospheric pollutants, including heavy metals, sulphur, fly ash (spheroidal carbonaceous particles – SCPs) and PAHs, related to fossil-fuel combustion (Figure 7.8.4). At Lilla Öresjön this is evident from an initial increase in Pb, Zn and benzo(a)pyrene at about 1900 and a more pronounced increase of these signals along with sulphur and SCPs during the 1960s (Renberg & Battarbee 1990). Heavy-metal and sulphur signals peaked in the 1960s and 1970s, whereas SCPs and PAH peaked in the 1970s. Analysis of diatoms, chrysophyte and cladocerans show changes reflecting decreasing trends in lake pH, while pollen shows 20th-century pine-spruce-forest expansion until clear felling started in the 1980s (Renberg & Battarbee 1990).

7.8.4.5 Peat Mires

Northern (boreal and subarctic) peatlands represent about 90% of global peat development, whereas the equivalent southern peatlands, mainly in Patagonia, cover about 2%, and tropical peatlands cover about

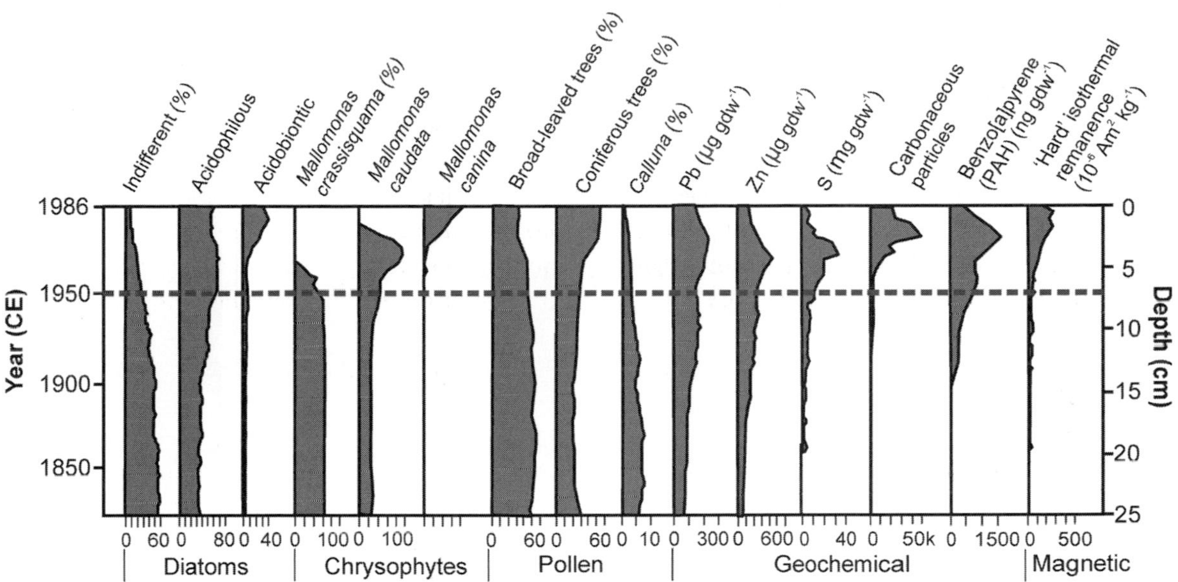

Figure 7.8.4 Physical, chemical and biological trends from Lilla Öresjön (southwest Sweden), a high sulphate deposition area. (from Renberg & Battarbee 1990)

8% (Yu et al. 2010). As these environments only receive input of geochemical signals through the atmosphere, they provide good records of atmospheric metal deposition (especially Pb, Ni, As and Cd; e.g., Shotyk et al. 1998) and fly ash (e.g., Swindles et al. 2015) with no delays due to settling through water. Pb signals in many European peats show pollution from early mining and smelting; in the case of the Jura Mountains of Switzerland, the earliest signal is about 3,000 years ago, but with the greatest Pb flux occurring in the late 20th century (Shotyk et al. 1998).

Peat deposits can accumulate relatively rapidly. However, the deposits are not varved and can only be dated by ^{14}C and ^{210}Pu methods, tied to known volcanic ashfall events, and so are not as highly resolved as those of other environments discussed here. Pu may be significantly mobile in these chemically reduced deposits (Quinto et al. 2013), and this may limit the suitability of a Pu fallout signal as a proxy for the base of the Anthropocene in this type of environment.

The lowest occurrence of SCPs in a peat succession in Malham, England, dates to the 1850s, but there was a marked increase in their abundance in the 1950s, then a peak in the 1970s (Swindles et al. 2015; Figure 7.8.5 herein). Atmospheric lead pollution from local industrial activity and additives in petrol, as well as increased soil erosion, reflected in the Fe and loss-on-ignition data, show comparable upturns in the mid-20th century. The Anthropocene there represents less than 10 cm of peat.

7.8.4.6 Anthropogenic Deposits

Anthropogenic deposits (see Section 2.5) include sedimentary successions that have accumulated through direct and deliberate human deposition (artificial ground) or by human influence on natural systems. They include a continuum from entirely natural through to entirely anthropogenic materials (Ford et al. 2014). Such deposits are extensive. Zalasiewicz et al. (2016b) estimated that ~55% of the terrestrial land surface has been affected by extensive human modification. In comparison with natural deposits, anthropogenic deposits may show remarkably high accumulation rates. The deposits can rapidly incorporate novel lithological, geochemical or biotic signatures, and the presence of artefacts (technofossils; see Section 4.2) and

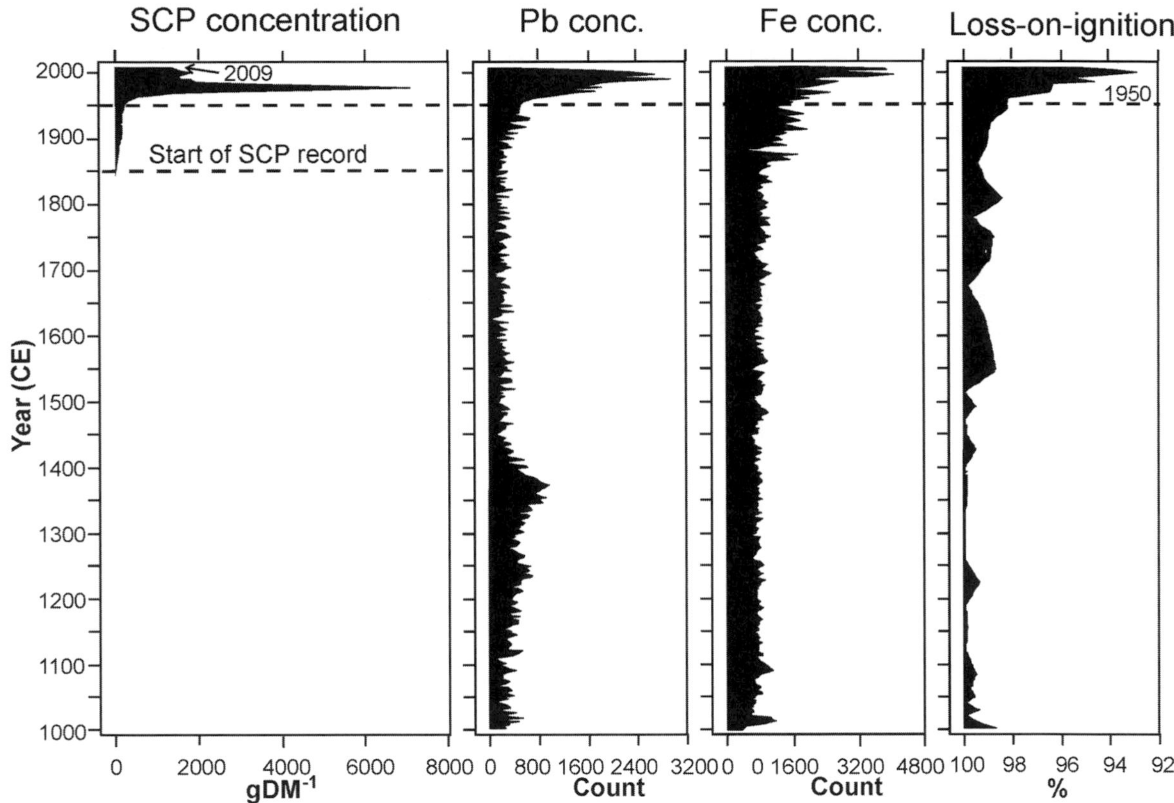

Figure 7.8.5 Spheroidal carbonaceous (fly-ash) particles (SCPs) and lead and iron concentrations from Malham Tarn Moss, England. (modified from Swindles et al. 2015; reprinted from figure 22 of Waters et al. (2018), ©2018, with permission from Elsevier)

novel minerals and materials (see Sections 2.1 and 2.2) can allow precise stratigraphic dating. However, such successions are not normally annually layered and commonly reflect intermittent deposition, often with internal erosive surfaces, and they commonly lack a simple stratiform geometry.

Examples of artificial ground include landfill sites (e.g., the Fresh Kills Landfill, New York) or building rubble (e.g., the Teufelsberg, Berlin, associated with WWII urban destruction, or the Schuttdecke of Vienna, Austria, associated with normal aggradation of urban landscapes). Perhaps the most stratigraphically coherent sections are those within natural deposits that include abundant artificial materials. For example, the Gorrondatxe-Tunelboca beachrock of northern Spain is composed of naturally redeposited blast-furnace waste that has distinct foraminiferal assemblages (Martínez-García et al. 2013). Nevertheless, even sections such as these are unlikely to provide the fine time resolution needed for an Anthropocene GSSP candidate.

7.8.4.7 Ice

Continental ice sheets, including those of Antarctica and Greenland, are equivalent to about 9.4% of the total land surface area. There is a precedent in using an ice core from Greenland to locate the GSSP for the Pleistocene-Holocene boundary (see Section 1.3.1.6), and polar ice sheets are likely to provide a relatively complete and permanent sedimentary record. Ice from glaciers occurring both marginal to the ice sheets and in alpine regions, which makes up a further ~0.5% of the Earth's land surface (Pfeffer et al. 2014), is more likely to be prone to significant seasonal melting and potential loss of laminae. Sea ice fluctuates in extent seasonally and does not represent a permanent ice record for consideration as a potential GSSP site.

Polar ice has annual layers, can also be radiometrically dated using ^{210}Pb and can include sulphur spikes that reflect known volcanic events (see Section 5.5.3). A problem with ice as a medium for a GSSP is that the mid-20th century increased CO_2, CH_4 and N_2O concentrations, and the $\delta^{13}C$ depletion signal (see Section 5.2) is acquired from air bubbles, and air in firn ice is in weak contact with the atmosphere until sufficient compaction has taken place to isolate the air pockets as sealed bubbles. As a consequence, at any one depth the ice is older than its trapped air bubbles, leading to a lag between those signals fixed in ice and those fixed in the bubbles. This potentially can lead to millennial-scale discrepancies in this parameter across those parts of Antarctica with very slow ice-accumulation rates.

Aside from the CO_2, CH_4 and N_2O concentrations and their $\delta^{13}C$ ratios, which are affected by the offset described above, the principal chemical signals in ice cores include ^{239}Pu and excess ^{14}C nuclear-fallout signals, starting in 1953–1954 (Gabrieli et al. 2011; Wolff 2014; Arienzo et al. 2016); clear though diachronous SO_4^{2-} spikes that start approximately during the mid-20th century (Wolff 2014); a NO_3^- increase (Fischer et al. 1998; Wolff 2014) and associated $\delta^{15}N$ depletion dating from ~1950 CE (Hastings et al. 2009); and increased black carbon, Pb, Zn, Cd and Cu accumulation, also dating from the 1950s (McConnell & Edwards 2008).

A clear contrast exists between ice sheets from Antarctica and Greenland as regards the most suitable location for a potential GSSP. The Greenland ice sheet shows the effects of greater local contamination from Northern Hemisphere industrial activities, while Antarctic land ice is more pristine, and many of its signals (NO_3^- and $\delta^{15}N$, SO_4^{2-}, ^{239}Pu, dust) occur at very low concentrations, potentially below levels of resolution.

The Law Dome ice core from Antarctica has a high-resolution multi-proxy record for the last 1,000–2,000 years that includes CO_2, CH_4, N_2O and $\delta^{13}C$ (Figures 5.2.2b, 5.4.1 and 7.8.6). This core shows many characteristics that make it suitable for a potential GSSP, such as high accumulation rates, low temperatures and small quantities of impurities (Etheridge et al. 1996; Francey et al. 1999; Ferretti et al. 2005; MacFarling Meure et al. 2006; Rubino et al. 2013). Ice accumulation rates at an average of ~1.4 m/yr result in a gas-age/ice-age difference of only ~31 years for CO_2 (Rubino et al. 2013).

7.8.4.8 Speleothems

Annually laminated calcareous speleothems, typically stalagmites, occur within natural cave systems in karst environments or within artificial tunnels where either the adjacent bedrock or the degradation of mortar in concrete linings of the tunnel contributes the calcium carbonate. There is precedent for using speleothems for the location of a GSSP, as the proposed stratotype for the Upper Holocene Subseries is in a stalagmite from Mawmluh Cave, India (Walker et al. 2012), and potential speleothem use for the Anthropocene is outlined by Fairchild and Frisia (2014) and Fairchild (2018). Speleothems have the advantage of being lithified and commonly being from undisturbed locations, but their growth rate is typically slow (tens to hundreds of microns per year). Sulphur concentrations, ^{34}S stable isotopes (Wynn et al. 2010) and radiogenic ^{14}C records within speleothems reflect atmospheric composition but are modified by prior travel through soil and rock, which can cause attenuation and decadal-scale lags of signals. However, some key global signals are not apparent, such as Pu fallout, increases in atmospheric CO_2 concentrations and the Suess effect on stable carbon isotopes ($\delta^{13}C$), while the impacts of changes to the nitrogen cycle are as yet unknown. Local, potentially diachronous signals may include variations in growth rates of laminae and $\delta^{18}O$, related to air temperature and humidity. Changes in trace pollutant metals (e.g., Pb, Zn, REE) and $\delta^{13}C$ are

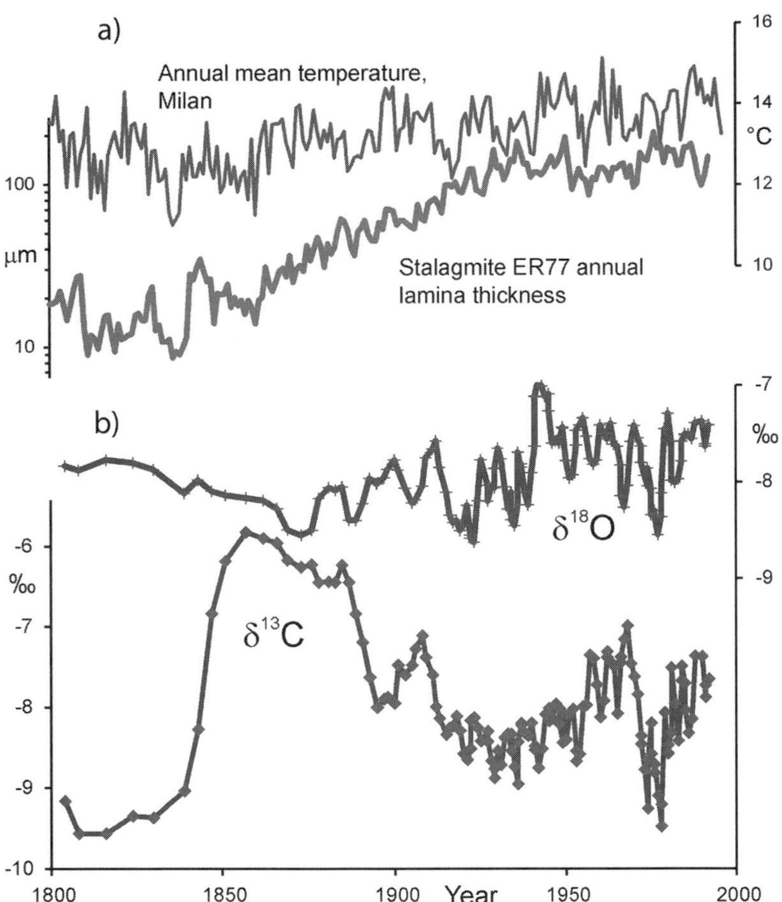

Figure 7.8.6 Ernesto Cave, Italy: (a) Age model based on lamina counting related to local mean air temperature (Frisia et al. 2003); (b) δ^{18}O and δ^{13}C profiles (Scholz et al. 2012); (c) Radiocarbon profile and comparable European atmospheric emissions (Fohlmeister et al. 2011); (d) S concentration and δ^{34}S (Frisia et al. 2005; Wynn et al. 2010). Reprinted from figure 29 of Waters et al. (2018). ©2018, with permission from Elsevier.

strongly linked to organic matter from soil, and their changes tend to reflect soil disturbance and deforestation.

Ernesto Cave in northern Italy has supplied various proxy-data series from annually laminated speleothems (Figure 7.8.7). Increases in laminae thickness, commencing about 1840, relate to a warming event at the end of the Little Ice Age (Frisia et al. 2003; Fairchild & Frisia 2014), and increased sulphate concentrations from 1880 broadly coincide with initial atmospheric pollution coincident with the start of the Industrial Revolution (Frisia et al. 2005). No appreciable change in stable isotopes has occurred across the mid-20th century. The ^{14}C bomb spike has a decade lag compared with atmospheric signals (Fohlmeister et al. 2011), whereas sulphur concentrations and ^{34}S stable isotopes show a 15–20-year lag (Wynn et al. 2010). Trace-element concentrations increased greatly during the early 20th century, reflecting extensive deforestation in the area (Borsato et al. 2007). Hence speleothems, while extraordinary and detailed archives of

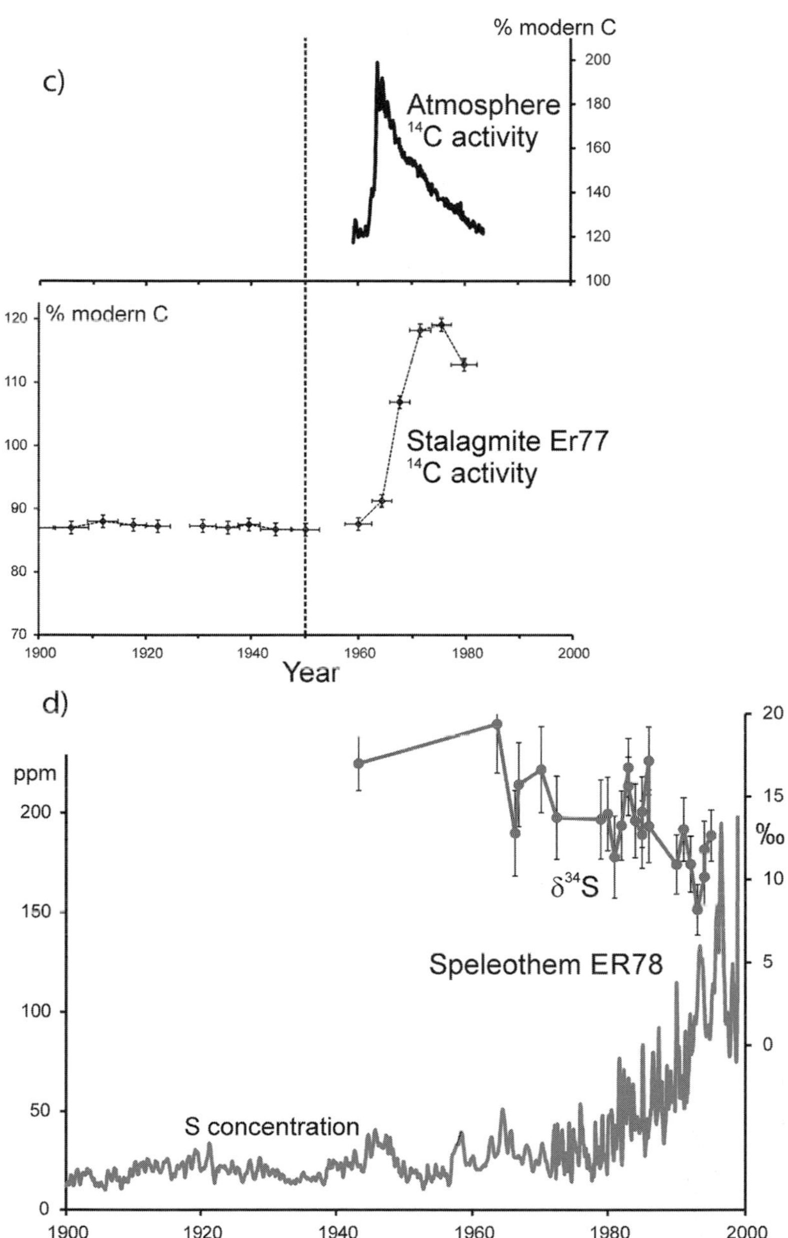

Figure 7.8.6 *(cont.)*

natural and anthropogenic change, are unlikely to provide GSSP candidates for the Anthropocene.

7.8.4.9 Tree Rings

Forests cover ~30% of the world's land area and are present on all continents except Antarctica. For potential usage as a GSSP, only trees from the temperate or boreal forests, including those of North America and Eurasia, consistently show tree rings. The rings contain an annual archive of local changes in palaeoclimate through proxies including isotopic ratios of carbon (sourced from air) and hydrogen and oxygen (sourced from soil water). Tree-ring width provides a clear warming signal in the Northern Hemisphere since 1850, associated with the end of the Little Ice Age, but does not provide the resolution for any warming event across the mid-20th century (Esper et al. 2002).

Key potential markers for the Anthropocene include: elevated sulphur commencing during the mid-20th century (Figure 7.8.7a, b); tree rings being typically more responsive to atmospheric pollution than are speleothems (Fairchild et al. 2009); the ^{14}C bomb spike being clearly defined (Rakowski et al. 2013; Figure 5.8.6b herein); $\delta^{13}C$ being highly sensitive to the Suess effect on atmospheric CO_2 and showing marked decline since 1940 (Loader et al. 2013; Figure 7.8.7d herein) to 1960 (Bukata & Kyser 2007); depleted $\delta^{15}N$ values being seen after 1945 (Bukata & Kyser 2007); and in the $\delta^{18}O$ and δD records, the latter showing positive shifts ~1960 in response to warming trends (Epstein et al. 1990). The $\delta^{34}S$ signal may show no clear excursions across the mid-20th century (Wynn et al. 2014; Figure 7.8.7c herein), but there are decreasing values from the 1950s to a minimum in the late 1970s, with an approximately five-year delay compared with atmospheric S concentrations (Kawamura et al. 2006). Historical changes in trace-metal levels (e.g., Cu, Ni, Cr, Zn, Cd, Co, As) can be recognised in some trees, but with a possible time lag between metal deposition and the passage through soils and root systems, and element concentrations for a given year differ markedly between trees from a single location (Watmough 1999).

7.8.5 Summary of GSSP Potential

There is a wide array of stratigraphic proxy signals present in recent strata that reflect major human impacts on the Earth System (Steffen et al. 2016; Zalasiewicz et al. 2017c) and that have the potential to be used both to define an Anthropocene chronostratigraphic unit and, if such a unit were defined, to correlate it widely across the globe in most environmental settings.

A number of these signals reflect the enormous, largely hydrocarbon-fuelled anthropogenic energy expenditure that has taken place since the mid-20th century (Section 7.5): for instance, the marked change in carbon isotopes related to the Suess effect and the worldwide dissemination of fly ash. This energy expenditure has been crucial in driving both the industrial and the agricultural development that produced further stratigraphic signals (nitrogen isotopes, persistent organic pollutants, plastics and so on), with knock-on impacts to the biosphere and a complex pattern of resulting biostratigraphic signals (see Chapter 3). Some or all of these signals extend in one way or another across many or all of the environments discussed above.

None of these signals is truly instantaneous in the manner of the K-T boundary meteorite impact – but that chronostratigraphic boundary is thus far unique in this respect (see Section 1.3.1.3). Had any of the major volcanic eruptions of the last few centuries (e.g., Tambora) been of sufficient scale to produce a widely detectable ash layer, found globally and in most of the environments described above, then that might have been considered a potential candidate event. There is no requirement that a boundary marker needs to relate to the main drivers of geological change in any boundary definition. All that is required is that such markers be effective in practice for correlation. Although large volcanic eruptions have left traces in their immediate surroundings and often in one or another of the major ice caps, they are rarely detectable beyond that.

The nearest thing to such an isochronous event in relation to a possible start date for the Anthropocene

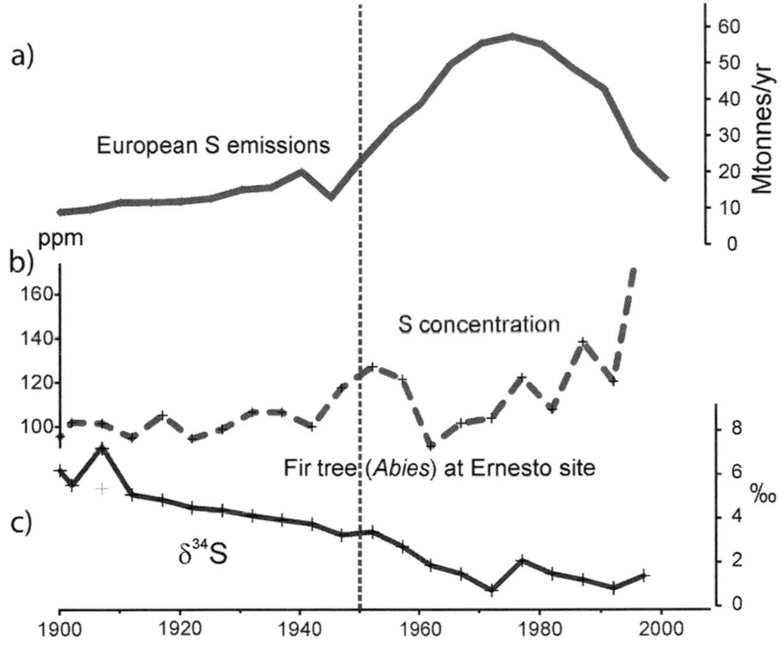

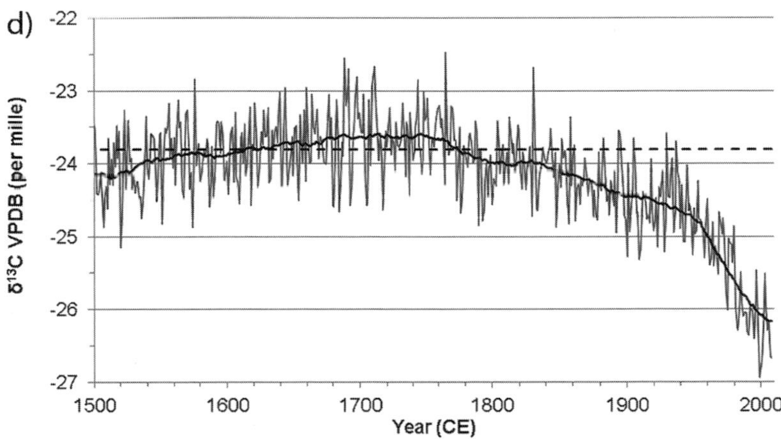

Figure 7.8.7 (a) European S emissions; (b) S concentrations; and (c) $\delta^{34}S$ in *Abies alba* from northeast Italy compared with European S emissions (Wynn et al. 2014); (d) δC variability, from Loader et al. (2013), for the period 1500–2008, measured in *P. sylvestris* tree rings from northern Fennoscandia. Fine line represents annually resolved $\delta^{13}C$ variability, thick solid line represents the annual data smoothed with a centrally weighted 51-year moving average, and dashed line represents the mean $\delta^{13}C$ value for the pre-industrial period 1500–1799 CE. Reprinted from figures 29 and 31 of Waters et al. (2018). ©2018, with permission from Elsevier.

is the start of the period of major thermonuclear atmospheric tests of the early 1950s that, within a year or two, had left a still detectable signal across much of the globe and within many environments. Whether this signal is expressed as a plutonium or excess-radiocarbon spike, it seems to be the most effective practical candidate for a primary marker. That does not necessarily mean that it has to be traceable worldwide – the deuterium-excess signal used to define the base of the Holocene is only discernible per se in Northern Hemisphere ice, for instance. But its *level* has to be widely traceable, using all available evidence. Given that the timing of the artificial-radionuclide signal coincides closely with the beginning of or major inflection in other major proxy signals (e.g., fly ash, plastics, nitrogen isotopes) and with the other signals from the Great Acceleration, this widespread traceability considerably helps make the case for the correlatability of an Anthropocene boundary level worldwide dating to approximately 1950.

Clearly, there is a large range of potential environments in which an appropriate GSSP might be found, datable to annual resolution. As we write, work is underway to narrow down the possibilities and to seek amongst these environments specific examples of potential GSSPs for further study. An optimum solution may be to follow the example of the Holocene GSSP and to suggest a primary stratotype augmented with half a dozen or so auxiliary stratotypes to provide effective global reference. This is work in progress.

7.8.6 Arguments for Use of Either a GSSP or a GSSA

Within the Phanerozoic, the current internationally accepted method for defining chronostratigraphic boundaries is using a GSSP, a physical reference point within geological strata that, once defined, can be correlated widely between sections of the same age. This option for a formally proposed definition is the one most likely to gain approval through the Subcommission on Quaternary Stratigraphy and International Commission on Stratigraphy. That is because it provides a definite fixed point to correlate to, with a stratal interface that is associated with evidence of elapsed events – fossils, geochemical patterns, magnetic properties and so forth. By using a GSSP for the definition, actual stratigraphic content is demonstrated. In contrast, GSSAs are agreed and ratified numerical dates chosen without any specific reference to the rock record (Plumb 1991). They were specifically introduced to subdivide Precambrian time where more conventional methods of correlation were considered lacking (but see Section 7.8.2). The recognition of these subdivisions within geological strata normally requires radiometric dating, which in turn necessitates an understanding of relevant radiometric-decay constants and their intrinsic analytical errors.

In determining whether the Anthropocene should be considered to be a new geological time unit, it is important to demonstrate the practical correlation potential of any candidate boundary and hence show that the reference section across the Holocene-Anthropocene boundary behaves similarly to existing formalised GSSPs throughout the geological column. Also, the task of identifying and choosing between GSSP candidates in distinct environments, as well as selecting primary and secondary signals from which the base of the Anthropocene is potentially defined, will help to deepen understanding of the nature of the Anthropocene as a chronostratigraphic unit.

Evidently, the Anthropocene is different from the rest of the geological column in being the first envisioned chronostratigraphic unit that spans written and instrumentally documented human history, as well as being one that covers an interval in which we have a more or less complete understanding of the operation of the many different parts of the Earth System. To some, this is an argument against defining the Anthropocene as a chronostratigraphic unit, because it is based more on direct human observation than on a stratigraphic record (Finney & Edwards 2016). In contrast, a response might be that by searching for a GSSP using standard stratigraphic protocols, the analysis of the Anthropocene is uniquely aided by the overlap of geological and historical time and by access to detailed instrumental records (Zalasiewicz et al. 2017d).

The origins of some signals evident in geological successions can be attributed to specific dated events. If such a datable event were used as the primary signal for the base of the chronostratigraphic Anthropocene, a GSSA definition might be considered a point in time in the human calendar for the start of the Anthropocene, as explored by Zalasiewicz et al. (2015b). It could provide an unambiguous and clear boundary when making correlations with the written, historical record, capable of being correlated too into strata. But it would not be *based* on stratal content in the manner of other Phanerozoic geological time boundaries – hence, the recommendation of the Anthropocene Working Group to prepare a proposal for formalisation based on a GSSP (Zalasiewicz et al. 2017c). Within this context, the process may require as flexible and creative an approach as the Holocene Working Group took in deciding the boundary for that epoch (Walker et al. 2009).

7.9 Epilogue and Forward Look for the Anthropocene

Jan Zalasiewicz, Colin N. Waters, Mark Williams, Colin P. Summerhayes and Martin J. Head

The evidence gathered together in this book shows that the Anthropocene is geologically real and represents a substantial change in the Earth System from Holocene conditions. This change is reflected in a distinct and globally near-synchronous body of strata characterised by a wide array of stratigraphic proxy markers, a unit that is most clearly recognisable as a globally near-synchronous unit with a boundary placed somewhere around the 1950s. This starting point coincides with the Great Acceleration of human population growth, industrialisation and globalisation. The duration of the Anthropocene, to date, is geologically brief – just the equivalent of one human lifetime – but its effects have already irrevocably changed the future course of history of this planet. Even the character of far-future strata will be influenced by what has happened in this geologically almost infinitesimally short interval, much as the nature of all strata of the Cenozoic was influenced by the geologically instantaneous perturbation that terminated the conditions of the Mesozoic Era.

On the basis of that realisation, the Anthropocene Working Group (AWG) has, by large majority (Zalasiewicz et al. 2017c), provisionally agreed to facilitate and vote on a formal proposal to add the Anthropocene to the Geological Time Scale, for consideration and voting on by, in succession, the Subcommission on Quaternary Stratigraphy and the International Commission on Stratigraphy, with the potential for eventual ratification by the Executive Committee of the International Union of Geological Sciences. A supermajority (>60%) is needed at any voting stage for a proposal to be passed, and there is no guarantee of a positive outcome. In any event, the work done towards this proposal will add to our collective understanding both of the current extraordinary and rapid phase of change in the Earth System towards what has been described as a no-analogue state and of the role of human impact in this change.

The AWG has provisionally agreed, also by large majority (Zalasiewicz et al. 2017c), that the proposal will be based on consideration of the Anthropocene as a potential series/epoch (meaning that, if accepted, the Holocene will have terminated at the point where the Anthropocene is taken to begin). The proposal would identify and document a candidate Global Boundary Stratotype Section and Point (GSSP), more colloquially known as a golden spike, together with auxiliary stratotypes representing the same level in different geological settings and identified by different proxy indicators around the world. This work has begun, initially by the identification of the range of settings within which a GSSP might be located (see Section 7.8.4). That is now being followed by the concurrent analysis of specific sections within these environments that might be considered as potential GSSPs. The work is aimed at being sufficiently advanced to vote upon and recommend a proposal within the coming few years, though this will depend on progress made by individual research groups investigating potential candidate sections as they become identified.

That now represents the main focus of the work programme, in response to the mandate placed on the AWG by the SQS. However, other lines of inquiry will take place (not counting the many studies of Anthropocene phenomena that are developing outside of – though many may well prove of very considerable relevance to – the confines and work of the AWG).

More generally, there is still a very great deal of scientific work to do in describing the nature, stratigraphic range and geographical extent of proxy indicators. This includes extending knowledge concerning known stratigraphic proxies (for instance, tracking the distribution of fly-ash particles from lacustrine environments, where they have been constrained well enough to show clear stratigraphic patterns, into the marine realm and cryosphere). New stratigraphic proxies will undoubtedly emerge (as, for instance, the recent discovery by Gałuszka and Migaszewski [2017] that industrial glass microspheres may be used as age markers in sediments). Areas that offer particular scope here include developing and refining the biostratigraphic information available from the recent complex, transglobal history of neobiota ('invasive species') that now dominate many areas both on land and in the sea. The AWG is also working closely with the archaeological community to exploit the use of artefacts/technofossils to help constrain the chronology and development of the extensive spreads of artificial ground around our burgeoning cities. New information and insights relevant to characterising and correlating the Anthropocene as a stratigraphic unit will undoubtedly emerge over the next few years.

Another necessary line of study is to explore more deeply the significance of formalising (and, conversely, not formalising) the Anthropocene in the geological sciences, for the more contemporary-focused Earth System science (within which the term emerged) and for a range of other disciplines both within the physical sciences and well beyond it (see Section 1.3), extending into the social sciences, humanities and arts, all of which have taken a keen interest in the Anthropocene and have interpreted it in various ways. The AWG's emphasis concerns the 'geological Anthropocene' essentially as originally intended rather than other interpretations. Even with this narrowing of focus, the question of the relevance for other disciplines, far from geology, of formalising a geological unit is a novel one in stratigraphy and needs careful consideration.

Overall, the Anthropocene has emerged as a concept that has provided not only a particular perspective on Earth history but also a remarkable and positive catalyst for cross-disciplinary discussions. These will continue to generate wider insights, even as the specific work of examining the Anthropocene in technical geological terms continues.

REFERENCES

Aarkrog, A. (2003). Input of anthropogenic radionuclides into the World Ocean. *Deep-Sea Research II*, **50**(17–21), 2597–2606.

Abram, N. J., McGregor, H. V., Tierney, J. E., et al. (2016). Early onset of Industrial-era warming across the oceans and continents. *Nature*, **536**, 411–418.

Abram, N. J., Mulvaney, R., Wolff, E. W., et al. (2013). Acceleration of snow melt in an Antarctic Peninsula ice core during the twentieth century. *Nature Geoscience*, **6**, 404–411.

Abram, N. J., Thomas, E. R., McConnell, J. R., et al. (2010). Ice core evidence for a 20th century decline of sea ice in the Bellingshausen Sea, Antarctica. *Journal of Geophysical Research*, **115**, D23101. doi:10.1029/2010JD014644.

ACIA (2005). *Arctic Climate Impact Assessment.* Cambridge University Press.

Aguirre, E., and Pasini, G. (1985). The Pliocene–Pleistocene boundary. *Episodes*, **8**, 116–120.

Ahmad, S. M., Padmakumari, V. M., Raza, W., et al. (2011). High-resolution carbon and oxygen isotope records from a scleractinian (*Porites*) coral of Lakshadweep Archipelago. *Quaternary International*, **238**, 107–114.

Ahn, J., and Brook, E. J. (2007). Atmospheric CO_2 and climate from 65 to 30 ka BP. *Geophysical Research Letters*, **34**, L10703. doi:10.1029/2007GL029551.

Aitken, A. R. A., Roberts, J. L., van Ommen, T. D., et al. (2016). Repeated large-scale retreat and advance of Totten Glacier indicated by inland bed erosion. *Nature*, **533**, 385–389.

Aldridge, D. C., Elliot, P., and Moggridge, G. D. (2004). The recent and rapid spread of the zebra mussel (*Dreissena polymorpha*) in Great Britain. *Biological Conservation*, 119, 253–261.

Allan, M., Fagel, N., Van Rampelbergh, M., et al. (2015). Lead concentrations and isotope ratios in speleothems as proxies for atmospheric metal pollution since the industrial revolution. *Chemical Geology*, **401**, 140–150.

Alley, R. B., Anandakrishnan, S., Jung, P., and Clough, A. (2001). Stochastic resonance in the North Atlantic: Further insights. In Seidov, D., Haupt, B. J., and Maslin, M., eds., *The Oceans and Rapid Climate Change: Past, Present and Future (Geophysical Monographs)*, **126**. Washington, DC: American Geophysical Union, pp. 57–68.

Allwood, A. C., Grotzinger, J. P., Knoll, A. H., et al. (2009). Controls on development and diversity of early Archean stromatolites. *Proceedings of the National Academy of Sciences (USA)*, **106**, 9548–9555.

Alonso-Hernández, C. M., Tolosa, I., Mesa-Albernas, M., et al. (2015). Historical trends of organochlorine pesticides in a sediment core from the Gulf of Batabanó, Cuba. *Chemosphere*, **137**, 95–100.

Al-Rousan, S., Pätzold, J., Al-Moghrabi, S., and Wefer, G. (2004). Invasion of anthropogenic CO_2 recorded in planktonic foraminifera from the northern Gulf of Aquaba. *International Journal of Earth Sciences*, **93**, 1066–1076.

Alvarez, L. W., Alvarez, W., Asaro, F., and Michel, H. (1984). Extra-terrestrial cause for the Cretaceous-Tertiary extinction: Experimental results and theoretical interpretation. *Science*, **208**, 1095–1108.

Amos, S. E., and Yalcin, B. (2015). *Hollow Glass Microspheres for Plastics, Elastomers, and Adhesives Compounds.* Oxford: Elsevier.

Amundson, R., and Jenny, H. (1991). The place of humans in the state factor theory of ecosystems and their soils. *Soil Science*, **151**, 99–109.

Anderegg, W. R. L., Kane, J. M., and Anderegg, L. D. L. (2012). Consequences of widespread tree mortality triggered by drought and temperature stress. *Nature Climate Change*, **3**, 30–36.

Andrady, A. (2011). Microplastics in the marine environment. *Marine Pollution Bulletin*, **62**, 1596–1605.

Angus, I. (2016). *Facing the Anthropocene: Fossil Capitalism and the Crisis of the Earth System.* New York: Monthly Review Press.

Araújo, D. F., Boaventura, G. R., Machado, W., et al. (2017). Tracing of anthropogenic zinc sources in coastal environments using stable isotope composition. *Chemical Geology*, **449**, 226–235.

Archer, D., Buffett, B., and Brovkin, V. (2009). Ocean methane hydrates as a slow tipping point in the global carbon cycle. *Proceedings of the National Academy of Sciences (USA)*, **106**, 20596–20601.

Archer, D., Eby, M., Brovkin, V., et al. (2009). Atmospheric lifetime of fossil fuel carbon dioxide. *Annual Review of Earth and Planetary Science*, **37**, 117–134.

Archer, D., and Ganopolski, A. (2005). A movable trigger: Fossil fuel CO_2 and the onset of the next glacial inception. *Geochemistry, Geophysics, Geosystems*, **6**, Q05003. doi:10.1029/2004GC000891.

Arduino, G. (1760). Due lettere del Sig. Giovanni Arduino sopra varie sue osservazioni naturali. *Nuova Raccolta d'Opuscoli Scientifici e Filologici*, **6**, 97–132, 133–180.

Arenillas, I., Alegret, L., Arz, J. A., et al. (2002). Cretaceous-Tertiary boundary planktic foraminiferal mass extinction and biochronology at Le Ceiba and Bochil, Mexico and El Kef, Tunisia. *Geological Society of America, Special Paper*, **356**, 253–263.

Arienzo, M. M., McConnell, J. R., Chellman, N., et al. (2016). A method for continuous ^{239}Pu determinations in Arctic and Antarctic ice cores. *Environmental Science and Technology*, **50**(13), 7066–7073.

Armour, K. C., Marshall, J., Scott, J. R., et al. (2016). Southern Ocean warming delayed by circumpolar upwelling and equatorward transport. *Nature Geoscience*, **9**, 549–554.

Arnold, R. W., Szabolcs, J., and Targulian, V. O. (1990). *Global Soil Change*. Laxenburg, Austria: International Institute for Applied Systems Analysis.

Arrhenius, S. (1896). On the influence of carbonic acid in the air upon the temperature of the ground. *London, Edinburgh, and Dublin Philosophical Magazine and Journal of Science*, Series 5, **41** (251), 237–276.

Arrhenius, S. (1908). *Worlds in the Making: The Evolution of the Universe*. New York: Harper and Bros.

Ashbee, P., Smith, I. F., and Evans, J. G. (1979). Excavation of three long barrows near Avebury, Wiltshire. *Proceedings of the Prehistoric Society*, **45**, 207–300.

Aubert, M., Brumm, A., Ramil, M., et al. (2014). Pleistocene cave art from Sulawesi, Indonesia. *Nature*, **514**, 223–227.

Aubry, M.-P., Ouda, K., Depuis, C., et al. (2007). The Global Standard Stratotype-section and Point (GSSP) for the base of the Eocene Series in the Dababiya section (Egypt). *Episodes*, **30**, 271–286.

Autin, W. J., and Holbrook, J. M. (2012). Is the Anthropocene an issue of stratigraphy or pop culture? *GSA Today*, **22**(7), 60–61.

Avio, C. M., Gorbi, S., and Regoli, F. (2017). Plastics and microplastics in the oceans: From emerging pollutants to emerged threat. *Marine Environmental Research*, **128**, 2–11.

Babcock, L. E., Peng, S., Zhu, M., Xiao, S., and Ahlberg, P. (2014). Proposed reassessment of the Cambrian GSSP. *Journal of African Earth Sciences*, **98**, 3–10.

Bacaloni, A., Insogna, S., and Zoccolillo, L. (2011). Remote zones air quality. Persistent organic pollutants: Sources, sampling and analysis. In Mazzeo, N., ed., *Air Quality Monitoring*,

Assessment and Management. Rijeka, Croatia: InTech, pp. 223–240.

Bacon, A. R., Richter, D. deB., Bierman, P. R., and Rood, D. H. (2012). Coupling meteoric ^{10}Be with pedogenic losses of ^{9}Be to improve soil residence time estimates on an ancient North American interfluve. *Geology*, **40**, 847–850.

Bader, J., Mesquita, D. S., Hodges, K. I., et al. (2011). A review on Northern Hemisphere sea-ice, storminess and the North Atlantic Oscillation: Observations and projected changes. *Atmospheric Research*, **101**(4), 809–834.

Baker, T. D., and Jikells, A. R. (2015). Biogeochemical cycles: Heavy metals. In North, G. R., Pyle, J. A., and Zhang, F., eds., *Encyclopedia of Atmospheric Sciences*, 2nd ed., vol. 1. Academic Press, pp. 201–204.

Bakker, P., Clark, P., Golledge, N. R., et al. (2017). Centennial-scale Holocene climate variations amplified by Antarctic Ice Sheet discharge. *Nature*, **541**, 72–76.

Balter, M. (2013). Archaeologists say the 'Anthropocene' is here–but it began long ago. *Science*, **340**, 261–262.

Bamber, J. L., Riva, R. E. M., Vermeersen, B. L. A., and LeBrocq, A. M. (2009). Reassessment of the potential sea-level rise from a collapse of the West Antarctic Ice Sheet. *Science*, **324**, 901–903.

Barbante, C., Spolaor, A., Cairns, W. R., and Boutron, C. (2017). Man's footprint on the Arctic environment as revealed by analysis of ice and snow. *Earth-Science Reviews*, **168**, 218–231.

Bard, E., Raisbeck, G., Yiou, F., and Jouzel, J. (2000). Solar irradiance during the last 1200 years based on cosmogenic nuclides. *Tellus B*, **52**, 985–992.

Barker, G. (2002). A tale of two deserts: Contrasting desertification histories in Rome's desert frontiers. *World Archaeology*, **33**, 488–507.

Barnes, D. K. A. (2005). Remote islands reveal rapid rise of Southern Hemisphere sea debris. *Directions in Science*, **5**, 915–921.

Barnes, D. K. A., Galgani, F., Thompson, R. C., and Barlaz, M. (2009). Accumulation and fragmentation of plastic debris in global environments. *Philosophical Transactions of the Royal Society*, **B364**, 1985–1998.

Barnett, R. L., Newton, T. L., Charman, D. J., and Gehrels, W. R. (2017). Salt-marsh testate amoebae as precise and widespread indicators of sea-level change. *Earth-Science Reviews*, **164**, 193–207.

Barnett, V., Payne, R., and Steiner, R. (1995). *Agricultural Sustainability*. New York: J. Wiley.

Barnola, J. M., Anklin, M., Porcheron, J., et al. (1995). CO_2 evolution during the last millennium as recorded by Antarctic and Greenland ice. *Tellus B*, **47**, 264–272.

Barnosky, A. D. (2005). Effects of Quaternary climatic change on speciation in mammals. *Journal of Mammalian Evolution*, **12**, 247–256.

Barnosky, A. D. (2008). Megafauna biomass tradeoff as a driver of Quaternary and future extinctions. *Proceedings of the National Academy of Sciences (USA)*, **105**, 11543–11548.

Barnosky, A. D. (2014). Palaeontological evidence for defining the Anthropocene. In Waters, C. N., Zalasiewicz, J., Williams, M., et al., eds., *A Stratigraphical Basis for the Anthropocene*. Geological Society, London, Special Publications, **395**, pp. 149–165.

Barnosky, A. D., Holmes, M., Kirchholtes, R., et al. (2014). Prelude to the Anthropocene: Two new North American Land Mammal Ages (NALMAs). *Anthropocene Review*, **1**(3), 225–242.

Barnosky, A. D., and Lindsey, E. L. (2010). Timing of Quaternary megafaunal extinction in South America in relation to human arrival and climate change. *Quaternary International*, **217**, 10–29.

Barnosky, A. D., Matzke, N., Tomiya, S., et al. (2011). Has the Earth's sixth mass extinction already arrived? *Nature*, **471**, 51–57.

Baroni, M., Savarino, J., Cole-Dai, J., et al. (2008). Anomalous sulfur isotope compositions of volcanic sulfate over the last millennium in Antarctic ice cores. *Journal of Geophysical Research*, **113**, D20112. doi:10.1029/2008JD010185.

Barrett, E. M. (1963). The California oyster industry. *Fish Bulletin*, **123**, 2–103.

Barshis, D. J., Ladner, J. T., Oliver, T. A., et al. (2013). Genomic basis for coral resilience to climate change. *Proceedings of the National Academy of Sciences (USA)*, **110**, 1387–1392.

Bar-Yosef, O. (2002). The Upper Paleolithic revolution. *Annual Reviews of Anthropology*, **31**, 363–393.

Baskin, J. (2015). Paradigm dressed as epoch: The ideology of the Anthropocene. *Environmental Values*, **24**, 9–29.

Bassis, J. N., Petersen, S. V., and MacCathles, L. (2017). Heinrich events triggered by ocean forcing and modulated by isostatic adjustment. *Nature*, **542**, 332–334.

Basson (2009). *Sedimentation and Sustainable Use of Reservoirs and River Systems.* Draft ICOLD Bulletin. http://icold-cigb.org/userfiles/files/CIRCULAR/CL1793Annex.pdf.

Battarbee, R. W. (1990). The causes of lake acidification, with special reference to the role of acid deposition. *Philosophical Transactions of the Royal Society of London Series B-Biological Sciences*, **327**, 339–347.

Battarbee, R. W., Jones, V. J., Flower, R. J., et al. (1996). Palaeolimnological evidence for the atmospheric contamination and acidification of high Cairngorm lochs with special reference to Lochnagar. *Botanical Journal of Scotland*, **48**, 79–87.

Bauer, A. M., and Ellis, E. C. (2018). The Anthropocene divide: Obscuring understanding of social-environmental change. *Current Anthropology*, 59 (2), 209–227.

Baustian, J. J., Mendelssohn, I. A., and Hester, M. W. (2012). Vegetation's importance in regulating surface elevation in a coastal salt marsh facing elevated rates of sea level rise. *Global Change Biology*, **18**, 3377–3382.

Bax, N., Williamson, A., Aguero, M., et al. (2003). Marine invasive alien species: A threat to global biodiversity. *Marine Policy*, **27**, 313–323.

Beasley, T. M., Carpenter, R., and Jennings, C. D. (1982). Plutonium, ^{241}Am and ^{137}Cs ratios, inventories and vertical profiles in Washington and Oregon continental shelf sediments. *Geochimica et Cosmochimica Acta*, **46**, 1931–1946.

Beerling, D. J. (2007). *The Emerald Planet.* Oxford University Press.

Beerling, D. J., and Royer, D. L. (2011). Convergent Cenozoic CO_2 history. *Nature Geoscience*, **4**, 418–420.

Bell, R. E., Chu, W., Kingslake, J., et al. (2017). Antarctic ice shelf potentially stabilized by export of meltwater in surface river. *Nature*, **544**, 344–348.

Bender, M. L. (2013). *Paleoclimate.* Princeton Primers in Climate. Princeton University Press.

Bennett, C. E., Thomas, R., Williams, M., et al. (in press). The broiler chicken as a signal of a human reconfigured biosphere. *Royal Society Open Science.*

Benninger L. K., and Dodge R. E. (1986). Fallout plutonium and natural radionuclides in annual bands of the coral *Montastrea annularis*, St. Croix, U.S. Virgin Islands. *Geochimica Cosmochimica Acta*, **50**, 2785–2797.

Berger, A., and Loutre, M. F. (2002). An exceptionally long interglacial ahead? *Science*, **297**, 1287–1288.

Berner, R. A. (2004). *The Phanerozoic Carbon Cycle: CO_2 and O_2.* Oxford University Press.

Berrojalbiz, N., Lacorte, S., Calbet, A., et al. (2009). Accumulation and cycling of polycyclic aromatic hydrocarbons in zooplankton. *Environmental Science & Technology*, **43**(7), 2295–2301.

Berry, E. W. (1925). The term Psychozoic. *Science*, **44**, 16.

Berry, K. L. E., Seemann, J., Dellwig, O., et al. (2013). Sources and spatial distribution of heavy metals in scleractinian coral tissues and sediments from the Bocas del Toro Archipelago, Panama. *Environmental Monitoring and Assessment*, **185** (11), 9089–9099.

Bertler, N. A. M., and Barrett, P. J. (2010). Vanishing polar ice sheets. In Dodson, J., ed., *Changing Climates, Earth Systems and Society*, International Year of Planet Earth, 49. Springer Science+Business Media B. V., pp. 49–84.

Besseling, E., Wang, B., Lürling, M., and Koelmans, A. A. (2014). Nanoplastic affects growth of *S. obliquus* and reproduction of *D. magna*. *Environmental Science and Technology*, **48**, 12336–12343.

Bhattacharya, D., Agrawa, S., Aranda, M., et al. (2016). Comparative genomics explains the evolutionary success of reef-forming corals. *eLife*, 5, e13288. doi:10.7554/eLife.13288.

Biber, E. (2016). Law in the Anthropocene Epoch. *UC Berkeley Public Law Research Paper No. 2834037* (September). https://papers.ssrn.com/sol3/papers.cfm?abstract_id=2834037.

Bidwell, O. W., and Hole, F. D. (1965). Man as a factor of soil formation. *Soil Science*, **99**, 65–72.

Bigalke, M., Weyer, S., Kobza, J., and Wilcke, W. (2010). Stable Cu and Zn isotope ratios as tracers of sources and transport of Cu and Zn in contaminated soil. *Geochimica et Cosmochimica Acta*, **74**(23), 6801–6813.

Bigalke, M., Weyer, S., and Wilcke, W. (2011). Stable Cu isotope fractionation in soils during oxic weathering and podzolization. *Geochimica et Cosmochimica Acta*, **75**(11), 3119–3134.

Bigler, M., Wagenbach, D., Fischer, H., et al. (2002). Sulphate record from a northeast Greenland ice core over the last 1200 years based on continuous flow analysis. *Annals of Glaciology*, **35**, 250–256.

Bigus, P., Tobiszewski, M., and Namieśnik, J. (2014). Historical records of organic pollutants in sediment cores. *Marine Pollution Bulletin*, **78**(1), 26–42.

Bindler, R. (2006). Mired in the past—looking to the future: Geochemistry of peat and the analysis of past environmental changes. *Global and Planetary Change*, **53**(4), 209–221.

Bing, H., Wu, Y., Zhou, J., Li, R., and Wang, J. (2016). Historical trends of anthropogenic metals in Eastern Tibetan Plateau as reconstructed from alpine lake sediments over the last century. *Chemosphere*, **148**, 211–219.

Bintanja, R., and Andry, O. (2017). Towards a rain-dominated Arctic. *Nature Climate Change*, **7**, 263–267.

Bintanja, R., van Oldenborgh, G. J., Drijfhout, S. S., et al. (2013). Important role for ocean warming and increased ice-shelf melt in Antarctic sea-ice expansion. *Nature Geoscience*, **6**, 376–379.

Birch, G. F., Gunns, T. J., and Olmos, M. (2015). Sediment-bound metals as indicators of anthropogenic change in estuarine environments. *Marine Pollution Bulletin*, **101**(1), 243–257.

Bird, E. C. F. (1993). *Submerging Coasts: The Effects of a Rising Sea Level on Coastal Environments.* Chichester: Wiley.

Blais, J. M., Macdonald, R. W., Mackay, D., et al. (2007). Biologically mediated transport of contaminants to aquatic systems. *Environmental Science & Technology*, **41**(4), 1075–1084.

Blasing, T. J. (2016). *Recent Greenhouse Gas Concentrations.* Carbon Dioxide Information Analysis Center. doi:10.3334/CDIAC/atg.032. http://cdiac.ornl.gov/pns/current_ghg.html.

Blöschl, G., Hall, J., Parajka, J., et al. (2017). Changing climate shifts timing of European floods. *Science*, **357**, 588–590.

Blunier, T., Chappellez, J., Schwander, J., et al. (1995). Variations in atmospheric methane concentration during the Holocene epoch. *Nature*, **374**, 46–49.

Bobrov, V. A., Bogush, A. A., Lenova, G. A., et al. (2011). Anomalous concentrations of zinc and copper in Highmoor Peat Bog, southeast coast of Lake Baikal. *Doklady Akademii Nauk*, **439**, 784–788.

Boerger, C. M., Lattin, G. L., Moore, S. L., and Moore, C. J. (2010). Plastic ingestion by planktivorous fishes in the North Pacific Central Gyre. *Marine Pollution Bulletin*, **60**, 2275–2278.

Bogdal, C., Schmid, P., Zennegg, M., et al. (2009). Blast from the past: Melting glaciers as a relevant source for persistent organic pollutants.

Environmental Science & Technology, **43**(21), 8173–8177.

Bollhöfer, A., and Rosman, K. J. R. (2001). Isotopic source signatures for atmospheric lead: The Northern Hemisphere. *Geochimica et Cosmochimica Acta*, **65**(11), 1727–1740.

Bolton, C. T., Hernández-Sánchez, M. T., Fuertes, M.-Á., et al. (2016). Decrease in coccolithophore calcification and CO_2 since the middle Miocene. *Nature Communications*, **7**, 10284. doi:10.1038/ncomms10284.

Bombelli, P., Howe, C. J., and Bertocchini, F. (2017a). Polyethylene bio-degradation by caterpillars of the wax moth *Galleria mellonella*. *Current Biology*, **27**(8), R292–293.

Bombelli, P., Howe, C. J., and Bertocchini, F. (2017b). Response to Weber et al. *Current Biology*, **27**(15), R745.

Bond, G., Heinrich, H., Broecker, W. S., et al. (1992). Evidence for massive discharges of icebergs into the North Atlantic Ocean during the last glacial period. *Nature*, **360**, 245–249.

Bond, G., Kromer, B., Beer, J., et al. (2001). Persistent solar influence on North Atlantic climate during the Holocene. *Science*, **294**, 2130–2136.

Bond, G., Showers, W., Cheseby, M., et al. (1997). A pervasive millennial-scale cycle in North Atlantic Holocene and glacial climates. *Science*, **278**, 1257–1266.

Bond, T. C., Bhardwaj, E., Dong, R., et al. (2007). Historical emissions of black and organic carbon aerosol from energy-related combustion, 1850–2000. *Global Biogeochemical Cycles*, **21**, GB2018. doi:10.1029/2006GB002840.

Bonhommet, N., and Zähringer, J. (1969). Paleomagnetism and potassium argon age determinations of the Laschamp geomagnetic polarity event. *Earth and Planetary Science Letters*, **6**, 43–46.

Bonneuil, C., and Fressoz, J.-B. (2016). *The Shock of the Anthropocene: The Earth, History and Us*. Translated from French by D. Fernbach. London, New York: Verso Books.

Borel, B. (2017). When the pesticides run out. *Nature*, **543**, 302–304.

Borsato, A., Frisia, S., Fairchild, I. J., et al. (2007). Trace element distribution in annual stalagmite laminae mapped by micrometer-resolution X-ray fluorescence: Implications for incorporation of environmentally significant species. *Geochimica et Cosmochimica Acta*, **71**, 1494–1512.

Boucher, J., and Friot D. (2017). *Primary Microplastics in the Oceans: A Global Evaluation of Sources*. Gland, Switzerland: IUCN. doi:10.2305/IUCN.CH.2017.01.en.

Boucher, O., Randall, D., Artaxo, P., et al. (2013). Clouds and aerosols. In Stocker, T. F., Qin, D., Plattner, G. K., et al., eds., *Climate Change 2013: The Physical Science Basis, Contribution of Working Group I to the Fifth Assessment Report of the Intergovernmental Panel on Climate Change*. New York: Cambridge University Press, pp. 571–657.

Boutron, C. F., Candelone, J. P., and Hong, S. (1995). Greenland snow and ice cores: Unique archives of large-scale pollution of the troposphere of the Northern Hemisphere by lead and other heavy metals. *Science of the Total Environment*, **160**, 233–241.

Bowell, R. J., Williams, K. P., Connelly, R. J., et al. (1999). Chemical containment of mine waste. In Metcalfe, R., and Rochelle, C. A., eds., *Chemical Containment of Waste in the Geosphere*. Geological Society, London, Special Publications, **157**, pp. 213–240.

Bowen, G. J., Maibauer, B. J., Kraus, M. J., et al. (2015). Two massive, rapid releases of carbon during the Paleocene-Eocene thermal maximum. *Nature Geoscience*, **8**, 44–47.

Boyce, J. I., Reinhardt, E. G., and Goodman, B. N. (2009). Magnetic detection of ship ballast mounds and anchorages at Caesarea Maritima, Israel. *Journal of Archaeological Science*, **36**, 1516–1526.

Bracegirdle, T. J., Connolley, W. M., and Turner, J. (2008). Antarctic climate change over the twenty first century. *Journal of Geophysical Research*, **113**, D03103. doi:10.1029/2007JD008933.

Brantley, S. L., McDowell, W. H., Dietrich, W. E., et al. (2017). Designing a network of critical zone observatories to explore the living skin of the terrestrial Earth. *Earth Surface Dynamics*. doi:10.5194/esurf-2017-36.

Braun, H., Christl, M., Rahmstorf, S., et al. (2005). Possible solar origin of the 1,470-year glacial climate cycle demonstrated in a coupled mode. *Nature*, **438**, 208–211.

Breitenlechner, E., Stöllner, T., Thomas, P., et al. (2014). An interdisciplinary study on the environmental reflection of prehistoric mining actvities at the Mitterberg main lode (Salzburg, Austria). *Archaeometry*, **56**(1), 102–128.

Bridgewater, A. V. (1986). Refuse composition projections and recycling technology. *Resources and Conservation*, **12**, 159–174.

Brigham-Grette, J., Melles, M., Minyuk, P., et al. (2013). Pliocene warmth, polar amplification, and stepped Pleistocene cooling recorded in NE Arctic Russia. *Science*, **340**(6139), 1421–1427.

Brimblecombe, P. (2005). The global sulfur cycle. In Schlesinger, W. H., ed., *Biogeochemistry: Treatise on Geochemistry*, vol. 8. Elsevier Science, pp. 645–682.

Brodie, J., Waterhouse, J., Schaffelke, B., et al. (2013). *Scientific Consensus Statement: Land Use Impacts on Great Barrier Reef Water Quality and Ecosystem Condition*. Reef Water Quality Protection Plan Secretariat, the State of Queensland. http://reefplan.qld.gov.au/about/assets/scientific-consensus-statement-2013.pdf.

Broecker, W. S. (1998). Paleocean circulation during the last deglaciation: A bipolar seesaw? *Paleoceanography*, **13**, 119–121.

Broecker, W. S., and Stocker, T. F. (2006) The Holocene CO_2 rise: Anthropogenic or natural? *EOS*, **87**, 27.

Brook, B. W., and Barnosky, A. D. (2012). Quaternary extinctions and their link to climate change. In Hannah, L., ed., *Saving a Million Species: Extinction Risk from Climate Change*. Washington, DC: Island Press, pp. 179–198.

Brown, A., Toms, P., Carey, C., and Rhodes, E. (2013). Geomorphology of the Anthropocene: Time-transgressive discontinuities of human-induced alluviation. *Anthropocene*, **1**, 3–13.

Browne, M. A., Crump, P., Niven, S. J., et al. (2011). Accumulation of microplastic on shorelines worldwide: Sources and sinks. *Environmental Science and Technology*, **45**, 9175–9179.

Browne, M. A., Galloway, T. S., and Thompson, R. C. (2010). Spatial patterns of plastic debris along estuarine shorelines. *Environmental Science and Technology*, **44**, 3404–3409.

Bryant, R. B., and Galbraith, J. M. (2002). Incorporating anthropogenic processes in soil classification. In Eswaran, H., Ahrens, R., Rice, T. J., and Stewart, B. A., eds., *Soil Classification: A Global Desk Reference*. Boca Raton, Florida: CRC Press, pp. 57–66.

Buddemeier, R. W., Baker, A. C., Fautin, D. G., and Jacobs, J. R., eds., (2004). The adaptive hypothesis of bleaching. *Coral Health and Disease*. Springer, pp. 427–444.

Buddemeier, R. W., and Fautin, D. (1993). Coral bleaching as an adaptive mechanism. A testable hypothesis. *Bioscience*, **43**, 320–326.

Buddemeier, R. W., and Smith, S. V. (1988). Coral reef growth in an era of rapidly rising sea level: Predictions and suggestions for long-term research. *Coral Reefs*, **7**, 51–56.

Budov, V. V. (1994). Hollow glass microspheres. Use, properties, and technology (Review). *Glass and Ceramics*, **51**(7), 230–235.

Budyko, M. I. (1977). On present-day climate changes. *Tellus*, **29**, 193–204.

Buesseler, K. O., Lamborg, C. H., Boyd, P. W., et al. (2007). Revisiting carbon flux through the ocean's twilight zone. *Science*, **316**, 567–570.

Buffon, G-L. L. de. (2018). *The Epochs of Nature*, ed. and trans. J. Zalasiewicz, A.-S. Milon, and M. Zalasiewicz. University of Chicago Press.

Buggisch, W. (1991). The global Frasnian-Famennian "Kellwasser Event." *Geologische Rundschau*, **80**, 49–72.

Buizert, C., and Schmittner, A. (2016). Southern Ocean control of glacial AMOC stability and Dansgaard-Oeschger interstadial duration. *Palaeoceanography*, **30**, 1595–1612.

Bukata, A. R., and Kyser T. K. (2007). Carbon and nitrogen isotope variations in tree-rings as records of perturbations in regional carbon and nitrogen cycles. *Environmental Science and Technology*, **41**(4), 1331–1338.

Buol, S. W., Southard, R. J., Graham, R. C., and McDaniel, P. A. (2011). *Soil Genesis and Classification*. New York: John Wiley & Sons.

Burghardt, T. E., Pashkevich, A., and Żakowska, L. (2016). Influence of volatile organic compounds emissions from road marking paints on ground-level ozone formation: Case study of Kraków, Poland. *Transportation Research Procedia*, **14**, 714–723.

Burke, L., Reytar, K., Spalding, M., et al. (2011). *Reefs at Risk Revisited*. World Resource Institute, p. 115. http://wri.org/publication/reefs-risk-revisited.

Burney, D. A., James, H. F., Pigott Burney, L., et al. (2001). Fossil evidence for a diverse biota from Kaua'i and its transformation since human arrival. *Ecological Monographs*, **7**, 615–641.

Burnley, S. J. (2007). A review of municipal waste composition in the United Kingdom. *Waste Management*, **27**, 1274–1285.

Burns, A., Pickering, M. D., Green, K. A., et al. (2017). Micromorphological and chemical investigation of late-Viking age grave fills at Hofstaðir, Iceland. *Geoderma*, **306**, 183–194. http://dx.doi.org/10.1016/j.geoderma.2017.06.021.

Butler, B. E. (1959). *Periodic Phenomena in Landscapes as a Basis for Soil Studies*. Melbourne, Australia: CSIRO.

Butler, C. D. (2016). Sounding the alarm: Health in the Anthropocene. *International Journal of Environmental Research and Public Health*, **13** (7), E665.

Butterfield, N. J. (2011). Animals and the invention of the Phanerozoic Earth system. *Trends in Ecology & Evolution*, **26**(2), 81–87.

Caiazzo, L., Baccolo, G., Barbante, C., et al. (2017). Prominent features in isotopic chemical and dust stratigraphies in a coastal East Antarctic ice sheet (Eastern Wilkes Land). *Chemosphere*, **176**, 273–287.

Callender, E. (2014). Heavy metals in the environment – historical trends. Reference Module in Earth Systems and Environmental Sciences. In Holland, H. D., and Turekian, K. K., eds., *Treatise on Geochemistry*, vol. 11, 2nd ed. Elsevier Ltd, pp. 59–89.

Campbell, J. E., Berry, J. A., Seibt, U., et al. (2017). Large historical growth in global terrestrial gross primary production. *Nature*, **544**, 84–87.

Campbell, J. E., Whelan, M. E., Seibt, U., et al. (2015). Atmospheric carbonyl sulfide sources from anthropogenic activity: Implications for carbon cycle constraints. *Geophysical Research Letters*, **42**, 3004–3010.

Canadell, J. G., Le Quéré, C., Raupach, M. R., et al. (2007). Contributions to accelerating atmospheric CO_2 growth from economic activity, carbon intensity, and efficiency of natural sinks. *Proceedings of the National Academy of Science (USA)*, **104**, 18866–1870.

Candelone, J. P., Hong, S., Pellone, C., and Boutron, C. F. (1995). Post-Industrial Revolution changes in large-scale atmospheric pollution of the northern hemisphere by heavy metals as documented in central Greenland snow and ice. *Journal of Geophysical Research: Atmospheres*, **100**(D8), 16605–16616.

Canfield, D. E., Glazer, A. N., and Falkowski, P. G. (2010). The evolution and future of Earth's nitrogen cycle. *Science*, **330**, 192–196.

Cantin, N. E., and Lough, J. (2014). Surviving coral bleaching events: *Porites* growth anomalies on the Great Barrier Reef. *PLoS One*, **9**(2), e88720.

Cao, Z. H., Ding, J. L., Hu, Z. H., et al. (2006). Ancient paddy soils from the Neolithic Age in China's

Yangtze River Delta. *Naturwissenschaften*, **93**, 232–236.

Capelotti, P. J. (2010). *The Human Archaeology of Space: Lunar, Planetary and Interstellar Relics of Exploration*. Jefferson: Mcfarland & Co.

Cappelletti, N., Skorupka, C. N., Migoya, M. C., et al. (2014). Behavior of dioxin like PCBs and PBDEs during early diagenesis of organic matter in settling material and bottom sediments from the sewage impacted Buenos Aires' coastal area, Argentina. *Bulletin of Environmental Contamination and Toxicology*, **93**(4), 388–392.

Capron, E., Govin, A., Stone, E. J., et al. (2014). Temporal and spatial structure of multi-millennial temperature changes at high latitudes during the Last Interglacial. *Quaternary Science Reviews*, **103**, 116–133.

Carlozo, L. R. (2008). Kicking butts. *Chicago Tribune*, 18 June 2008. http://articles.chicagotribune.com/2008-06-18/features/0806170174_1_cigarette-butts-secondhand-beach-house/2.

Carlton, J. T. (1979). History, biogeography, and ecology of the introduced marine and estuarine invertebrates of the Pacific coast of North America. Unpublished PhD dissertation, University of California, Davis.

Carlton, J. T. (1992). Introduced marine and estuarine mollusks of North America: An end-of-the-20th-century perspective. *Journal of Shellfish Research*, **11**, 489–505.

Carlton, J. T., and Cohen, A. N. (1998). Periwinkle's progress: The Atlantic snail *Littorina saxatilis* (Mollusca: Gastropoda) establishes a colony on a Pacific shore. *Veliger*, **41**, 333–338.

Carlton, J. T., Thompson, J. K., Schemel, L. E., and Nichols, F. H. (1990). Remarkable invasion of San Francisco Bay (California, USA) by the Asian clam *Potamocorbula amurensis*. *Marine Ecology Progress Series*, **66**, 81–94.

Carver, M. (1987). *Underneath English Towns: Interpreting Urban Archaeology*. London: Batsford.

Caseldine, C. J., Turney, C., and Long, A. J. (2010). IPCC and Palaeoclimate: An Evolving Story? *Journal of Quaternary Science*, **25**, 1–4.

Ceballos, G., Ehrlich, P. R., Barnosky, A. D., et al. (2015). Accelerated modern human-induced species losses: Entering the sixth mass extinction. *Scientific Advances*, **1**(5), e1400253. doi:10.1126/sciadv.1400253.

Certini, G., and Scalenghe, R. (2011). Anthropogenic soils are the golden spikes for the Anthropocene. *Holocene*, **21**, 1269–1274.

Chakrabarty, D. (2009). The climate of history: Four theses. *Critical Inquiry*, **35**(2), 197–222.

Chakrabarty, D. (2014). Conjoined histories. *Critical Inquiry*, **41**(Autumn), 1–23.

Chappell, J., and Shackleton, N. J. (1986). Oxygen isotopes and sea level. *Nature*, **6093**, 137–140.

Charlson, R. J., Anderson, T. L., and McDuff, R. E. (1992). The sulfur cycle. In Butcher, S. S., Charlson, R. J., Orians, G. H., and Wolfe, G. V., eds., *Global Biogeochemical Cycles*. London: Academic Press, pp. 285–300.

Charman, D. J., Roe, H. M., and Gehrels, W. R. (1998). The use of testate amoebae in studies of sea-level change: A case study from the Taf Estuary, South Wales, UK. *Holocene*, **8**, 209–218.

Chavez, F. P., Ryan, J., Lluch-Cota, S. E., and Ñiquen, M. (2003). From anchovies to sardines and back: Multidecadal change in the Pacific Ocean. *Science*, **299**, 217–221.

Chawla, F., Steinmann, P., Loizeau, J.-L., et al. (2010). Binding of ^{239}Pu and ^{90}Sr to organic colloids in soil solutions: Evidence from a field experiment. *Environmental Science & Technology*, **44**(22), 8509–8514.

Chemekov, Y. F. (1983). Technogenic deposits. In INQUA Congress, 11, Moscow, Abstracts 3, p. 62.

Chen, C. W., Chen, C. F., and Dong, C. D. (2012). Copper contamination in the sediments of Salt River Mouth, Taiwan. *Energy Procedia*, **16**, 901–906.

Church, J. A., Clark, P. U., Cazenave, A., et al. (2013). Sea level change. In Stocker, T. F., Qin, D.,

Plattner, G. K., et al., eds., *Climate Change 2013: The Physical Science Basis, Contribution of Working Group I to the Fifth Assessment Report of the Intergovernmental Panel on Climate Change*. New York: Cambridge University Press, pp. 1137–1216.

Church, J. A., and White, N. J. (2011). Sea-level rise from the late 19th to the early 21st Century. *Surveys in Geophysics*, **32**, 585–602.

Church, J. A., White, N. J., Aarup, T., et al. (2008). Understanding global sea levels: Past, present and future. *Sustainability Science*, **3**, 9–22.

Clark, J. D., Beyene, Y., WoldeGabriel, G., et al. (2003). Stratigraphic, chronological and behavioural contexts of Pleistocene *Homo sapiens* from Middle Awash, Ethiopia. *Nature*, **423**, 747–752.

Clark, J. D., de Heinzelin, J., Schick, K. D., et al. (1994). African *Homo erectus*: Old radiometric ages and young Oldowan assemblages in the Middle Awash Valley, Ethiopia. *Science*, **264**, 1907–1910.

Clark, N., and Yusoff, K. (2017). Geosocial formations and the Anthropocene. *Theory, Culture & Society*, **34**(2–3), 1–275.

Clark, P. U., Shakun, J. D., Marcott, S. A., et al. (2016). Consequences of twenty-first-century policy for multi-millennial climate and sea-level change. *Nature Climate Change*, **6**, 360–369.

Clark, W. C. (1986). Sustainable development of the biosphere: Themes for a research program. In Clark, W. C., and Munn R. E., eds., *Sustainable Development of the Biosphere*. IIASA, Laxenburg: Cambridge University Press, pp. 5–48.

Clarkson, C., Jacobs, Z., Marwick, B., et al. (2017). Human occupation of northern Australia by 65,000 years ago. *Nature*, **547**, 306–310.

Clarkson, O., Kasemann, S. A., Wood, R. A., et al. (2015). Ocean acidification and the Permo-Triassic mass extinction. *Science*, **348**(6231), 229–232.

Clette, F., Svalgaard, L., Vaquero, J. M., and Cliver, E. W. (2014). Revisiting the sunspot number. *Space Science Review*, **186**, 35–103.

Cocroft, W., and Schofield, G. (2012). The secret hill: Cold War archaeology of the Teufelsberg. *British Archaeology*, **126**, 38–43.

Codling, G., Vogt, A., Jones, P. D., et al. (2014). Historical trends of inorganic and organic fluorine in sediments of Lake Michigan. *Chemosphere*, **114**, 203–209.

Coe, M. T., and Foley, J. A. (2001). Human and natural impacts on the water resources of the Lake Chad basin. *Journal of Geophysical Research*, **106**, 3349–3356.

Cofaigh, C. O., Davies, B. J., Livingstone, S. J., et al. (2014). Reconstruction of ice-sheet changes in the Antarctic Peninsula since the Last Glacial Maximum. *Quaternary Science Reviews*, **100**, 87–110.

Cohen, A. N. (2004). Invasions in the sea. *Park Science*, **22**, 37–41.

Cohen, A. N. (2011). *The Exotics Guide: Non-native Marine Species of the North American Pacific Coast*. Oakland, CA: Center for Research on Aquatic Bioinvasions, Richmond, CA, and San Francisco Estuary Institute. http://exoticsguide.org.

Cohen, A. N., and Carlton, J. T. (1998). Accelerating invasion rate in a highly invaded estuary. *Science*, **279**, 555–557.

Cohen, D. J. (2013). The advent and spread of early pottery in East Asia: New dates and new considerations for the world's earliest ceramic vessels. *Journal of Austronesian Studies*, **4**(2), 55–90.

Cohen, S., Kettner, A. J., Syvitski, J. P. M., and Fekete, B. M. (2013). WBMsed: A distributed global-scale daily riverine sediment flux model-model description and validation. *Computers & Geosciences*, **53**, 80–93.

Cohen, T. J., Jansen, J. D., Gliganic, L. A., et al. (2015). Hydrological transformation coincided with megafaunal extinction in central Australia. *Geology*, **43**, 195–198.

Cole, J. E. (1996). Coral records of climate change: Understanding past variability in the tropical ocean-stmosphere. In Jones, P. D., Bradley, R. S., and Jouzel, J., eds., *Climatic Fluctuations and Forcing: Mechanisms of the Last 2000 Years*. Berlin: Springer, pp. 333–355.

Cole, M., Lindeque, P., Fileman, E., et al. (2013). Microplastic ingestion by zooplankton. *Environmental Science and Technology*, **47**, 6646–6655.

Collins, M., Knutti, R., Arblaster, J. M., et al. (2013). Long-term climate change: Projections, commitments and irreversibility. In Stocker, T. F., Qin, D., Plattner, G. K., et al., eds., *Climate Change 2013: The Physical Science Basis, Contribution of Working Group I to the Fifth Assessment Report of the Intergovernmental Panel on Climate Change*. New York: Cambridge University Press, pp. 1029–1136.

Committee on Nonnative Oysters in the Chesapeake Bay, National Research Council (2004). *Non-native Oysters in the Chesapeake Bay.*

Conard, N. J. (2003). Palaeolithic ivory sculptures from southwestern Germany and the origin of figurative art. *Nature*, **426**, 830–832.

Conard, N. J. (2009). A female figurine from the basal Aurignacian of Hohle Fels Cave in southwestern Germany. *Nature*, **459**, 248–252.

Conway, K. W., Krautter, M., Barrie, J. V., and Neuweiler, M. (2001). Hexactinellid sponge reefs on the Canadian continental shelf: A unique "living fossil." *Geoscience Canada*, **28**, 71–78.

Cook, A. J., Fox, A. J., Vaughan, D. G., and Ferrigno, J. G. (2005). Retreating glacier fronts on the Antarctic Peninsula over the past half-century. *Science*, **308**, 541–544.

Cook, A. J., Holland, P. R., Meredith, M. P., et al. (2016). Ocean forcing of glacier retreat in the western Antarctic Peninsula. *Science*, **353**(6296), 283–286.

Cook, A. J., van der Flierdt, T., Williams, T., et al. (2013). Dynamic behaviour of the East Antarctic ice sheet during Pliocene warmth. *Nature Geoscience*, **6**, 765–769.

Cook, A. J., and Vaughan, D. G. (2010). Overview of areal changes of the ice shelves on the Antarctic Peninsula over the past 50 years. *Cryosphere*, **4** (10), 77–98.

Cook, D. E., and Gale, S. J. (2005). The curious case of the date of introduction of leaded fuel to Australia: Implications for the history of Southern Hemisphere atmospheric lead pollution. *Atmospheric Environment*, **39**(14), 2553–2557.

Cook, P. J., and Shergold, J. H. (eds.) (1986). *Phosphate Deposits of the World: Volume 3 - Neogene to Modern Phosphorites.* Cambridge University Press.

Cooke, C. A., and Bindler, R. (2015). Lake sediment records of preindustrial metal pollution. In Blais, J. M., Rosen, M. R., and Smol, J. P., eds., *Environmental Contaminants: Using Natural Archives to Track Sources and Long-Term Trends of Pollution*. Dordrecht: Springer, pp. 101–119.

Cooper, A. H., Brown, T. J., Price, S. J., et al. (2018, in press). Humans are the most significant global geological driving force of the 21st century. *Anthropocene Review*.

Copper, P. (2001). Evolution, radiations, and extinctions in Proterozoic to mid-Paleozoic reefs. In Stanley, G. D., Jr., ed., *The History and Sedimentology of Ancient Reef Systems*. Topics in Geobiology, **17**, pp. 89–119. New York: Academic/Plenum Publishers.

Corcoran, P. L. (2015). Benthic plastic debris in marine and fresh water environments. *Environmental Science Processes & Impacts*, **17**, 1363–1369.

Corcoran, P. L., Moore, C. J., and Jazvac, K. (2014). An anthropogenic marker horizon in the rock record. *Geological Society of America Today*, **24** (6), 4–8.

Corcoran, P. L., Norris, T., Ceccanese, T., et al. (2015). Hidden plastics of Lake Ontario, Canada and their potential preservation in the sediment record. *Environmental Pollution*, **204**, 17–25.

Cordell, D., Drangert, J.-O., and White, S. (2009). The story of phosphorus: Global food security and food for thought. *Global Environmental Change*, **19**, 292–305.

Correggiari, A., Cattaneo, A., and Trincardi, F. (2005). The modern Po Delta system: Lobe switching and asymmetric prodelta growth. *Marine Geology*, **222**, 49–74.

Craig, O. E., Saul, H., Lucquin, A., et al. (2013). Earliest evidence for the use of pottery. *Nature*, **496**, 351–354.

Crucifix, M. (2012). Oscillators and relaxation phenomena in Pleistocene climate theory. *Philosophical Transactions of the Royal Society A*, **370**, 1140–1165.

Crutzen, P. J. (2002). Geology of Mankind. *Nature*, **415**(January), 23.

Crutzen, P. J. (2006). Albedo enhancement by stratospheric sulfur injections: A contribution to resolve a policy dilemma? *Climate Change*, **77**, 211–219.

Crutzen, P. J., and Stoermer, E. F. (2000). The "Anthropocene." *IGBP Global Change Newsletter*, **41**, 17–18.

Cuffey, K. M., Clow, G. D., Steig, E. J., et al. (2016). Deglacial temperature history of West Antarctica. *Proceedings of the National Academy of Science (USA)*, **113**(50), 14249–14254.

Cui, Y., Kump, L. R., Ridgwell, A. J., et al. (2011). Slow release of fossil carbon during the Palaeocene-Eocene Thermal Maximum. *Nature Geoscience*, **4** (7), 481–485.

Dachs, J., and Méjanelle, L. (2010). Organic pollutants in coastal waters, sediments, and biota: A relevant driver for ecosystems during the Anthropocene? *Estuaries and Coasts*, **33**(1), 1–14.

Dalby, S. (2009). *Security and Environmental Change*. Cambridge/Boston: Polity.

Darwin, C. R. (1881). *The Formation of Vegetable Mould*. London: John Murray.

Davies, J. (2016). *The Birth of the Anthropocene*. Oakland: University of California Press.

Davis, H., and Turpin, E. (2015). *Art in the Anthropocene: Encounters among Aesthetics, Politics, Environments and Epistemologies*. London: Open Humanities Press.

Davis, R. V. (2011). Inventing the present: Historical roots of the Anthropocene. *Earth Sciences History*, **30**(1), 63–84.

Dawson, A. G., and Stewart, I. (2007). Tsunami deposits in the geological record. *Sedimentary Geology*, **200**(3–4), 166–183.

Day, J. W., Jr., Gunn, J. D., Folan, W., and Yanez-Arancibia, A. (2007). Emergence of complex societies after sea level stabilized. *EOS, Transactions of the American Geophysical Union*, **88**, 169–176.

De Beer, J., Price, S. J., and Ford, J. R. (2012). 3D modelling of geological and anthropogenic deposits at the World Heritage Site of Bryggen in Bergen, Norway. *Quaternary International*, **251**, 107–116.

de Bruyn, A. M. H., and Gobas, F. A. P. C. (2004). Modelling the diagenetic fate of persistent organic pollutants in organically enriched sediments. *Ecological Modelling*, **179**(3), 405–416.

Deacon, T. W. (2012). *Incomplete Nature: How Mind Emerged from Matter*. New York: Norton.

Dean, J. R., Leng, M. J., and Mackay, A. W. (2014). Is there an isotopic signature of the Anthropocene? *Anthropocene Review*, **1**(3), 276–287.

Dearing, J. A., and Jones, R. T. (2003). Coupling temporal and spatial dimensions of global sediment flux through lake and marine sediment records. *Global and Planetary Change*, **39**, 147–168.

Dearing, J. A., Yang, X.-D., Dong, X.-H., et al. (2012). Extending the timescale and range of ecosystem services through paleoenvironmental analyses, exemplified in the lower Yangtze basin. *Proceedings of the National Academy of Sciences (USA)*, **109**, E1111–E1120.

DeCarlo, T. M., and Cohen, A. L. (2017). Dissepiments, density bands and signatures of thermal stress in *Porites* skeletons. *Coral Reefs*, **6**, 749–761.

DeConto, R. M., and Pollard, D. (2016). Contribution of Antarctica to past and future sea-level rise. *Nature*, **531**, 591–597.

Dedkov, A. P., and Mozzherin, V. I. (1992). Erosion and sediment yield in mountain regions of the world. In Walling, D. E., Davies, T. R., and Hasholt, B., eds., *Erosion, Debris Flows and Environment in Mountain Regions*, **209**. Wallingford, UK: International Association of Hydrological Sciences (IAHS) Publication, pp. 29–36.

DeFelice, R. C., Eldredge, L. G., and Carlton, J. T. (2001). Nonindigenous invertebrates. In Eldredge, L. G., and Smith, C. M., eds., *A Guidebook of Introduced Marine Species in Hawaii*. Bishop Museum Technical Report, p. 21.

Dehaut, A., Cassone, A.-L., Frère, L., et al. (2016). Microplastics in seafood: Benchmark protocol for their extraction and characterization. *Environmental Pollution*, **215**, 223–233.

Dekiff, J. H., Remy, D., Kasmeier, J., and Fries, E. (2014). Occurrence and spatial distribution of microplastics in sediments from Norderney. *Environmental Pollution*, **186**, 248–256.

Della Torre, C., Bergami, E., Salvati, A., et al. (2014). Accumulation and embryotoxicity of polystyrene nanoparticles at early stage of development of sea urchin embryos *Paracentrotus lividus*. *Environmental Science and Technology*, **48**, 12302–12311.

DeLong, K. L., Maupin, C. R., Flannery, J. A., et al. (2016). Refining temperature reconstructions with the Atlantic coral *Siderastrea siderea*. *Palaeogeography, Palaeoclimatology, Palaeoecology*, **462**, 1–15.

Department for Environment Food and Rural Affairs (2009). *Municipal Waste Composition: A Review of Municipal Waste Component Analyses* (Project WR0119).

D'Errico, F., Henshilwood, C., Vanhaeren, M., and van Niekerk, K. (2005). *Nassarius kraussianus* shell beads from Blombos Cave: Evidence for symbolic behaviour in the Middle Stone Age. *Journal of Human Evolution*, **48**, 3–24.

Deschamps, P., Durand, N., Bard, E., et al. (2012). Ice-sheet collapse and sea-level rise at the Bølling warming 14,600 years ago. *Nature*, **483**, 559–564.

Desnoyers, J. (1829). Observations sur un ensemble de dépôts marins plus récents que les terrains tertiaires du Bassin de la Seine et constituant une formation géologique distincte: Précédés d'un apercu de la nonsimultanéité des bassins tertiaires. *Annales Scientifiques Naturelles*, **16** (171–214), 402–419.

Diamond, J., and Bellwood, P. (2003). Farmers and their languages: The first expansions. *Science*, **300**, 597–603.

Diaz, R. J., and Rosenberg, R. (2008). Spreading dead zones and consequences for marine ecosystems. *Science*, **321**, 926–929.

Ding, Q., Schweiger, A., L'Heureux, M., et al. (2017). Influence of high-latitude atmospheric circulation changes on summertime Arctic sea ice. *Nature Climate Change*, **7**, 289–295.

Dodd, M. S., Papineau, D., Grenne, T., et al. (2017). Evidence for early life in Earth's oldest hydrothermal vent precipitates. *Nature*, **543**, 60–64.

Dokuchaev, V. V. (1883). *Russian chernozem*. In *Selected Works of V. V. Dokuchaev*, vol. 1, trans. N. Kaner. Jerusalem: Israel Program for Scientific Translations Ltd (for USDA-NSF), pp. 14–419.

Dolan, A. M., Haywood, A. M., Hill, D. J., et al. (2011). Sensitivity of Pliocene ice sheets to orbital forcing. *Palaeogeography, Palaeoclimatology, Palaeoecology*, **309**, 98–110.

D'Olivo, J. P., and McCulloch, M. T. (2017). Response of coral calcification and calcifying fluid composition to thermally induced bleaching stress. *Scientific Reports*, **7**(2207).

Domack, E. W., Amblas, D., Gilbert, R., et al. (2006). Subglacial morphology and glacial evolution of the Palmer deep outlet system, Antarctic Peninsula. *Geomorphology*, **75**, 125–142.

Donner, S. D., Rickbeil, G. J. M., and Heron, S. F. (2017). A new, high-resolution global mass coral bleaching database. *PLoS ONE* **12**(4), e0175490.

Döring, M. (2007). Wasser für Gadara: 94 km langer Tunnel antiker Tunnel im Norden Jordaniens entdeckt. *Querschnitt*, **21**, 24–35

Douglas, I. (1993). Sediment transfer and siltation. In Turner, B. L., Clark, W. C., Kates, R. W., et al., eds., *The Earth as Transformed by Human Action*. Cambridge University Press, pp. 215–234.

Douglas, I., and Lawson, N. (2001). The human dimensions of geomorphological work in Britain. *Journal of Industrial Ecology*, **4**(2), 9–33.

Dounas, C., Davies, I., Triantafyllou, G., et al. (2007). Large-scale impacts of bottom trawling on shelf primary productivity. *Continental Shelf Research*, **27**, 2198–2210.

Drage, D., Mueller, J. F., Birch, G., et al. (2015). Historical trends of PBDEs and HBCDs in sediment cores from Sydney estuary, Australia. *Science of the Total Environment*, **512**, 177–184.

Drijfhout, S., Bathiany, S., Brovkin, V., et al. (2015). Catalogue of abrupt shifts in Intergovernmental Panel on Climate Change climate models. *Proceedings of the National Academy of Sciences (USA)*, **112**(43), E5777–E5786.

Druffel, E. R. M. (1996). Post-bomb radiocarbon records of surface corals from the tropical Atlantic Ocean. *Radiocarbon*, **38**(3), 563–572.

Duan, K., Thompson, L. G., Yao, T., et al. (2007). A 1000 year history of atmospheric sulfate concentrations in southern Asia as recorded by a Himalayan ice core. *Geophysical Research Letters*, **34**, L01810. doi:10.1029/2006GL027456.

Dudal, R. (2005). The sixth factor of soil formation. *Eurasian Soil Science*, **38**, S60.

Dunham, A. C. (1992). Developments in industrial mineralogy: I. The mineralogy of brick-making. *Proceedings of the Yorkshire Geological Society*, **49**(2), 95–104.

Ebbesson, J. (2014). Social-ecological security and international law in the Anthropocene. In Ebbesson, J., et al., eds., *International Law and Changing Perceptions of Security: Liber Amicorum Said Mahmoudi*. Boston/Leiden: Brill/ Martinus Nijhoff, pp. 71–92.

Eby, N., Hermes, R., Charnley, N., et al. (2010). Trinitite–the atomic rock. *Geology Today*, **26**(5), 180–185.

Edgeworth, M. (2011). *Fluid Pasts: Archaeology of Flow*. London: Bloomsbury Academic.

Edgeworth, M. (2013). Scale. In Graves-Brown, P., Harrison, R., and Piccini, A., eds., *The Oxford Handbook of the Archaeology of the Contemporary World*. Oxford University Press.

Edgeworth, M. (2014). The relationship between archaeological stratigraphy and artificial ground and its significance to the Anthropocene. In Waters, C. N., Zalasiewicz, J., Williams, M., et al., eds., *A Stratigraphical Basis for the Anthropocene*. Geological Society, London, Special Publications, **395**, pp. 91–108.

Edgeworth, M., Richter, D. deB., Waters, C. N., et al. (2015). Diachronous beginnings of the Anthropocene: The lower bounding surface of anthropogenic deposits. *Anthropocene Review*, **2** (1), 33–58.

Edinburgh, T., and Fay, J. J. (2016). Estimating the extent of Antarctic summer sea ice during the Heroic Age of Antarctic Exploration. *Cryosphere*, **10**, 2721–2730.

Edwards, K. J., and Whittington, G. (2001). Lake sediments, erosion and landscape change during the Holocene in Britain and Ireland. *Catena*, **42**, 143–173.

Eerkes-Medrano, D., Thompson, R. C., and Aldridge, D. C. (2015). Microplastics in freshwater systems: A review of the emerging threats, identification of knowledge gaps and prioritisation of research needs. *Water Resources*, **75**, 63–82.

Eichler, A., Tobler, L., Eyrikh, S., et al. (2014). Ice-core based assessment of historical anthropogenic heavy metal (Cd, Cu, Sb, Zn) emissions in the Soviet Union. *Environmental Science & Technology*, **48**(5), 2635–2642.

Eighmy, J. L., and Sternberg, R. S. (eds.) (1990). *Archaeomagnetic Dating*. Tucson: University of Arizona Press.

Ekaykin, A. E., Vladimirova, D. O., Lipenkov, V. Y., and Masson-Delmotte, V. (2016). Climatic variability in Princess Elizabeth Land (East Antarctica) over the last 350 years. *Climate of the Past*, **13**, 61–71.

Ekdahl, E. J., Teranes, J. L., Guilderson, T. P., et al. (2004). Prehistorical record of cultural eutrophication from Crawford Lake, Canada. *Geology*, **32**, 745–748.

Ellis, E. C. (2011). Anthropogenic transformation of the terrestrial biosphere. *Philosophical Transactions of the Royal Society A*, **369**, 1010–1035.

Ellis, E. C. (2015). Ecology in an anthropogenic biosphere. *Ecological Monographs*, **85**, 287–331.

Ellis, E. C., Maslin, M., Boivin, N., and Bauer, A. (2016). Involve social scientists in defining the Anthropocene. *Nature*, **540**, 192–193.

Elsig, J., Schmitt, J., Leuenberger, D., et al. (2009). Stable isotope constraints on Holocene carbon cycle changes from an Antarctic ice core. *Nature*, **461**(7263), 507–510.

Emerson, T. E., Hedman, K. M., and Simon, M. L. (2005). Marginal horticulturalists or maize agriculturalists? Archaeobotanical, paleopathological, and isotopic evidence relating to Langford Tradition Maize Consumption. *Midcontinent Journal of Archaeology*, **30**, 67–118.

Emmett, R., and Lekan, T. (2016). *Whose Anthropocene? Revisiting Dipesh Chakrabarty's "Four Theses."* Transformations in Environment and Society, 2016/2, Munich: Rachel Carson Center. http://environmentandsociety.org/sites/default/files/2015_new_final.pdf.

EPICA Community Members (2006). One-to-one coupling of glacial climate variability in Greenland and Antarctica. *Nature*, **444**, 195–198.

Epstein, S., Krishnamurthy, R. V., Oeschger, H., et al. (1990) Environmental information in the isotopic record in trees [and discussion]. *Philosphical Transactions of the Royal Society of London A*, **330**, 427–439.

Ericson, J. P., Vorosmarty, C. J., Dingman, S. L., et al. (2005). Effective sea-level rise and deltas: Causes of change and human dimension implications. *Global and Planetary Change*, **50**, 63–82.

Eriksen, M., Masin, S., Wilson, S., et al. (2013). Microplastic pollution in surface waters of the Laurentian Great Lakes. *Marine Pollution Bulletin*, **77**, 177–182.

Erisman, J. W., Sutton, M. A., Galloway, J., et al. (2008). How a century of ammonia synthesis changed the world. *Nature Geoscience*, **1**, 636–639.

Erlandson, J. (2013). Shell middens and other anthropogenic soils as global stratigraphic signatures of the Anthropocene. *Anthropocene*, **4**, 24–32.

Erwin, D. H., Laflamme, M., Tweedt, S. M., et al. (2011). The Cambrian conundrum: Early divergence and later ecological success in the early history of animals. *Science*, **334**, 1091–1097.

Esper, J., Cook, E. R., and Schweingruber, F. H. (2002). Low-frequency signals in long tree ring chronologies for reconstructing past temperature variability. *Science*, **295**, 2250–2253.

Etheridge, D. M., Steele, L. P., Langenfelds, R. L., et al. (1996). Natural and anthropogenic changes in atmospheric CO_2 over the last 1000 years from air in Antarctic ice and firn. *Journal of Geophysical Research*, **101**(D2), 4115–4128.

Evans, D., Stephenson, M., and Shaw, R. (2009). The present and future use of "land" below ground. *Land Use Policy*, **26S**, S302–S316.

Evans, M. E., and Heller, F. (2003). *Environmental Magnetism: Principles and Applications of Enviromagnetics*. Amsterdam, Boston: Academic Press.

Fairbanks, R. G., and Matthews, R. K. (1978). The marine oxygen isotope record in Pleistocene coral, Barbados, West Indies. *Quaternary Research*, **10**, 181–196

Fairbridge, R. W. (1961). Eustatic changes in sea level. In Ahrens, L. H., Press, F., Rankama, K., and

Runcorn, S. K., eds., *Physics and Chemistry of the Earth*, **4**. New York: Pergamon, pp. 99–185.

Fairchild, I. J. (2018). Geochemical records in speleothems. In DellaSala, D., and Goldstein, M. I., eds., *Encyclopedia of the Anthropocene*, vol. 1. Oxford: Elsevier. doi:10.1016/B978-0-12-809665-9.09775 -5.

Fairchild, I. J., and Frisia, S. (2014). Definition of the Anthropocene: A view from the underworld. In Waters, C. N., Zalasiewicz, J., Williams, M., et al., eds., *A Stratigraphical Basis for the Anthropocene*. Geological Society Special Publication, **395**, pp. 239–254.

Fairchild, I. J., Loader, N. J., Wynn, P. M., et al. (2009). Sulfur fixation in wood mapped by synchrotron X-ray studies: Implications for environmental archives. *Environmental Science and Technology*, **43**, 1310–1315.

Fairchild, I. J., Quest, M., Tucker, M. E., and Hendry, G. L. (1988). Chemical analysis of sedimentary rocks. In Tucker, M. E., ed., *Techniques in Sedimentology*. Oxford: Blackwells, pp. 274–354.

Falcon, A. (2015). Aristotle on causality. In Zalta, E. N., ed., *The Stanford Encyclopedia of Philosophy* (Spring 2015 edition). https://plato.standord.edu/archives/spr2015/entries/aristotle-causality/.

Farbstein, R., Radić, D., Brajković, D., and Miracle, P. T. (2012). First Epigravettian ceramic figurines from Europe (Vela Spila, Croatia). *PLoS ONE*, **7**, e41437.

Fekiacova, Z., Cornu, S., and Pichat, S. (2015). Tracing contamination sources in soils with Cu and Zn isotopic ratios. *Science of the Total Environment*, **517**, 96–105.

Feldpausch, T. R., Phillips, O. M., Brienen, T. J. W., et al. (2016). Amazon forest response to repeated droughts. *Global Biogeochemical Cycles*, **30**, 964–982.

Felis, T., Merkel, U., Asami, R., et al. (2012). Pronounced interannual variability in tropical South Pacific temperatures during Heinrich Stadial 1. *Nature Communications*, **3**, 965.

Ferré, B., Durrieu de Madron, X., Estournel, C., et al. (2008). Impact of natural (waves and currents) and anthropogenic (trawl) resuspension on the export of particulate matter to the open ocean: Application to the Gulf of Lion (NW Mediterranean). *Continental Shelf Research*, **28**, 2071–2091.

Ferretti, D. F., Miller, J. B., White, J. W. C., et al. (2005). Unexpected changes to the global methane budget over the last 2,000 years. *Science*, **309**, 1714–1717.

Ferse, S. (2008). *Artificial Reefs and Coral Transplantation: Fish Community Responses and Effects on Coral Recruitment in North Sulawesi/Indonesia*. Dissertation. University of Bremen.

Fey, F., Dankers, N., Steenbergen, J., and Goudswaard, K. (2010). Development and distribution of the non-indigenous Pacific oyster (*Crassostrea gigas*) in the Dutch Wadden Sea. *Aquaculture International*, **18**, 45–59.

Field, D. B., Baumgartner, T. R., Charles, C. D., Ferreira-Bartrina, V., and Ohman, M. D. (2006). Planktonic foraminifera of the California Current reflect 20th-century warming. *Science*, **311**, 63–66.

Field, L. P., Milodowski, A. E., Shaw, R. P., et al. (2017). Unusual morphologies and the occurrence of ikaite (CaCO3 · 6H2O) pseudomorphs in rapid growth, hyperalkaline speleothem. *Mineralogical Magazine*, **81**(3), 565–589.

Fielding, C. R., Browne, G. H., Field, B., et al. (2011). Sequence stratigraphy of the ANDRILL AND-2A drillcore, Antarctica: A long-term, ice-proximal record of Early to Mid-Miocene climate, sea-level and glacial dynamism. *Palaeogeography, Palaeoclimatology, Palaeoecology*, **305**, 337–351.

Figueres, C., Schellnhuber, H. J., Whiteman, G., et al. (2017). Three years to safeguard our climate. *Nature*, **546**, 593–595.

Filippelli, G. (2002). The global phosphorus cycle. *Reviews in Mineralogy and Geochemistry*, **48**, 391–425.

Finlayson, B., Mithen, S. J., Najjar, M., et al. (2011). Architecture, sedentism, and social complexity at Pre-Pottery Neolithic A WF16, Southern Jordan. *Proceedings of the National Academy of Sciences (USA)*, **108**(20), 8183–8188.

Finney, S. C. (2014). The "Anthropocene" as a ratified unit in the ICS International Chronostratigraphic Chart: Fundamental issues that must be addressed by the Task Group. In Waters, C. N., Zalasiewicz, J. A., Williams, M., et al., eds., *A Stratigraphical Basis for the Anthropocene*. Geological Society, London, Special Publications, **395**, pp. 23–28.

Finney, S. C., and Edwards, L. E. (2016). The "Anthropocene" epoch: Scientific decision or political statement? *GSA Today*, **26**(2–3), 4–10.

Firestone, R. B., West, A., and Kennett, J. P., et al. (2007). Evidence for an extraterrestrial impact 12,900 years ago that contributed to the megafaunal extinctions and the younger Dryas cooling. *Proceedings of the National Academy of Sciences (USA)*, **104**, 16016–16021.

Fischer, H., Wagenbach, D., and Kipfstuhl, J. (1998). Sulfate and nitrate firn concentrations on the Greenland ice sheet: 2. Temporal anthropogenic deposition changes. *Journal of Geophysical Research*, **103**, 21935–21942.

Fischer, V., Elsner, N. O., Brenke, N., et al. (2015). Plastic pollution of the Kuril–Kamchatka Trench area (NW Pacific). *Deep Sea Research, Part II*, **111**, 399–405.

Fischer-Kowalski, M., Krausmann, F., and Pallua, I., 2014. A sociometabolic reading of the Anthropocene: Modes of subsistence, population size and human impact on Earth. *Anthropocene Review*, **1**, 8–33.

Fisher, A. T., Mankoff, K. D., Tulaczyk, S. M., et al. (2015). High geothermal heat flux measured below the West Antarctic Ice Sheet. *Science Advances*, **1**(6), e1500093.

FitzGerald, D. M., Fenster, M. S., Argow, B. A., and Buynevich, I. V. (2008). Coastal impacts due to sea-level rise. *Annual Review of Earth and Planetary Sciences*, **36**, 601–647.

Flügel, E., and Senowbari-Daryan, B. (2001). Triassic reefs of the Tethys. In Stanley, G. D., Jr., ed., *The History and Sedimentology of Ancient Reef Systems*. Topics in Geobiology, **17**. New York: Academic/Plenum Publishers, pp. 217–249.

Fofonoff, P. W., Ruiz, G. M., Steves, B., et al. (2017). *National Exotic Marine and Estuarine Species Information System*. http://invasions.si.edu/nemesis/ (accessed 2 September 2017).

Fohlmeister, J., Kromer, B., and Mangini, A. (2011). The influence of soil organic matter age spectrum on the reconstruction of atmospheric ^{14}C levels via stalagmites. *Radiocarbon*, **53**, 99–115.

Foley, R. A., and Lahr, M. M. (2015). Lithic landscapes: Early human impact from stone tool production on the central Saharan environment. *PLo SONE*, **10**(3), e0116482. doi:10.1371/journal.pone.0116482.

Foley, S. F., Gronenborn, D., Andreae, M. O., et al. (2013). The Palaeoanthropocene: The beginnings of anthropogenic environmental change. *Anthropocene*, **3**, 83–88.

Food and Agriculture Organization (FAO) of the United Nations, World Food Situation, monthly update. http://fao.org/worldfoodsituation/csdb/en/ (accessed 22 June 2018).

Food and Agriculture Organization of the United Nations (FAOSTAT) (2017). FAOSTAT database. Available from FAO, Rome. http://faostat.fao.org/.

Forbes, D., and Syvitski, J. P. M. (1995). Paraglacial coasts. In Woodruffe, C., and Carter, R. W. G., eds., *Coastal Evolution*. Cambridge University Press, pp. 373–424.

Forbes, R. J. (1958). *Man the Maker: A History of Technology and Engineering*. New York: Abelard-Schuman.

Ford, J. R., Kessler, H., Cooper, A. H., et al. (2010). An enhanced classification of artificial ground. British Geological Survey, Open Report OR/10/036.

Ford, J. R., Price, S. J., Cooper, A. H., and Waters, C. N. (2014). An assessment of lithostratigraphy for anthropogenic deposits. In Waters, C. N., Zalasiewicz, J., Williams, M., et al., eds., *A Stratigraphical Basis for the Anthropocene.* Geological Society, London, Special Publications, **395**, pp. 55–89.

Foster, G. L., Lear, C. H., and Rae, J. W. B. (2012). The evolution of pCO2, ice volume and climate during the middle Miocene. *Earth and Planetary Science Letters*, **341–344**, 243–254.

Foster, G. L., Royer, D. L., and Lunt, D. J. (2017). Future climate forcing potentially without precedent in the last 420 million years. *Nature Communications*, **8**, 14845. doi:10.1038/ncomms14845.

Francey, R. J., Allison, C. E., Etheridge, D. M., et al. (1999). A 1000-year high precision record of $\delta^{13}C$ in atmospheric CO_2. *Tellus B*, **51**, 170–193.

Free, C. M., Jensen, O. P., Mason, S. A., et al. (2014). High-levels of microplastic pollution in a large, remote, mountain lake. *Marine Pollution Bulletin*, **85**, 156–163.

Freestone, D., Vidas, D., and Torres Camprubi, A. (2017). Sea level rise and impacts on maritime zones and limits: The work of the ILA Committee on International Law and Sea Level Rise. *Korean Journal of International and Comparative Law*, **5** (1), 5–35.

Friedman, G. M. (1950). Benthos of Lake Sevan. *Works of the Sevan Hydrobiological Station*, **11**, 7–93.

Friedman, T. (2016). *Thank You for Being Late.* Allen Lane/Penguin Books.

Frisia, S., Borsato, A., Fairchild, I. J., and Susini, J. (2005). Variations in atmospheric sulphate recorded in stalagmites by synchrotron micro-XRF and XANES analyses. *Earth and Planetary Science Letters*, **235**, 729–740.

Frisia, S., Borsato, A., Preto, N., and McDermott, F. (2003). Late Holocene annual growth in three Alpine stalagmites records the influence of solar activity and the North Atlantic Oscillation on winter climate. *Earth and Planetary Science Letters*, **216**, 411–424.

Fudge, T. J., Steig, E. J., Markle, B. R., et al. (2013). Onset of deglacial warming in West Antarctica driven by local orbital forcing. *Nature*, **500**, 440–446.

Fujii, T., Moynier, F., Abe, M., et al. (2013). Copper isotope fractionation between aqueous compounds relevant to low temperature geochemistry and biology. *Geochimica et Cosmochimica Acta*, **110**, 29–44.

Gabrieli, J., Cozzi, G., Vallelonga, P., et al. (2011). Contamination of Alpine snow and ice at Colle Gnifetti, Swiss/Italian Alps, from nuclear weapons tests. *Atmospheric Environment*, **45**, 587–593.

Gallardo, B., and Aldridge, D. C. (2013). Evaluating the combined threat of climate change and biological invasions on endangered species. *Biological Conservation*, **160**, 225–233.

Gallet, Y., Genevey, A., and Courtillot, V. (2003). On the possible occurrence of archeomagnetic jerks in the geomagnetic field over the past three millennia. *Earth and Planetary Science Letters*, **214**, 237–242.

Galloway, T. S., Cole, M., and Lewis, C. (2017). Interactions of microplastic debris throughout the marine ecosystem. *Nature Ecology and Evolution*, **1**. doi:10.1038/s41559-017-0116.

Gałuszka, A., and Migaszewski, Z. M. (2017). Glass microspheres as a potential indicator of the Anthropocene: A first study in an urban environment. *Holocene*, **28**(2), 323–329.

Gałuszka, A., and Migaszewski, Z. M. (2018). Chemical signals of the Anthropocene. In DellaSala, D., and Goldstein, M. I., eds., *Encyclopedia of the Anthropocene*, vol. 1. Oxford: Elsevier.

Gałuszka, A., Migaszewski, Z. M., and Zalasiewicz, J. (2014). Assessing the Anthropocene with geochemical methods. In Waters, C. N., Zalasiewicz, J., Williams, M., et al., eds., *A Stratigraphical Basis for the Anthropocene.*

Geological Society, London, Special Publications, **395**, pp. 221–238.

Gambaryan, M. G. (1979). Temperature regime of Lake Sevan. *Trudy Sevanskoi Gidrobiologicheskoi Stantsii*, **17**, 123–129 [in Russian].

Ganopolski, A., Winkelmann, R., and Schellnhuber, H. J. (2016) Critical insolation–CO_2 relation for diagnosing past and future glacial inception. *Nature*, **529**, 200–203.

Garcia, T., Féraud, G., Falguères, C., et al. (2010). Earliest human remains in Eurasia: New $^{40}Ar/^{39}Ar$ dating of the Dmanisi hominid-bearing levels, Georgia. *Quaternary Geochronology*, **5**, 443–451.

García-Alix, A., Jimenez-Espejo, F. J., Lozano, J. A., et al. (2013). Anthropogenic impact and lead pollution throughout the Holocene in Southern Iberia. *Science of the Total Environment*, **449**, 451–460.

García-Artola, A., Cearreta, A., and Leorri, E. (2015). Relative sea-level changes in the Basque coast (northern Spain, Bay of Biscay) during the Holocene and Anthropocene: The Urdaibai estuary case. *Quaternary International*, **364**, 172–180.

García-Ruiz, J. M., Beguería, S., Nadal-Romero, E., González-Hidalgo, J. C., Lana-Renault, N., and Sanjuán, Y. (2015). A meta-analysis of soil erosion rates across the world. *Geomorphology*, **239**, 160–173.

Garrett, R. G. (2000). Natural sources of metals to the environment. *Human and Ecological Risk Assessment*, **6**, 945–963.

Garrett, T. L. (2014). Long-run evolution of the global economy: 1. Physical basis. *Earth's Future*, **2**, 127–151.

Gasperi, J., Dris, R., Mirande-Bret, C., et al. (2015). First overview of microplastics in indoor and outdoor air. In 15th EuCheMS International Conference on Chemistry and the Environment. https://hal-enpc.archives-ouvertes.fr/hal-01195546/.

Gattuso, J.-P., Magnan, A., Billé, R., et al. (2015). Contrasting futures for ocean and society from different anthropogenic CO_2 emissions scenarios. *Science*, **349**(6243).

Gauthier-Lafaye, F, Holliger, P., and Blanc, P. L. (1996). Natural fission reactors in the Franceville basin, Gabon: A review of the conditions and results of a "critical event" in a geologic system. *Geochimica et Cosmochimica Acta*, **60**, 4831–4852.

GBRMA (2017). Significant coral decline and habitat loss on the Great Barrier Reef. http://gbrmpa.gov.au/media-room/latest-news/coral-bleaching/2017/significant-coral-decline-and-habitat-loss-on-the-great-barrier-reef.

Gehrels, R. (2010a). Sea-level changes since the Last Glacial Maximum: An appraisal of the IPCC Fourth Assessment Report. *Journal of Quaternary Science*, **25**, 26–38.

Gehrels, W. R. (2010b). Late Holocene land- and sea-level changes in the British Isles: Implications for future sea-level predictions. *Quaternary Science Reviews*, **29**, 1648–1660.

Gehrels, W. R., Roe, H. M., and Charman, D. J. (2001). Foraminifera, testate amoebae and diatoms as sea-level indicators in UK saltmarshes: A quantitative multiproxy approach. *Journal of Quaternary Science*, **16**, 201–220.

Gehrels, W. R., and Shennan, I. (2015). Sea level in time and space: Revolutions and inconvenient truths. *Journal of Quaternary Science*, **30**, 131–143.

Gehrels, W. R., and Woodworth, P. L. (2013). When did modern rates of sea-level rise start? *Global and Planetary Change*, **100**, 263–277.

Gemmell, N. J., Schwartz, M. K., and Robertson, B. C. (2004). Moa were many. *Proceedings of the Royal Society London B*, **271**, S430–S432.

Genty, D., and Massault, M. (1999). Carbon transfer dynamics from bomb-^{14}C and $\delta^{13}C$ time series of a laminated stalagmite from SW France: Modelling and comparison with other stalagmite records. *Geochimica et Cosmochimica Acta*, **63**, 1537–1548.

Georgescu-Roegen, N. (1975). Energy and economic myths. *Southern Economic Journal*, **41**(3),

347–381. (reprinted as chapter 1 [1972] in *Energy and Economic Myths: Institutional and Analytical Economic Essays.* New York: Pergamon.)

Gerasimov, I. P. (1979). Anthropogene and its major problem. *Boreas*, **8**, 23–30.

Gerig, B. S., Chaloner, D. T., Janetski, D. J., et al. (2015). Congener patterns of persistent organic pollutants establish the extent of contaminant biotransport by pacific salmon in the Great Lakes. *Environmental Science & Technology*, **50** (2), 554–563.

GESAMP (2016). *Sources, Fate and Effects of Microplastics in the Marine Environment: Part Two of a Global Assessment*, ed. P. J. Kershaw and C. M. Rochman. (IMO/FAO/ UNESCO-IOC/ UNIDO/WMO/IAEA/UN/UNEP/UNDP Joint Group). Rep. Stud. GESAMP No. 93. http://gesamp.org/data/gesamp/files/file_element/0c50c023936f7ffd16506be330b43c56/rs93e.pdf.

Gevao, B., Boyle, E. A., Carrasco, G. G., et al. (2016). Spatial and temporal distributions of polycyclic aromatic hydrocarbons in the Northern Arabian Gulf sediments. *Marine Pollution Bulletin*, **112** (1), 218–224.

Geyer, R., Jambeck, J. R., Lavender Law, K. (2017). Production, use, and fate of all plastics ever made. *Science Advances*, **3**, e1700782. doi:10.1126/sciadv.1700782.

GFDL (Geophysical Fluid Dynamics Laboratory) (2017). Global warming and hurricanes: An overview of current research results. https://gfdl.noaa.gov/global-warming-and-hurricanes/ (accessed June 2017).

Gherardi, F., Audigane, P., and Gaucher, E. C. (2012). Predicting long-term geochemical alteration of wellbore cement in a generic geological CO_2 confinement site: Tackling a difficult reactive transport modelling challenge. *Journal of Hydrology*, **420–421**, 340–359.

Gibbard, P. L., and Head, M. J. (2009). IUGS ratification of the Quaternary System/Period and the Pleistocene Series/Epoch with a base at 2.58 Ma. *Quaternaire*, **20**(4), 411–412.

Gibbard, P. L., and Head, M. J. (2010). The newly-ratified definition of the Quaternary System/Period and redefinition of the Pleistocene Series/Epoch, and comparison of proposals advanced prior to formal ratification. *Episodes*, **33**, 152–158.

Gibbard, P. L., Head, M. J., and Walker, M. J. C. (2010). Formal ratification of the Quaternary system/period and the Pleistocene series/Epoch with a base at 2.58 Ma. *Journal of Quaternary Science*, **25**, 96–102.

Gibbard, P. L., Smith, A. G., Zalasiewicz, J. A., et al. (2005). What status for the Quaternary? *Boreas*, **34**, 1–6.

Gibbard, P. L., and Walker, M. J. C. (2014). The term "Anthropocene" in the context of formal geological classification. In Waters, C. N., Zalasiewicz, J. A., Williams, M., et al., eds., *A Stratigraphical Basis for the Anthropocene.* Geological Society, London, Special Publications, **395**, pp. 29–37.

Gibbs, S. J., Bown, P. R., Sessa, J. A., et al. (2006). Nannoplankton extinction and origination across the Paleocene-Eocene Thermal Maximum. *Science*, **314**(5806), 1770–1773.

Gilbert, A., Flowers, G. E., Miller, G. H., et al. (2017). Projected demise of Barnes Ice Cap: Evidence of an unusually warm 21st century Arctic. *Geophysical Research Letters*, **44**(6), 2810–2816.

Gilbert, G. K. (1877). *The Geology of the Henry Mountains.* Washington, DC: US Government Printing Office.

Gillings, M. R., and Paulsen, I. T. (2014). Microbiology of the Anthropocene. *Anthropocene*, **5**, 1–8.

Gingerich, P. D. (2006). Environment and evolution through the Paleocene–Eocene thermal maximum. *Trends in Ecology & Evolution*, **21**(5), 246–253.

Giosan, L., Syvitski, J., Constantinescu, S., and Day, J. (2014). Climate change: Protect the world's deltas. *Nature*, **516**, 31–33.

Giuliani, S., Bellucci, L. G., Çağatay, M. N., et al. (2017). The impact of the 1999 Mw 7.4 event in the İzmit Bay (Turkey) on anthropogenic contaminant (PCBs, PAHs and PBDEs) concentrations recorded in a deep sediment core. *Science of the Total Environment*, **590–591**, 799–808.

Glacken, C. J. (1956). Changing ideas of the habitable world. In Thomas, W. L., Jr., ed., *Man's Role in Changing the Face of the Earth*. University of Chicago Press, pp. 79–92.

Glasser, N. F., Harrison, S., Jansson, K. N., et al. (2011). Global sea-level contribution from the Patagonian Icefields since the Little Ice Age maximum. *Nature Geoscience*, **4.5**(May 2011), 303–307.

Glikson, A. (2013). Fire and human evolution: The deep-time blueprints of the Anthropocene. *Anthropocene*, **3**, 89–92.

Goldberg, E. D. (1997). Plasticizing the seafloor: An overview. *Environmental Science and Technology*, **18**, 195–201.

Golledge, N. R., Levy, R. H., McKay, R. M., and Naish, T. R. (2017). East Antarctic ice sheet most vulnerable to Weddell Sea warming. *Geophysical Research Letters*, **44**, 2343–2351.

Gomez, B., Cui, Y., Kettner, A. J., et al. (2009). Simulating changes to the sediment transport regime of the Waipaoa River driven by climate change in the twenty-first century. *Global and Planetary Change*, **67**, 153–166.

Gonzalez, R. O., Strekopytov, S., Amato, F., et al. (2016). New insights from zinc and copper isotopic compositions into the sources of atmospheric particulate matter from two major European cities. *Environmental Science & Technology*, **50**(18), 9816–9824.

Goodman, D., and Chant, C. (eds.) (1999). *European Cities & Technology: Industrial to Post-Industrial City*. London: Routledge.

Gori, G. (1965). *La Forestal: La tragedia del quebracho Colorado*. Buenos Aires: Platina/Stilcograf.

Goto-Azuma, K., and Koerner, R. M. (2001). Ice core studies of anthropogenic sulfate and nitrate trends in the Arctic. *Journal of Geophysical Research*, **106**(D5), 4959–4969.

Gowlett, J. A. J. (2016). The discovery of fire by humans: A long and convoluted process. *Philosophical Transactions of the Royal Society B*, **371**, 20150164.

Gradstein, F. M., Ogg, J. G., Schmitz, M., and Ogg, G. (eds.) (2012). *The Geological Time Scale 2012*. Elsevier.

Gradstein, F. M., Ogg, J. G., Smith, A. G., et al. (2004). A new geologic time scale with special reference to the Precambrian and Neogene. *Episodes*, **27**, 83–100.

Gradstein, F. M., Ogg, J. G., Smith, A. G. (eds.) (2005). *A Geologic Time Scale 2004*. Cambridge University Press.

Grasso, D. N. (2000). Geological effects of underground nuclear testing. United States Geological Survey Open-File Report 00–176.

Greb, L., Saric, B., Seyfried, H., and Leinfelder, R. R. (1996). Ökologie und Sedimentologie eines rezenten Rampensystems an der Karibikküste von Panamá. *Profil*, **10**.

Greenbaum, J. S., Blankenship, D. D., Young, D. A., et al. (2015). Ocean access to a cavity beneath Totten Glacier in East Antarctica. *Nature Geoscience*, 8, 294–298.

Greenop, R., Foster, G. L., Wilson, P. A., and Lear, C. H. (2014). Middle Miocene climate instability associated with high-amplitude CO_2 variability. *Paleoceanography*, **29**, 845–853.

Gregory, M. R. (2009). Environmental implications of plastic debris in marine settings—entanglement, ingestion, smothering, hangers-on, hitch-hiking and alien invasions. *Philosophical Transactions of the Royal Society B*, **364**, 2013–2025.

Gregory, M. R., and Andrady, A. L. (2003). *Plastics in the marine environment*. In Andrady, A. L., ed., *Plastics and the Environment*. New Jersey: Wiley & Sons, pp. 379–401.

Griffin, E. (2010). *A Short History of the British Industrial Revolution*. London: Palgrave.

Grimalt, J. O., Fernandez, P., Berdie, L., et al. (2001). Selective trapping of organochlorine compounds in mountain lakes of temperate areas. *Environmental Science & Technology*, **35**, 2690–2697.

Grinevald, J. (1987). On a holistic concept for deep and global ecology: The Biosphere. *Fundamenta Scientiae*, **8**(2), 197–226.

Grinevald, J. (1988). Sketch for a history of the idea of the Biosphere. In Bunyard, P., and Goldsmith, E., eds., *GAIA, the Thesis, the Mechanisms, and the Implications*. Camelford, Cornwall, UK: Wadebridge Ecological Centre, pp. 1–34. (reprinted in Bunyard, P., ed., *Gaia in Action: Science of the Living Earth*. Edinburgh: Floris Books, 1996, pp. 34–53.)

Grinevald, J. (2007). *La Biosphère de l'Anthropocène: Climat et pétrole, la double menace. Repères transdisciplinaires (1824–2007)*. Geneva, Switzerland: Georg/Editions Médecine & Hygiène.

Grossman, E. L. (2012). Oxygen isotope stratigraphy. In Gradstein, F. M., Ogg, J. G, Schmitz, M., and Ogg, G., eds., *The Geological Time Scale 2012*. Elsevier, pp. 181–206.

Grunwald, S., Thompson, J. A., and Boettinger, J. L. (2011). Digital soil mapping and modeling at continental scales: Finding solutions for global issues. (SSSA 75th Anniversary Special Paper). *Soil Science Society of America Journal*, **75**(4), 1201–1213.

Gutjahr, M., Ridgwell, A., Sexton, P. F., et al. (2017). Very large release of mostly volcanic carbon during the Palaeocene-Eocene Thermal Maximum. *Nature*, **548**(7669), 573–777.

Haff, P. K. (2012). Technology and human purpose: The problem of solids transport on the Earth's surface. *Earth System Dynamics*, **3**, 149–156.

Haff, P. K. (2014a). Technology as a geological phenomenon: Implications for human well-being. In Waters, C. N., Zalasiewicz, J., Williams, M., eds., *A Stratigraphical Basis for the Anthropocene*. Geological Society London, Special Publication, **395**, pp. 301–309.

Haff, P. K. (2014b). Humans and technology in the Anthropocene: Six rules. *Anthropocene Review*, **1**, 126–136.

Haff, P. K. (2016). Purpose in the Anthropocene: Dynamical role and physical basis. *Anthropocene*, **16**, 54–60.

Haigh, J. D., and Cargill, P. (2015). *The Sun's Influence on Climate*. Princeton Primers in Climate. New Jersey: Princeton University Press.

Hall, N. M., Berry, K. L. E., Rintoul, L., and Hoogenboom, M. O. (2015). Microplastic ingestion in scleractinian corals. *Marine Biology*, **162**, 725–732.

Halpern, B. S., Walbridge, S., Selkoe, K. A., et al. (2008). A global map of human impact on marine ecosystems. *Science*, **319**, 948–952.

Hamilton, C. (2017). *Defiant Earth: The Fate of Humans in the Anthropocene*. Cambridge, UK: Polity Books.

Hamilton, C., and Grinevald, J. (2015). Was the Anthropocene anticipated? *Anthropocene Review*, **2**(1), 59–72.

Hammarlund, E. U., Dahl, T. W., Harper, D. A. T., et al. (2012). A sulfidic driver for the end-Ordovician mass extinction. *Earth and Planetary Science Letters*, **331–332**, 128–139.

Han, Y. M., An, Z. S., and Cao, J. J. (2017). The Anthropocene: A potential stratigraphic definition based on black carbon, char, and soot records. In DellaSala, D., and Goldstein, M. I., eds., *Encyclopedia of the Anthropocene*, vol. 1. Oxford: Elsevier. doi:10.1016/B978-0-12-409548-9.10001-6.

Han, Y. M., Wei, C., Huang, R. J., et al. (2016). Reconstruction of atmospheric soot history in inland regions from lake sediments over the past 150 years. *Scientific Reports*, **6**, 19151.

Hancock, G. J., Leslie, C., Everett, S. E., Tims, S. G., Brunskill, G. J., and Haese, R. (2011). Plutonium as a chronomarker in Australian and New Zealand sediments: A comparison with ^{137}Cs. *Journal of Environmental Radioactivity*, **102**, 919–929.

Hancock, G. J., Tims, S. G., Fifield, L. K., and Webster, I. T. (2014). The release and persistence of radioactive anthropogenic nuclides. In Waters, C. N.,

Zalasiewicz, J., Williams, M., et al., eds., *A Stratigraphical Basis for the Anthropocene.* Geological Society, London, Special Publications, **395**, pp. 265–281.

Hancock, T. (2016). Healthcare in the Anthropocene: Challenges and opportunities. *Health Care Quarterly*, **19**(3), 17–22.

Hancock, T. (2017). Population health promotion in the Anthropocene. In Rootman, I., et al., eds., *Health Promotion in Canada*, 4th ed. Toronto: Canadian Scholars Press.

Hancock, T., Capon, A., Dietrich, U., and Patrick, R. (2016). Governance for health in the Anthropocene. *International Journal of Health Governance*, **21**(4), 1–20.

Hancock, T., Spady, D. W., and Soskolne, C. (eds.) (2015). Global change and public health: Addressing the ecological determinants of health. http://cpha.ca/uploads/policy/edh-brief.pdf.

Hanebuth, T. J. J., Lantzsch, H., and Nizou, J. (2015). Mud depocenters on continental shelves—appearance, initiation times, and growth dynamics. *Geo-Marine Letters*, **35**(6), 487–503.

Hanna, E., Navarro, F. J., Pattyn, F., et al. (2013). Ice-sheet mass balance and climate change. *Nature*, **498**, 51–59.

Hansen, J., Sato, M., Hearty, P., et al. (2016). Ice melt, sea level rise and superstorms: Evidence from paleoclimate data, climate modeling, and modern observations that 2°C global warming could be dangerous. *Atmospheric Chemistry and Physics*, **16**, 3761–3812.

Hansen, P. H. (2013). *The Summits of Modern Man: Mountaineering after the Enlightenment.* Cambridge, MA: Harvard University Press.

Hanvey, J. S., Lewis, P. J., Lavers, J. L., et al. (2017). A review of analytical techniques for quantifying microplastics in sediments. *Analytical Methods*, **9**, 1369–1383.

Haraway, D. (2015). Anthropocene, Capitalocene, Plantationocene, Chthulucene: Making kin. *Environmental Humanities*, **6**, 159–165.

Harmand, S., Lewis, J. E., Feibel, C. S., et al. (2015). 3.3-million-year-old stone tools from Lomekwi 3, west Turkana, Kenya. *Nature*, **521**, 310–315.

Harris, E. C. (2014). Archaeological stratigraphy as a paradigm for the Anthropocene. *Journal of Contemporary Archaeology*, **1**(1), 105–109.

Harris, T. D., and Smith, V. H. (2016). Do persistent organic pollutants stimulate cyanobacterial blooms? *Inland Waters*, **6**(2), 124–130.

Harshvardhan, K., and Jha, B. (2013). Biodegradation of low-density polyethylene by marine bacteria from pelagic waters, Arabian Sea, India. *Marine Pollution Bulletin*, **77**, 100–106.

Hart, A. B., and Lawn, C. J. (1977). Combustion of coal and oil power station boilers. *CEGB Research*, **5**, 4–17.

Harvey, M. C., Brassel, S. C., Belcher, C. M., and Montanari, A. (2008) Combustion of fossil organic matter at the Cretaceous-Paleogene (K-P) boundary. *Geology*, **36**, 355–358.

Harwood, T. D., Tomlinson, I., Potter, C. A., and Knight, J. D. (2011). Dutch elm disease revisited: Past, present and future management in Great Britain. *Plant Pathology*, **60**, 545–555.

Hastings, M. G., Jarvis, J. C., and Steig, E. J. (2009). Anthropogenic impacts on nitrogen isotopes of ice-core nitrate. *Science*, **324**, 1288.

Haughton, S. (1865). *Manual of Geology.* Dublin, London: Longman & Co.

Hautmann, M. (2012). *Extinction: End-Triassic Mass Extinction.* In *Encyclopedia of Life Sciences.* Chichester: John Wiley & Sons Ltd. doi:10.1002/9780470015902.a0001655.pub3.

Hawkins, E., Ortega, P., Suckling, E., et al. (2017) Estimating changes in global temperature since the pre-industrial period. *Bulletin of the American Meteorological Society.* doi:10.1175/BAMS-D-16-0007.1.

Hawkins, W., and Wohletz, K. (1996). Visual inspection for CTBT verification. Los Alamos National Laboratory, OAC Project Number: ST484A.

Hay, W. W. (1994). Pleistocene-Holocene fluxes are not the earth's norm. In Hay, W., ed., *Global*

Sedimentary Geofluxes, **599**. Washington: National Academy of Sciences Press, pp. 15–27.

Hay, W. W. (2013). *Experimenting on a Small Planet: A Scholarly Entertainment.* London: Springer.

Haygarth, P. M., and Jones, K. C. (1992). Atmospheric deposition of metals to agricultural surfaces. In *Biogeochemistry of Trace Elements*. Boca Raton: Lewis Publishers, pp. 249–276.

Hazen, R. M., Papineau, D., Bleeker, W., et al. (2008). Mineral evolution. *American Mineralogist*, **93**, 1639–1720.

Hazen, R. M., and Ferry, J. M. (2010). Mineral evolution: Mineralogy in the fourth dimension. *Elements*, **6**, 9–12.

Hazen, R. M., Grew, E. S., Origlieri, M. J., and Downs, R. T. (2017). On the mineralogy of the "Anthropocene Epoch." *American Mineralogist*, **102**, 595–611.

Head, M. J., Aubry, M.-P., Walker, M., et al. (2017). A case for formalizing subseries (subepochs) of the Cenozoic Era. *Episodes*, **40**(1), 22–27.

Head, M. J., and Gibbard, P. L. (2015). Formal subdivision of the Quaternary System/Period: Past, present, and future. *Quaternary International*, **383**, 4–35.

Heaney, P. J. (2017). Defining minerals in the age of humans. *American Mineralogist*, **102**, 925–926.

Heard, B. P., Brook, B. W., Wigley, T. M. L., and Bradshaw, C. J. A. (2017). Burden of proof: A comprehensive review of the feasibility of 100% renewable-electricity systems. *Renewable and Sustainable Energy Reviews*, **76**, 1122–1133.

Heim, S., and Schwarzbauer, J. (2013). Pollution history revealed by sedimentary records: A review. *Environmental Chemistry Letters*, **11** (3), 255–270.

Heim, S., Schwarzbauer, J., Littke, R., et al. (2004). Geochronology of anthropogenic pollutants in riparian wetland sediments of the Lippe River (Germany). *Organic Geochemistry*, **35**(11–12), 1409–1425.

Heinrich, H. (1988). Origin and consequences of cyclic ice rafting in the northeast Atlantic Ocean during the past 130,000 years. *Quaternary Research*, **29**, 142–142.

Heiss, G., and Leinfelder, R. R. (2008). "Fünf vor Zwölf"– Verschwinden die Riffe? In Leinfelder, R. R., Heiss, G., and Moldrzyk, U., eds., *"Abgetaucht": Begleitbuch zur Sonderausstellung zum Internationalen Jahr des Riffes*. Leinfelden-Echterdingen: Konradin-Verlag, pp. 182–197.

Hemming, N. G., Guilderson, T. P., and Fairbanks, R. G. (1998). Seasonal variations in the boron isotopic composition of coral: A productivity signal? *Global Biogeochemical Cycles*, **12**, 581–586.

Hemming, N. G., and Hanson, G. N. (1992). Boron isotope composition and concentration in modern marine carbonates. *Geochimica Cosmochimica Acta*, **56**, 537–543.

Hempelmann, A., and Weber, W. (2012). Correlation between the sunspot number, the total solar irradiance, and the terrestrial insolation. *Solar Physics*, **277**, 417–430.

Hendy, I. L., Napier, T. J., and Schimmelmann, A. (2015). From extreme rainfall to drought: 250 years of annually resolved sediment deposition in Santa Barbara Basin, California. *Quaternary International*, **387**, 3–12.

Hennissen, J. A. I., Head, M. J., De Schepper, S., and Groeneveld, J. (2014). Palynological evidence for a southward shift of the North Atlantic Current at ~2.6 Ma during the intensification of late Cenozoic Northern Hemisphere glaciation. *Paleoceanography*, **28**. doi:10.1002/2013PA002543.

Hennissen, J. A. I., Head, M. J., De Schepper, S., and Groeneveld, J. (2015). Increased seasonality during the intensification of Northern Hemisphere glaciation at the Pliocene–Pleistocene boundary ~2.6 Ma. *Quaternary Science Reviews*, 129, 321–332.

Hennissen, J. A. I., Head, M. J., De Schepper, S., and Groeneveld, J. (2017). Dinoflagellate cyst paleoecology during the Pliocene–Pleistocene climatic transition in the North Atlantic.

Palaeogeography, Palaeoclimatology, Palaeoecology, 470, 81–108.

Henshilwood, C. S., d'Errico, F., and Watts, I. (2009). Engraved ochres from the Middle Stone Age levels at Blombos Cave, South Africa. *Journal of Human Evolution*, **57**, 27–47.

Henshilwood, C. S., d'Errico, F., Yates, R., et al. (2002). Emergence of modern human behaviour: Middle Stone Age engravings from South Africa. *Science*, **295**, 1278–1280.

Heringman, N. (2015). Deep time and the dawn of the Anthropocene. *Representations*, **129**, 56–85.

Herz-Fischler, R. (2009). *The Shape of the Great Pyramid*. Waterloo, Canada: Wilfrid Laurier University Press.

Herzschuh, U., Birks, J. B., Laepple, T., et al. (2016). Glacial legacies on interglacial vegetation at the Pliocene-Pleistocene transition in NE Asia. *Nature Communications*, **7**, 11967. doi:10.1038/ncomms11967.

Hetzinger, S., Pfeiffer, M., Dullo, W.-C., et al. (2010). Rapid 20th century warming in the Caribbean and impact of remote forcing on climate in the northern tropical Atlantic as recorded in a Guadeloupe coral. *Palaeogeography, Palaeoclimatology, Palaeoecology*, **296**, 111–124.

Hey, E. (2016). International law and the Anthropocene. *European Society of International Law (ESIL) Reflections*, **5**(10), 1–7.

Hibbard, K. A., Crutzen, P. J., Lambin, E. F., et al. (2006). Decadal interactions of humans and the environment. In Costanza, R., Graumlich, L., and Steffen, W., eds., *Integrated History and Future of People on Earth*. Dahlem Workshop Report, **96**, pp. 341–375.

Hicks, S., and Isaksson, E. (2006). Assessing source areas of pollutants from studies of fly ash, charcoal, and pollen from Svalbard snow and ice. *Journal of Geophysical Research*, **111**(D02113). doi:10.1029/2005JD006167.

Higgins, S. (2016). Review: Advances in delta-subsidence research using satellite methods. *Hydrogeology Journal*, **24**, 587–600.

Higgins, S., Overeem, I., Tanaka, A., and Syvitski, J. P. M. (2013). Land subsidence at aquaculture facilities in the Yellow River delta, China. *Geophysical Research Letters*, **40**, 3898–3902.

Hilgard, E. W. (1860). *Report of the Geology and Agriculture of the State of Mississippi*. Jackson, MS: E Barksdale, State Printer.

Hilgen, F. J., Lourens, L. J., and Van Dam, J. A. (2012). The Neogene Period. In Gradstein, F. M., Ogg, J. G., Schmitz, M., and Ogg, G., eds., *The Geological Time Scale 2012*. Elsevier, pp. 923–978.

Hillenbrand, C.-D., Smith, J. A., Hodell, D. A., et al. (2017). West Antarctic Ice Sheet retreat driven by Holocene warm water incursions. *Nature*, **547**, 43–48.

Historic UK. http://historic-uk.com/HistoryMagazine/DestinationsUK/LondonPlaguePits/ (accessed January 2017).

Hobbs, W. R., Massom, R. A., Stammerjohn, S., et al. (2016). A review of recent changes in Southern Ocean sea ice, their drivers and forcings. *Global Planetary Change*, **143**, 228–250.

Hoegh-Guldberg, O. (1999). Climate change, coral bleaching and the future of the world's coral reefs. *Marine and Freshwater Research*, **50**, 839–66.

Hoegh-Guldberg, O., and Bruno, J. F. (2010). The impact of climate change on the world's marine ecosystems. *Science*, **328**, 1523–1528.

Hoegh-Guldberg, O., Hoegh-Guldberg, H., Veron, J. E. N., et al. (2009). *The Coral Triangle and Climate Change: Ecosystems, People and Societies at Risk*. Brisbane: WWF Australia. http://wwf.de/fileadmin/fm-wwf/Publikationen-PDF/climate_change_coral_triangle_summary_report.pdf.

Hoegh-Guldberg, O., Mumby, P. J., Hooten, A. J., et al. (2007). Coral reefs under rapid climate change and ocean acidification. *Science*, **318**, 1737–1742.

Hoesly, R. M., Smith, S. J., Feng, L., et al. (2018). Historical (1750–2014) anthropogenic emissions of reactive gases and aerosols from the

Community Emissions Data System (CEDS). *Geosicence Model Development Discussions*, **11** (1), 369–408.

Hoffmann, G., and Reicherter, K. (2014). Reconstructing Anthropocene extreme flood events by using litter deposits. *Global Planet Change*, **123**, 22–28.

Hoffman, J., Clark, P. U., Parnell, A. C., and He, F. (2017). Regional and global sea-surface temperatures during the last interglaciation. *Science*, **335**(6322), 276–279.

Hofmann, A. (1981). The ecostratigraphic paradigm. *Lethaia*, **14**, 1–7.

Holland, P. R., and Kwok, R. (2012). Wind-driven trends in Antarctic sea-ice drift. *Nature Geoscience*, **5**, 872–875.

Holliday, V. T. (2004). *Soils in Archaeological Research*. Oxford University Press.

Holmes, A. (1913). *The Age of the Earth*, 1st ed. Harper & Brothers.

Holtgrieve, G. W., Schindler, D. E., Hobbs, W. O., et al. (2011). A coherent signature of anthropogenic nitrogen deposition to remote watersheds of the northern hemisphere. *Science*, **334**, 1545–1548.

Hom, W., Risebrough, R. W., Soutar, A., and Young, D. R. (1974). Deposition of DDE and polychlorinated biphenyls in dated sediments of the Santa Barbara Basin. *Science*, **184**, 1197–1199.

Hong, S., Candelone, J. P., Patterson, C. C., and Boutron, C. F. (1994). Greenland ice evidences of hemispheric pollution for lead two millennia ago by Greek and Roman civilizations. *Science*, **265**, 1841–1843.

Hong, S., Candelone, J. P., Patterson, C. C., and Boutron, C. F. (1996). History of ancient copper smelting pollution during Roman and medieval times recording in Greenland ice. *Science*, **272** (5259), 246.

Hong, S., Lee, J., Kang, D., et al. (2014). Quantities, composition and sources of beach debris in Korea from the results of nationwide monitoring. *Marine Pollution Bulletin*, **84**, 27–34.

Hönisch, B., Hemming, N. G., Archer, D., et al. (2009). Atmospheric carbon dioxide concentration across the Mid-Pleistocene transition. *Science*, **324**, 1551–1554.

Hönisch, B., Ridgwell, A., Schmidt, D. N., et al. (2012). The geological record of ocean acidification. *Science*, **335**, 1058–1063.

Hooke, R. L. (2000). On the history of humans as geomorphic agents. *Geology*, **28**(9), 843–846.

Hooke, R. L., Martín-Ducque, J. F., and Pedraza, J. (2012). Land transformation by humans: A review. *GSA Today*, **22**(12), 4–10.

Hori, K., Tanabe, S., Saito, Y., et al. (2004). Delta initiation and Holocene sea-level change: Example from the Song Hong (Red River) delta, Vietnam. *Sedimentary Geology*, **164**, 237–249.

Horng, C.-S., Huh, C.-A., Chen, K.-H., et al. (2009). Air pollution history elucidated from anthropogenic spherules and their magnetic signatures in marine sediments offshore of Southwestern Taiwan. *Journal of Marine Systems*, **76**, 468–478.

Horton, R. (2013). Offline: Planetary health—a new vision for the post-2015 era. *The Lancet*, **382**, 1012.

Hou, X.-G., Siveter, D. J., Siveter, D. J., et al. (2017). *The Cambrian Fossils of Chengjiang, China: The Flowering of Early Animal Life*, 2nd ed. Oxford: Wiley Blackwell.

Hounslow, M. W. (2018). Magnetic particulates as markers of fossil fuel burning. In DellaSala, D., and Goldstein, M. I., eds., *Encyclopedia of the Anthropocene*, vol. 1. Oxford: Elsevier.

Hovhannissian, R. H. (1994). *Lake Sevan, Yesterday, Today*. Erevan: Armenia National Academy of Sciences [in Russian, with extended summaries in English and Armenian].

Hovhannissian, R. H. (1996). Evolution of eutrophication processes of Lake Sevan and how to control them. In *Lake Sevan: Problems and Strategies of Action*, pp. 81–85. Proceeding of the international conference, Sevan, Armenia, 13–16 October 1996, Yeravan.

Howard, J. L. (2014). Proposal to add anthrostratigraphic and technostratigraphic units to the stratigraphic code for classification of anthropogenic Holocene deposits. *Holocene*, **24**(12), 1856–1861.

Hu, B., Yang, Z., Wang, H., et al. (2009). Sedimentation in the Three Gorges Dam and the future trend of Changjiang (Yangtze River) sediment flux to the sea. *Hydrology of Earth System Science*, **13**, 2253–2264.

Hu, Z., and Gao, S. (2008). Upper crustal abundances of trace elements: A revision and update. *Chemical Geology*, **253**(3), 205–221.

Hua, Q., Barbetti, M., and Rakowski, A. Z. (2013). Atmospheric radiocarbon for the period 1950–2010. *Radiocarbon*, **55**(4), 2059–2072.

Hublin, J.-J., Ben-Ncer, A., Bailey, S. E., et al. (2017). New fossils from Jebel Irhoud, Morocco and the pan-African origin of *Homo sapiens*. *Nature*, **546**, 289–292.

Hughes, T. P., Baird, A. H., and Bellwood, D. R. (2003). Climate change, human impacts, and the resilience of coral reefs. *Science*, **301**(5635), 929–933.

Hughes, T. P., Barner, M. K., Bellwood, D. R., et al. (2017a). Coral reefs in the Anthropocene. *Nature*, **546**, 82 90.

Hughes, T. P., Kerry, J. T., Alvarez-Noriega, M., et al. (2017b). Global warming and recurrent mass bleaching of corals. *Nature*, **543**(7645), 373–377.

Hume, B. C. C., D'Angelo, C., Smith, E. G., et al. (2015). *Symbiodinium thermophilum* sp. nov., a thermotolerant symbiotic alga prevalent in corals of the world's hottest sea, the Persian/Arabian Gulf. *Scientific Reports*, **5**(8562). doi:10.1038/srep08562.

Humphreys, G. S., and Wilkinson, M. T. (2007). The soil production function: A brief history and its rediscovery. *Geoderma*, **139**, 73–78.

Hupy, J. P., and Schaetzl, R. J. (2006). Introducing "bombturbation," a singular type of soil disturbance and mixing. *Soil Science*, **171**(11), 823–836.

Hussain, I., and Hamid, H. (2003). Plastics in agriculture. In Andrady, A. L., ed., *Plastics and the Environment*. Hoboken: Wiley, pp. 185–209.

Hutton, J. (1795). *Theory of the Earth with Proofs and Illustrations (in Four Parts)*, Edinburgh vol. 1, vol. 2, vol. 3. London: Geological Society, 1899.

Huybers, P., and Denton, G. (2008). Antarctic temperature at orbital timescales controlled by local summer duration. *Nature Geoscience*, **1**, 787–792.

Huybers, P., and Langmuir, C. (2009). Feedback between deglaciation, volcanism, and atmospheric CO_2. *Earth and Planetary Science Letters*, **286**(3–4), 479–491.

ICOLD (2009). Sedimentation and sustainable use of reservoirs and river systems. *ICOLD bulletin*. CIGB-ICOLD.

ICOLD (2017). http://icold-cigb.net/article/GB/world_register/general_synthesis/general-synthesis.

Imhof, H. K., Ivleva, N. P., Schmid, J., et al. (2013). Contamination of beach sediments of a subalpine lake with microplastic particles. *Current Biology*, **23**, R867–868.

Intergovernmental Panel on Climate Change (IPCC) (2013). *Climate Change 2013: The Physical Science Basis, Contribution of Working Group I to the Fifth Assessment Report of the Intergovernmental Panel on Climate Change*. New York: Cambridge University Press.

International Council on Mining and Metals (ICMM) (2012). Trends in the mining and metals industry: Mining's contribution to sustainable development, October 2012. London.

International Energy Agency (IEA) (2017). Key World Energy Statistics. http:// iea.org/statistics/.

International Law Association (ILA) (2016). *Committee on International Law and Sea Level Rise: Interim Report*. London: International Law Association.

International Union for the Conservation of Nature and Natural Resources (IUCN) (2014). The IUCN Red List of Threatened Species. Available at iucnredlist.org/.

Iozza, S., Müller, C. E., Schmid, P., et al. (2008). Historical profiles of chlorinated paraffins and polychlorinated biphenyls in a dated sediment core from Lake Thun (Switzerland). *Environmental Science & Technology*, **42**(4), 1045–1050.

Irabien, M. J., García-Artola, A., Cearreta, A., and Leorri, E. (2015). Chemostratigraphic and lithostratigraphic signatures of the Anthropocene in estuarine areas from the eastern Cantabrian coast (N. Spain). *Quaternary International*, **364**, 196–205.

Ivar do Sul, J. A., and Costa, M. F. (2014). The present and future of microplastic pollution in the marine environment. *Environmental Pollution*, **185**, 352–364.

Ivar do Sul, J. A., Costa, M. F., Silva-Cavalcanti, J., and Araújo, M. C. B. (2014). Plastic debris retention and exportation by a mangrove forest patch. *Marine Pollution Bulletin*, **78**, 252–257.

Jackson, J. B. C. (1997). Reefs since Columbus. *Corals Reefs*, **16**, 23–32.

Jackson, M. D., Landis, E. N., Brune, P. F., et al. (2014). Mechanical resilience and cementitious processes in Imperial Roman architectural mortar. *Proceedings of the National Academy of Sciences (USA)*, **111**(52), 18484–18489.

Jahan, K., Axe, L. B., Sandhu, N. K., et al. (2011). *Heavy Metal Contamination in Highway Marking Glass Beads*. New Jersey Department of Transportation.

Jambeck, J. R., Geyer, R., Wilcox, C., et al. (2015). Plastic waste inputs from land into the ocean. *Science*, **347**, 768–771.

James, L. A. (2013). Legacy sediment: Definitions and processes of episodically produced anthropogenic sediment. *Anthropocene*, **2**, 16–26.

Jamieson, A. J., Malkocs, T., Piertney, S. B., et al. (2017). Bioaccumulation of persistent organic pollutants in the deepest ocean fauna. *Nature Ecology & Evolution*, **1**(0051).

Jansen, E., Overpeck, J., Briffa, K. R., et al. (2007). Palaeoclimate. In Solomon, S., Qin, D., Manning, M., et al., eds., *Climate Change 2007: The Physical Science Basis; Contribution of Working Group I to the Fourth Assessment Report of the Intergovernmental Panel on Climate Change*. New York: Cambridge University Press, pp. 433–497.

Jaspers, V. L. B., Megson, D., and O'Sullivan, G. (2014). POPs in the terrestrial environment. In O'Sullivan, G., and Sandau, C. D., eds., *Environmental Forensics for Persistent Organic Pollutants*. Amsterdam: Elsevier, pp. 291–356.

Jeandel, C. (1981). *Comportement du Plutonium dans les mileux naturels (Lacustre, Fluvial et Estuarien)*. Thèse 3° Cycle, Université de Paris VII.

Jeandel, C., and Oelkers, E. H. (2015). The influence of terrigenous particulate material dissolution on ocean chemistry and global element cycles. *Chemical Geology*, **395**, 50–66.

Jeker, P., and Krahenbühl, U. (2001). Sulfur profiles of the twentieth century in peat bogs of the Swiss Midlands measured by ICP-OES and by IC. *Chimia*, **55**, 1029–1032.

Jenkyn, T. W. (1854a). Lessons in geology XLVI, chapter IV: On the effects of organic agents on the Earth's crust. *Popular Educator*, **4**, 139–141.

Jenkyn, T. W. (1854b). Lessons in geology XLIX, chapter V: On the classification of rocks (section IV: On the tertiaries). *Popular Educator*, **4**, 312–316.

Jennings, N. S. (2011). Mining and quarrying. In Armstrong, J. R., and Menon, R., eds., *Encyclopedia of Occupational Health and Safety*. Geneva: International Labor Organization.

Jeong, S., Howat, I. M., and Bassis, J. N. (2016). Accelerated ice shelf rifting and retreat at Pine Island Glacier, West Antarctica. *Geophysical Research Letters*, **43**, 11720–11725.

Jevrejeva, S., Grinsted, A., Moore, J. C., and Holgate, S. J. (2006). Nonlinear trends and multiyear cycles in sea level records. *Journal of Geophysical Research*, **111**, C09012.

Jevrejeva, S., Moore, J. C., Grinsted, A., and Woodworth, P. L. (2008). Recent global sea-level acceleration started over 200 years ago? *Geophysical Research Letters*, **35**, L08715.

Jimenez, H., and Ruiz, G. M. (2016). Contribution of non-native species to soft-sediment marine community structure of San Francisco Bay, California. *Biological Invasions*, **18**, 2007–2016.

Jinhui, L., Yuan, C., and Wenjing, X. (2017). Polybrominated diphenyl ethers in articles: A review of its applications and legislation. *Environmental Science and Pollution Research*, **24**, 4312–4321.

Johannsson, O. E., Mills, E. L., and O'Gorman, R. (1991). Changes in the nearshore and offshore zooplankton communities in Lake Ontario: 1981–88. *Canadian Journal of Fisheries and Aquatic Sciences*, **48**, 1546–1557.

Johnsen, S. I., and Taugbøl, T. (2010). NOBANIS – invasive alien species fact sheet: *Pacifastacus leniusculus. Online Database of the European Network on Invasive Alien Species – NOBANIS.* http://nobanis.org (accessed 23 June 2017).

Johnson, B. (2018). The reputed plague pits of London. *The History Magazine.* https://historic-uk.com/HistoryMagazine/DestinationsUK/LondonPlaguePits/ (accessed June 2018).

Johnson, L. (2017). Fisheries minister to announce protection for ancient glass sponge reefs: 9,000 year old reefs expected to be protected by new Marine Protected Area on BC Coast. *CBCNews*, 15 February 2017. http://cbc.ca/news/canada/british-columbia/leblanc-sponge-announcement-1.3984590

Johnson, L. C. (1995). China's Pompeii: Twelfth century Dongjing. *Historian*, **58**, 49–68.

Jomelli, V., Favier, V., Vuille, M., et al. (2014). A major advance of tropical Andean glaciers during the Antarctic cold reversal. *Nature*, **513**, 224–228.

Joos, F., Prentice, I. C., Sitch, S., et al. (2001). Global warming feedbacks on terrestrial carbon uptake under the Intergovernmental Panel on Climate Change (IPCC) emission scenarios. *Global Biogeochemical Cycles*, **15**, 891–908.

Juarrero, A. (1999). *Dynamics in Action: Intentional Behavior as a Complex System.* Cambridge: MIT Press.

Juniper, T. (2013). *What Has Nature Ever Done for Us?* London: Profile Books.

Kabata-Pendias, A. (2010). *Trace Elements in Soils and Plants*, 4th ed. Boca Raton: CRC Press.

Kakonyi, G., and Ahmed, I. (2013). Cadmium and lead: Contamination. In Jørgensen, S. E., ed., *Encyclopedia of Environmental Management.* Boca Raton: CRC Press.

Kallenborn, R., Halsall, C., Dellong, M., and Carlsson, P. (2012). The influence of climate change on the global distribution and fate processes of anthropogenic persistent organic pollutants. *Journal of Environmental Monitoring*, **14**(11), 2854–2869.

Kannan, K., Johnson-Restrepo, B., Yohn, S. S., et al. (2005). Spatial and temporal distribution of polycyclic aromatic hydrocarbons in sediments from Michigan inland lakes. *Environmental Science & Technology*, **39**(13), 4700–4706.

Kaplan, M. R., Schaefer, J. M., Denton, G. H., et al. (2010). Glacier retreat in New Zealand during the Younger Dryas Stadial. *Nature*, **467**, 194–197.

Karkanas, P., Shahack-Gross, R., Ayalon, A., et al. (2007). Evidence for habitual use of fire at the end of the Lower Palaeolithic: Site-formation processes at Qesem Cave, Israel. *Journal of Human Evolution*, **53**, 197–212.

Kaser, G., Cogley, J. G., Dyurgerov, M. B., et al. (2006). Mass balance of glaciers and ice caps: Consensus estimates for 1961–2004. *Geophysical Research Letters*, **33**, L19501. doi:10.1029/2006GL027511.

Kasirajan, S., and Ngouajio, M. (2012). Polyethylene and biodegradable mulches for agricultural applications: A review. *Agronomy for Sustainable Development*, **32**, 501–529.

Kaspari, S. D., Schwikowski, M., Gysel, M., et al. (2011). Recent increase in black carbon concentrations from a Mt. Everest ice core spanning 1860–2000 AD. *Geophysical Research Letters*, **38**, L04703.

Kawamura, H., Matusoka, N., Momoshima, N., et al. (2006). Isotopic evidence in tree rings for historical changes in atmospheric sulfur sources. *Environmental Science and Technology*, **40**, 5750–5754.

Keeling, C. D. (1960). The concentration and isotopic abundance of carbon dioxide in the atmosphere. *Tellus*, **12**(2), 200–203.

Keenan, T. F., Prentice, C., Canadell, J. G., et al. (2016). Recent pause in the growth rate of atmospheric CO_2 due to enhanced terrestrial carbon uptake. *Nature Communications*, **7**, 13428. doi:10.1038/ncomms13428.

Kelleher, J. (2016). Studying Anthropocene sedimentation behind a 19th century dam in western Connecticut. *Keck Geology Consortium, Short Contributions*, 29th Annual Symposium Volume.

Keller, G., Adatte, T., Stinnesbeck, W., et al. (2004). Chicxulub impact predates the K–T boundary mass extinction. *Proceedings of the National Academy of Sciences (USA)*, **101**, 3753–3758.

Kelly, A. E., Reuer, M. K., Goodkin, N. F., and Boyle, E. A. (2009). Lead concentrations and isotopes in corals and water near Bermuda, 1780–2000. *Earth and Planetary Science Letters*, **283**, 93–100.

Kelly, J. M., Scarpino, P., Berry, H., et al. (eds.) (2017). *Rivers of the Anthropocene*. University of California Press.

Kemp, A. C., Hawkes, A. D., Donnelly, J. P., et al. (2015). Relative sea-level change in Connecticut (USA) during the last 2200 yrs. *Earth and Planetary Science Letters*, **428**, 217–229.

Kennett, J. P. (1982). *Marine Geology*. New York: Prentice-Hall.

Kettner, A. J., Gomez, B., and Syvitski, J. P. M. (2007). Modeling suspended sediment discharge from the Waipaoa River system, New Zealand: The last 3000 years. *Water Resources Research*, **43**, W07411. doi:10.1029/2006WR005570.

Kettner, A. J., and Syvitski, J. P. M. (2009). Fluvial responses to environmental perturbations in the Northern Mediterranean since the Last Glacial Maximum. *Quaternary Science Reviews*, **28**, 2386–2397.

Key, R. M., Kozyr, A., Sabine, C. L., et al. (2004). A global ocean carbon climatology: Results from Global Data Analysis Project (GLODAP). *Global Biogeochemical Cycles*, **18**, GB4031. doi:10.1029/2004GB002247.

Khan, A., and Lemmen, C. (2014). Bricks and urbanism in the Indus Valley rise and decline. Cornell University Library. *arXiv:1303.1426v2 [physics.hist-ph]*. Last revised 24 July 2014.

Khan, S. A., Sasgen, I., Bevis, M., et al. (2016). Geodetic measurements reveal similarities between post–Last Glacial Maximum and present-day mass loss from the Greenland ice sheet. *Science Advances*, **2**, e1600931.

Kim, R. E., and Bosselmann, K. (2013). International environmental law in the Anthropocene: Towards a purposive system of multilateral environmental agreements. *Transnational Environmental Law*, **2**, 285–309.

Kinnard, C., Zdanowicz, C. M., Fisher, D. A., et al. (2011). Reconstructed changes in Arctic sea ice over the past 1450 years. *Nature*, **479**, 509–512.

Kinnard, C., Zdanowicz, C. M., Koerner, R. M., and Fisher, D. A. (2008). A changing Arctic seasonal ice zone: Observations from 1870–2003 and possible oceanographic consequences. *Geophysical Research Letters*, **35**, L02507.

Kintisch, E. (2013). Can coastal marshes rise above it all? *Science*, **341**, 480–481.

Klee, R. J., and Graedel, T. E. (2004). Elemental cycles: A status report on human or natural dominance. *Annual Review of Environment and Resources*, **29**, 69–107.

Klein, G. D. (2015). The “ANTHROPOCENE”: What is its geological utility? (Answer: It has none!). *Episodes*, September 2015, 218.

Knoll, A. H., Walter, M. R., Narbonne, G. M., and Christie-Blick, N. (2006). The Ediacaran Period:

A new addition to the geologic time scale. *Lethaia*, **39**, 13–30.

Knowlton, N. (2001). Sea urchin recovery from mass mortality: New hope for Caribbean coral reefs? *Proceedings of the National Academy of Sciences (USA)*, **98**(9), 4822–4824.

Knudsen, M. F., Seidenkrantz, M.-S., Jacoben, B. H., and Kuijpers, A. (2011). Tracking the Atlantic Multidecadal Oscillation through the last 8,000 years. *Nature Communications*, **2**, 178.

Kobashi, T., Shindell, D. T., Kodera, K., et al. (2013). On the origin of multidecadal to centennial Greenland temperature anomalies over the past 800 yr. *Climate of the Past*, **9**, 583–596.

Koch, P. L., and Barnosky, A. D. (2006). Late Quaternary extinctions: State of the debate. *Annual Review of Ecology, Evolution, and Systematics*, **37**, 215–250.

Koch, P. L., Zachos, J. C., and Gingerich, P. D. (1992). Correlation between isotope records in marine and continental carbon reservoirs near the Paleocene/Eocene boundary. *Nature*, **358**, 319–322.

Koide, M., Griffin, J. J., and Goldberg, E. D. (1975). Records of plutonium fallout in marine and terrestrial samples. *Journal of Geophysical Research*, **80**, 4153–4162.

Kominz, M. A., Browning, J. V., Miller, K. G., et al. (2008). Late Cretaceous to Miocene sea-level estimates from the New Jersey and Delaware coastal plain coreholes: An error analysis. *Basin Research*, **20**(2), 211–226.

Kondolf, G. M., Gao, Y., Annandale, G. W., et al. (2014). Sustainable sediment management in reservoirs and regulated rivers: Experiences from five continents. *Earth's Future*, **2**, 256–280.

Konrad, H., Gilbert, L., Cornford, S. L., et al. (2017). Uneven onset and pace of ice dynamic imbalance in the Amundsen Sea Embayment, West Antarctica. *Geophysical Research Letters*. doi:10.1002/2016GL070733.

Kopp, R. E., Horton, B. P., Kemp, A. C., and Tebaldi, C. (2015). Past and future sea-level rise along the coast of North Carolina, USA. *Climatic Change*, **132**, 693–707.

Kopp, R. E., Kemp, A. C., Bittermann, K., et al. (2016). Temperature-driven global sea-level variability in the Common Era. *Proceedings of the National Academy of Sciences (USA)*, **113**, E1434–E1441.

Kopp, R. E., Simons, F. J., Mitrovica, J. X., et al. (2009). Probabilistic assessment of sea level during the last interglacial stage. *Nature*, **462**, 863–868.

Kotzé, L. J. (2014). Rethinking global environmental law and governance in the Anthropocene. *Journal of Energy & Natural Resources Law*, **32**, 121–156.

Krachler, M., Zheng, J., Fisher, D., and Shotyk, W. (2009). Global atmospheric As and Bi contamination preserved in 3000 year old Arctic ice. *Global Biogeochemical Cycles*, **23**, GB3011.

Kramer, N., Wohl, E. E., and Harry, D. L. (2011). Using ground penetrating radar to "unearth" buried beaver dams. *Geology*, **40**, 43–46.

Krause, J. C., Diesing, M., and Arlt, G. (2010). The physical and biological impact of sand extraction: A case study of the western Baltic Sea. *Journal of Coastal Research*, **51**, 215–226.

Krotkov, N. A., McLinden, C. A., Li, C., et al. (2016). Aura OMI observations of regional SO_2 and NO_2 pollution changes from 2005 to 2015. *Atmospheric Chemistry and Physics*, **16**, 4605–4629.

Kubo, Y., Syvitski, J. P. M., and Tanabe, S. (2006). An application of the hydrological model HYDROTREND to the paleo-Tonegawa: Numerical estimates of sediment discharge for the last 13,000 years. *Journal of the Geological Society of Japan*, **112**, 719–729.

Kubota, K., Yokoyama, Y., Ishikawa, T., and Suzuki, A. (2015). A new method for calibrating a boron isotope paleo-pH proxy using massive *Porites* corals. *Geochemistry, Geophysics, Geosystems*, **16**(9), 3333–3342.

Kummu, M., and Varis, O. (2011). The world by latitudes: A global analysis of human population, development level and environment across the north-south axis over the past half century. *Applied Geography*, **31**, 495–507.

Kunasek, S. A, Alexander, B., Steig, E. J., et al. (2010). Sulfate sources and oxidation chemistry over the past 230 years from sulfur and oxygen isotopes of sulfate in a West Antarctic ice core. *Journal of Geophysical Research*, **115**, D18313. doi:10.1029/2010JD013846.

Kuoppamaa, M., Goslar, T., and Hicks, S. (2009). Pollen accumulation rates as a tool for detecting land-use changes in a sparsely settled boreal forest. *Vegetation History and Archaeobotany*, **18**, 205–217.

Kurashov, E. A., and Belyakov, V. P. (1987). Role of meiofauna in benthic communities of various types in the Lattgalian lakes. *Hydrobiological Journal*, **23**, 46–50.

Kurz, W. A., and Apps, M. J. (1999). A 70-year retrospective analysis of carbon fluxes in the Canadian forest sector. *Ecological Applications*, **9**, 526–547.

Lal, R., Reicosky, D. C., and Hanson, J. D. (2007). Evolution of the plow over 10,000 years and the rationale for no-till farming. *Soil and Tillage Research*, **93**(1), 1–12.

Lambeck, K., Esat, T., and Potter, E. (2002). Links between climate and sea levels for the past three million years. *Nature*, **419**, 199–206.

Landing, E. (1994). Precambrian–Cambrian global stratotype ratified and a new perspective of Cambrian time. *Geology*, **22**, 179–182.

Landing, E., Geyer, G., Brasier, M. D., and Bowring, S. A. (2013). Cambrian evolutionary radiation: Context, correlation, and chronostratigraphy-overcoming deficiencies of the first appearance datum (FAD) concept. *Earth Science Reviews*, **123**, 133–172.

Landrigan, P. J., Fuller, R., Acosta, N. J. R., et al. (2017). The Lancet Commission on pollution and health. *The Lancet*, **391**, 407–408.

Larson, G., Karlsson, E. K., Perri, A., et al. (2012). Rethinking dog domestication by integrating genetics, archeology, and biogeography. *Proceedings of the National Academy of Sciences (USA)*, **109**, 8878–8883.

Latour, B. (2015). *Face à Gaïa*. Les Empêcheurs de Penser en Rond/La Découverte, Paris.

Latti, S. J. (1932). The lake bed of Sevan: Materials on the investigation of Lake Sevan and its basin. Tiflis.

Lavanchy, V. M. H., Gäggler, H. W., Schotterer, U., Schwikowski, M., and Baltensperger, U. (1999). Historical record of carbonaceous particle concentrations from a European high-alpine glacier (Colle Gnifetti, Switzerland). *Journal of Geophysical Research*, **104**(D17), 21227–21236.

Lavers, J. L., and Bond, A. L. (2017). Exceptional and rapid accumulation of anthropogenic debris on one of the world's most remote and pristine islands. *Proceedings of the National Academy of Sciences (USA)*, **114**(23), 6052–6055.

Law, K. L., Morét-Ferguson, S. E., Goodwin, D. S., et al. (2014). Distribution of surface plastic debris in the Eastern Pacific Ocean from an 11-year data set. *Environmental Science and Technology*, **48**, 4732–4738.

Lay, M. G. (1992). *Ways of the World: A History of the World's Roads and of the Vehicles That Used Them*. New Bunswick, New Jersey: Rutgers University Press.

Le Quéré, C., Andrew, R. M., Canadell, J. P., et al. (2016). Global carbon budget 2016. *Earth System Science Data*, **8**, 605–649.

Le Quéré, C., Andrew, R. M., Friendlingstein, P., et al. (2018). Global carbon budget 2017. *Earth System Science Data*, **10**, 405–448.

Le Quéré, C., Moriarty, R., Andrew, R. M., et al. (2015). Global carbon budget 2014. *Earth Systems Science Data*, **7**, 47–85.

Leão, Z. M. A. N. (1982). *Morphology, Geology and Developmental History of the Southernmost Coral Reefs of Western Atlantic, Abrolhos Bank, Brazil*. PhD Dissertation, Rosenstiel School of Marine and Atmospheric Sciences. Florida: University of Miami.

Leão, Z. M. A. N., and Kikuchi, R. (2001). The Abrolhos reefs of Brazil. *Ecological Studies*, **144**, 83–92.

Lechner, A., Keckeis, H., Lumesberger-Loisl, F., and Zens, B. (2014). The Danube so colourful: A potpourri of plastic litter outnumbers fish larvae in Europe's second largest river. *Environmental Pollution*, **188**, 177–181.

Lee, J.-M., Eltgroth, S. F., Boyle, E. A., and Adkins, J. F. (2017). The transfer of bomb radiocarbon and anthropogenic lead to the deep North Atlantic Ocean observed from a deep sea coral. *Earth and Planetary Science Letters*, **458**, 223–232.

Lee, S.-H., Povinec, P. P., Wyse, E., et al. (2005). Distribution and inventories of ^{90}Sr, ^{137}Cs, ^{241}Am and Pu isotopes in sediments of the Northwest Pacific Ocean. *Marine Geology*, **216** (4), 249–263.

Legovich, N. A. (1979). "Blooms" in the water of Lake Sevan (Observations 1964–1972). *Materials of Sevan Hydrobiological Station*, **17**, 51–74 [in Russian].

Leinfelder, R. R. (1997). Coral reefs and carbonate platforms within a siliciclastic setting: General aspects and examples from the Late Jurassic of Portugal. *Proceedings of the 8th International Coral Reef Symposium*, **2**, 1737–1742.

Leinfelder, R. R. (2001). Jurassic reef ecosystems. In Stanley, G. D., Jr., ed., *The History and Sedimentology of Ancient Reef Systems*. Topics in Geobiology Series, **17**, pp. 251–309.

Leinfelder, R. R. (2017). Das Zeitalter des Anthropozäns und die Notwendigkeit der grossen Transformation: Welche Rollen spielen Umweltpolitik und Umweltrecht? *Zeitschrift für Umweltrecht (ZUR)*, **28**(5), 259–266, Nomos. http://zur.nomos.de/fileadmin/zur/doc/Aufsatz_ZUR_17_05.pdf.

Leinfelder, R. R., and Haum, R. (2015). Ozeane. In Kersten, J., ed., *Inwastement: Abfall in Umwelt und Gesellschaft*. Bielefeld: Transcript-Verlag, pp. 153–179.

Leinfelder, R. R., and Leão, Z. M. A. N. (2000). *Increasing Reef Complexity-Decreasing Reef Flexibility Through Time, and a Unique Exception – The Evolutionary Relic Reefs of Brazil*. Rio de Janeiro, Brazil: 31st International Geological Congress, abstract vol.

Leinfelder, R. R., and Nose, M. (1999). Increasing complexity - decreasing flexibility: A different perspective of reef evolution through time. *Profil*, **17**, 135–147 (Univ. Stuttgart).

Leinfelder, R. R., and Schmid, D. U. (2000). Mesozoic reefal thrombolites and other microbolites. In Riding, R., ed., *Microbial Sediments*. Berlin: Springer, pp. 289–294.

Leinfelder, R. R., Schmid, D. U., Nose, M., and Werner, W. (2002). Jurassic reef patterns: The expression of a changing globe. In Flügel, E., Kiessling W., and Golonka, J., eds., *Phanerozoic Reef Patterns*. Tulsa: SEPM Special Publication, **72**, pp. 465–520.

Leinfelder, R. R., Seemann, J., Heiss, G. A., and Struck, U. (2012). Could "ecosystem atavisms" help reefs to adapt to the Anthropocene? *Proceedings of the 12th International Coral Reef Symposium, Cairns, Australia*, 9–13 July 2012. 2B Coral reefs: Is the past the key to the future? http://icrs2012.com/proceedings/manuscripts/ICRS2012_2B_2.pdf.

Leinfelder, R. R., and Wilson, R. C. L. (1998). Third order sequences in an Upper Jurassic rift-related second order sequence, Central Lusitanian Basin, Portugal. In Graciansky, P.-C. de, Hardenbol, J., Jacquin, T., and Vail, P., eds., *Mesozioc-Cenozoic Sequence Stratigraphy of European Basins*. Tulsa: SEPM Special Publication, **60**, pp. 507–525.

Lenton, T. (2016). *Earth System Science: A Very Short Introduction*. Oxford University Press.

Lenton, T., and Watson, A. (2011). *Revolutions That Made the Earth*. Oxford University Press.

Letcher, T. M. (2016). *Climate Change: Observed Impacts on Planet Earth*, 2nd ed. Elsevier.

Levy, R., Harwood, D., Florindo, F., et al. (2016). Antarctic ice sheet sensitivity to atmospheric CO_2

variations in the early to mid-Miocene. *Proceedings of the National Academy of Sciences (USA)*, **113**(13), 3453–3458.

Lewin, J. (2012). Enlightenment and the GM floodplain. *Earth Surface Processes and Landforms*, **38**, 17–29.

Lewis, S. L., and Maslin, M. A. (2015). Defining the Anthropocene. *Nature*, **519**, 171–180.

Lindahl, P., Asami, R., Iryu, Y., et al. (2011). Sources of plutonium to the tropical Northwest Pacific Ocean (1943–1999) identified using a natural coral archive. *Geochemica et Cosmochimica Acta*, **75**, 1346–1356.

Lirman, D., Schopmeyer, S., Galvan, V., et al. (2014). Growth dynamics of the threatened Caribbean staghorn coral *Acropora cervicornis*: Influence of host genotype, symbiont identity, colony size, and environmental setting. *PLoS ONE*, **9**(9), e107253. http://doi.org/10.1371/journal.pone.0107253.

Lisiecki, L., and Raymo, M. E. (2005). A Pliocene-Pleistocene stack of 57 globally distributed benthic $\delta^{18}O$ records. *Paleoceanography*, **20**, PA1003. doi:10.1029/2004PA001071.

Lithner, D., Larsson, A., Dave, G. (2011). Environmental and health hazard ranking and assessment of plastic polymers based on chemical composition. *Science of the Total Environment*, **409**, 3309–3324.

Liu, W., Xie, S.-P., Liu, A., and Zhu, J. (2017). Overlooked possibility of a collapsed Atlantic Meridional Overturning Circulation in warming climate. *Science Advances*, 3(1), e1601666. doi:10.1126/sciadv.1601666.

Livingston, H. D., and Povinec, P. P. (2000). Anthropogenic marine radioactivity. *Ocean & Coastal Management*, **43**(8–9), 689–712.

Livingston, H. D., Povinec, P. P., Ito, T., et al. (2001). The behaviour of plutonium in the Pacific Ocean. In Kudo, A., ed., *Radioactivity in the Environment*, vol. 1, Plutonium in the Environment. Amsterdam: Elsevier, pp. 267–292.

Loader, N. J., Young, G. H. F., Grudd, H., and McCarroll, D. (2013). Stable carbon isotopes from Torneträsk, northern Sweden provide a millennial length reconstruction of summer sunshine and its relationship to Arctic circulation. *Quaternary Science Reviews*, **62**, 97–113.

Lobelle, D., and Cunliffe, M. (2011). Early microbial biofilm formation on marine plastic debris. *Marine Pollution Bulletin*, **62**, 197–200.

Loch, K., Loch, W., and Anlauf, H. (2007). Der Zustand der Steinkorallen in maledivischen Riffen und die Regeneration nach dem 1998er Korallenbleichen. *Bufus*, **37**. http://bufus.sbg.ac.at/Info/Info37/Info37-2.htm.

Loch, K., Loch, W., Schuhmacher, H., and See, W. R. (2002). Coral recruitment and regeneration on a Maldivian reef 21 months after the coral bleaching event of 1998. *Marine Ecology*, **23**, 219–236

Long, A. J., Barlow, N. L. M., Gehrels, W. R., et al. (2014). Contrasting records of sea-level change in the eastern and western North Atlantic during the last 300 years. *Earth and Planetary Science Letters*, **388**, 110–122.

Lopes-Lima, M., Sousa, R., Geist, J., et al. (2016). Conservation status of freshwater mussels in Europe: State of the art and future challenges. *Biological Reviews*, **92**, 572–607.

Lorgeoux, C., Moilleron, R., Gasperi, J., et al. (2016). Temporal trends of persistent organic pollutants in dated sediment cores: Chemical fingerprinting of the anthropogenic impacts in the Seine River basin, Paris. *Science of the Total Environment*, **541**, 1355–1363.

Lough, J. M., and Barnes, D. J. (2000). Environmental controls on growth of the massive coral *Porites*. *Journal of Experimental Marine Biology and Ecology*, **245**(2), 225–243.

Loulergue, L., Schilt, A., Spahni, R., et al. (2008). Orbital and millennial-scale features of atmospheric CH_4 over the past 800,000 years. *Nature*, **453**, 383–386.

Lovejoy, S. (2014a). Scaling fluctuation analysis and statistical hypothesis testing of anthropogenic warming, *Climate Dynamics*, **42**, 2339–2351.

Lovejoy, S. (2014b). Return periods of global climate fluctuations and the pause. *Geophysical Research Letters*, **41**, 4704–4710.

Lovejoy, S. (2015). Climate Closure. *EOS*, **96**. doi:10.1029/2015EO037499.

Lovelock, J. E., and Margulis, L. (1974). Atmospheric homeostasis by and for the biosphere: The Gaia hypothesis. *Tellus*, **26**, 2–10.

Lowenthal, D. (2000). *George Perkins Marsh: Prophet of Conservation*. Seattle & London: University of Washington Prcss.

Lu, Z., Letcher, R. J., Chu, S., et al. (2015). Spatial distributions of polychlorinated biphenyls, polybrominated diphenyl ethers, tetrabromobisphenol A and bisphenol A in Lake Erie sediment. *Journal of Great Lakes Research*, **41**(3), 808–817.

Lüdecke, C., and Summerhayes, C. P. (2012). *The Third Reich in Antarctica: The Third German Antarctic Expedition 1938–39*. Bluntisham Books & Erskine Press.

Lundstrom, M. (2003). Moore's law forever? *Science*, **299**, 210–211.

Lunt, D. J., Haywood, A. M., Schmidt, G. A., et al. (2012). On the causes of mid-Pliocene warmth and polar amplification. *Earth and Planetary Science Letters*, **321–322**, 128–138.

Lüthi, D., Le Floch, M., Bereiter, B., et al. (2008). High-resolution carbon dioxide concentration record 650,000–800,000 years before present. *Nature*, **453**, 379–382.

Lyon, T. L., and Buckman, H. O. (1946). *The Nature and Properties of Soils*. New York: Macmillan Co.

MA (Millennium Ecosystem Assessment) (2005) *Ecosystems and Human Well-Being: Synthesis*. Washington, DC: Island Press.

Ma, Z., Melville, D. S., Liu, J., et al. (2014). Rethinking China's new great wall. *Science*, **346**(6212), 912–914.

MacAyeal, D. R. (1993). Binge/purge oscillations of the Laurentide ice sheet as a cause of the North Atlantic's Heinrich Events. *Paleoceanography*, **8** (6), 775–784.

MacFarling Meure, C., Etheridge, D. E., Trudinger, C., et al. (2006). Law Dome CO_2, CH_4 and N_2O ice core records extended to 2000 years BP. *Geophysical Research Letters*, **33**(14), L14810. doi:10.1029/2006GL026152.

Mackintosh, A. N., Anderson, B. M., Lorrey, A. M., et al. (2017). Regional cooling caused recent New Zealand glacier advances in a period of global warming. *Nature Communications*. doi:10.1038/ncomms14202.

Maher, B. A. (1986). Characterisation of soils by mineral magnetic measurements. *Physics of the Earth and Planetary Interiors*, **42**, 76–92.

Maher, B. A., and Thompson, R. (1992). Paleoclimatic significance of the mineral magnetic record of the Chinese loess and paleosols. *Quaternary Research*, **37**(2), 155–170.

Mahiques, M. M., Hanebuth, T. J. J., Martins, C. C., et al. (2016). Mud depocentres on the continental shelf: A neglected sink for anthropogenic contaminants from the coastal zone. *Environmental Earth Sciences*, **75**(44). doi:10.1007/s12665-015-4782-z.

Mahmood, K. (1987). Reservoir sedimentation: Impact, extent and mitigation. *World Bank Technical Paper*, **71**.

Malakoff, D. (2002). Trawling's a drag for marine life, say studies. *Science*, **298**, 2123.

Malm, A., and Hornborg, A. (2014). The geology of mankind? A critique of the Anthropocene narrative. *Anthropocene Review*, **1**, 62–69.

Mann, M. E., Miller, S. K., Rahmstorf, S., et al. (2017). Record temperature streak bears anthropogenic fingerprint. *Geophysical Research Letters*, **44**. doi:10.1002/2017GL074056.

Mao, J., Ribes, A., Yan, B., et al. (2016). Human-induced greening of the northern extratropical land surface. *Nature Climate Change*, **6**, 959–964.

Marchant, J. (2017). Biologists rush to study creatures living beneath Larsen C ice shelf before they disappear. *Nature*, **549**, 443.

Marcott, S. A., Bauska, T. K., Buizert, C., et al. (2014). Centennial-scale changes in the global carbon cycle during the last deglaciation. *Nature*, **514**, 616–619.

Marcott, S. A., Shakun, J. D., Clark, P. U., and Mix, A. (2013). A reconstruction of regional and global temperature for the past 11,300 years. *Science*, **339**(6124), 1198–1201.

Mari, M., and Domingo, J. L. (2010). Toxic emissions from crematories: A review. *Environment International*, **36**(1), 131–137.

Marie, B., Zanella-Cléon, I., Guichard, N., et al. (2011). Novel proteins from the calcifying shell matrix of the Pacific oyster *Crassostrea gigas*. *Marine Biotechnology*, **13**, 1159–1168.

Marini, F. (2003). Natural microtektites versus industrial glass beads: An appraisal of contamination problems. *Journal of Non-Crystalline Solids*, **323**(1), 104–110.

Markosyan, A. G. (1959). Benthos production in Lake Sevan. In *Trudy VI Soveshchaniya po Problemam Biologii Vnutrennikh Vod*. Moscow, Leningrad, pp. 139–145 [in Russian].

Marsh, G. P. (1864). *Man and Nature; Or, Physical Geography as Modified by Human Action*. New York: Charles Scribner (reprinted by Lowenthal, D., ed., Cambridge, MA: Belknap Press/Harvard University Press, 1965).

Marsh, G. P. (1874). *The Earth as Modified by Human Action: A New Edition of "Man and Nature" Charles Scribner*. New York: Armstrong & Co.

Martín, J., Puig, P., Palanques, A., et al. (2015). Commercial bottom trawling as a driver of sediment dynamics and deep seascape evolution in the Anthropocene. *Anthropocene*, **7**, 1–15.

Martin, K., and Sauerborn, J. (2013). *Agroecology*. Netherlands: Springer.

Martinez, A., Hadnott, B. N., Awad, A. M., et al. (2017). Release of airborne polychlorinated biphenyls from New Bedford Harbor results in elevated concentrations in the surrounding air. *Environmental Science & Technology Letters*, **4** (4), 127–131.

Martinez, J. M., Guyot, J. L., Filizola, N., and Sondag, F. (2009). Increase in suspended sediment discharge of the Amazon River assessed by monitoring network and satellite data. *Catena*, **79**, 257–264.

Martínez-García, B., Pascual, A., Baceta, J. I., and Murelaga, X. (2013). Estudio de los foraminíferos bentónicos del "beach-rock" de Azkorri (Getxo, Bizkaia). *Geogaceta*, **53**, 29–32.

Martinez-Porchas, M., and Martinez-Cordova, L. R., 2012. World aquaculture: Environmental impacts and troubleshooting alternatives. *Scientific World Journal*, **2012**, 389623.

Marx, S. K., Rashid, S., and Stromsoe, N. (2016). Global-scale patterns in anthropogenic Pb contamination reconstructed from natural archives. *Environmental Pollution*, **213**, 283–298.

Maselli, V., and Trincardi, F. (2013). Man made deltas. *Scientific Reports*, **3**, 1926.

Masiokas, M. H., Rivera, A., Espizua, L. E., et al. (2009). Glacier fluctuations in extratropical South America during the past 1000 years. *Palaeogeography, Palaeoclimatology, Palaeoecology*, **281**, 242–268.

Masson-Delmotte, V., Schulz, M., Abe-Ouchi, A., et al. (2013). Information from paleoclimate archives. In Stocker, T. F., Qin, D., Plattner, G. K., et al., eds., *Climate Change 2013: The Physical Science Basis, Contribution of Working Group I to the Fifth Assessment Report of the Intergovernmental Panel on Climate Change*. New York: Cambridge University Press, pp. 383–464.

Masson-Delmotte, V., Swingedouw, D., Landais, A., et al. (2012). Greenland climate change: From the past to the future. *Wiley Interscience Reviews: Climate Change*, 3(5), 427–449.

Massos, A., and Turner, A. (2017). Cadmium, lead and bromine in beached microplastics. *Environmental Pollution*, **227**, 139–145.

Mathew, W. M. (1970). Peru and the British guano market, 1840–1870. *Economic History Review* 23 (1), 112–128.

Matisoo-Smith, E., Roberts, R. M., Irwin, G. J., et al. (1998). Patterns of prehistoric human mobility in Polynesia indicated by mtDNA from the Pacific rat. *Proceedings of the National Academy of Sciences (USA)*, **95**, 15145–15150.

Mattey, D. M., Lowry, D., Duffet, J., et al. (2008). A 53 year seasonally resolved oxygen and carbon isotope record from a modern Gibraltar speleothem: Reconstructed drip water and relationship to local precipitation. *Earth and Planetary Science Letters*, **269**, 80–95.

Mayewski, P. A., Carleton, A. M., Birkel, S. D., et al. (2017). Ice core and climate reanalysis analogs to predict Antarctic and southern hemisphere climate changes. *Quaternary Science Reviews*, **155**, 50–66.

McCauley, D. J., Pinsky, M. L., Palumbi, S. R., et al. (2015). Marine defaunation: Animal loss in the global ocean. *Science*, **347**, 1255641.

McConnell, J. R., and Edwards, R. (2008). Coal burning leaves toxic heavy metal legacy in the Arctic. *Proceedings of the National Academy of Sciences (USA)*, **105**(34), 12140–12144.

McConnell, J. R., Edwards, R., Kok, G. L., et al. (2007). 20th-century industrial black carbon emissions altered Arctic climate forcing. *Science*, **317**, 1381–1384.

McConnell, J. R., Kipfstuhl, S., and Fischer, H. (2006). The NGT and PARCA shallow ice core arrays in Greenland: A brief overview. *PAGES News*, **14**(1), 13–14.

McCormick, A., Hoellin, T. J., Mason, S. A., et al. (2014). Microplastic is an abundant and distinct microbial habitat in an urban river. *Environmental Science and Technology*, **48**, 11863–11871.

McDougall, I., Brown, F. H., and Fleagle, J. G. (2005). Stratigraphic placement and age of modern humans from Kibish, Ethiopia. *Nature*, **433**, 733–736.

McFarlane, D. A., Lundberg, J., and Neff, H. (2014). A speleothem record of early British and Roman mining at Charterhouse, Mendip, England. *Archaeometry*, **56**(3), 431–443.

McGann, M., Grossman, E. E., Takesue, R. K., et al. (2012). Arrival and expansion of the invasive foraminifera *Trochammina hadai* Uchio in Padilla Bay, Washington. *Northwest Science*, **86**, 9–26.

McGann, M., Sloan, D., and Cohen, A. N. (2000). Invasion by a Japanese marine microorganism in western North America. *Hydrobiologia*, **421**, 25–30.

McGann, M., Sloan, D., and Wan, E. (2002). Biostratigraphy beneath central San Francisco Bay along the San Francisco-Oakland Bay Bridge transect. In *Crustal Structure of the Coastal and Marine San Francisco Bay Region, California*, pp. 11–28.

McInerney, F. A., and Wing, S. L. (2011). The Paleocene-Eocene Thermal Maximum: A perturbation of carbon cycle, climate, and biosphere with implications for the future. *Annual Review of Earth and Planetary Sciences*, **39**, 489–516.

McKee, K., Rogers, K., and Saintilan, H. (2012). Response of salt marsh and mangrove wetlands to changes in atmospheric CO_2, climate, and sea level. In Middleton, B. A., ed., *Global Change and the Function and Distribution of Wetlands*, Dordrecht: Springer, pp. 63–96.

McLean, D. (1991). Magnetic spherules in recent lake sediments. *Hydrobiologica*, **214**, 91–97.

McLinden, C. A., Fioletov, V., Shephard, M. W., et al. (2016). Space-based detection of missing sulfur dioxide sources of global air pollution. *Nature Geoscience*, **9**, 496–500.

McMichael, A. (1993). *Planetary Overload*. Cambridge University Press.

McMillan, A. A., and Powell, J. H. (1993). BGS Rock Classification Scheme: The classification of artificial (man-made) ground and natural superficial deposits. Version 2. *British Geological Survey Technical Report*, WG/93/46/R.

McMillan, A. A., and Powell, J. H. (1999). BGS Rock Classification Scheme: Classification of artificial (man-made) ground and natural superficial deposits; Applications to geological maps and datasets in the UK, vol. 4. *British Geological Survey Research Report*, RR/99/004.

McMinn, A., Hallegraeff, G. M., Thomson, P., et al. (1997). Cyst and radionucleotide evidence for the recent introduction of the toxic dinoflagellate *Gymnodinium catenatum* into Tasmanian waters. *Marine Ecology Progress Series*, **161**, 165–172.

McNeely, J. (2001). Invasive species: A costly catastrophe for native biodiversity. *Land Use and Water Resources Research*, **1**, 1–10.

McNeill, J., Barrie, F. R., Buck, W. R., et al. (2012). International Code of Nomenclature for algae, fungi, and plants (Melbourne Code). Regnum Vegetabile 154. Koeltz Scientific Books. ISBN 978-3-87429-425-6.

McNeill, J. R. (2001). *Something New under the Sun: An Environmental History of the Twentieth-Century World*. New York: W. W. Norton & Company.

McNeill, J. R., and Engelke, P. (2016). *The Great Acceleration: An Environmental History of the Anthropocene since 1945*. Cambridge, MA: Harvard University Press.

McPherron, S., Alemseged, Z., Marean, C. W., et al. (2010). Evidence for stone-tool-assisted consumption of animal tissues before 3.39 million years ago at Dikika, Ethiopia. *Nature*, **466**, 857–860.

Mee, L. (2012). Between the devil and the deep blue sea: The coastal zone in an era of globalisation. *Estuarine, Coastal and Shelf Science*, **96**, 1–8.

Meehl, G. A., Arblaster, J. M., Bitz, C. M., et al. (2016). Antarctic sea-ice expansion between 2000 and 2014 driven by tropical Pacific decadal climate variability. *Nature Geoscience*, **9**, 590–595.

Meharg, A. A., and Killham, K. (2003). A pre-industrial source of dioxins and furans. *Nature*, **421**, 909–910.

Mehta, P. K., and Monteiro, P. J. M. (2006). *Concrete: Microstructure, properties, and materials*, 3rd ed. New York: McGraw-Hill.

Meier, K. J. S., Beaufort, L., Heussner, S., and Ziveri, P. (2014). The role of ocean acidification in *Emiliania huxleyi* coccolith thinning in the Mediterranean Sea. *Biogeosciences*, **11**, 2857–2869.

Melchin, M. J., Sadler, P. M., and Cramer, B. D. (2012). The Silurian Period. In Gradstein, F. M., Ogg, J. G., Schmitz, M., and Ogg, G., eds., *The Geological Time Scale 2012*. Elsevier, pp. 526–558.

Mernild, S. H., Lipscomb, W. H., Bahr, D. B., et al. (2013). Global glacier retreat: A revised assessment of committed mass losses and sampling uncertainties. *The Cryosphere Discussions*, **7**, 1987–2005.

Merrill, G. P. (1897). *A Treatise on Rocks: Rock-Weathering and Soils*. New York: Macmillan Co.

Merritts, D., Walter, R., Rahnis, M., et al. (2011). Anthropocene streams and base-level controls from historic dams in the unglaciated mid-Atlantic region, USA. *Philosophical Transactions of the Royal Society A*, **369**, 976–1009.

Meshkova, T. M. (1976). Eutrophication of Lake Sevan. *Biological Journal of Armenia*, **29**, 14–22 [in Russian].

Messager, E., Lordkipanidze, D., Kvavadze, E., et al. (2010). Palaeoenvironmental reconstruction of Dmanisi site (Georgia) based on palaeobotanical data. *Quaternary International*, **223–224**, 20–27.

Meybeck, M. (2003). Global analysis of river systems: From Earth System controls to Anthropocene syndromes. *Philosophical Transactions of the Royal Society B*, **358**, 1935–1955.

Meybeck, M., Laroche, L., Darr, H. H., and Syvitski, J. P. M. (2003). Global variability of total suspended solids and their fluxes in rivers. *Global and Planetary Change*, **39**(1/2), 65–93.

Michelutti, N., Simonetti, A., Briner, J. P., et al. (2009). Temporal trends of pollution Pb and other metals in east-central Baffin Island inferred from lake sediment geochemistry. *Science of the Total Environment*, **407**(21), 5653–5662.

Mighall, T. M., Timberlake, S., Foster, I. D., et al. (2009). Ancient copper and lead pollution records from a raised bog complex in Central Wales, UK. *Journal of Archaeological Science*, **36**(7), 1504–1515.

Miles, B. W. J., Stokes, C. R., Veili, A., and Cox, N. J. (2013). Rapid, climate-driven changes in outlet glaciers on the Pacific coast of East Antarctica. *Nature*, **500**, 563–566.

Miller, G., Magee, J., Smith, M., et al. (2016). Human predation contributed to the extinction of the Australian megafunal bird *Genyornis newtoni* ~ 47 ka. *Nature Communications*, **7**, 10496. doi:10.1038/ncomms10496.

Miller, G. H., Brigham-Grette, J., Alley, R. B., et al. (2010a). Temperature and precipitation history of the Arctic. *Quaternary Science Reviews*, **29**, 1679–1715.

Miller, G. H., Brigham-Grette, J., Alley, R. B., et al. (2010b). Arctic amplification: Can the past constrain the future? *Quaternary Science Reviews*, **29**(15–16), 1779–1790.

Miller, G. H., Geirsdottir, A., Zhong, Y., et al. (2012). Abrupt onset of the Little Ice Age triggered by volcanism and sustained by sea-ice/ocean feedbacks. *Geophysical Research Letters*, **39**, L02708. doi:10.1029/2011GL050168.

Miller, G. H., Lehman, S. J., Refsnider, K. A., et al. (2013). Unprecedented recent summer warmth in Arctic Canada. *Geophysical Research Letters*, **40**, 5745–5751.

Miller, K. G., Kopp, R. E., Horton, B. P., et al. (2013). A geological perspective on sea-level rise and its impacts along the U.S. mid-Atlantic coast. *Earth's Future*, **1**, 3–18.

Miller, K. G., and Wright, J. D. (2017). Success and failure in Cenozoic global correlations using golden spikes: A geochemical and magnetostratigraphic perspective. *Episodes*. 10 .18814/epiiugs/2017/v40i1/017003.

Miller, K. G., Wright, J. D., Browning, J. V., et al. (2012). High tide of the warm Pliocene: Implications of global sea level for Antarctic deglaciation. *Geology*, **49**, 407–421.

Milliman, J. D., and Syvitski, J. P. M. (1992). Geomorphic/tectonic control of sediment discharge to the ocean: The importance of small mountainous rivers. *Journal of Geology*, **100**, 525–544.

Milne, G. A., Gehrels, W. R., Hughes, C. W., and Tamisiea, M. E. (2009). Identifying the causes of sea-level change. *Nature Geoscience*, **2**, 471–478.

Molina, E., Alegret, L., Arenillas, I., et al. (2006). The Global Boundary Stratotype Section and Point for the base of the Danian Stage (Paleocene, Paleogene, "Tertiary", Cenozoic) at El Kef, Tunisia: Original definition and revision. *Episodes*, **29**, 263–273.

Monge, G., Jimenez-Espejo, F. J., García-Alix, A., et al. (2015). Earliest evidence of pollution by heavy metals in archaeological sites. *Scientific Reports*, **5**, 14252.

Monnin, E., Indermühle, A., Dällenbach, A., et al. (2001). Atmospheric CO_2 concentrations over the last glacial termination. *Science*, **291**, 112–114.

Monteith, D. T., Evans, C. D., Henrys, P. A., et al. (2014). Trends in the hydrochemistry of acid-sensitive surface waters in the UK 1988–2008. *Ecological Indicators*, **37**, 287–303.

Montzka, S. A., Aydin, M., Battle, M., et al. (2004). A 350-year atmospheric history for carbonyl sulfide inferred from Antarctic firn air and air trapped in ice. *Journal of Geophysical Research*, **109**, D22302. doi:10.1029/2004JD004686.

Moore, A. M. T., Hillman, G. C., and Legge, A. J. (2000). *Village on the Euphrates: From Foraging to Farming at Abu Hureyra*. Oxford University Press.

Moore, C. J., Moore, S. L., Leecaster, M. K., and Weisberg, S. B. (2001). A comparison of plastic and plankton in the North Pacific Central Gyre. *Marine Pollution Bulletin*, **42**, 1297–1300.

Mora, C., Dousset, B., Caldwell, I. R., et al. (2017). Global risk of deadly heat. *Nature Climate Change*, **7**, 501–506.

Morehead, M. D., Syvitski, J. P. M., Hutton, E. W. H., and Peckham, S. D. (2003). Modeling the inter-annual and intra-annual variability in the flux of sediment in ungauged river basins. *Global and Planetary Change*, **39**(1/2), 95–110.

Morf, L. S., Tremp, J., Gloor, R., et al. (2005). Brominated flame retardants in waste electrical and electronic equipment: Substance flows in a recycling plant. *Environmental Science & Technology*, **39**(22), 8691–8699.

Morritt, D., Stefanoudis, P. V., Pearce, D., et al. (2014). Plastic in the Thames: A river runs through it. *Marine Pollution Bulletin*, **78**, 196–200.

Morton, O. (2015). *The Planet Remade: How Geoengineering Could Change the World*. London: Granta Publications.

Moura, R. L., Amado-Filho, G. M., Moraes, F. C., et al. (2016). An extensive reef system at the Amazon River mouth. *Science Advances*, **2**(4), e1501252. doi:10.1126/sciadv.1501252.

Muir, D. C. G., and de Wit, C. A. (2010). Trends of legacy and new persistent organic pollutants in the circumpolar arctic: Overview, conclusions and recommendations. *Science of the Total Environment*, **408**, 3044–3051.

Muir, D. C. G., and Howard, P. H. (2006). Are there other persistent organic pollutants? A challenge for environmental chemists. *Environmental Science & Technology*, **40**(23), 7157–7166.

Muir, D. C. G., and Rose, N. L. (2007). Persistent organic pollutants in the sediments of Lochnagar. In Rose, N. L., ed., *Lochnagar: The Natural History of a Mountain Lake, Developments in Paleoenvironmental Research*, 12. Dordrecht: Springer, pp. 375–402.

Mulder, T., and Syvitski, J. P. M. (1996). Climatic and morphologic relationships of rivers: Implications of sea level fluctuations on river loads. *Journal of Geology*, **104**, 509–523.

Muri, G., Wakeham, S., and Rose, N. L. (2006). Records of atmospheric delivery of pyrolysis-derived pollutants in recent mountain lake sediments of the Julian Alps (NW Slovenia). *Environmental Pollution*, **139**, 461–468.

Murphy, B. H., Farley, K. A., and Zachos, J. C. (2010). An extraterrestrial ^{3}He-based timescale for the Paleocene-Eocene thermal maximum (PETM) from Walvis Ridge, IODP Site 1266. *Geochimica et Cosmochimca Acta*, **74**, 5098–5108.

Muscheler, R., Beer, J., and Kubik, P. W. (2004). Long-term solar variability and climate change based on radionuclide data from ice cores. In Pap, J. M., and Fox, P., eds., *Solar Variability and Its Effects on Climate*. Geophysical Monograph, 114. American Geophysical Union, pp. 221–235.

Myers, R. A., and Worm, B. (2003). Rapid worldwide depletion of predatory fish communities. *Nature*, **423**, 280–283.

Naish, T. R., Powell, R., Levy, R., et al. (2009). Obliquity-paced Pliocene West Antarctic oscillations. *Nature*, **458**, 322–328.

Naish, T. R., and Wilson, G. S. (2009). Constraints on the amplitude of mid-Pliocene (3.6–2.4 Ma) eustatic sea-level fluctuations from the New Zealand shallow-marine sediment record. *Philosophical Transactions of the Royal Society of London A*, **367**, 169–187.

Naredo, J. M., and Gutiérrez, L. (eds.) (2005). *La Incidencia de la Especie Humana sobre la Faz de la Tierra (1955–2005)*. Fundacion César Manrique, Universidad de Granada.

NASA (2007). GISS data at https://data.giss.nasa.gov/modelforce/ghgases/Fig1A.ext.txt.

National Academies of Sciences, Engineering, and Medicine (NASEM) (2015). *A Strategic Vision for NSF Investment in Antarctic and Southern Ocean Research*. Washington, DC: The National Academies Press.

National Academies of Sciences, Engineering, and Medicine (NASEM) (2017). *Antarctic Sea Ice Variability in the Southern Ocean-Climate*

System. Washington, DC: The National Academies Press. doi:10.17226/24696.

National Funeral Directors Association (NFDA) (2016). *The 2016 NFDA Cremation and Burial Report: Research, Statistics and Projections.*

National Snow and Ice Data Center (NSIDC) (2017). http://nsidc.org/.

NEEM Community Members (2013). Eemian interglacial reconstructed from a Greenland folded ice core. *Nature*, **493**, 489–494.

Négrel, P., Kloppmann, W., Garcin, M., and Giot, D. (2004). Lead isotope signatures of Holocene fluvial sediments from the Loire River valley. *Applied Geochemistry*, **19**, 957–972.

Netz, R., and Noel, W. (2007). *The Archimedes Codex*. New York: DaCapo Press.

Neukom, R., Gergis, J., Karoly, D. J., et al. (2014). Interhemispheric temperature variability over the past millennium. *Nature Climate Change*, **4**, 362–367.

Newman, L., Wanner, H., and Kiefer, T. (2009). Towards a global synthesis of the climate of the last two millennia: Workshop of the pages 2k regional network, Corvallis, USA, 7 July 2009. *PAGES News*, **17**(3), 130–131.

Nickel, E. H. (1995a). Definition of a mineral. *Canadian Mineralogist*, **33**, 689–690.

Nickel, E. H. (1995b). Mineral names applied to synthetic substances. *Canadian Mineralogist*, **33**, 1335.

Nickel, E. H., and Grice, J. D. (1998). The IMA Commission on New Minerals and Mineral Names: Procedures and guidelines on mineral nomenclature. *Canadian Mineralogist*, **36**, 913–926.

Nir, D. (1983). *Man, a Geomorphological Agent: An Introduction to Anthropic Geomorphology*. Keter Publication Jerusalem.

Nirei, H., Furuno, K., Osamu, K., et al. (2012). Classification of man made strata for assessment of geopollution. *Episodes*, **35**(2), 333–336.

Nisbet, E. G., Dlugokencky, E. J., Manning, M. R., et al. (2016). Rising atmospheric methane: 2007–2014 growth and isotopic shift. *Global Biogeochemical Cycles*, **30**, 1356–1370.

NOAA (National Oceanic and Atmospheric Administration) (2017). Global analysis: Annual 2016. https://ncdc.noaa.gov/sotc/global/201613.

Nobre, C. A., and de Simone Borma, L. (2009). "Tipping points" for the Amazon forest. *Current Opinion in Environmental Sustainability*, **1**, 28–36.

Nogués-Bravo, D., Rodríguez, J., Hortal, J., et al. (2008). Climate change, humans, and the extinction of the woolly mammoth. *PLoS Biology*, **6**(4), e79. doi:10.1371/journal.pbio.0060079.

Nordhaus, W. D. (1996). Do real-output and real-wage measures capture reality? The history of lighting suggests not. In Bresnahan, T. F., and Gordon, R. J., eds., *The Economics of New Goods*. http://nber.org/chapters/c6064.

North American Commission on Stratigraphic Nomenclature (NACSN) (1983). North American Stratigraphic Code. *American Association of Petroleum Geologists B*, **67**, 841–875.

North American Commission on Stratigraphic Nomenclature (NACSN) (2005). North American Stratigraphic Code. *American Association of Petroleum Geologists B*, **89**, 1547–1591.

Novak, M., Adamová, M., Wieder, R. K., and Bottrell, S. H. (2005). Sulfur mobility in peat. *Applied Geochemistry*, **20**, 673–681.

Novak, M., Sipkova, A., Chrastny, V., et al. (2016). Cu-Zn isotope constraints on the provenance of air pollution in Central Europe: Using soluble and insoluble particles in snow and rime. *Environmental Pollution*, **218**, 1135–1146.

Novakov, T., Ramanathan, V., Hansen, J. E., et al. (2003). Large historical changes of fossil-fuel black carbon aerosols. *Geophysical Research Letters*, **30**(6), 1324, 57-1–57-4.

Nuelle, M.-T., Dekiff, J. H., Remy, D., and Fries, E. (2014). A new analytical approach for monitoring microplastics in marine sediments. *Environmental Pollution*, **184**, 161–169.

Obbard, R. W., Sadri, S., Wong, Y. Q., et al. (2014). Global warming releases microplastic legacy frozen in Arctic Sea ice. *Earth's Future*, **2**, 315–320.

Oberle, F. K. J., Storlazzi, C. D., and Hanebuth, T. J. J. (2016). What a drag: Quantifying the global impact of chronic bottom trawling on continental shelf sediment. *Journal of Marine Systems*, **159**, 109–119.

Oberle, F. K. J., Swarzenski, P. W., Reddy, C. M., et al. (2015). Deciphering the lithological consequences of bottom trawling to sedimentary habitats on the shelf. *Journal of Marine Systems*, **159**, 120–131.

O'Connor, B. (2001). The origins and development of the British coprolite industry. *Mining History: The Bulletin of the Peak District Mines Historical Society*, **14**(5), 46–57.

Officer, C. B., Hallam, A., Drake, C. L., and Devine, J. D. (1987). Late Cretaceous and paroxysmal Cretaceous/Tertiary extinctions. *Nature*, **326**, 143–149.

Ogg, J. G., and Hinnov, L. A. (2014). Cretaceous. In Gradstein, F. M., Ogg, J. G., Schmitz, M. D., and Ogg, G., eds., *The Geologic Time Scale*. Elsevier, pp. 793–853.

Oldfield, F. (2015). Can the magnetic signatures from inorganic fly ash be used to mark the onset of the Anthropocene? *Anthropocene Review*, **2**(1), 3–13.

Oldfield, F., and Richardson, N. (1990). Lake sediment magnetism and atmospheric deposition. *Philosophical Transactions of the Royal Society of London B*, **327**, 325–330.

Oldfield, F., Thompson, R., and Barber, K. E. (1978). Changing atmospheric fallout of magnetic particles recorded in recent ombrotrophic peat. *Science*, **199**, 679–680.

Olóriz, F., Caracuel, J. E., and Rodríguez-Tovar, F. J. (1995). Using ecostratigraphic trends in sequence stratigraphy. In Haq, B. U., ed., *Sequence Stratigraphy and Depositional Response to Eustatic, Tectonic and Climatic*. Dordrecht: Springer-Publ., pp. 59–85.

Oppenheimer, M., and Alley, R. B. (2016). How high will the seas rise? *Science*, **354**, 1375–1377.

Oreskes, N. (1999). *The Rejection of Continental Drift: Theory and Method in American Earth Science*. New York: Oxford University Press.

Oreskes, N. (2003). *Plate Tectonics: An Insider's History of the Modern Theory of the Earth*. Boulder, Westview Press.

Orio, A. A., and Botkin, D. B. (eds.) (1986). Man's role in changing the global environment (papers presented at an international conference, Venice, Italy, 21–26 October 1985). *Science of Total Environment*, **55**, 10–399; 56, 1–415.

Orsi, A. J., Kawamura, K., Masson-Delmotte, V., et al. (2017). The recent warming trend in North Greenland. *Geophysical Research Letters*. doi:10.1002/2016GL072212.

Osborn, F. (1948). *Our Plundered Planet*. New York: Little, Brown and Company.

Osterman, L. E., Poore, R. Z., and Swarzenski, P. W. (2008). The last 1000 years of natural and anthropogenic low-oxygen bottom-water on the Louisiana shelf, Gulf of Mexico. *Marine Micropaleontology*, **66**(3–4), 291–303.

Ostrovsky, I. S. (1983). Productivity of the common zoobenthos species and their role in Lake Sevan ecosystem. *Moscow* [in Russian].

O'Sullivan, G., and Megson, D. (2014). Brief overview: Discovery, regulation, properties, and fate of POPs. In O'Sullivan, G., and Sandau, C. D., eds., *Environmental Forensics for Persistent Organic Pollutants*. Amsterdam: Elsevier, pp. 1–20.

Our World in Data (OWID) (2017). University of Oxford, *OurWorldInData.org*.

Overeem, I., Syvitski, J. P. M., Hutton, E. W. H., and Kettner, A. J. (2005). Stratigraphic variability due to uncertainty in model boundary conditions: A case study of the New Jersey Shelf over the last 21,000 years. *Marine Geology*, **224**, 23–41.

Overland, J. E., Wood, K. R., and Wang, M. (2011). Warm Arctic – cold continents: Climate impacts of the newly open Arctic Sea. *Polar Research*, **30**. doi:10.3402/polar.v30i0.15787.

Pacyna, J. M., and Pacyna, E. G. (2001). An assessment of global and regional emissions of trace metals to the atmosphere from anthropogenic sources worldwide. *Environmental Reviews*, **9**(4), 269–298.

Pacyna, J. M., and Pacyna, E. G. (2016). *Environmental Determinants of Human Health.* Molecular and Integrative Toxicology. Switzerland: Springer International Publishing.

Padilla, D. K. (2010). Context-dependant impacts of a non-native ecosystem engineer, the Pacific oyster *Crassostrea gigas*. *Integrative and Comparative Biology*, **50**, 213–225.

Pagani, M., Lemarchand, D., Spivack, A., and Gaillardet, J. A. (2005). Critical evaluation of the boron isotope pH-proxy: The accuracy of ancient ocean pH estimates. *Geochimica Cosmochimica Acta*, **69**, 953–961.

Palanques, A., Guillén, J., and Puig, P. (2001). Impact of bottom trawling on water turbidity and muddy sediment of an unfished continental shelf. *Limnology and Oceanography*, **46**, 1100–1110.

Palmieri, A., Shah, F., Annandale, G. W., and Dinar, A. (2003). *Reservoir Conservation Vol. 1: The RESCON Approach: Economic and Engineering Evaluation of Alternative Strategies for Managing Sedimentation in Storage Reservoirs.* International Bank for Reconstruction and Development, the World Bank.

Pandolfi, J. M. (2015). Incorporating uncertainty in predicting the future response of coral reefs to climate change. *Annual Review of Ecology, Evolution, and Systematics*, **46**, 281–303.

Paolo, F. S., Fricker, H. A., and Padman, L. (2015). Volume loss from Antarctic ice shelves is accelerating. *Science*, **348**(6232), 327–331.

Parrenin, F., Masson-Delmotte, V., Köhler, P., et al. (2013). Synchronous change of atmospheric CO_2 and Antarctic temperature during the last deglacial warming. *Science*, **339**(6123), 1060–1063.

Patris, N., Delmas, R. J., Legrand, M., et al. (2002). First sulfur isotope measurements in central Greenland ice cores along the preindustsrial and industrial periods. *Journal of Geophysical Research*, **107**(D11), 4115. doi:10.1029/2001JD000672.

Patton, H., Hubbard, A., Andreassen, K., et al. (2017). Deglaciation of the Eurasian ice sheet complex. *Quaternary Science Reviews*, **169**, 148–172.

Paull, C. K., Ussler, W., III, Mitts, P. J., et al. (2006). Discordant ^{14}C-stratigraphies in upper Monterey Canyon: A signal of anthropogenic disturbance. *Marine Geology*, **233**, 21–36.

Pauly, D. (2010). *5 Easy Pieces: The Impact of Fisheries on Marine Systems*. Island Press.

Pauly, D., Christensen, V., Guenette, S., et al. (2002). Towards sustainability in world fisheries. *Nature*, **418**, 689–695.

Pearce, F. (2017). A year on thin ice. *New Scientist*, **234**(3120), 33–37.

Pearson, A. J., Pizzuto, J. E., and Vargas, R. (2015). Influence of run of river dams on floodplain sediments and carbon dynamics. *Geoderma*, **272**, 51–63.

Pearson, P. N., and Palmer, M. R. (2000). Atmospheric carbon dioxide concentrations over the past 60 million years. *Nature*, **406**, 695–699.

Pedro, J. B., Rasmussen, S. O., and Van Ommen, T. D (2012). Tightened constraints on the time-lag between Antarctic temperature and CO_2 during the last deglaciation. *Climate of the Past*, **8**, 1213–1221.

Peloggia, A. U. G., Oliveira, A. M. S., Oliveira, A. A., et al. (2014). Technogenic geodiversity: A proposal on the classification of artificial ground. *Quaternary and Environmental Geosciences*, **5**(1), 28–40.

Penman, D. E., Hönisch, B., Zeebe, R. E., et al. (2014). Rapid and sustained ocean acidification during the Paleocene-Eocene Thermal Maximum. *Paleoceanography*, **29**, 357–369.

Pereira, N. S., Sial, A. N., Kikuchi, R. K. P., et al. (2015). Coral-based climate records from tropical South Atlantic: 2009/2010 ENSO event in C and O isotopes from *Porites* corals (Rocas Atoll, Brazil). *Anais da Academia Brasileira de Ciência*, **87**(4). doi:10.1590/0001-3765201520150072.

Periman, R. D. (2006). Visualizing the Anthropocene: Human land use history and environmental management. In Aguirre-Bravo, C., Pellicane, P. J., Burns, D. P., and Draggan, S., eds., *Monitoring Science and Technology Symposium: Unifying Knowledge for Sustainability in the Western Hemisphere*. Fort Collins: US Department of Agriculture, pp. 558–564.

Perry, W. E. (1992). The utilization of by-products and waste products in the production of commercial fertilizers. *Fertilizer Research*, **32**, 111–114.

Petit, J. R., Jouzel, J., Raynaud, D., et al. (1999). Climate and atmospheric history of the past 420,000 years from the Vostok ice core, Antarctica. *Nature*, **399**, 429–436.

Petoukhov, V., and Semenov, V. A. (2010). A link between reduced Barents-Kara sea ice and cold winter extremes over northern continents. *Journal of Geophysical Research*, **115**, D21111. doi:10.1029/2009JD013568.

Pfeffer, W. T., Arendt, A. A., Bliss, A., et al. (2014). The Randolph Glacier Inventory: A globally complete inventory of glaciers. *Journal of Glaciology*, **60** (221), 537–552.

Pham, C. K., Ramirez-Llodra, E., Alt, C. H. S., et al. (2014). Marine litter density and distribution in European seas, from shelves to deep basins. *PLoS ONE*, **9**, e95839.

Pillans, B., and Gibbard, P. (2012). The Quaternary Period. In Gradstein, F. M., Ogg, J. G., Schmitz, M., and Ogg, G., eds., *The Geological Time Scale 2012*. Elsevier, pp. 979–1010.

Pillans, B., and Naish, T. (2004). Defining the Quaternary. *Quaternary Science Reviews*, **23**, 2271–2282.

Piperno, D. R. (2011). The origins of plant cultivation and domestication in the New World tropics: Patterns, process, and new developments. *Current Anthropology*, **52**, S453–S470.

Piruzyan, L. A., Malenkov, A. G., Barenboim, G. M., and Precoda, N. (1980). Chemistry and the Biosphere. *Environment*, **22**(10), 25–30.

Pla, S., Monteith, D., Flower, R., and Rose, N. (2009). The recent palaeolimnology of a remote Scottish loch with special reference to the relative impacts of regional warming and atmospheric contamination. *Freshwater Biology*, **54**, 505–523.

Planchon, F. A., Boutron, C. F., Barbante, C., et al. (2002). Changes in heavy metals in Antarctic snow from Coats Land since the mid-19th to the late-20th century. *Earth and Planetary Science Letters*, **200**(1), 207–222.

Plass, G. N. (1961). The influence of infrared absorptive molecules on the climate. *Annals of the New York Academy of Sciences*, **95**, 61–71.

PlasticsEurope (2016). Plastics – the facts 2016: An analysis of European latest plastics production, demand and waste data. http://plasticseurope.org/Document/plastics—the-facts-2016-15787.aspx?FolID=2.

Plater, A. J., Ridgway, J., Appleby, P. J., et al. (1998). Historical contaminant fluxes in the Tees Estuary, UK: Geochemical, magnetic and radionuclide evidence. *Marine Pollution Bulletin*, **37**, 343–360.

Plumb, K. A. (1991). New Precambrian time scale. *Episodes*, **14**(2), 139–140.

Pohl, M. E. D., Piperno, D. R., Pope, K. O., and Jones, J. G. (2007). Microfossil evidence for pre-Columbian maize dispersals in the neotropics from San Andrés, Tabasco, Mexico. *Proceedings of the National Academy of Sciences (USA)*, 104, 6870–6875.

Pohl, T., Al-Muqdadi, S. W., Ali, M. H., et al. (2013). Discovery of a living coral reef in the coastal waters of Iraq. *Scientific Reports*, **4**, 4250. doi:10.1038/srep04250.

Polato, N. R., Voolstra, C. R., and Schnetzer, J. (2010). Location-specific responses to thermal stress in larvae of the reef-building coral *Montastraea faveolata*. *PLoS ONE*, 5(6), e11221. doi:10.1371/journal.pone.0011221.

Pollard, D., and DeConto, R. (2005). Hysteresis in Cenozoic Antarctic ice-sheet variations. *Global and Planetary Change*, **45**, 9–21.

Pollard, D., and DeConto, R. (2009). Modelling West Antarctic ice sheet growth and collapse through the past five million years. *Nature*, **458**, 329–333.

Pollard, D., DeConto, R. M., and Alley, R. B. (2015). Potential Antarctic Ice Sheet retreat driven by hydrofracturing and ice cliff failure. *Earth and Planetary Science Letters*, **412**, 112–121.

Poloczanska, E. S., Burrows, M. T., Brown, C. J., et al. (2016). Responses of marine organisms to climate change across oceans. *Frontiers in Marine Science*, 3(62). doi:10.3389/fmars.2016.00062.

Pompeani, D. P., Abbott, M. B., Bain, D. J., et al. (2015). Copper mining on Isle Royale 6500–5400 years ago identified using sediment geochemistry from McCargoe Cove, Lake Superior. *Holocene*, **25**, 253–262.

Pompeani, D. P., Abbott, M. B., Steinman, B. A., and Bain, D. J. (2013). Lake sediments record prehistoric lead pollution related to early copper production in North America. *Environmental Science & Technology*, **47**(11), 5545–5552.

Pongratz, R., and Heumann, K. G. (1999). Production of methylated mercury, lead, and cadmium by marine bacteria as a significant natural source for atmospheric heavy metals in polar regions. *Chemosphere*, **39**(1), 89–102.

Porter, S. (2011). The rise of predators. *Geology*, **39**, 607–608.

Prado, J., Martinez-Maza, C., and Alberdi, M. T. (2015). Megafauna extinction in South America: A new chronology for the Argentine Pampas. *Palaeogeography, Palaeoclimatology, Palaeoecology*, **425**, 41–49.

Preunkert, S., and Legrand, M. (2013). Towards a quasi-complete reconstruction of past atmospheric aerosol load and composition (organic and inorganic) over Europe since 1920 inferred from Alpine ice cores. *Climates of the Past*, **9**, 1403–1416.

Price, S. J., Ford, J. R., Cooper, A. H., and Neal, C. (2011). Humans as major geological and geomorphological agents in the Anthropocene: The significance of artificial ground in Great Britain. In Zalasiewicz, J. A., Williams, M., Haywood, A., and Ellis, M., eds., *The Anthropocene: A New Epoch of Geological Time*. Philosophical Transactions of the Royal Society (Series A), **369**, pp. 1056–1084.

Pritchard, H. D., Ligtenberg, S. R. M., Fricker, H. A., et al. (2012). Antarctic ice-sheet loss driven by basal melting of ice shelves. *Nature*, **484**, 502–505.

Pugh, D., and Woodworth, P. (2014). *Sea-Level Science: Understanding Tides, Surges, Tsunamis and Mean Sea-Level Changes*. Cambridge University Press.

Puig, P., Canals, M., Company, J. B., et al. (2012). Ploughing the deep sea floor. *Nature*, **489**, 286–289.

Punmia, B. C., Jain, A. K., and Jain, A. K. (2004). *Basic Civil Engineering*. New Delhi, India: Laxmi Publications (P) Ltd.

Qiu, Y. W., Zhang, G., Liu, G. Q., et al. (2009). Polycyclic aromatic hydrocarbons (PAHs) in the water column and sediment core of Deep Bay, South China. *Estuarine, Coastal and Shelf Science*, **83**(1), 60–66.

Quinto, F., Hrnecek, E., Krachler, M., et al. (2013). Determination of ^{239}Pu, ^{240}Pu, ^{241}Pu and ^{242}Pu at femtogram and attogram levels—evidence for the migration of fallout plutonium in an ombrotrophic peat bog profile. *Environmental Science: Processes & Impacts*, **15**(4), 839–847.

Rackow, T., Wesche, C., Timmermann, R., et al. (2016). A simulation of small to giant Antarctic iceberg evolution: Differential impact on climatology estimates. *Journal of Geophysical Research – Oceans*, **122**(4), 3170–3190.

Radivojević, M., Rehren, T., Pernicka, E., et al. (2010). On the origins of extractive metallurgy: New evidence from Europe. *Journal of Archaeological Science*, **37**, 2775–2787.

Raes, F., Van Dingenen, R., Vignati, E., et al. (2000). Formation and cycling of aerosols in the global troposphere. *Atmospheric Environment*, **34**(25), 4215–4240.

Rahmstorf, S., Foster, G., and Cahill, N. (2017). Global temperature evolution: Recent trends and some pitfalls. *Environmental Research Letters*, **12**(5).

Rakowski, A. Z., Nadeau, M.-J., Nakamura, T., et al. (2013). Radiocarbon method in environmental monitoring of CO_2 emission. *Nuclear Instruments and Methods in Physics Research B*, **294**, 503–507.

Ramirez-Llodra, E., Tyler, P. A., Baker, M. C., et al. (2011). Man and the last great wilderness: Human impact on the deep sea. *PLoS ONE*, **6**, e22588. doi.org/10.1371/journal.pone.0022588.

Rasmussen, B., and Buick, R. (1999). Redox state of the Archaean atmosphere: Evidence from detrital heavy minerals in ca 3250–2750 Ma sandstones from the Pilbara Craton, Australia. *Geology*, **27**, 115–118.

Rathje, W. L., and Murphy, C. (2001). *Rubbish!: The Archaeology of Garbage*. Tucson: University of Arizona Press.

Rauch, J. N. (2010). Global spatial indexing of the human impact on Al, Cu, Fe and Zn mobilization. *Environmental Science and Technology*, **44**, 5728–5734.

Rauch, J. N. (2012). The present understanding of Earth's global anthrobiogeochemical metal cycles. *Mineral Economics*, **25**(1), 7–15.

Rauch, J. N., and Graedel, T. E. (2007). Earth's anthrobiogeochemical copper cycle. *Global Biogeochemical Cycles*, **21**(2). doi:10.1029/2006GB002850.

Rauch, J. N., and Pacyna, J. M. (2009). Earth's global Ag, Al, Cr, Cu, Fe, Ni, Pb, and Zn cycles. *Global Biogeochemical Cycles*, **23**(2). doi:10.1029/2008GB003376.

Rausch, N., Nieminen, T., Ukonmaanaho, L., et al. (2005). Comparison of atmospheric deposition of copper, nickel, cobalt, zinc, and cadmium recorded by Finnish peat cores with monitoring data and emission records. *Environmental Science & Technology*, **39**(16), 5989–5998.

Razali, M. N., Effendi, M. L. H. M, Musa, M., and Yunus, R. M. (2016). Formulation of bitumen from industrial waste. *ARPN Journal of Engineering and Applied Sciences*, **11**(8), 5244–5250.

Reading, H. G. (ed.) (1996). *Sedimentary Environments: Processes, Facies and Stratigraphy*, 3rd ed. Malden, MA: Blackwell Publishing.

Reager, J. T., Gardner, A. S., Famiglietti, J. S., et al. (2016). A decade of sea level rise slowed by climate-driven hydrology. *Science*, **351**, 699–703.

Reed, A. J., Mann, M. E., Emanuel, K. A., et al. (2015). Increased threat of tropical cyclones and coastal flooding to New York City during anthropogenic era. *Proceedings of the National Academy of Sciences (USA)*, **112**(41), 12610–12615.

Reed, C. (2015). Dawn of the Plasticene age. *New Scientist*, **225** (3006, 31 January 2015), 28–32.

Reis, L. (2014). Occurrence of macro- and microplastics in sediments from Fehmarm. BSc. Unpublished thesis, Freie Universität, Berlin.

Remane, J. (1997). Foreword: Chronostratigraphic standards: How are they defined and when should they be changed? *Quaternary International*, **40**, 3–4.

Remane, J. (2003). Chronostratigraphic correlations: Their importance for the definition of geochronologic units. *Palaeogeography, Palaeoclimatology, Palaeoecology*, **196**, 7–18.

Remane, J., Bassett, M. G., Cowie, J. W., et al. (1996). Revised guidelines for the establishment of global chronostratigraphic standards by the International Commission on Stratigraphy (ICS). *Episodes*, **19**, 77–81.

Renaud, F., Syvitski, J. P. M., Sebesvari, Z., et al. (2013). Tipping from the Holocene to the Anthropocene: How threatened are major world deltas? *Current Opinion in Environmental Sustainability*, **5**, 644–654.

Renberg, I., and Battarbee, R. W. (1990). The SWAP Palaeolimnology Programme: A synthesis. In Mason, B. J (ed). *The Surface Waters Acidification Programme*. Cambridge University Press, pp. 281–300.

Renberg, I., Brännvall, M.-L., Bindler, R., and Emteryd, O. (2000). Atmospheric lead pollution history during four millennia (2000 BC to 2000 AD) in Sweden. *Ambio*, **29**(3), 150–156.

Renberg, I., Persson, M., and Emteryd, O. (1994). Pre-industrial atmospheric lead contamination detected in Swedish lake sediments. *Nature*, **368**, 323–326.

Restrepo, J. D., and Syvitski, J. P. M. (2006). Assessing the effect of natural controls and land use change on sediment yield in a major Andean river: The Magdalena drainage basin, Colombia. *Ambio*, **35**, 65–74.

Revkin, A. C. (1992). *Global Warming: Understanding the Forecast* (American Museum of Natural History, Environmental Defense Fund). New York: Abbeville Press.

Reynolds, P., Planke, S., Millett, J. M., et al. (2017). Hydrothermal vent complexes offshore Northeast Greenland: A potential role in driving the PETM. *Earth and Planetary Science Letters*, **467**, 72–78.

Reznikov, S. A. (1984). Biogenic elements of sediments in Lake Sevan. *Trudy Sevanskoi Gidrobiologicheskoi Stantsii*, **19**, 5–17 [in Russian].

Rhoads, B. L., Quinn, W., and Lewis, W. A. (2016). Historical changes in channel network extent and channel planform in an intensively managed landscape: Natural versus human-induced effects. *Geomorphology*, **252**, 17–31.

Ribeiro, S., Amorim, A., Abrantes, F., and Ellegaard, M. (2016). Environmental change in the Western Iberia Upwelling Ecosystem since the preindustrial period revealed by dinoflagellate cyst records. *Holocene*, **26**, 874–889.

Ribeiro, S., Amorim, A., Andersen, T. J., et al. (2012). Reconstructing the history of an invasion: The toxic phytoplankton species *Gymnodinium catenatum* in the Northeast Atlantic. *Biological Invasions*, **14**, 969–985.

Richards, Z. T., and Beger, M. (2011). A quantification of the standing stock of macro-debris in Majuro lagoon and its effect on hard coral communities. *Marine Pollution Bulletin*, **62**, 1693–1701.

Richter, D. deB. (2007). Humanity's transformation of Earth's soil: Pedology's new frontier. *Soil Science*, **172**, 957–967.

Richter, D. deB., Bacon, A. R., Megan, L. M., et al. (2011). Human–soil relations are changing rapidly: Proposals from SSSA's cross-divisional Soil Change Working Group. *Soil Science Society of America Journal*, **75**(6), 2079–2084.

Richter, D. deB., Grün, R., Joannes-Boyau, R., et al. (2017). The age of the hominin fossils from Jbel Irhoud, Morocco, and the origins of the Middle Stone Age. *Nature*, **546**, 293–296.

Richter, D. deB., and Markewitz, D. (1995). How deep is soil? *BioScience*, **45**, 600–609.

Richter, D. deB., and Yaalon, D. H. (2012). "The changing model of soil" revisited. *Soil Science Society of America Journal*, **76**, 766–778.

Richter-Menge, J., Overland, J. E., and Mathis, J. T. (eds.) (2016). *Arctic Report Card 2016*. http://arctic.noaa.gov/Report-Card.

Rietbroek, R., Brunnabend, S.-E., Kusche, J., et al. (2016). Revisiting the contemporary sea-level budget on global and regional scales. *Proceedings of the National Academy of Sciences (USA)*, **113**, 1504–1509.

Rigaud, J.-P., Texier, P.-J., Parkington, J., and Poggenpoel, C. (2006). South African Middle Stone Age chronology: New excavations at Diepkloof Rock Shelter; Preliminary results. *Comptes Rendus Palevol*, **5**, 839–849.

Rignot, E., Casassa, G., Gogineni, P., et al. (2004). Accelerated ice discharge from the Antarctic Peninsula following the collapse of Larsen B ice shelf. *Geophysical Research Letters*, **31**, L18401. doi:10.1029/2004GL020697.

Rignot, E., Velicogna, I., van den Broeke, M. R., et al. (2011). Acceleration of the contribution of the Greenland and Antarctic ice sheets to sea level rise. *Geophysical Research Letters*, **38**, L05503. doi:10.1029/2011GL046583.

Rijsdijk, K. F, Hume, J. P., Bunnik, F., et al. (2009). Mid-Holocene vertebrate bone concentration-Lagerstätte on oceanic island Mauritius provides a window into the ecosystem of the dodo *(Taphus cucullatus). Quaternary Science Reviews*, **28**, 14–24.

Rillig, M. C. (2012). Microplastic in terrestrial ecosystems and the soil? *Environmental Science and Technology*, **46**, 6453–6454.

Rintoul, S. R., Silvano, A., Pena-Molino, B., et al. (2016). Ocean heat drives rapid basal melt of the Totten Ice Shelf. *Science Advances*, **2**, e1601610.

Rio, D., Sprovieri, R., Castradori, D., and Di Stefano, E. (1998). The Gelasian Stage (Upper Pliocene): A new unit of the global standard chronostratigraphic scale. *Episodes*, **21**(2), 82–87.

Rivas, V., Cendero, A., Hurtado, M., et al. (2006). Geomorphic consequences of urban development and mining activities: An analysis of study areas in Spain and Argentina. *Geomorphology*, **73**, 185–206.

Roberts, C. (2007). *The Unnatural History of the Sea: The Past and Future of Humanity and Fishing (Gaia Thinking)*. Washington, DC: Island Press.

Roberts, C. (2013). *Ocean of Life*. London: Penguin.

Roberts, J., Moy, A., Plummer, C., et al. (2017). A revised Law Dome age model (LD2017) and implications for last glacial climate. *Climate of the Past Discussions*. https://doi.org/10.5194/cp-2017-96.

Roberts, R. G., Flannery, T. F., Ayliffe, L. K., et al. (2001). New ages for the last Australian megafauna: Continent-wide extinction about 46,000 years ago. *Science*, **292**, 1888–1892.

Robertson, J. (2017). Runoff pollution from Cyclone Debbie flooding sweeps into Great Barrier Reef. *The Guardian*, 11 April 2017. https://theguardian.com/australia-news/2017/apr/11/run-off-pollution-from-cyclone-debbie-flooding-sweeps-into-great-barrier-reef (accessed June 2017).

Robinson, N. (2012). Beyond sustainability: Environmental management for the Anthropocene epoch. *Journal of Public Affairs*, **12**, 181–194.

Robison, B. H., Reisenbichler, K. R., and Sherlock, R. E. (2005). Giant larvacean houses: Rapid carbon transport to the deep sea floor. *Science*, **308**, 1609–1611.

Rochman, C., Browne, M. A., Halpern, B., et al. (2013). Classify plastic waste as hazardous. *Nature*, **494**, 169–171.

Rockström, J., Gaffney, O., Rogelj, J., et al. (2017). A roadmap for rapid decarbonization. *Science*, **355**, 1269–1271.

Rockström, J., Steffen, W., Noone, K., et al. (2009). A safe operating space for humanity. *Nature*, **461**, 472–475.

Roebroeks, W., and Villa, P. (2011). On the earliest evidence for habitual use of fire in Europe. *Proceedings of the National Academy of Sciences (USA)*, **108**, 5209–5214.

Roger, J. (ed.) (1962). *Buffon: Les Époques de la Nature*. Edition critique, Mémoires du Muséum National d'Histoire Naturelle. Nouvelle Série, Série C, Sciences de la Terre, Tome X, Paris.

Röhl, U., Westerhold, T., Bralower, T. J., and Zachos, J. C. (2007). On the duration of the Paleocene-Eocene thermal maximum (PETM). *Geochemistry, Geophysics, Geosystems*, **8**(12), Q12002. doi:10.1029/2007GC001784.

Rohling, E. J., Haigh, I. D., Foster, G. L., et al. (2013). A geological perspective on potential future sea-level rise. *Scientific Reports*, **3**, 3461.

Rohwer, F., and Youle, M. (2010). *Coral reefs in the microbial seas*. San Francisco, CA: Plaid Press.

Rolison, J. M., Landing, W. M., Luke, W., et al. (2013). Isotopic composition of species-specific atmospheric Hg in a coastal environment. *Chemical Geology*, **336**, 37–49.

Roosevelt, A. C. (2013). The Amazon and the Anthropocene: 13,000 years of human influence in a tropical rainforest. *Anthropocene*, **4**, 69–87.

Rose, N. L. (1996). Inorganic ash spheres as pollution tracers. *Environmental Pollution*, **91**, 245–252.

Rose, N. L. (2001). Fly-ash particles. In Last, W. M., and Smol, J. P., eds., *Tracking Environmental Change Using Lake Sediments: Volume 2. Physical and Chemical Techniques*. Dordrecht, Netherlands: Kluwer Academic Publishers, pp. 319–349.

Rose, N. L. (2015). Spheroidal carbonaceous fly-ash particles provide a globally synchronous stratigraphic marker for the Anthropocene. *Environmental Science & Technology*, **49**, 4155–4162.

Rose, N. L. (2018). Spheroidal carbonaceous fly ash particles in the Anthropocene. In DellaSala, D. A., and Goldstein, M. I., eds., *Encyclopaedia of the Anthropocene*, vol. 1. Oxford: Elsevier.

Rose, N. L., and Appleby, P. G. (2005). Regional applications of lake sediment dating by spheroidal carbonaceous particle analysis I. *Journal of Paleolimnology*, **34**, 349–361.

Rose, N. L., Backus, S., Karlsson, H., and Muir, D. C. G. (2001). An historical record of toxaphene and its congeners in a remote lake in western Europe. *Environmental Science & Technology*, **35**, 1312–1319.

Rose, N. L., Jones, V. J., Noon, P. E., et al. (2012). Long-range transport of pollutants to the Falkland Islands and Antarctica: Evidence from lake sediment fly-ash particle records. *Environmental Science & Technology*, **46**, 9881–9889.

Rose, N. L., and Ruppel, M. (2015). Environmental archives of contaminant particles. In Blais, J. M., Rosen, M. R., and Smol, J. P., eds., *Environmental Contaminants: Using Natural Archives to Track Sources and Long-Term Trends of Pollution*. Developments in Paleoenvironmental Research, **18**. Dordecht, Netherlands: Springer, pp. 187–221.

Rosenbaum, M. S., McMillan, A. A., Powell, J. H., et al. (2003). Classification of artificial (man-made) ground. *Engineering Geology*, **69**, 399–409.

Rosman, K. J., Chisholm, W., Hong, S., et al. (1997). Lead from Carthaginian and Roman Spanish mines isotopically identified in Greenland ice dated from 600 BC to 300 AD. *Environmental Science & Technology*, **31**(12), 3413–3416.

Rosswall, T., Liss, P., Rapley, C., et al. (2015). Reflections on Earth-System Science. *Global Change*, **84**, 8–13.

Rothman, A. (2007). *Slave Country: American Expansion and the Origins of the New South*. Cambridge, MA: Harvard University Press.

Royal Society (2009). *Geoengineering the Climate: Science, Governance and Uncertainty. Report 10/9 RS1636*. London: Royal Society.

Royer, D. L. (2006). CO_2-forced climate thresholds during the Phanerozoic. *Geochimica Cosmochimica Acta*, **70**, 5665–5675.

Royer, D. L., Berner, R. A., Montañez, I. P., et al. (2004). CO_2 as a primary driver of Phanerozoic climate. *GSA Today*, **14**(3), 4–10.

Rubino, M., Etheridge, D. M., Trudinger, C. M., et al. (2013). A revised 1000 year atmospheric $\delta^{13}C$-CO_2 record from Law Dome and South Pole, Antarctica. *Journal of Geophysical Research*, **118**, 8482–8499.

Rubino, M., Etheridge, D. M., Trudinger, C. M., et al. (2016). Low atmospheric CO_2 levels during the Little Ice Age due to cooling-induced terrestrial uptake. *Nature Geoscience*, **9**, 691–694.

Ruddiman, W. F. (1977). Late Quaternary deposition of ice-rafted sand in the subpolar North Atlantic (Lat 40° to 65°N). *Bulletin of the Geological Society of America*, **88**, 1813–1827.

Ruddiman, W. F. (2003). The anthropogenic Greenhouse Era began thousands of years ago. *Climatic Change*, **61**, 261–293.

Ruddiman, W. F. (2005). *Plows, Plagues and Petroleum*. Princeton University Press.

Ruddiman, W. F. (2013). Anthropocene. *Annual Review of Earth and Planetary Sciences*, **41**, 45–68.

Ruddiman, W. F. (2014). *Earth's Climate: Past and Future*, 3rd ed. New York: W. H. Freeman.

Ruddiman, W. F., Ellis, E. C., Kaplan, J. O., and Fuller, D. Q. (2015a). Defining the epoch we live in. *Science*, **348**, 38–39.

Ruddiman, W. F., Fuller, D. Q., Kutzbach, J. E., et al. (2015b). Late Holocene climate: Natural or anthropogenic? *Reviews of Geophysics*, **54**(1), 93–118.

Rudwick, M. J. S. (2005). *Bursting the Limits of Time: The Reconstruction of Geohistory in the Age of Revolution*. Chicago University Press.

Rudwick, M. J. S. (2008). *Worlds before Adam: The Reconstruction of Geohistory in the Age of Reform*. Chicago University Press.

Ruhe, R. V., and Scholtes, W. H. (1956). Age and development of soil landscapes in relation to climate and vegetational changes in Iowa. *Soil Science Society of America Proceedings*, **20**, 264–273.

Ruppel, C. D., and Kessler, J. D. (2017). The interaction of climate change and methane hydrates. *Reviews of Geophysics*, **55**(1), 126–168.

Rushton, A. W. A., Brück, P. M., Molyneux, S. G., et al. (2011). A revised correlation of the Cambrian rocks in the British Isles. *Geological Society of London, Special Report*, **25**.

Ryan, P. G., Moore, C. J., van Franeker, J. A., and Moloney, C. L. (2009). Monitoring the abundance of plastic debris in the marine environment. *Philosophical Transactions of the Royal Society B*, **364**, 1999–2012.

Sabine, C. L., Feely, R. A., Gruber, N., et al. (2004). The oceanic sink for anthropogenic CO_2. *Science*, **305**, 367–371.

Sadri, S. S., and Thompson, R. C. (2014). On the quantity and composition of floating plastic debris entering and leaving the Tamar Estuary, Southwest England. *Marine Pollution Bulletin*, **81**, 55–60.

Sagan, C. (1994). *Pale Blue Dot: A Vision of the Human Future in Space*, 1st ed. New York: Random House.

Sagnotti, L., Scardia, G., Giaccio, B., et al. (2014). Extremely rapid directional change during Matuyama-Brunhes geomagnetic polarity reversal. *Geophysical Journal International*, **199**, 1110–1124.

Saito, Y., Chaimanee, N., Jarupongsakul, T., and Syvitski, J. P. M. (2007). Shrinking megadeltas in Asia: Sea-level rise and sediment reduction impacts from case study of the Chao Phraya delta. *Imprint Newsletter of the IGBP/IHDP Land Ocean Interaction in the Coastal Zone*, **2007**(2), 3–9.

Sakurai, T., Ohno, H., Motoyama, H., and Uchida, T. (2016). Micro-droplets containing sulfate in the Dome Fuji deep ice core, Antarctica: Findings using micro-Raman spectroscopy. *Journal of Raman Spectroscopy*, **48**, 448–452.

Salomons, W., and Förstner, U. (2012). *Metals in the Hydrocycle*. Berlin: Springer Science & Business Media.

Salvador, A. (ed.) (1994). *International Stratigraphic Guide: A Guide to Stratigraphic Classification, Terminology, and Procedure*, 2nd ed. Boulder, Colorado: International Subcommission on Stratigraphic Classification of IUGS International Commission on Stratigraphy and the Geological Society of America.

Salzmann, U., Williams, M., Haywood, A. M., et al. (2011) Climate and environment of a Pliocene warm world. *Palaeogeography, Palaeoclimatology, Palaeoecology*, **309**, 1–8.

Samways, M. (1999). Translocating fauna to foreign lands: Here comes the Homogenocene. *Journal of Insect Conservation*, **3**, 65–66.

Sandom, C., Faurby, S., Sandel, B., and Svenning, J.-C. (2013). Global late Quaternary megafauna extinctions linked to humans, not climate change. *Proceedings of the Royal Society B*, **281**. doi:10.1098/rspb.2013.3254.

Sarg, J. F. (1988). Carbonate sequence stratigraphy. In Wilgus, C. K., Hastings, B. S., Posamentier, H., et al., eds., *Sea-Level Changes: An Integrated Approach*. Tulsa: SEPM Special Publication, **42**, pp. 155–181.

Saunois, M., Jackson, R. B., Bousquet, P., et al. (2016). The growing role of methane in anthropogenic climate change. *Environmental Research Letters*, **11**. doi:10.1088/1748-9326/11/12/120207.

Schaefer, J. M., Finkel, R. C., Balco, G., et al. (2016). Greenland was nearly ice-free for extended periods during the Pleistocene. *Nature*, **540**, 252–255.

Schaetzl, R. J., and Thompson, M. L. (2015). *Soils: Genesis and Geomorphology*. Cambridge University Press.

Scheffers, B. R., De Meester, L., Bridge, T. C. L., et al. (2016). The broad footprint of climate change from genes to biomes to people. *Science*, **354**. doi:10.1126/science.aaf7671.

Schellnhuber, H. J., Rahmstorf, S., and Winkelmann, R. (2016). Why the right climate target was agreed in Paris. *Nature Climate Change*, **6**, 649–653.

Scherer, R. P., DeConto, R. M., Pollard, D., and Alley, R. B. (2016). Windblown Pliocene diatoms and East Antarctic Ice Sheet retreat. *Nature Communications*, **7**. doi:10.1038/ncomms12957.

Schimmelmann, A., Hendy, I. L., Dunn, L., et al. (2013). Revised ~2000-year chronostratigraphy of partially varved marine sediment in Santa Barbara Basin, California. *GFF*, **135**(3–4), 258–264.

Schimmelmann, A., Lange, C. B., Schieber, J., et al. (2016). Varves in marine sediments: A review. *Earth-Science Reviews*, **159**, 215–246.

Schimper, W. P. (1874). *Traité de Paléontologie Végétale*. Paris: J. B. Ballière, **3**.

Schlager, W. (1992). Sedimentology and sequence stratigraphy of reefs and carbonate platforms. *American Association of Petroleum Geologists Contin. Educ. Course Note Series*, **34**, Tulsa.

Schlining, K., von Thun, S., Kuhnz, L., et al. (2013). Debris in the deep: Using a 22-year video annotation database to survey marine litter in Monterey Canyon, central California, USA. *Deep-Sea Research Part I: Ocean Research Paper*, **79**, 96–105.

Schmid, P., Bogdal, C., Blüthgen, N., et al. (2010). The missing piece: Sediment records in remote mountain lakes confirm glaciers being secondary sources of persistent organic pollutants. *Environmental Science & Technology*, **45**(1), 203–208.

Schmidt, H., and Reimers, C. E. (1991). The recent history of trace metal accumulation in the Santa Barbara Basin, southern California borderland. *Estuarine, Coastal and Shelf Science*, **33**, 485–500.

Schmitt, J., Schneider, R., Elsig, J., et al. (2012). Carbon isotope constraints on the deglacial CO_2 rise from ice cores. *Science*, **336**, 711–714.

Schoepf, V., Stat, M., Falter, J. L., and McCulloch, M. T. (2015). Limits to the thermal tolerance of corals adapted to a highly fluctuating, naturally extreme temperature environment. *Scientific Reports*, **5**. doi:10.1038/srep17639.

Schofield, P. J. (2009). Geographic extent and chronology of the invasion of non-native lionfish (*Pterois volitans* [Linnaeus 1758] and *P. miles* [Bennett 1828]) in the Western North Atlantic and Caribbean Sea. *Aquatic Invasions*, **4** (3), 473–479.

Scholz, D., Frisia, S., Borsato, A., et al. (2012). Holocene climate variability in north-eastern Italy: Potential influence of the NAO and solar activity recorded by speleothem data. *Climate of the Past*, **8**, 1367–1383.

Schulte, P., Alegret, L., Arenillas, I., et al. (2010). The Chicxulub asteroid impact and mass extinction at the Cretaceous-Paleogene boundary. *Science*, **327**, 1214–1218.

Scott, D. B., and Medioli, F. S. (1980). Quantitative studies of marsh foraminiferal distributions in Nova Scotia: Implications for sea level studies. *Cushman Foundation for Foraminiferal Research, Special Publication* **17**, 1–58.

Scott, D. B., Medioli, F. S., and Schafer, C. T. (2001). *Monitoring in Coastal Environments Using Foraminifera and Thecamoebian Indicators.* Cambridge University Press.

Scott, K. (2013). International law in the Anthropocene: Responding to the geoengineering challenge. *Michigan Journal of International Law*, **34**, 309–358.

Scrivener, K. L., and Kirkpatrick, R. J. (2008). Innovation in use and research on cementitious material. *Cement and Concrete Research*, **38**, 128–136.

Seemann, J. (2013). The use of ^{13}C and ^{15}N isotope labeling techniques to assess heterotrophy of corals. *Journal of Experimental Marine Biology and Ecology*, **442**(88). doi:10.1016/j.jembe.2013.01.004.

Seemann, J., Carballo-Bolaños, R., Berry, K. L., et al. (2012). Importance of heterotrophic adaptations of corals to maintain energy reserves. *Proceedings of the 12th International Coral Reef Symposium, Cairns, Australia*, 9–13 July 2012. 19A Human impacts on coral reef. http://icrs2012.com/proceedings/manuscripts/ICRS2012_19A_4.pdf.

Sen, I. S., and Peuckner-Ehrenbrink, B. (2012). Anthropogenic disturbance of element cycles at the Earth's surface. *Environmental Science and Technology*, **46**, 8601–8609.

Setälä, O., Fleming-Lehtinen, V., and Lehtiniemi, M. (2014). Ingestion and transfer of microplastics in the planktonic food web. *Environmental Pollution*, **185**, 77–83.

Shah, A. A., Hasan, F., Hameed, A., and Ahmed, S. (2008). Biological degradation of plastics: A comprehensive review. *Biotechnology Advances*, **26**, 246–265.

Shakhova, N., Semiletov, I., Leifer, I., et al. (2013). Ebullition and storm induced methane release from the East Siberian Arctic shelf. *Nature Geoscience*, **7**(1), 64–70.

Shakhova, N., Semiletov, I., Salyk, A., and Yusupov, V. (2010). Extensive methane venting to the atmosphere from sediments of the East Siberian Arctic shelf. *Science*, **327**, 1246.

Shakun, J. D., Clark, P. U., He, F., et al. (2012). Global warming preceded by increasing carbon dioxide concentrations during the last deglaciation. *Nature*, **383**, 49–54.

Shaler, N. S. (1905). *Man and the Earth.* New York: Fox, Duffield & Co.

Shamberger, K. E. F., Cohen, A. L., Golbuu, Y., et al. (2014). Diverse coral communities in naturally acidified waters of a Western Pacific reef. *Geophysical Research Letters*, **41**(2), 499–504.

Shaviv, N. J. (2002). The spiral structure of the Milky Way, cosmic rays, and ice age epochs on Earth. *New Astronomy*, **8**, 39–77.

Shaviv, N. J., and Veizer, J. (2003). Celestial driver of Phanerozoic climate? *GSA Today*, **13**(7), 4–10.

Shell International BV (2013). New Lens Scenarios. A Shift in Perspective for a World in Transition. http://shell.com/content/dam/royaldutchshell/documents/corporate/scenarios-newdoc.pdf.

Shen, B., Wu, J., and Zhao, Z. (2017). A ~150-year record of human impact in the Lake Wuliangsu (China) watershed: Evidence from polycyclic aromatic hydrocarbon and organochlorine pesticide distributions in sediments. *Journal of Limnology*, **76**(1), 129–136.

Shepherd, A., Ivins, E. R., Geruo, A., et al. (2012). A reconciled estimate of ice-sheet mass balance. *Science*, **338**, 1183–1189.

Sheppard, C. (2006). Trawling the sea bed. *Marine Pollution Bulletin*, **52**, 831–835.

Sherlock, R. L. (1922). *Man as a Geological Agent: An Account of His Action on Inanimate Nature.* London: H. F. & G. Witherby.

Sherwood, O. A., Heikoop, J. M., Scott, D. B., et al. (2005a). Stable isotopic composition of deep-sea gorgonian corals *Primnoa* spp.: A new archive of surface processes. *Marine Ecology Progress Series*, **301**, 135–148.

Sherwood, O. A., Lehmann, M. F., Schubert, C. J., et al. (2011). Nutrient regime shift in the western North Atlantic indicated by compound-specific $\delta^{15}N$ of deep-sea gorgonian corals. *Proceedings of the National Academy of Sciences (USA)*, **108**(3), 1011–1015.

Sherwood, O. A., Scott, D. B., Risk, M. J., and Guilderson, T. P. (2005b). Radiocarbon evidence for annual growth rings in a deep sea octocoral (*Primnoa resedaeformis*). *Marine Ecology Progress Series*, **301**, 129–134.

Shevenell, A. E. (2016). Drilling and modeling studies expose Antarctica's Miocene secrets. *Proceedings of the National Academy of Sciences (USA)*, **113** (13), 3419–3421.

Shields, W. J., Ahn, S., Pietari, J., et al. (2014). Atmospheric fate and behavior of POPs. In O'Sullivan, G., and Sandau, C. D., eds., *Environmental Forensics for Persistent Organic Pollutants*. Amsterdam: Elsevier, pp. 199–289.

Shields-Zhou, G. A., Porter, S., and Halverson, G. P. (2016). A new rock-based definition for the Cryogenian Period. *Episodes*, **39**, 3–8.

Shimada, I., and Cavallaro, R. (1985). Monumental adobe architecture of the late prehispanic northern north coast of Peru. *Journal de la Société des Américanistes*, **71**, 41–78.

Sholkovitz, E. R., and Mann, D. R. (1984). The pore water chemistry of ^{239,240}Pu and ^{137}Cs in sediments of Buzzards Bay, Massachusetts. *Geochimica et Cosmochimica Acta*, **48**, 1107–1114.

Shotyk, W., Appleby, P. G., Bicalho, B., et al. (2016). Peat bogs in northern Alberta, Canada reveal decades of declining atmospheric Pb contamination. *Geophysical Research Letters*, **43** (18), 9964–9974.

Shotyk, W., Krachler, M., Martinez-Cortizas, A., et al. (2002). A peat bog record of natural, pre-anthropogenic enrichments of trace elements in atmospheric aerosols since 12370 ^{14}C yr BP, and their variation with Holocene climate change. *Earth and Planetary Science Letters*, **199**(1), 21–37.

Shotyk, W., Weiss, D., Appleby, P. G., et al. (1998). History of atmospheric lead deposition since 12,370 ^{14}C yr BP from a peat bog, Jura Mountains, Switzerland. *Science*, **281**, 1635–1640.

Sigl, M., Winstrup, M., McConnell, J. R., et al. (2015). Timing and climate forcing of volcanic eruptions for the past 2,500 years. *Nature*, **523**, 543–549.

Simonyan, A. A. (1988). Zooplankton in changing conditions of a water body (case study: Lake Sevan). Unpublished PhD Thesis, Leningrad [in Russian].

Simonyan, A. A. (1991). Zooplankton of the Lake Sevan. *Yerevan* [in Russian].

Simpson, I. A. (1997). Relict properties of anthropogenic deep top soils as indicators of infield management in Marwick, West Mainland, Orkney. *Journal of Archaeological Science*, **24**(4), 365–380.

Sinkkonen, S., and Paasivirta, J. (2000). Degradation half-life times of PCDDs, PCDFs and PCBs for environmental fate modeling. *Chemosphere*, **40** (9–11), 943–949.

Skinner, L. C., Fallon, S., Waelbroeck, C., et al. (2010). Ventilation of the deep Southern Ocean and deglacial CO_2 rise. *Science*, **328**, 1147–1151.

Sloan, D. (1992). The Yerba Buena mud: Record of the last-interglacial predecessor of San Francisco Bay, California. *GSA Bulletin*, **104**, 716–727.

Slowikowski, A. (1995). "The greatest depository of archaeological material": The role of pottery in ploughzone archaeology. In Shepherd, E., ed., *Interpreting Stratigraphy*. Hunstanton: Witley Press, **5**, pp. 15–20.

Smil, V. (2010). *Energy Transitions: History, Requirements, Prospects*. Cambridge: MIT Press.

Smil, V. (2015). It's too soon to call this the Anthropocene. *IEEE Spectrum*, **52**(6), 28.

Smith, A. G., Barry, T., Bown, P., et al. (2014). GSSPs, global stratigraphy and correlation. In Smith, D. G., Bailey, R. J., Burgess, P. M., and Fraser, A. J., eds., *Strata and Time: Probing the Gaps in Our Understanding*. Geological Society, London, Special Publications, **404**, pp. 37–67.

Smith, B. D., and Zeder, M. A. (2013). The onset of the Anthropocene. *Anthropocene*, **4**, 8–13.

Smith, C. L., Fairchild, I. J., Spötl, C., et al. (2009). Chronology-building using objective identification of annual signals in trace element profiles of stalagmites. *Quaternary Geochronology*, **4**, 11–21.

Smith, D. M., Zalasiewicz, J. A., Williams, M., et al. (2010). Holocene drainage of the English Fenland: Roddons and their environmental significance. *Proceedings of the Geologists' Association*, **121**, 256–269.

Smith, J. A., Andersen, T. J., Shortt, M., et al. (2016). Sub-ice-shelf sediments record twentieth century retreat of Pine Island Glacier. *Nature*, **541**, 77–80.

Smith, J. N., and Levy, E. M. (1990). Geochronology for polycyclic aromatic hydrocarbon contamination in sediments of the Saguenay Fjord. *Environmental Science & Technology*, **24**, 874–879.

Smith, S. D. A., and Markic, A. (2013). Estimates of marine debris accumulation on beaches are strongly affected by the temporal scale of sampling. *PloS ONE*, **8**, 1-6.

Smith, S. J., Van Aardenne, J., Klimont, Z., et al. (2011). Anthropogenic sulfur dioxide emissions: 1850–2005. *Atmospheric Chemistry and Physics*, **11**, 1101–1116.

Smol, J. P. (2008). *Pollution of Lakes and Rivers: An Environmental Perspective*, 2nd ed. Wiley-Blackwell.

Snir, A., Nadel, D., Groman-Yaroslavski, I., et al. (2015). The origin of cultivation and proto-weeds, long before Neolithic farming. *PLoS ONE*, **10**(7), e0131422. doi:10.1371/journal.pone.0131422.

Snowball, I., Hounslow, M. W., and Nilsson, A. (2014). Geomagnetic and mineral magnetic characterisation of the Anthropocene. In Waters, C. N., Zalasiewicz, J., Williams, M., et al., eds., *A Stratigraphical Basis for the Anthropocene*. Geological Society, London, Special Publications, **395**, pp. 119–141.

Snyder, N. P., Rubin, D. M., Alpers, C. N., et al. (2004). Estimating accumulation rates and physical properties of sediment behind a dam: Englebright Lake, Yuba River, northern California. *Water Resources Research*, **40**, W11301. doi:10.1029/2004WR003279.

Sokolov, B. S. (1986). Ekostratigrafiya, yeye mesto i rol' v sovremennoy stratigrafii. In Kaljo, D. L., and Klaamann, E. R., eds., *Teoriya i opyt ekostratigrafii (The Theory and Practice of Ecostratigraphy)*, pp. 9–18; Inst. Geol. AN Eston. SSR, Valgus Press, Tallinn, 1986. (Using translated version: Ecostratigraphy, its place and role in modern stratigraphy, International Geology Review, 30[1], 3–10, 1988. doi:10.1080/00206818809465980.)

Solomon, S., Plattner, G.-K. Knutti, R., and Friedlingstein, P. (2009). Irreversible climate change due to carbon dioxide emissions. *Proceedings of the National Academy of Sciences (USA)*, **106**, 1704–1709.

Sousa, R., Pilotto, F., Aldridge, D. C. (2011). Fouling of European freshwater bivalves (Unionidae) by the invasive zebra mussel (*Dreissena polymorpha*). *Freshwater Biology*, **56**, 867–876.

Spencer, K., and O'Shea, F. T. (2014). The hidden threat of historical landfills on eroding and low-lying coasts. *ECSA Bulletin*, **Summer 2014**, 16–17.

Stanley, D. J., and Warne, A. G. (1997). Holocene sea-level change and early human utilization of deltas. *GSA Today*, **7**, 1–7.

Stanley, G. D., Jr. (2001a). Introduction to reef ecosystems and their evolution. In Stanley, G. D., Jr., ed., *The History and Sedimentology of Ancient Reef Systems*. Topics in Geobiology, **17**, pp. 1–39.

Stanley, G. D., Jr. (ed.) (2001b). *The History and Sedimentology of Ancient Reef Systems*. Topics in Geobiology, **17**.

Stefani, M., and Vincenzi, M. (2005). The interplay of eustasy, climate and human activity in the late Quaternary depositional evolution and sedimentary architecture of the Po Delta system. *Marine Geology*, **222–223**, 19–48.

Steffen, W., Broadgate, W., Deutsch, L., et al. (2015). The trajectory of the Anthropocene: The Great Acceleration. *Anthropocene Review*, **2**(1), 81–98.

Steffen, W., Crutzen, P. J., and McNeill, J. R. (2007). The Anthropocene: Are humans now overwhelming the great forces of Nature? *Ambio*, **36**, 614–621.

Steffen, W., Grinevald, J., Crutzen, P., and McNeill, J. (2011). The Anthropocene: Conceptual and historical perspectives. *Philosophical Transactions of the Royal Society A*, **369**, 842–867.

Steffen, W., Leinfelder, R., Zalasiewicz, J., et al. (2016). Stratigraphic and Earth System approaches in defining the Anthropocene. *Earth's Future*, **8**, 324–345.

Steffen, W., Rockström, J., Richardson, K., et al. (2018). Trajectories of the Earth System in the Anthropocene. *Proceedings of the National Academy of Sciences (USA)*.

Steffen, W., Sanderson, A., Tyson, P. D., et al. (2004). *Global Change and the Earth System: A Planet under Pressure*. The IGBP Book Series. Berlin, Heidelberg, New York: Springer-Verlag.

Steffensen, J. P., Andersen, K. K., Bigler, M., et al. (2008). High-resolution Greenland ice core data show abrupt climate change happens in a few years. *Science*, **321**, 680–683.

Steig, E. J., Ding, Q., White, J. C., et al. (2013). Recent climate and ice-sheet changes in West Antarctica compared with the past 2,000 years. *Nature Geoscience*, **6**, 372–375.

Stein, R., Fahl, K., Schade, I., et al. (2017). Holocene variability in sea ice cover, primary production, and Pacific-Water inflow and climate change in the Chukchi and East Siberian Seas (Arctic Ocean). *Journal of Quaternary Science*, **32**(3), 362–379.

Steinhilber, F., Abreu, J. A., Beer, J., et al. (2012). 9,400 years of cosmic radiation and solar activity from ice cores and tree rings. *Proceedings of the National Academy of Sciences*, **109**(16), 5967–5971.

Stenni, B., Curran, M. A. J., Abram, N. J., et al. (2017). Antarctic climate variability on regional and continental scales over the last 2000 years. *Climate of the Past*, **13**, 1609–1634.

Stickley, C. E. (2014). The sea ice thickens. *Nature Geoscience*, **7**, 165–166.

Stocker, T. F. (1998). The seesaw effect. *Science*, **282**, 61–62.

Stoppani, A. (1873). *Corso di Geologia*, vol. 2. Geologia Stratigrafica. Milan: G. Bernardoni e G. Brigola.

Stordeur, D., and Khawam, R. (2007). Les crânes surmodelés de Tell Aswad (PPNB, Syrie): Premier regard sur l'ensemble, premières réflexions. *Syria*, **84**, 5–32.

Stow, D. A. V. (2001). Deep sea sediment drifts. Steele. Steele, J. H., Thorpe, S. A., and Turekian, K. K. (eds.), Encyclopaedia of Ocean Sciences. London: Academic Press.

Strasser, T. F., Panagopoulou, E., Runnels, C. N., et al. (2010). Stone Age seafaring in the Mediterranean: Evidence from the Plakias Region for Lower Palaeolithic and Mesolithic habitation of Crete. *Hesperia*, **79**, 145–190.

Strecker, A. L., and Arnott, S. E. (2010). Complex interactions between regional dispersal of native taxa and an invasive species. *Ecology*, **91**, 1035–1047.

Stuart, A. J., Kosintsev, P. A., Higham, T. F. G., and Lister, A. M. (2004). Pleistocene to Holocene extinction dynamics in giant deer and woolly mammoth. *Nature*, **431**, 684–689.

Stuart, A. J., Sulerzhitsky, L. D., Orlova, L. A., et al. (2002). The latest woolly mammoths (*Mammuthus primigenius* Blumenbach) in Europe and Asia: A review of the current evidence. *Quaternary Science Reviews*, **21**, 1559–1569.

Sucharovà, J., Suchara, I., Hola, M., et al. (2012). Top-/bottom-soil ratios and enrichment factors: What do they really show? *Applied Geochemistry*, **27**(1), 138–145.

Summerhayes, C. P. (2010). Climate change: A creeping catastrophe. *Bulletin of the World Health Organization*, **88**(6). http://dx.doi.org/10.1590/S0042-96862010000600007.

Summerhayes, C. P. (2015). *Earth's Climate Evolution*. Chichester: Wiley.

Summerhayes, C. P., Ellis, J. P., and Stoffers, P. (1985). Estuaries as sinks for sediment and industrial waste: A case history from the Massachusetts coast. *Contributions to Sedimentology*, **14**, 1–47.

Surovell, T. A., Holliday, V. T., Gingerich, J. A. M., et al. (2009). An independent evaluation of the Younger Dryas extraterrestrial impact hypothesis. *Proceedings of the National Academy of Sciences (USA)*, **106**, 18155–18158.

Svendsen, J. I., Alexanderson, H., Astakhov, V. I., et al. (2004). Late Quaternary ice sheet history of northern Eurasia. *Quaternary Science Reviews*, **23**, 1229–1271.

Swart, N. (2017). Natural causes of Arctic sea-ice loss. *Nature Climate Change*, **7**, 239–241.

Swart, P. K., Greer, L., Rosenheim, B. E., et al. (2010). The ^{13}C Suess effect in scleractinian corals mirror changes in the anthropogenic CO_2 inventory of the surface oceans. *Geophysical Research Letters*, **37**, L05604.

Swindles, G. T., Watson, E., Turner, T. E., et al. (2015). Spheroidal carbonaceous particles are a defining stratigraphic marker for the Anthropocene. *Scientific Reports*, **5**, 10264. doi:10210.11038/srep10264.

Syvitski, J. P. M. (1993). Glaciomarine environments in Canada: An overview. *Canadian Journal of Earth Sciences*, **30**, 354–371.

Syvitski, J. P. M. (Editor) (2003a). The supply and flux of sediment along hydrological pathways: Anthropogenic influences at the global scale. *Global and Planetary Change*, **39**(1/2), 1–199.

Syvitski, J. P. M. (2003b). Sediment fluxes and rates of sedimentation. In Middleton, G. V., ed., *Encyclopedia of Sediments and Sedimentary Rocks*. Dordrecht, Netherlands: Kluwer Academic Publishers, pp. 600–606.

Syvitski, J. P. M. (2008). Deltas at Risk. *Sustainability Science*, **3**, 23–32.

Syvitski, J. P. M., Burrell, D. C., and Skei, J. M. (1987). *Fjords: Processes & Products*. New York: Springer-Verlag.

Syvitski, J. P. M., Harvey, N., Wollanski, E., et al. (2005a). Dynamics of the coastal zone. In Crossland, C. J., Kremer, H. H., Lindeboom, H. J., et al., eds., *Global Fluxes in the Anthropocene*. Berlin: Springer, pp. 39–94.

Syvitski, J. P. M., and Kettner, A. (2011). Sediment flux and the Anthropocene. *Philosophical Transactions of the Royal Society A*, **369**(1938), 957–975.

Syvitski, J. P. M., Kettner, A. J., Correggiari, A., and Nelson, B. W. (2005b). Distributary channels and their impact on sediment dispersal. *Marine Geology*, **222–223**, 75–94.

Syvitski, J. P. M., Kettner, A. J., Overeem, I., et al. (2009). Sinking deltas due to human activities. *Nature Geoscience*, **2**, 681–689.

Syvitski, J. P., Kettner, A. J., Overeem, I., et al. (2017). Latitudinal controls on siliciclastic sediment production and transport. *SEPM Special Issue No. 108, Latitudinal Controls on Stratigraphic Models and Sedimentary Concepts, 1–15, Tulsa OK.*

Syvitski, J. P. M., Kettner, A., Peckham, S. D., and Kao, S. J. (2005c). Predicting the flux of sediment to the coastal zone: Application to the Lanyang watershed, northern Taiwan. *Journal of Coastal Research*, **21**, 580–587.

Syvitski, J. P. M., and Milliman, J. D. (2007). Geology, geography and humans battle for dominance over the delivery of sediment to the coastal ocean. *Journal of Geology*, **115**, 1–19.

Syvitski, J. P. M., and Saito, Y. (2007). Morphodynamics of deltas under the influence of Humans. *Global and Planetary Changes*, **57**, 261–182.

Syvitski, J. P. M., Vörösmarty, C., Kettner, A. J., and Green, P. (2005d). Impact of humans on the flux of terrestrial sediment to the global coastal ocean. *Science*, **308**, 376–380.

Szabó, J. (2010). Anthropogenic geomorphology: Subject and system. In Szabó, J., Dávid, L., and Lóczy, D., eds., *Anthropogenic Geomorphology: A guide to Man-Made Landforms*. Dordecht, Heidelberg, London, New York: Springer, pp. 3–10.

Ta, T. K. O., Nguyen, V. L., Tateishi, M., et al. (2002). Sediment facies and late Holocene progradation of the Mekong River delta in Bentre Province, southern Vietnam: An example of evolution from a tide-dominated to a tide- and wave- dominated delta. *Sedimentary Geology*, **152**, 313–325.

Tagliabue, A., Aumont, O., and Bopp, L. (2014). The impact of different external sources of iron on the global carbon cycle. *Geophysical Research Letters*. doi:10.1002/2013GL059059.

Tanabe, S., Saito, Y., Vu, Q. L., et al. (2006). Holocene evolution of the Song Hong (Red River) delta system, northern Vietnam. *Sedimentary Geology*, **187**, 29–61.

Tansel, B., and Yildiz, B. S. (2011). Goal-based waste management strategy to reduce persistence of contaminants in leachate at municipal solid waste landfills. *Environment, Development and Sustainability*, **13**, 821–831.

Targulian, V. O., and Goryachkin, S. V. (2008). Soil memory: Types of records, carriers, hierarchy and diversity. *Revista Mexicana Ciencias Geologica*, **21**, 1–8.

Tarolli, P., Preti, F., and Romano, N. (2014). Terraced landscapes: From an old best practice to a potential hazard for soil degradation due to land abandonment. *Anthropocene*, **6**, 10–25.

Tarolli, P., and Sofia, G. (2016). Human topographic signatures and derived geomorphic processes. *Geomorphology*, **255**, 140–161.

Terrington, R. L., Silva, É. C. N., Waters, C. N., et al. (2018). Quantifying anthropogenic modification of the shallow geosphere in central London, UK. *Geomorphology*, **319**, 15–34.

Tessler, Z., Vörösmarty, C., Grossberg, M., et al. (2015). Profiling risk and sustainability in coastal deltas of the world. *Science*, **349**(6248), 638–643.

Tesson, M., Labaune, C., and Gensous, B. (2005). Small rivers contribution to the Quaternary evolution of a Mediterranean littoral system: The western gulf of Lion, France. *Marine Geology*, 222–223, 313–334.

Thevenon, F., Guédron, S., Chiaradia, M., et al. (2011). (Pre-) historic changes in natural and anthropogenic heavy metals deposition inferred from two contrasting Swiss Alpine lakes. *Quaternary Science Reviews*, **30**(1), 224–233.

Thiede, J., Jessen, C., Knutz, P., et al. (2010). Millions of years of Greenland Ice Sheet history recorded in ocean sediments. *Polarforschung*, **80**(3), 141–159.

Thiengo, S. C., Faraco, F. A., Salgado, N. C., et al. (2007). Rapid spread of an invasive snail in South America: The giant African snail, *Achatina fulica* in Brasil. *Biological Invasions*, **9**, 693–702.

Thomas, E. (2003). Benthic foraminiferal record across the Initial Eocene Thermal Maximum, Southern Ocean Site 690: Causes and consequences of globally warm climates in the Early Paleogene. *Geological Society of America Special Paper*, **369**, 319–331.

Thomas, E. R., Hosking, J. S., Tuckwell, R. R., et al. (2015). Twentieth century increase in snowfall in coastal West Antarctica. *Geophysical Research Letters*, **42**, 9387–9393.

Thomas, E. R., Van Wessem, J. M., Roberts, J., et al. (2017). Regional Antarctic show accumulation over the past 1000 years. *Climate of the Past*, **13**, 1491–1513.

Thomas, W. L., Jr. (ed.) (1956). *Man's Role in Changing the Face of the Earth*. Wenner-Gren Foundation for Anthropological Research Symposium, Princeton, June 1955. University of Chicago Press.

Thompson, L. G. (2010). Climate change: The evidence and our options. *Behavior Analyst*, **33**(2), 153–170.

Thompson, L. G., Mosley-Thompson, E., Brecher, H., et al. (2006). Abrupt tropical climate change: Past and present. *Proceedings of the National Academy of Sciences (USA)*, **103**(28), 10536–10543.

Thompson, R., Battarbee, R. W., O'Sullivan, P. E., and Oldfield, F. (1975). Magnetic susceptibility of lake sediments. *Limnology and Oceanography*, **20**, 687–698.

Thompson, R. C., Moore, C., vom Saal, F. S., and Swan, S. H. (2009). Plastics, the environment and human health: Current consensus and future trends. *Philosophical Transactions of the Royal Society B*, **364**, 2153–2166.

Thompson, R., and Morton, D. J. (1978). Magnetic susceptibility and particle-size distribution in recent sediments of the Loch Lomond drainage basin, Scotland. *Journal of Sedimentary Petrology*, **49**, 801–811.

Thompson, R., and Oldfield, F. (1986). *Environmental Magnetism*. London: Allen & Unwin.

Thomson, J., Brown, L., Nixon, S., et al. (2000). Bioturbation and Holocene sediment accumulation fluxes in the north-east Atlantic Ocean (Benthic Boundary Layer experiment sites). *Marine Geology*, **169**, 21–39.

Tickell, C. (2011). Societal responses to the Anthropocene. *Philosophical Transactions of the Royal Society A*, **369**, 926–932.

Tierney, J. E., Abram, N. J., Anchukaitis, K. J., et al. (2015). Tropical sea surface temperatures for the past four centuries reconstructed from coral archives. *Paleoceanography*, **30**, 226–252.

Tobiszewski, M., and Namieśnik, J. (2012). PAH diagnostic ratios for the identification of pollution emission sources. *Environmental Pollution*, **162**, 110–119.

Toggweiler, R. (2008). Origin of the 100,000-year timescale in Antarctic temperatures and atmospheric CO_2. *Paleoceanography*, **23**, PA2211.

Tollefson, J. (2017). Larsen C's big divide: Collapse of nearby Antarctic ice shelves offers a glimpse of the future. *Nature*, **542**, 402–403.

Torres Camprubi, A. (2016). *Statehood under Water: Challenges of Sea-Level Rise to the Continuity of Pacific Island States*. Boston/Leiden: Brill/ Martinus Nijhoff.

Toynbee, J. (1996). *Death and Burial in the Roman World*. Baltimore, MD: John Hopkins University Press, reprint.

Tubau, X., Canals, M., Lastras, G., and Rayo, X. (2015). Marine litter on the floor of deep submarine canyons of the Northwestern Mediterranean Sea: The role of hydrodynamic processes. *Progress in Oceanography*, **134**, 379–403.

Tully, J. (2009). A Victorian ecological disaster: Imperialism, the telegraph, and Gutta-Percha. *Journal of World History*, **20**, 559–579.

Turner, B. L., II, Clark, W. C., Kates, R. W., et al., (eds.) (1990). *The Earth as Transformed by Human Action: Global and Regional Changes in the Biosphere over the Past 300 Years*. Cambridge University Press.

Turner, J., Binschadler, R., Convey, P., et al. (eds.) (2009a). *Antarctic Climate Change and the Environment*. Cambridge, UK: Scientific Committee on Antarctic Research, pp. 389–393.

Turner, J., Comiso, J. C., Marshall, G. J., et al. (2009b). Non-annular atmospheric circulation change induced by stratospheric ozone depletion and its role in the recent increase of Antarctic sea ice extent. *Geophysical Research Letters*, **36**, L08502. doi:10.1029/2009GL037524.

Turner, J., Lu, H., White, I., et al. (2016). Absence of 21st century warming on Antarctic Peninsula consistent with natural variability. *Nature*, **535**, 411–415.

Turner, J., Summerhayes, C. P., Sparrow, M. D., et al. (2017). Antarctic climate change and the environment – 2017 update. *Information Paper* **80**, Antarctic Treaty Consultative Meeting 40, Beijing, China.

Turra, A., Manzano, A. B., Dias, R. J. S., et al. (2014). Three-dimensional distribution of plastic pellets in sandy beaches: Shifting paradigms. *Scientific Reports*, **4**, 4435.

Turvey, S., Pitman, R. L., Taylor, B. L., et al. (2007). First human-caused extinction of a cetacean species? *Biology Letters*, **3**, 537–540.

Tyndall, J. (1868). *On Radiation*, the 1865 Rede Lecture. New York: D. Appleton and Co. Also included in his 1871 book *Fragments of Science*, available as a Project Gutenberg e-Book.

Tyrrell, T. (2011). Anthropogenic modification of the oceans. *Philosophical Transactions of the Royal Society A*, **369**, 887–908.

Tyrrell, T., and Zeebe, R. E. (2004). History of carbonate ion concentration over the last 100 million years. *Geochimica Cosmochimica Acta*, **68**, 3521–3530.

Tzedakis, P. C., Crucifix, M., Mitsui, T., and Wolff, E. W. (2017). A simple rule to determine which insolation cycles lead to interglacials. *Nature*, **452**, 427–432.

UBA (Umweltbundesamtes) (2016). Kunststoffabfälle. Umweltbundesamt Data. https://umweltbundesamt.de/daten/abfall-kreislaufwirtschaft/entsorgung-verwertung-ausgewaehlter-abfallarten/kunststoffabfaelle#textpart-1 (accessed 18 May 2017).

Uhrqvist, O., and Linnér, B.-O. (2015). Narratives of the past for Future Earth: The historiography of global environmental change research. *Anthropocene Review*, **2**(2), 159–173.

Ulanowicz, R. E. (1997). *Ecology, the Ascendent Perspective*. New York: Columbia University Press.

Underwood, J. R. (2001). Anthropic rocks as a fourth basic class. *Environmental & Engineering Geoscience*, **7**(1), 104–110.

UNEP (United Nations Environment Programme) (2013). *Environmental Risks and Challenges of Anthropogenic Metals Flows and Cycles: A Report of the Working Group on the Global Metal Flows to the International Resource Panel.*

UNESCO (2011). Sediment issues and sediment management in large river basins: Interim case study synthesis report. *International Sediment Initiative Technical Documents in Hydrology*, IRTCES 2011.

United Nations Convention on the Law of the Sea (1982). *United Nations Treaty Series*, **1833**, 3.

United Nations, Department of Economic and Social Affairs, Population Division (2014). World Urbanization Prospects: The 2014 Revision, Highlights (ST/ESA/SER. A/352).

United Nations, Department of Economic and Social Affairs, Population Division (2015). *World Urbanization Prospects: The 2015 Revision, Key Findings and Advance Tables*. Working Paper No. ESA/P/WP.241.

United Nations Scientific Committee on the Effects of Atomic Radiation (UNSCEAR) (2000). *Sources and Effects of Ionizing Radiation*, 2000 Report, vol. 1. New York: United Nations.

United States Geological Survey (USGS) (2010). Aluminium statistics. In Kelly, T. D., and Matos, G. R., eds., *Historical Statistics for Mineral and Material Commodities in the United States*. US Geological Survey Data Series, 140. http://pubs.usgs.gov/ds/2005/140/ (accessed 16 December 2012).

United States Geological Survey (USGS) (2016). http://minerals.usgs.gov/minerals/pubs/commodity/cement/.

US CLIVAR Project Office (2012). *Understanding the Dynamic Response of Greenland's Marine Terminating Glaciers to Oceanic and Atmospheric Forcing*. A whitepaper by the US CLIVAR Working Group on Greenland Ice Sheet-Ocean Interactions (GRISO), Report 2012-2, US CLIVAR Project Office, Washington, DC, 20006.

Vai, G. B. (2007). A history of chronostratigraphy. *Stratigraphy*, **4**, 83–97.

Van Cappellen, P., and Wang, Y. (1995). Metal cycling in surface sediments: Modeling the interplay of transport and reaction. In Allen, H. E., ed., *Metal Contaminated Aquatic Sediments*. Chelsea, MI: Ann Arbor Press, pp. 21–64.

Van Cauwenberghe, L., Vanreusel, A., Mees, J., and Janssen, C. R. (2013). Microplastic pollution in deep-sea sediments. *Environmental Pollution*, **182**, 495–499.

van der Gon, H. D., van het Bolscher, M., Visschedijk, A., and Zandveld, P. (2007). Emissions of persistent organic pollutants and eight candidate POPs from UNECE–Europe in 2000, 2010 and 2020 and the emission reduction resulting from the implementation of the UNECE POP protocol. *Atmospheric Environment*, **41**(40), 9245–9261.

Van der Velde, G., and Rajagopal, S. (2010). *Zebra Mussels in Europe*. Leiden: Backhuys Publishers.

van der Voet, E., Salminen, R., Eckelman, M., et al. (1995). Metal cycling in surface sediments: Modeling the interplay of transport and reaction. In Allen, H. E., ed., *Metal Contaminated Aquatic Sediments*. Chelsea: Ann Arbor Press, pp. 21–64.

Van Kranendonk, M. J., Altermann, W., Beard, B. L., et al. (2012). *A Chronostratigraphic Division of the Precambrian*. In Gradstein, F. M., Ogg, J. G., Schmitz, M., and Ogg, G., eds., *The Geologic Time Scale 2012*, vol. 1. Elsevier, pp. 299–392.

Van Oppen, M. J. H., Oliver, J. K., Putnam, H. M., and Gates, R. D. (2015). Building coral reef resilience through assisted evolution. *Proceedings of the National Academy of Sciences (USA)*, **112**(8), 2307–2313.

Vandenberghe, N., Hilgen, F. J., and Speijer, R. P. (2012). The Paleogene Period. In Gradstein, F. M., Ogg, J. G., Schmitz, M., and Ogg, G., eds., *The Geological Time Scale 2012*. Elsevier, pp. 853–922.

Vandiver, P. B., Soffer, O., Klima, B., and Svoboda, J. (1989). The origins of ceramic technology at Dolni Věstonice, Czechoslovakia. *Science*, **246** (4933), 1002–1008.

Vane, C. H., Chenery, S. R., Harrison, I., et al. (2011). Chemical signatures of the Anthropocene in the Clyde estuary, UK: Sediment-hosted Pb, 207/206Pb, total petroleum hydrocarbon, polyaromatic hydrocarbon and polychlorinated biphenyl pollution records. *Philosophical Transactions of the Royal Society A*, **369**, 1085–1111.

Velicogna, I., Sutterley, T. S., and van den Broeke, M. R. (2014). Regional acceleration in ice mass loss from Greenland and Antarctica using GRACE time-variable gravity data. *Journal of Geophysical Research Space Physics*, **41**, 8130–8137.

Vella, C., Fleury, T.-J., Raccasi, G., et al. (2005). Evolution of the Rhône delta plain in the Holocene. *Marine Geology*, **222–223**, 235–265.

Velzboer, I., Kwadijk, C. J. A. F., and Koelmans, A. A. (2014). Strong sorption of PCBs to nanoplastics, microplastics, carbon tubules, and fullerenes. *Environmental Science and Technology*, **48**, 4869–4876.

Vernadsky, V. I. (1924). *La Géochimie*. Paris: Librairie Félix Acan. (Lectures at the Sorbonne in 1922–1923).

Vernadsky, V. I. (1945). The Biosphere and the Noosphere. *American Scientist*, **33**(1), 1–12.

Vernadsky, V. I. (1997). *Scientific Thought as a Planetary Phenomenon*. Translated from the Russian [1938, 1977, 1991] by B. A. Starostin. Moscow: Nongovernmental Ecological V. I. Foundation.

Vernadsky, V. I. [1926](1998). *The Biosphere*. Translated from the Russian by D. B. Langmuir, revised and annotated by M. A. S. McMenamin. New York: Copernicus (Springer-Verlag).

Veron, J. E. N. (1995). *Corals in Space and Time: The Biogeography & Evolution of the Scleractinia*. Ithaca, NY: Cornell University Press.

Veron, J. E. N. (2008). Mass extinctions and ocean acidification: Biological constraints on geological dilemmas. *Coral Reefs*, **27**, 459–472.

Verpoorter, C., Kutser, T., Seekell, D. A., and Tranvik, L. J. (2014). A global inventory of lakes based on high-resolution satellite imagery. *Geophysical Research Letters*, **41**(18), 6396–6402.

Vidas, D. (2010). Responsibility for the seas. In Vidas, D., ed., *Law, Technology and Science for Oceans in Globalization*. Boston/Leiden: Brill/Martinus Nijhoff, pp. 3–40.

Vidas, D. (2011). The Anthropocene and the international law of the sea. *Philosophical Transactions of the Royal Society A*, **369**, 909–925.

Vidas, D. (2015). The Earth in the Anthropocene – and the world in the Holocene? *European Society of International Law (ESIL) Reflections*, **4**(6), 1–7.

Vidas D., Fauchald, O. K., Jensen, Ø., and Tvedt, M. W. (2015a). International law for the Anthropocene? Shifting perspectives in regulation of the oceans, environment and genetic resources. *Anthropocene*, **9**, 1–13.

Vidas D., Zalasiewicz, J., and Williams, M. (2015b). What is the Anthropocene: And why is it relevant for international law? *Yearbook of International Environmental Law*, **25**, 3–23.

Vienna Convention on the Law of Treaties (1969). *United Nations Treaty Series*, **1155**, 331.

Viers, J., Dupré, B., and Gaillardet, J. (2009). Chemical composition of suspended sediments in world rivers: New insights from a new database. *Science of the Total Environment*, **407**(2), 853–868.

Vigney, J.-D. (2011). The origins of animal domestication and husbandry: A major change in the history of humanity and the biosphere. *Comptes Rendus Biologies*, **334**, 171–181.

Villavicencio, N. A., Lindsey, E. L., Martin, F. M., et al. (2015). Combination of humans, climate, and vegetation change triggered Late Quaternary megafauna extinction in the Última Esperanza region, southern Patagonia, Chile. *Ecography*, **38**, 1–16.

Villmoare, B., Kimbel, W. H., Seyoum, C., et al. (2015). Early *Homo* at 2.8 Ma from Ledi-Geraru, Afar, Ethiopia. *Science*, **347**, 1352–1355.

Vinther, B. M., Buchardt, S. L., Clausen, H. B., et al. (2009). Holocene thinning of the Greenland ice sheet. *Nature*, **461**, 385–388.

Vinuales, J. E. (2016). Law and the Anthropocene. *C-EENRG Working Papers*, 2016–4 (September), University of Cambridge, 1–72.

Vogel, J. C. (1970). Groningen radiocarbon dates IX. *Radiocarbon*, **12**(2), 444–471.

von Glasow, R., Bobrowski, N., and Kern, C. (2009). The effects of volcanic eruptions on atmospheric chemistry. *Chemical Geology*, **263**, 131–142.

Vörösmarty, C., Meybeck, M., Fekete, B., et al. (2003). Anthropogenic sediment retention: Major global-scale impact from the population of registered impoundments. *Global and Planetary Change*, **39**, 169–190.

Wacey, D., Kilburn, M. R., Saunders, M., et al. (2011). Microfossils of sulphur-metabolizing cells in 3.4-billion-year-old rocks of Western Australia. *Nature Geoscience*, **4**, 698–702.

Wadhams, P. (2016). *A Farewell to Ice: A Report from the Arctic*. London: Allen Lane.

Wagreich, M., and Draganits, E. (2018). Early mining and smelting lead anomalies in geological archives as potential stratigraphic markers for the base of an early Anthropocene. *The Anthropocene Review*. doi:10.1177/2053019618756682.

Walker, M. J. C., Berkelhammer, M., Björck, S., et al. (2012). Formal subdivision of the Holocene Series/Epoch: A discussion paper by a working group of INTIMATE (integration of ice-core, marine and terrestrial records) and the Subcommission on Quaternary Stratigraphy (International Commission on Stratigraphy). *Journal of Quaternary Science*, **27**, 649–659.

Walker, M. J. C., Berkelhammer, M., Björck, S., et al. (2016). Formal subdivision of the Holocene Series/Epoch: Three proposals by a working group of members of INTIMATE (integration of ice-core, marine and terrestrial records) and the Subcommission on Quaternary Stratigraphy. Unpublished proposal submitted to the ICS Subcommission on Quaternary Stratigraphy.

Walker, M. J. C., Johnsen, S., Rasmussen, O. S., et al. (2009). Formal definition and dating of the GSSP (Global Stratotype Section and Point) for the base of the Holocene using the Greenland NGRIP ice core, and selected auxiliary records. *Journal of Quaternary Science*, **24**, 3–17.

Walling, D. E., and Fang, D. (2003). Recent trends in the suspended sediment loads of the world's rivers. *Global and Planetary Change*, **39**, 111–126.

Walsh, J. E. (2009). A comparison of Arctic and Antarctic climate change, present and future. *Antarctic Science*, **21**(3), 179–188.

Walsh, J. E., and Chapman, W. L. (2001). 20th-century sea-ice variations from observational data. *Annals of Glaciology*, **33**, 444–448.

Walter, R. C., and Merritts, D. J. (2008). Natural streams and the legacy of water-powered mills. *Science*, **319**, 299–304.

Wang, H., Saito, Y., Zhang, Y., et al. (2011). Recent changes of sediment flux to the western Pacific Ocean from major rivers in East and Southeast Asia. *Earth Science Reviews*, **108**, 80–100.

Wang, H., Yang, Z., Saito, Y., et al. (2007). Stepwise decreases of the Huanghe (Yellow River) sediment load (1950–2004): Impacts from climate changes and human activities. *Global Planetary Change*, **57**, 331–354.

Wang, M., and Overland, J. E. (2009). A sea ice free summer Arctic within 30 years? *Geophysical Research Letters*, **36**(7). doi:10.1029/2009GL037820.

Wania, F. (2003). Assessing the potential of persistent organic chemicals for long-range transport and accumulation in polar regions. *Environmental Science & Technology*, **37**(7), 1344–1351.

Wanner, H., Beer, J., Butikofer, J., et al. (2008). Mid- to Late Holocene climate change: An overview. *Quaternary Science Reviews*, **27**, 1791–1828.

Ward, C. V., Tocheri, M. W., Plavcan, J. M., et al. 2014. Early Pleistocene third metacarpal from Kenya and the evolution of modern human-like hand morphology. *Proceedings of the National Academy of Sciences (USA)*, **111**, 121–124.

Ward, L. (2015). *The London County Council Bomb Damage Maps*. Thames & Hudson.

Waste Online (2004). History of waste and recycling information sheet. http://dl.dropbox.com/u/21130258/resources/informationsheets/historyofwaste.htm.

Waters, C. N., Graham, C., Tapete, D., et al. (2018, in press). Recognising anthropogenic modification of the subsurface in the geological record. *Quarterly Journal of Engineering Geology and Hydrogeology*.

Waters, C. N., Northmore, K., Prince, G., et al. (1996). Volume 2: A technical guide to ground conditions. In Waters, C. N., Northmore, K., Prince, G., and Marker, B. R., eds., *A Geological Background for Planning and Development in the City of Bradford Metropolitan District*. British Geological Survey Technical Report WA/96/1.

Waters, C. N., Syvitski, J. P. M., Gałuszka, A., et al. (2015). Can nuclear weapons fallout mark the beginning of the Anthropocene Epoch? *Bulletin of the Atomic Scientists*, **71**(3), 46–57.

Waters, C. N., and Zalasiewicz, J. (2018). Concrete: The most abundant novel rock type of the Anthropocene. In DellaSala, D., and Goldstein, M. I., eds., *Encyclopedia of the Anthropocene*, vol. 1. Oxford: Elsevier. https://doi.org/10.1016/B978-0-12-809665-9.09775 -5.

Waters, C. N., Zalasiewicz, J., Summerhayes, C., et al. (2016). The Anthropocene is functionally and stratigraphically distinct from the Holocene. *Science*, **351**(6269), 137.

Waters, C. N., Zalasiewicz, J., Summerhayes, C., et al. (2018). Global Boundary Stratotype Section and Point (GSSP) for the Anthropocene Series: Where and how to look for a potential candidate. *Earth-Science Reviews*, **178**, 379–429.

Waters, C. N., Zalasiewicz, J., Williams, M., et al. (eds.) (2014). *A Stratigraphical Basis for the Anthropocene. Geological Society, London, Special Publications*, **395**.

Watling, L., and Norse, E. A. (1998). Disturbance of the seabed by mobile fishing gear: A comparison to forest clearcutting. *Conservation Biology*, **12**, 1180–1197.

Watmough, S. A. (1999). Monitoring historical changes in soil and atmospheric trace metal levels by dendrochemical analysis. *Environmental Pollution*, **106**, 391–403.

Watters, D. L., Yoklavich, M. M., Love, M. S., and Schroeder, D. M. (2010). Assessing marine debris in deep seafloor habitats off California. *Marine Pollution Bulletin*, **60**, 131–138.

Weaver, P. P. E., Wynn, R. B., Kenyon, N. H., and Evans, J. (2000). Continental margin sedimentation, with special reference to the north-east Atlantic margin. *Sedimentology*, **47** (1), 239–256.

Weaver, T. D., and Roseman, C. C. (2008). New developments in the genetic evidence for modern human origins. *Evolutionary Anthropology*, **17**, 69–80.

Webby, B. D. (1998). Steps toward a global standard for Ordovician stratigraphy. *Newsletters on Stratigraphy*, **36**(1), 1–33.

Weber, C., Pusch, S., and Opatz, T. (2017). Correspondence: Polyethylene bio-degradation by caterpillars? *Current Biology*, **27**(15), R744–R745.

Wei, G., McCulloch, M. T., Mortimer, G., et al. (2009). Evidence for ocean acidification in the Great Barrier Reef of Australia. *Geochimica et Cosmochimica Acta*, **73**, 2332–2346.

Wei, S., Wang, Y., Lam, J. C., et al. (2008). Historical trends of organic pollutants in sediment cores from Hong Kong. *Marine Pollution Bulletin*, **57** (6), 758–766.

Weidman, C. R., and Jones, G. A. (1993). A shell-derived time history of bomb ^{14}C on Georges Bank and its Labrador Sea implications. *Journal of Geophysical Research*, **98**(C8), 14577–14588.

Weightman, G. (2007). *The Industrial Revolutionaries: The Creation of the Modern World, 1776–1914*. New York: Grove Press.

Wellman, C., Osterloff, P. L., and Mohiuddin, U. (2003). Fragments of the earliest land plants. *Nature*, **425**, 282–285.

Wells, S., and Hanna, R. (1992). *The Greenpeace Book of Coral Reefs*. London: Cameron.

West, G. (2017). *Scale*. New York: Penguin Press.

Westbrook, G. K., Thatcher, K. E., Rohling, E. J., et al. (2009). Escape of methane gas from the seabed along the West Spitzbergen continental margin. *Geophysical Research Letters*, **36**(15). doi:10.1029/2009GL039191.

Wetzel, R. G. (2001). *Limnology: Lake and River Systems*. San Diego: Academic Press.

White, S. J. O., and Hemond, H. F. (2012). The anthrobiogeochemical cycle of indium: A review of the natural and anthropogenic cycling of indium in the environment. *Critical Reviews in Environmental Science and Technology*, **42**(2), 155–186.

Whiteman, G., Hope, C., and Wadhams, P. (2013). Vast costs of Arctic change. *Nature*, **499**, 401–403.

Whitmee, S., Haines, A., Beyrer, C., et al. (2015). Safeguarding human health in the Anthropocene epoch: Report of The Rockefeller Foundation–Lancet Commission on planetary health. *The Lancet*, **386**, 1973–2028.

Wikipedia (2017). World population estimates. https://en.wikipedia.org/wiki/World_population_estimates.

Wilkinson, B. H., and McElroy, B. J. (2007). The impact of humans on continental erosion and sedimentation. *Geological Society of America Bulletin*, **119**, 140–156.

Wilkinson, I. P., and Gulakyan, S. Z. (2010). Holocene to recent Ostracoda of Lake Sevan, Armenia: Biodiversity and ecological controls. *Stratigraphy*, **7**, 301–315.

Wilkinson, I. P., Poirier, C., Head, M. J., et al. (2014). Microbiotic signatures of the Anthropocene in marginal marine and freshwater palaeoenvironments. In Waters, C. N., Zalasiewicz, J. A., Williams, M., et al., eds., *A Stratigraphical Basis for the Anthropocene*. Geological Society, London, Special Publications, **395**, pp. 185–219.

Wilkinson, T. J. (2003). *Archaeological Landscapes of the Near East*. Tucson: University of Arizona Press.

Williams, A. T., and Simmons, S. L. (1996). The degradation of plastic litter in rivers: Implications for beaches. *Journal of Coastal Conservation*, **2**, 63–72.

Williams, M., Edgeworth, M., Zalasiewicz, J., et al. (2019, in press). Underground metro systems: A durable geological proxy of rapid urban population growth and energy consumption during the Anthropocene. *Anthropocene*. In Benjamin, C., Quaedakers, E., and Baker, D., eds., *The Routledge Handbook of Big History (Routledge Companions)*.

Williams, M., Zalasiewicz, J., Davies, N., et al. (2015a). Humans as the third evolutionary stage of biosphere engineering of rivers. *Anthropocene*, **7**, 57–63.

Williams, M., Zalasiewicz, J., Haff, P. K., et al. (2015b). The Anthropocene biosphere. *Anthropocene Review*, **2**, 196–219.

Williams, M., Zalasiewicz, J., Haywood, A., and Ellis, M. (eds.) (2011). The Anthropocene: A new epoch of geological time? *Philosophical Transactions of the Royal Society A*, **369**, 833–1112.

Williams, M., Zalasiewicz, J., and Waters, C. N. (2017). The Anthropocene: A geological perspective. In Heikkurinen, P., ed., *Sustainability and Peaceful Coexistence for the Anthropocene*. Routledge Series on Transnational Law and Governance. Oxon.: Taylor & Francis, pp. 16–30.

Williams, M., Zalasiewicz, J., Waters, C. N., and Landing, E. (2014). Is the fossil record of complex animal behaviour a stratigraphical analogue for the Anthropocene? In Waters, C. N., Zalasiewicz, J., Williams, M., et al., eds., *A Stratigraphical Basis for the Anthropocene*. Geological Society, London, Special Publications, **395**, pp. 143–148.

Williams, M., Zalasiewicz, J., Waters, C. N., et al. (2016). The Anthropocene: A conspicuous stratigraphical signal of anthropogenic changes in production and consumption across the biosphere. *Earth's Future*, **4**, 34–53.

Willmore, P. L. (1949). Seismic experiments on the North German explosions, 1946 to 1947. *Philosophical Transactions of the Royal Society A*, **242**(843), 123–151.

Wilmshurst, J., Hunt, T. L., Lipo, C. P., and Anderson, A. J. (2011). High-precision radiocarbon dating shows recent and rapid initial colonization of East Polynesia. *Proceedings of the National Academy of Sciences (USA)*, **108**, 1815–1820.

Wilson, M. J., and Bell, N. (1996). Acid deposition and heavy metal mobilization. *Applied Geochemistry*, **11**(1), 133–137.

Wing, S. L., and Currano, E. D. (2013). Plant response to a global greenhouse 56 million years ago. *American Journal of Botany*, **100**, 1234–1254.

Wolfe, A. P., Hobbs, W. O., Birks, H. H., et al. (2013). Stratigraphic expressions of the Holocene–Anthropocene transition revealed in sediments from remote lakes. *Earth-Science Reviews*, **116**, 17–34.

Wolff, E. W. (2011). Greenhouse gases in the Earth System: A palaeoclimate perspective. *Philosophical Transactions of the Royal Society A*, **369**, 2133–2147.

Wolff, E. W. (2013). Ice sheets and nitrogen. *Philosophical Transactions of the Royal Society B*, **368**, 20130127. http://dx.doi.org/10.1098/rstb.2013.0127.

Wolff, E. W. (2014). Ice sheets and the Anthropocene. In Waters, C. N., Zalasiewicz, J. A., Williams, M., et al., eds., *A Stratigraphical Basis for the Anthropocene*. Geological Society, London, Special Publications, **395**, pp. 255–263.

Wolff, E. W., Fischer, H., and Rõthlisberger, R. (2009). Glacial terminations as southern warmings without northern control. *Nature Geoscience*, **2**, 206–209.

Wolff, E. W., and Suttie, E. D. (1994). Antarctic snow record of southern hemisphere lead pollution. *Geophysical Research Letters*, **21**, 781–784.

Wolff, E. W., Suttie, E. D., and Peel, D. A. (1999). Antarctic snow record of cadmium, copper, and zinc content during the twentieth century. *Atmospheric Environment*, **33**(10), 1535–1541.

Wood, K. (2008). Microspheres: Fillers filled with possibilities. *Composites World*, **14**(2), 28–32.

Wood, K. R., and Overland, J. E. (2010). Early 20th century Arctic warming in retrospect. *International Journal of Climatology*, **30**, 1269–1279.

Wood, R. (1999). *Reef Evolution*. Oxford University Press.

Woodall, L. C., Sanchez-Vidal, A., Canals, M., et al. (2014). The deep sea is a major sink for microplastic debris. *Royal Society Open Science*, **1**, 140317.

Woodroffe, S. A., and Horton, B. P. (2005). Holocene sea-level changes in the Indo-Pacific. *Journal of Asian Earth Sciences*, **25**, 29–43.

Woods, W. I. (2008). *Amazonian Dark Earths: Wim Sombroek's Vision*. New York: Springer.

Woodworth, P. L., Gehrels, W. R., and Nerem, R. S. (2011). Nineteenth and twentieth century changes in sea level. *Oceanography*, **24**, 80–93.

Wright, G. R. H. (1985). *Ancient Building in South Syria and Palestine*, vol. 1. Leiden-Köln: E. J. Brill.

Wright, J. D., and Schaller, M. F. (2013). Evidence for a rapid release of carbon at the Paleocene-Eocene thermal maximum. *Proceedings of the National Academy of Sciences (USA)*, **110**(40), 15908 15913.

Wright, S. L., Thompson, R. C., and Galloway, T. S. (2013). The physical impacts of microplastics on marine organisms: A review. *Environmental Pollution*, **178**, 483–492.

Wrigley, E. A. (2010). *Energy and the English Industrial Revolution*. Cambridge University Press.

Wroe, S., and Field, J. (2006). A review of the evidence for a human role in the extinction of Australian megafauna and an alternative interpretation. *Quaternary Science Reviews*, **25**, 2692–2703.

Wroe, S., Field, J. H., Archer, M., et al. (2013). Climate change frames debate over the extinction of megafauna in Sahul (Pleistocene Australia-New Guinea). *Proceedings of the National Academy of Sciences (USA)*, **110**, 8777–8781.

Wynn, P. M., Borsato, A., Baker, A., et al. (2013). Biogeochemical cycling of sulphur in karst and transfer into speleothem archives at Grotta di Ernesto, Italy. *Biogeochemistry*, **114**, 255–267.

Wynn, P. M., Fairchild, I. J., Baker, A., et al. (2008). Isotopic archives of sulphate in speleothems. *Geochimica Cosmochimica Acta*, **72**, 2465–2477.

Wynn, P. M., Fairchild, I. J., Frisia, C., et al. (2010). High-resolution sulphur isotope analysis of speleothem carbonate by secondary ionisation mass spectrometry. *Chemical Geology*, **271**, 101–107.

Wynn, P. M., Loader, N. J., and Fairchild, I. J. (2014). Interrogating trees for isotopic archives of atmospheric sulphur deposition and comparison to speleothem records. *Environmental Pollution*, **187**, 98–105.

Wypych, G. (2016). Fillers – origin, chemical composition, properties, and morphology. In Wypych, G., ed., *Handbook of Fillers*. Amsterdam: Elsevier, pp. 13–266.

Wu, X.-H., Zhang, C., Goldberg, P., Cohen, D., et al. (2013). Early pottery at 20,000 years ago in Xianrendong Cave, China. *Science*, **336**, 1696–1700.

Xu, B. Q., Cao, J. J., Hansen, J., Yao, T. D., et al. (2009). Black soot and the survival of Tibetan glaciers. *Proceedings of the National Academy of Sciences (USA)*, **106**, 22114–22118.

Xu, J. (2003). Naturally and anthropogenically accelerated sedimentation in the Lower Yellow River, China, over the past 13.000 years. *Geografiska Annaler*, Series A, **80**(1), 67–78.

Yaalon, D. H., and Yaron, B. (1966). Framework for man-made soil changes – an outline of metapedogenesis. *Soil Science*, **102**, 272–277.

Yan, H., Sun, L., Wang, Y., et al. (2010). A 2000-year record of copper pollution in South China Sea derived from seabird excrements: A potential indicator for copper production and civilization of China. *Journal of Paleolimnology*, **44**(2), 431–442.

Yan, X.-H., Boyer, T., Trenberth, K., Karl, T. R., et al. (2016). The global warming hiatus: Slowdown or redistribution? *Earth's Future*, **4**. doi:10.1002/2016EF000417.

Yang, C., Rose, N. L., Turner, S. D., et al. (2016). Hexabromocyclododecanes, polybrominated diphenyl ethers, and polychlorinated biphenyls in radiometrically dated sediment cores from

English lakes, ~ 1950–present. *Science of the Total Environment*, **541**, 721–728.

Yang, H., Engstrom, D. R., and Rose, N. L. (2010). Recent changes in atmospheric mercury deposition recorded in the sediments of remote equatorial lakes in the Rwenzori Mountains, Uganda. *Environmental Science & Technology*, **44**, 6570–6575.

Yang, H., Lohmann, G., Wei, W., et al. (2016). Intensification and poleward shift of subtropical western boundary currents in a warming climate. *Journal of Geophysical Research Oceans*, **121**(7), 4928–4945.

Yang, H., Rose, N. L., and Battarbee, R. W. (2001). Dating of recent catchment peats using spheroidal carbonaceous particle (SCP) concentration profiles with particular reference to Lochnagar, Scotland. *Holocene*, **11**, 593–597.

Yang, H., Rose, N. L., Battarbee, R. W., and Boyle, J. F. (2002). Mercury and lead budgets for Lochnagar, a Scottish mountain lake and its catchment. *Environmental Science & Technology*, **36**, 1383–1388.

Yang, J., Yang, Y., Wu, W. M., Zhao, J., and Jiang, L. (2014). Evidence of polyethylene biodegradation by bacterial strains from the guts of plastic-eating waxworms. *Environmental Science and Technology*, **48**, 13776–13784.

Yang, Y., Dong, G., Zhang, S., et al. (2017). Copper content in anthropogenic sediments as a tracer for detecting smelting activities and its impact on environment during prehistoric period in Hexi Corridor, Northwest China. *Holocene*, **27**(2), 282–291.

Yang, Z., Wang, H., Saito, Y., et al. (2006). Dam impacts on the Changjiang (Yangtze River) sediment discharge to the sea: The past 55 years and after the Three Gorges Dam. *Water Resources Research*, **42**, W04407. doi:10.1029/2005WR003970.

Yasuhara, M., Hunt, G., Breitburg, D., et al. (2012). Human-induced marine ecological degradation: Micropaleontological perspectives. *Ecology and Evolution*, **2**, 3242–3268.

Yin, Z., Maoyan, Z., Davidson, E. H., et al. (2015). Sponge grade body fossil with cellular resolution dating 60 Myr before the Cambrian. *Proceedings of the National Academy of Sciences (USA)*, **112**, E1453–1460.

Yoshida, S., Hiraga, K., Takehana, T., et al. (2016). A bacterium that degrades and assimilates poly (ethylene terephthalate). *Science*, **351**(6278), 1196–1199.

Yu, Z., Loisel, J., Brosseau, D. P., and Beilman, D. W. (2010). Global peatland dynamics since the Last Glacial Maximum. *Geophysical Research Letters*, **37**, L13402. doi:10.1029/2010GL043584.

Zachos, J. C., Bohaty, S. M., John, C. M., et al. (2007). The Palaeocene-Eocene carbon isotope excursion: Constraints from individual shell planktonic foraminifer records. *Philosophical Transactions of the Royal Society A*, **365**, 1829–1842.

Zachos, J. C., Dickens, G. R., and Zeebe, R. E. (2008). An early Cenozoic perspective on greenhouse warming and carbon-cycle dynamics. *Nature*, **451**, 279–283.

Zachos, J. C., Pagani, M., Sloan, M., et al. (2001). Trends, rhythms and aberrations in global climate 65 Ma to present. *Science*, **292**, 686–693.

Zachos, J. C., Röhl. U., Schellenberg, S. A., et al. (2005). Rapid acidification of the ocean during the Paleocene–Eocene thermal maximum. *Science*, **308**, 1611–161.

Zachos, J. C., Wara, M. W., Bohaty, S., et al. (2003). A transient rise in tropical sea surface temperature during the Paleocene-Eocene thermal maximum. *Science*, **302**(5650), 1551–1554.

Zalasiewicz, J., Cita, M. B., Hilgen, F., et al. (2013). Chronostratigraphy and geochronology: A proposed realignment. *GSA Today*, **23**(3), 4–8.

Zalasiewicz, J., Kryza, R., and Williams, M. (2014a). The mineral signature of the Anthropocene. In Waters, C. N., Zalasiewicz, J. A., Williams, M., et al., eds., *A Stratigraphical Basis for the Anthropocene.* Geological Society of London Special Publication, **395**, pp. 109–117.

Zalasiewicz, J., Steffen, W., Leinfelder, R., et al. (2017a). Petrifying Earth process: The stratigraphic imprint of key Earth System parameters in the Anthropocene. In Clark, N., and Yusoff, K., eds., *Theory Culture & Society, Special Issue: Geosocial Formations and the Anthropocene; Theory, Culture & Society*, **34**, pp. 83–104.

Zalasiewicz, J., Waters, C. N., Barnosky, A. D., et al. (2015a). Colonization of the Americas, "Little Ice Age" climate, and bomb-produced carbon: Their role in defining the Anthropocene. *Anthropocene Review*, **2**(2), 117–127.

Zalasiewicz, J., Waters, C. N., and Head, M. J. (2017b). Anthropocene: Its stratigraphic basis. *Nature*, **541**(7637), 289.

Zalasiewicz, J., Waters, C. N., Head, M. J., et al. (2018). The geological and Earth System reality of the Anthropocene: Reply to Bauer, A. M. and Ellis, E. C. The Anthropocene Divide: Obscuring understanding of social-environmental change. *Current Anthropology*, **59**(2), 220–223.

Zalasiewicz, J., Waters, C. N., Ivar do Sul, J., et al. (2016a). The geological cycle of plastics and their use as a stratigraphic indicator of the Anthropocene. *Anthropocene*, **13**, 4–17.

Zalasiewicz, J., Waters, C. N., Summerhayes, C. P., et al. (2017c). The Working Group on the Anthropocene: Summary of evidence and interim recommendations. *Anthropocene*, **19**, 55–60.

Zalasiewicz, J., Waters, C. N., and Williams, M. (2014b). Human bioturbation, and the subterranean landscape of the Anthropocene. *Anthropocene*, **6**, 3–9.

Zalasiewicz, J., Waters, C. N., Williams, M., et al. (2015b). When did the Anthropocene begin? A mid-twentieth century boundary level is stratigraphically optimal. *Quaternary International*, **383**, 196–203.

Zalasiewicz, J., Waters, C. N., Wolfe, A. P., et al. (2017d). Making the case for a formal Anthropocene: An analysis of ongoing critiques. *Newsletters on Stratigraphy*, **50**, 205–226.

Zalasiewicz, J., and Williams, M. (2012). *The Goldilocks Planet: An Earth History of Climate Change.* Oxford University Press.

Zalasiewicz, J., and Williams, M. (2014). The Anthropocene: A comparison with the Ordovician-Silurian boundary. *Rendiconti Lincei – Scienze Fisiche e Naturali*, **25**(1), 5–12.

Zalasiewicz, J., Williams, M., Fortey, R., et al. (2011a). Stratigraphy of the Anthropocene. In Zalasiewicz, J. A., Williams, M., Haywood, A., and Ellis, M., eds., *The Anthropocene: A New Epoch of Geological Time.* Philosophical Transactions of the Royal Society A, **369**, pp. 1036–1055.

Zalasiewicz, J., Williams, M., Haywood, A., and Ellis, M. (2011b). The Anthropocene: A new epoch of geological time? *Philosophical Transactions of the Royal Society*, **A369**, 835–841.

Zalasiewicz, J., Williams, M., Smith, A., et al. (2008). Are we now living in the Anthropocene? *GSA Today*, **18**(2), 4–8.

Zalasiewicz, J., Williams, M., and Waters, C. N. (2014c). Can an Anthropocene Series be defined and recognized? In Waters, C. N., Zalasiewicz, J. A., Williams, M., et al., eds., *A Stratigraphical Basis for the Anthropocene.* Geological Society of London Special Publication, **395**, pp. 39–54.

Zalasiewicz, J., Williams, M., Waters, C. N., et al. (2014d). The technofossil record of humans. *Anthropocene Review*, **1**, 34–43.

Zalasiewicz, J., Williams, M., Waters, C. N., et al. (2016b). Scale and diversity of the physical technosphere: A geological perspective. *Anthropocene Review*. **4**(1), 9–22.

Zalasiewicz, J., and Zalasiewicz, M. (2015). Battle scars. *New Scientist*. 36–39.

Zannoni, D., Valotto, G., Visin, F., and Rampazzo, G. (2016). Sources and distribution of tracer elements in road dust: The Venice mainland case of study. *Journal of Geochemical Exploration*, **166**, 64–72.

Zareitalabad, P., Siemens, J., Hamer, M., and Amelung, W. (2013). Perfluorooctanoic acid (PFOA) and perfluorooctanesulfonic acid (PFOS) in surface waters, sediments, soils and wastewater: A review on concentrations and distribution coefficients. *Chemosphere*, **91**, 725–732.

Zarfl, C., and Matthies, M. (2010). Are marine plastic particles transport vectors for organic pollutants to the Arctic? *Marine Pollution Bulletin*, **60**(10), 1810–1814.

Zayasu, T. Y., and Shinzato, C. (2016). Hope for coral reef rehabilitation: Massive synchronous spawning by outplanted corals in Okinawa, Japan. *Coral Reefs*, **35**, 1295–1295.

Zbyszewski, M., Corcoran, P. L., and Hockin, A. (2014). Comparison of the distribution and degradation of plastic debris along the shorelines of the Great Lakes, North America. *Journal of Great Lakes Research*, **40**, 288–299.

Zeder, M. A. (2011). The origins of agriculture in the Near East. *Current Anthropology*, **52**, 221–235.

Zeebe, R. E., Dickens, G. R., Ridgwell, A., et al. (2014). Onset of carbon isotope excursion at the Paleocene-Eocene thermal maximum took millennia, not 13 years. *Proceedings of the National Academy of Sciences*, **111**(12), E1062–E1063.

Zeebe, R. E., Ridgwell, A., and Zachos, J. C. (2016). Anthropogenic carbon release rate unprecedented during the past 66 million years. *Nature Geoscience*, **9**, 325–329.

Zeebe, R. E., Zachos, J. C., and Dickens, G. R. (2009). Carbon dioxide forcing alone insufficient to explain Palaeocene–Eocene Thermal Maximum warming. *Nature Geoscience*, **2**(8), 576–580.

Zemp, D. C., Schleussner, C. F., Barbosa, H. M. J., and Rammig, A. (2017). Deforestation effects on Amazon forest resilience. *Geophysical Research Letters*, **44** (12), 6182–6190.

Zemp, M., Frey, H., Gärtner-Roer, I., et al. (2015). Historically unprecedented global glacier decline in the early 21st century. *Journal of Glaciology*, **61**(228), 745–762.

Zettler, E. R., Mincer, T. J., and Amaral-Zettler, L. A. (2013). Life in the "Plastisphere": Microbial communities on plastic marine debris. *Environmental Science and Technology*, **47**, 7137–7146.

Zhao, S., Zhu, L., Wang, T., and Li, D. (2014). Suspended microplastics in the surface water of the Yangtze estuary system, China: First observations on occurrence, distribution. *Marine Pollution Bulletin*, **86**, 562–568.

Zheng, J., Shotyk, W., Krachler, M., and Fisher, D. A. (2007). A 15,800-year record of atmospheric lead deposition on the Devon Island ice cap, Nunavut, Canada: Natural and anthropogenic enrichments, isotopic composition and predominant sources. *Global Biogeochemical Cycles*, **21**, GB2027.

Zhu, R. X., Potts, R., Xie, F., et al. (2004). New evidence on the earliest human presence at high northern latitudes in northeast Asia. *Nature*, **431**, 559–562.

Zhu, Z., Piao, S., Myneni, R. B., et al. (2016). Greening of the Earth and its drivers. *Nature Climate Change*, **6**, 791–795.

Zielinski, G. A., Mayewski, P. A., Meeker, L. D., et al. (1994). Record of volcanism since 7000 B.C. from the GISP2 Greenland ice core and implications for the volcano-climate system. *Science*, **264**, 948–952.

Zinke, J., Loveday, B. R., Reason, C. J. C., et al. (2014). Madagascar corals track sea surface temperature variability in the Agulhas Current core region over the past 334 years. *Scientific Reports*, **4**, 4393. doi:10.1038/srep04393.

Zolitschka, B., Francus, P., Ojala, A. E. K., and Schimmelmann, A. (2015). Varves in lake sediments: A review. *Quaternary Science Reviews*, **117**, 1–41.

Zong, Y., and Horton, B. P. (1999). Diatom-based tidal-level transfer functions as an aid in reconstructing Quaternary history of sea-level movements in the UK. *Journal of Quaternary Science*, **14**, 153–167.

Zushi, Y., Tamada, M., Kanai, Y., and Masunaga, S. (2010). Time trends of perfluorinated compounds from the sediment core of Tokyo Bay, Japan (1950s–2004). *Environmental Pollution*, **158**, 756–763.

INDEX